LINEAR ALGEBRA

LINEAR ALGEBRA

TERRY LAWSON
Tulane University

JOHN WILEY & SONS, INC.
New York • Chichester • Brisbane
Toronto • Singapore

COVER ILLUSTRATION	Roy Wiemann
ACQUISITIONS EDITOR	Barbara Holland
MARKETING MANAGER	Debra Riegert
SENIOR PRODUCTION EDITOR	Tony VenGraitis
COVER AND TEXT DESIGN	Karin Kincheloe
MANUFACTURING MANAGER	Mark Cirillo
OUTSIDE PRODUCTION MANAGEMENT	Ingrao Associates

This book was set in Times Roman by Publication Services and
printed and bound by Courier/Westford. The cover was printed by Phoenix Color.

Recognizing the importance of preserving what has been written, it is a
policy of John Wiley & Sons, Inc. to have books of enduring value published
in the United States printed on acid-free paper, and we exert our best
efforts to that end.

The paper in this book was manufactured by a mill whose forest management programs include
sustained yield harvesting of its timberlands. Sustained yield harvesting principles ensure that
the number of trees cut each year does not exceed the amount of new growth.

ISBN 0-471-30897-8

Printed in the United States of America

10 9 8 7 6 5 4 3 2 1

CONTENTS

LIST OF FIGURES

PREFACE

This book approaches the subject matter of linear algebra from a blend of both computational and theoretical ideas. Students covering this material typically range from talented freshmen to graduate students in other areas, particularly science, business, and engineering. It is assumed that most students taking this course will have had a basic three semester sequence in calculus, and perhaps even an introduction to differential equations. This assumption is largely for the mathematical maturity gained through this background rather than specific material which is being assumed, although some examples and applications may build on such material. Some material on differential equations is treated here as an application of material on eigenvalues and eigenvectors.

The expectation is that the book should serve well both mathematics majors as well as students in other areas such as science, business, and engineering that use linear algebra. Its main characteristic is the manner in which it handles the interaction between the computational and theoretical aspects of the material. We have tried to make each aspect reinforce the other, and show the student the balanced view of both areas, which is necessary for developing a good understanding of linear algebra. In particular, a thorough knowledge of basic Gaussian elimination is used repeatedly in developing and illustrating the theoretical concepts. On the other hand, we repeatedly apply the theory to solve computational problems.

One aspect of our approach is that we treat many topics in matrix theory as special cases of statements about linear transformations. For example, the main theorems concerning solving the eigenvalue–eigenvector problem are treated conceptually in terms of linear transformations, and then the results are applied to specific types of matrices, such as symmetric matrices. On the other hand, we do give completely matrix oriented approaches to certain results when we feel that this approach provides the easiest way to understand the result, and then show what this implies in terms of linear transformations.

Our viewpoint on computation is that most computations will ultimately be done using some form of computer program. We feel it is important for the student to have experience working with small examples by hand to develop a feeling for the basic algorithms. The student should also be introduced to some software which implements the computations, even if this only means using a calculator with substantial matrix capabilities. For this book we provide supplementary materials for each of the programs MATLAB®, Maple®, and Mathematica®. These are in the form of labs which are designed to help the student learn to effectively use these programs, as well as to reinforce the main material in the text.

Each of these programs is available in a student version. Although it is not absolutely necessary to use one of these programs (or another similar program) in conjunction with the book, we do recommend doing so. We recommend that after spending some time working some small problems by hand to help develop an understanding for various algorithms, the student should use one of these programs to actually do the computational work. This also facilitates the student spending more time focusing on conceptual aspects of the course. We especially encourage students to use the computer to check their answers on exercises. The book does include a small amount of MATLAB code, as well as various graphics that were prepared using MATLAB Handle Graphics™. Handle Graphics is the trademark of The MathWorks, Inc.

Exercises provide an essential part of the book. There are two types of exercises. The first type is exercises within each section, which we call Embedded Exercises. It is intended that students attempt most of these, and these can provide a focus for class discussion. In my own class, I spend a significant portion of class time in discussion focused on these exercises. Answers to all of these exercises are provided in the appendix. Even if an embedded exercise is not worked, the student should read over the exercise and its solution as later sections may refer to material in these exercises. In particular, many routine verifications are left for these exercises.

The second type of exercise occurs as the last section of each chapter. These exercises range in level from routine computations to fairly difficult theoretical problems. Answers are not provided here, and these are intended to be used in graded written assignments in the course. It is expected that the student have available computational tools such as MATLAB, Maple, or Mathematica in working both types of exercises. Typically, I will put some restrictions on the use of these tools for certain of the assigned exercises to make sure that the student has mastered basic algorithms presented in class.

The text contains a number of sections devoted to applications of linear algebra. All of these sections are optional, although we recommend including at least a few of them in the course. Rather than locating these sections at the end of the text, we have put them at points where we felt they would be most naturally covered. There is more material in the text than can be comfortably covered in a semester in a first course. For this reason, I recommend that the applications sections only be covered selectively if one wants to treat the full range of conceptual material in the course. An alternative course featuring applications could mainly concentrate on the first four chapters with only enough treatment of the Spectral Theorem in Chapter 5 to allow a discussion of some of its applications.

ORGANIZATION BY CHAPTER

Chapter 1 discusses the basic matrix algebra involved in solving systems of linear equations. This is formulated in terms of the basic matrix equation $A\mathbf{x} = \mathbf{b}$. We discuss thoroughly the Gaussian elimination algorithm to solve this equation. We then interpret it in terms of elementary matrices, and apply these ideas to discuss inverses of matrices. The topic of determinants is also developed in this chapter. Although the chapter is lengthy, most of its length comes from giving many detailed examples of Gaussian elimination. The last two sections of the chapter are optional. Section 8 gives a short discussion of some of the computational issues involved in solving equations on a computer. Section 9 briefly introduces applications of Gaussian elimination to an open Leontief input–ouput model and to simple electrical circuits—these topics are revisited in later chapters. Many students have been exposed already to ideas in this chapter. Our approach is different enough from other ones and concepts such as elementary matrices and decompositions $OA = R$ are used in later chapters so that this chapter should be covered, even if at an accelerated pace, by all students.

Chapter 2 provides an introduction to the main conceptual ideas in linear algebra. This involves the basic concepts of vector spaces and linear transformations. We provide copious examples to illustrate each new concept introduced. Our work in matrix algebra is used both in providing methods for proving theorems as well as providing illustrative examples concerning these concepts. In particular, the null space and range of both a matrix A and its transpose A^t play a critical role. A basic theme of the chapter is the interaction between the theoretical concepts and computational methods. The chapter ends with an optional section applying the material of the first two chapters to elementary graph theory, where the topic of electric circuits is revisited.

Chapter 3 introduces the notion of an inner product, which is just the appropriate generalization of the dot product in Euclidean space. This is related to the solvability of $A\mathbf{x} = \mathbf{b}$. The Gram-Schmidt algorithm for providing an orthonormal basis is discussed, as well as a related $A = QR$ matrix decomposition. The least squares solution of $A\mathbf{x} = \mathbf{b}$ is also discussed and applied to data analysis. Finally, the chapter ends with a discussion of the singular value decomposition and pseudoinverse, leading to the introduction to the eigenvalue–eigenvector problem.

In Chapters 4 and 5, the eigenvalue–eigenvector problem is studied and applied to a variety of problems. Chapter 4 starts by discussing the eigenvalue–eigenvector problem and its interpretation in terms of diagonalizability. Section 3 gives a discussion of the complex numbers and complex vector spaces, which is essential for full analysis when there are complex eigenvalues. The chapter ends with two longer optional applications sections. The first covers difference equations, with special emphasis on Markov chains and recurrence relations. The second covers topics in differential equations. The focus of these applications is to feature the role of eigenvalue–eigenvector analysis.

Chapter 5 focuses on the Spectral Theorem, which is given in a variety of useful forms. This theorem is then applied to a number of different applications. These include singular value decomposition and pseudoinverse, and applications to geometry, both in terms of understanding orthogonal matrices and in studying the geometry of solutions of quadratic equations. Applications to calculus are also discussed here.

Chapter 6 discusses further normal forms for quadratic forms and for general square matrices. The first section gives a proof that is closely related to the Gaussian elimination algorithm discussed in Chapter 1 that a symmetric matrix is congruent to a diagonal matrix with diagonal entries ± 1 and 0. Tests for when a matrix is positive definite are given, and there is a proof of Sylvester's law of inertia. The second section discusses the Jordan canonical form of a matrix from a linear transformation point of view. This is then used to give solutions of differential equations when the matrix involved is not diagonalizable.

The appendix gives the solutions to all of the embedded exercises.

SUPPLEMENTS

As mentioned above, there are supplements available from Wiley for use of MATLAB, Maple, or Mathematica in conjunction with the text. We will describe the MATLAB supplement in detail. The other two supplements are based on the MATLAB supplement and have a similar structure. The MATLAB supplement is entitled *MATLAB Labs for Linear Algebra.* It consists of 13 labs, which are tied closely to the presentation in the text. After an initial lab, which introduces the basic features of MATLAB, all the other labs introduce the student to relevant features of MATLAB to use it as an effective tool to understand the concepts of linear algebra and work problems. Students will first be told what commands to input into MATLAB through examples, and later on will be asked questions about output as well as work problems similar to exercises within the text using MATLAB. These labs use a teaching software toolbox, which uses the same notation as in the text. It is available free through The MathWorks, Inc., or directly from the author. These labs may be used in a separate laboratory section of the class, as homework, or independently. We recommend highly using one of the versions of these labs, as it allows the student to easily learn how to use software tools in studying linear algebra, and can free class time to be used more on mastering the underlying concepts rather than working through by hand computations best left to the computer.

A second supplement is the *Student Solutions Manual to Accompany Linear Algebra.* This book is intended to aid the student in studying from this text. It consists of seven chapters. Each of the first six chapters corresponds to a chapter in the book. These chapters consist of four parts. The first part gives an overview of the main ideas in the chapter. It is intended that the student lightly read this before beginning the chapter, then read carefully the overview of a section before reading that section. After reading the section and working some of the embedded exercises, the student should reread the overview and use it as a guide to see if the main concepts have been mastered. A second section lists major terms in the chapter, as well as major concepts. This is intended to be used as discussion items for groups of students to test their mastery of the section. A third section gives detailed solutions to a selection of embedded exercises. Although all of the embedded exercises have their answers in the appendix, many of these just have an answer with no explanation as to how to obtain it. Here we provide complete explanations of these solutions. Finally, there is a section that provides detailed solutions for about one-fifth of the exercises at the end of each chapter. The last chapter provides some sample examinations with solutions to be used by students in preparing for their own examinations in the course.

The final supplement is the *Instructor's Manual to Accompany Linear Algebra.* This provides suggestions of topics to emphasize and timing in developing a syllabus. It gives notes on the text by section, as well as suggested syllabi for different courses. It also contains solutions to all of the chapter end exercises.

ACKNOWLEDGMENTS

Any book is influenced strongly by ideas from the many books that have preceded it. In my case, two books have had a particularly strong influence on my ultimate conception of how to best present the ideas of linear algebra to the diverse group of students that take it. The first was by my dissertation adviser, Hans Samelson, whose treatment emphasizes a conceptual approach of vector spaces and linear transformations. The second was by Gilbert Strang, whose approach emphasized a matrix oriented approach with computational aspects such as important matrix decompositions. While developing a balanced approach emphasizing the interrelation of the most important aspects of each approach, important user-friendly software became easily available, which made this new approach more feasible. In my case, the software of choice is MATLAB, but other packages can be used as well. This software made it possible to spend less time on routine computations and illustrate key ideas from both the theoretical and computational aspects of the subject, as well as their interrelationship.

I have taught out of preliminary forms of this book as I developed it for almost 10 years. These started with extensive notes to supplement other texts and developed into a full-fledged book prepared in AMS-LATEX. During the last few years, other faculty at Tulane have also used these preliminary forms and have given me useful feedback on its improvement. These are Lisa Fauci, Laszlo Fuchs, Ronald Knill, Victor Moll, Frank Quigley, Steve Rosencrans, and Dagang Yang. There have also been a number of graduate students at Tulane who have served as teaching assistants here using the preliminary versions whose input I wish to acknowledge: Dean Bottino, George Boros, Ken Brumer, Mike Cusac, Earl Packard, and Jeff Seifert. Perhaps the strongest source of feedback has been from the many students at Tulane who have used the book, whom I wish to thank for their help, particularly those who endured its rough early manifestations. I also wish to acknowledge the invaluable help of the Mathematics Department staff at Tulane: Geralyn Caradona, Susan Lam, and Meredith Mickel. The development of the book has received valuable input from the book's reviewers: Robert L. Borrelli, Harvey Mudd College, Ken W. Bosworth, Idaho State University, Jim Bruening, Southeast Missouri State University, Carl C. Cowen, Purdue University, Richard H. Elderkin, Pomona College, Sidney Graham, Michigan Technological University, Lynn Kiser, Rose-Hulman Institute of Technology, Stanley O. Kochman, York University, Ana M. Mantilla, University of Arizona, Steven C. McKelvey, Saint Olaf College, Frank D. Quigley, Tulane University, William W. Smith, University of North Carolina, and Avi Vardi, Drexel University. I would also like to thank the editorial and production staff at Wiley involved with the book, with particular thanks to Barbara Holland, Cindy Rhoads, and Ingrao Associates, who handled the production process.

Finally, I would like to thank my wife Barbara, and daughters Amy and Katherine, for their support, encouragement, and love while I have worked on this book.

LINEAR ALGEBRA

CHAPTER 1

MATRIX ALGEBRA

1.1. FINDING THE GENERAL SOLUTION OF ONE EQUATION IN n UNKNOWNS

We will start off by learning the Gaussian elimination algorithm for solving a matrix equation $A\mathbf{x} = \mathbf{b}$. First, however, we need to introduce a few basic definitions concerning matrices and relate matrices to linear equations. By a single linear equation we mean an equation of the form

$$a_1x_1 + a_2x_2 + \cdots + a_nx_n = b$$

Here the x_i are called the **variables,** and the numbers a_i are called the **coefficients** of the equation. The x_i's determine a point $\mathbf{x} = (x_1, \ldots, x_n)$ in Euclidean n-space $\mathbb{R}^n$ as do the coefficients $\mathbf{a} = (a_1, \ldots, a_n)$. A point in $\mathbb{R}^n$ is called a **vector** in $\mathbb{R}^n$. Our convention will be to denote vectors in boldface to distinguish them from real numbers, which we will call **scalars.** When $n = 3$ the space where our solutions will lie is usual 3-space.

In $\mathbb{R}^n$ there are two operations on vectors, addition and scalar multiplication. To add two vectors, we just add their coordinates:

$$(x_1, \ldots, x_n) + (y_1, \ldots, y_n) = (x_1 + y_1, \ldots, x_n + y_n)$$

We achieve scalar multiplication by multiplying the scalar by each entry of the vector:

$$c(x_1, \ldots, x_n) = (cx_1, \ldots, cx_n)$$

For example,

$$(2,3,1,4)+(2,-1,0,-3) = (4,2,1,1), \quad 3(2,3,1,4) = (6,9,3,12)$$

We illustrate these operations in Figure 1.1.

For two vectors $\mathbf{a} = (a_1, \ldots, a_n), \mathbf{x} = (x_1, \ldots, x_n)$ in $\mathbb{R}^n$, there is the **dot product,** denoted $\langle \mathbf{a}, \mathbf{x} \rangle$,

$$\langle \mathbf{a}, \mathbf{x} \rangle = a_1 x_1 + \cdots + a_n x_n$$

The equation $a_1x_1 + a_2x_2 + \cdots + a_nx_n = b$ is just the equation saying that the dot product of the vectors **a** and **x** is the number b. In several variable calculus the dot product is shown to be related to the geometric concepts of length and angle. The dot product of a vector with itself just gives the square of the length of the vector:

$$\langle \mathbf{x}, \mathbf{x} \rangle = x_1^2 + \ldots + x_n^2 = \|\mathbf{x}\|^2$$

Here we are denoting the length of **x** by $\|\mathbf{x}\|$. The angle θ between two vectors $\mathbf{x}, \mathbf{y}$ is given through

$$\langle \mathbf{x}, \mathbf{y} \rangle = \|\mathbf{x}\| \, \|\mathbf{y}\| \cos\theta$$

This formula is derived from applying the law of cosines to the triangle with edges $\mathbf{x}, \mathbf{y}, \mathbf{x} - \mathbf{y}$. The condition that two vectors $\mathbf{x}, \mathbf{y}$ are perpendicular is thus given by

$$\mathbf{x} \perp \mathbf{y} \text{ iff } \langle \mathbf{x}, \mathbf{y} \rangle = 0$$

In solving $a_1x_1 + a_2x_2 + \cdots + a_nx_n = b$ we think of **a** and b as being given and then try to solve for all such vectors **x**. When **a** is not the zero vector **0** and $b = 0$, then the solution is what is called a hyperplane in geometry. For $n = 3$, this will just be a plane through the origin whose normal vector is the vector **a**. For $n = 2$, this will be a line through the origin which has **a** as a normal vector. When b is not zero, the plane and line involved will be parallel to the ones given when $b = 0$. For example,

$$2x + 3y - 4z = 0$$

describes a plane through the origin in 3-space with normal vector $(2, 3, -4)$. If we look at the solutions to the equation

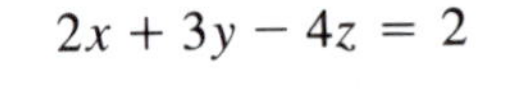

$$2x + 3y - 4z = 2$$

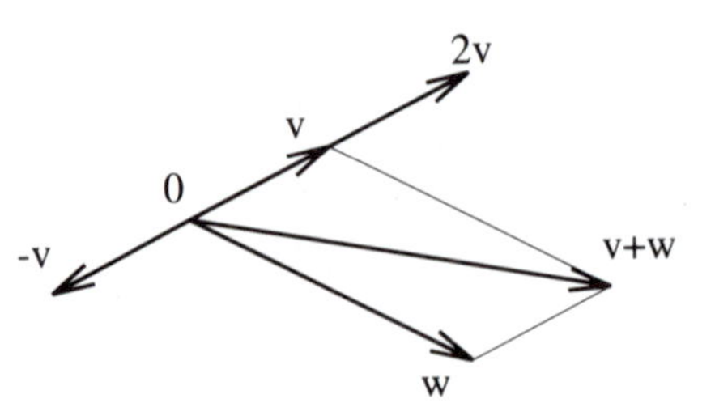

Figure 1.1. Addition and scalar multiplication of vectors

Figure 1.2. Parallel planes

this will be a plane which is parallel to the preceding plane. Two such parallel planes are illustrated in Figure 1.2.

Since these planes each contain an infinite number of points, there are an infinite number of solutions to either of these equations. We will want to describe all the solutions in an efficient manner, and this will lead us into some of the primary definitions in vector space theory. For the moment, however, let us note that the first equation has the property that if we have any solution and we multiply each of its entries by a scalar, then we will still get a solution. For example, $(x, y, z) = (3, -2, 0)$ is a solution and so is $c(3, -2, 0)$ for any real number c. For

$$\langle \mathbf{a}, c(3, -2, 0)\rangle = c\langle \mathbf{a}, (3, -2, 0)\rangle = c(0) = 0$$

Moreover, if we take two solutions such as $(3, -2, 0)$ and $(2, 0, 1)$ and add them to get the vector $(5, -2, 1)$, then it is easy to check that this is also a solution. These two statements depend on the fact that the dot product satisfies the following **linearity properties** for arbitrary vectors $\mathbf{a}, \mathbf{v}, \mathbf{w}$ and real number c:

$$\begin{aligned} \langle \mathbf{a}, \mathbf{v} + \mathbf{w}\rangle &= \langle \mathbf{a}, \mathbf{v}\rangle + \langle \mathbf{a}, \mathbf{w}\rangle \\ \langle \mathbf{a}, c\mathbf{v}\rangle &= c\langle \mathbf{a}, \mathbf{v}\rangle \end{aligned}$$

This implies that if $\mathbf{v}$ and $\mathbf{w}$ are solutions (and so $\langle \mathbf{a}, \mathbf{v}\rangle = 0 = \langle \mathbf{a}, \mathbf{w}\rangle$), then so are $\mathbf{v}+\mathbf{w}$ and $c\mathbf{v}$.

Given two vectors $\mathbf{v}, \mathbf{w}$, and two numbers c, d, we can form a new vector $c\mathbf{v} + d\mathbf{w}$ which is called a **linear combination** of $\mathbf{v}, \mathbf{w}$. In general, any linear combination $c\mathbf{v} + d\mathbf{w}$ of two solutions will be a solution. The two equations can then be combined to show

$$\langle \mathbf{a}, c\mathbf{v} + d\mathbf{w}\rangle = c\langle \mathbf{a}, \mathbf{v}\rangle + d\langle \mathbf{a}, \mathbf{w}\rangle = c(0) + d(0) = 0$$

We will show later that every solution is a linear combination

$$c(3, -2, 0) + d(2, 0, 1)$$

of the two solutions. The solutions form a plane through the origin which is determined by these two vectors.

To see how we can arrive at the general form of the solution, we will move the terms involving the variables y and z to the right-hand side and solve for x in terms of them. We will call y and z the free variables and x the basic variable. The only crucial fact used in selecting to solve for x is that the coefficient of x is nonzero; in this particular example we could have solved for y or z as well. We get $2x = -3y + 4z$, or $x = -1.5y + 2z$. We can then write our solution

$$(x, y, z) = (-1.5y + 2z, y, z) = y(-1.5, 1, 0) + z(2, 0, 1)$$

and thus we see that every solution is a linear combination of the two solutions $(-1.5, 1, 0)$ and $(2, 0, 1)$. Since $(-1.5, 1, 0) = -.5(3, -2, 0)$, this also establishes that all solutions are linear combinations of the two solutions $(3, -2, 0)$ and $(2, 0, 1)$, as we claimed earlier. When we discuss linear independence later, we will also show that these two solutions are linearly independent. This means that there is no way to write the general solution as multiples of just one vector (geometrically this is just saying that the solution space is not a line). Note that the solution $(-1.5, 1, 0)$ corresponds to assigning the free variables $(y, z) = (1, 0)$ and solving for x. Similarly, the solution $(2, 0, 1)$ arises by assigning $(y, z) = (0, 1)$ and solving for x.

Now let us look at the equation $2x + 3y - 4z = 2$. A way to get one solution to this is to assign the free variables y and z to be 0 and solve for $x = 1$, giving the solution $(1, 0, 0)$. If (x, y, z) is any other solution to this equation, then we can check that the linearity of the dot product implies that the difference $(x, y, z) - (1, 0, 0)$ is a solution to the (associated homogeneous) equation

$$\langle \mathbf{a}, \mathbf{v} \rangle = 0$$

But we already know all the solutions of this equation, so

$$(x, y, z) - (1, 0, 0) = y(-1.5, 1, 0) + z(2, 0, 1)$$

This gives the general solution

$$(x, y, z) = (1, 0, 0) + y(-1.5, 1, 0) + z(2, 0, 1)$$

Of course we could have just started with the equation $2x + 3y - 4z = 2$, moved the terms involving y and z to the right-hand side, solved for $x = 1 - 1.5y + 2z$, and then written $(x, y, z) = (1 - 1.5y + 2z, y, z)$, which then factors in the preceding combination. The form $(1, 0, 0) + y(-1.5, 1, 0) + z(2, 0, 1)$ makes more explicit that $(1, 0, 0)$ is one solution (called a **particular solution**) found by assigning 0 to each of the free variables y and z and that all other solutions occur from adding on linear combinations of the two solutions $(-1.5, 1, 0)$ and $(2, 0, 1)$ of the associated homogeneous equation. These two solutions are found by setting the free variables (y, z) equal to $(1, 0)$ and $(0, 1)$, respectively, when solving the associated homogeneous equation.

We can formulate an algorithm for writing down the general solution of one equation in n variables, based on the foregoing procedure. Informally, what we would do is select one of the variables which has a nonzero coefficient to be the basic variable and call all the other variables free variables. For definiteness we will choose as the basic variable the first one whose coefficient is nonzero in the equation. Then we subtract from each side of the equation the terms involving the free variables to get a new equation that expresses a multiple of the basic variable in terms of the free variables. Next we divide each side by the coefficient of the basic variable to solve for the basic variable in terms of the free variables. Finally, we substitute this expression for the basic variable into the vector $\mathbf{x} = (x_1, \ldots, x_n)$ to write $\mathbf{x}$ in terms of the free variables. When the free variables are all set to 0, this gives one solution, which is called a particular solution. Separating this out from the right-hand side and decomposing the remaining terms as a linear combination where the coefficients involved are the free variables writes the general solution in the form

$$\mathbf{x} = \mathbf{v}_0 + c_1\mathbf{v}_1 + \cdots + c_k\mathbf{v}_k$$

Here $\mathbf{v}_0$ is a particular solution found by setting all the free variables equal to 0, $k = n - 1$, and $\mathbf{v}_i$ is the solution of the associated homogeneous equation (coming from replacing b by 0) found by setting the ith free variable equal to 1 and all the other free variables equal to 0 and then solving for the basic variable. The coefficients c_i will occur as the free variables themselves in this method, but the important fact about them is that they are arbitrary real numbers.

Let's look at some more examples.

■ ***Example 1.1.1*** Consider the equation $5x - 4y + z = 3$. Since the coefficient of x is nonzero, it will serve as the basic variable. We rewrite the equation as $5x = 3 + 4y - z$. We divide by 5 to give $x = .6 + .8y - .2z$. Thus the solution is

$$(x, y, z) = (.6 + .8y - .2z, y, z) = (.6, 0, 0) + y(.8, 1, 0) + z(-.2, 0, 1)$$

The particular solution when each of the free variables is set to 0 is $(.6, 0, 0)$, and the solution to the associated homogeneous equation $5x - 4y + z = 0$ is the general linear combination of the solutions $(.8, 1, 0)$ and $(-.2, 0, 1)$, each of which could be found by setting one of the free variables equal to 1 and the others equal to 0 and then solving for the basic variable x. ■

■ ***Example 1.1.2*** Consider the equation in the five variables $x_1, \ldots, x_5$ given by

$$2x_2 - 3x_4 + x_5 = 6$$

Then we take x_2 as the basic variable and the other variables as the free variables. We solve

$$\begin{aligned} 2x_2 &= 6 + 3x_4 - x_5 \\ x_2 &= 3 + 1.5x_4 - .5x_5 \end{aligned}$$

Then

$$\begin{aligned}\mathbf{x} &= (x_1, 3 + 1.5x_4 - .5x_5, x_3, x_4, x_5)\\ &= (0, 3, 0, 0, 0) + x_1(1, 0, 0, 0, 0) + x_3(0, 0, 1, 0, 0)\\ &\quad + x_4(0, 1.5, 0, 1, 0) + x_5(0, -.5, 0, 0, 1)\end{aligned}$$

We have written the solution as the sum of the particular solution $(0, 3, 0, 0, 0)$ and the general solution to the associated homogeneous problem. The latter solution is a general linear combination of four solutions, one for each free variable. Each is found from the homogeneous equation by assigning that free variable equal to 1, the other free variables equal to 0, and then solving for the basic variables. ■

So far it has not been very difficult to solve a single linear equation in n variables. There is one case, which may seem very trivial at this point but which will arise later, that we have ignored. That is the case when the vector **a** is the zero vector. Then our equation takes the form $\langle \mathbf{0}, \mathbf{x} \rangle = b$. This has no solution unless $b = 0$; in that case any **x** is a solution. Let us summarize the discussion so far.

Consider the equation

$$\langle \mathbf{a}, \mathbf{x} \rangle = a_1 x_1 + \cdots + a_n x_n = b$$

■ **Case 1.** **a** is not the zero vector.
The general solution is of the form

$$\mathbf{v}_0 + c_1 \mathbf{v}_1 + \cdots + c_k \mathbf{v}_k$$

where $k = n - 1$, the c_i are arbitrary scalars. The first variable with nonzero coefficient is the basic variable, and the others are free variables. $\mathbf{v}_0$ is a solution which is found by assigning all the free variables equal to 0 and solving for the basic variable. $\mathbf{v}_i$ is the solution of the associated homogeneous equation which is found by assigning the ith free variable equal to 1, the other free variables equal to 0, and solving for the basic variable.

■ **Case 2.** $\mathbf{a} = \mathbf{0}$, b is not equal to 0.
Then there is no solution.

■ **Case 3.** $\mathbf{a} = \mathbf{0}$, $b = 0$.
Then every **x** is a solution. ■

Exercise 1.1.1

Find the general solution of each of the following equations.

1. $2x - 3y = 5$.
2. $4y - z = 7$ (variables are x, y, z).
3. $5x_1 - 4x_3 + x_4 = 1$ (variables $x_1, \ldots, x_5$).

1.2. MATRICES AND SYSTEMS OF EQUATIONS

It was relatively easy to find the solution of a single linear equation in n variables. We now want to reduce the general problem of solving m equations in n unknowns to a situation where we can apply the same type of algorithm to read off the general solution. We will need some more complicated notation even to write down the general system of equations. Since there are m equations, we will have to index the vectors which occur. We will call them $\mathbf{A}_1, \ldots, \mathbf{A}_m$. We will also have m right-hand sides, which we will make into a vector $\mathbf{b} = (b_1, \ldots, b_m)$. Then our m equations are

$$\begin{aligned} \langle \mathbf{A}_1, \mathbf{x} \rangle &= b_1 \\ \langle \mathbf{A}_2, \mathbf{x} \rangle &= b_2 \\ &\vdots \\ \langle \mathbf{A}_m, \mathbf{x} \rangle &= b_m \end{aligned}$$

To simplify our expression for these equations, we introduce matrix notation.

DEFINITION 1.2.1. By an m **by** n **matrix** of real numbers we mean an arrangement of mn real numbers into m rows and n columns. A 1 by n matrix is also called a **row vector,** and it can be identified with a vector in $\mathbb{R}^n$. An m by n matrix has m rows and so determines m row vectors, which then can be identified with m vectors in $\mathbb{R}^n$. Similarly, an m by 1 matrix is called a **column vector** and can be identified with a vector in $\mathbb{R}^m$. An m by n matrix contains n columns and so determines n column vectors and thus n vectors in $\mathbb{R}^m$.

For example,

$$\begin{pmatrix} 3 & 4 & 8 \end{pmatrix}$$

is a 1 by 3 matrix,

$$\begin{pmatrix} 1 & 9 & 3 \\ 3 & 2 & 0 \\ 1 & 0 & 8 \end{pmatrix}$$

is a 3 by 3 matrix,

$$\begin{pmatrix} 9 \\ 0 \\ 2 \end{pmatrix}$$

is a 3 by 1 matrix, and

$$\begin{pmatrix} 1 & 2 & 3 & 4 & 5 \\ 2 & 3 & 4 & 5 & 6 \\ 3 & 4 & 5 & 6 & 7 \end{pmatrix}$$

is a 3 by 5 matrix.

Row vectors (column vectors) may be added by adding the corresponding entries (i.e., identifying them with vectors in $\mathbb{R}^n$ ($\mathbb{R}^m$) and doing usual vector addition there). We can also multiply a (row or column) vector by a scalar by multiplying each entry by that scalar.

Let us introduce some notation concerning matrices.

DEFINITION 1.2.2. The entry in a matrix that is in the ith row and jth column is called the **ij-entry** of the matrix. If the matrix has the same number of rows as columns, the matrix is called **square.** Two m by n matrices A, B may be added by adding their corresponding entries: if $C = A + B$, then $c_{ij} = a_{ij} + b_{ij}$. A scalar r may be multiplied by an m by n matrix by multiplying the scalar by each entry in the matrix: if $D = rA$, then $d_{ij} = ra_{ij}$. These operations on matrices are consistent with the operations on row and column vectors given above.

For example, the 12-entry of

$$A = \begin{pmatrix} 2 & 3 & -1 \\ 1 & 0 & 1 \\ 3 & 1 & -1 \\ 0 & 2 & 1 \end{pmatrix}$$

is 3 and the 21-entry is 1.

If we denote the whole matrix by A, then we will denote its rows by $\mathbf{A}_1, \ldots, \mathbf{A}_m$ and its columns by $\mathbf{A}^1, \ldots, \mathbf{A}^n$. Note that when we have m equations in n variables, we can construct an m by n matrix, which is called the **coefficient matrix** of the system, so that the vectors $\mathbf{A}_1, \ldots, \mathbf{A}_m$ occurring in the equations are just the rows of this matrix. We can also make the vector $\mathbf{b}$ with the right-hand sides into a column vector. We now describe how to multiply a matrix times a vector so that our equations can just be written as $A\mathbf{x} = \mathbf{b}$. We first describe how to multiply a 1 by n row vector $\mathbf{a}$ by an n by 1 column vector $\mathbf{x}$. We just identify each with the corresponding vector in $\mathbb{R}^n$ and then take the dot product. For example,

$$\begin{pmatrix} 2 & 1 & 3 \end{pmatrix} \begin{pmatrix} 2 \\ -9 \\ 2 \end{pmatrix} = \langle (2, 1, 3), (2, -9, 2) \rangle = 4 - 9 + 6 = 1$$

In general,

$$\mathbf{ax} = \begin{pmatrix} a_1 & \ldots & a_n \end{pmatrix} \begin{pmatrix} x_1 \\ \vdots \\ x_n \end{pmatrix} = \langle \mathbf{a}, \mathbf{x} \rangle = a_1 x_1 + \cdots + a_n x_n$$

When we multiply an m by n matrix by an n by 1 column vector, the result is supposed to be an m by 1 column vector. It thus has m entries and the ith one is found by multiplying the ith row of the matrix (considered as a row vector) by the given column vector. For example, when we multiply the matrix

$$A = \begin{pmatrix} 2 & 3 & -1 \\ 1 & 0 & 1 \\ 3 & 1 & -1 \\ 0 & 2 & 1 \end{pmatrix}$$

by the column vector $\mathbf{x} = \begin{pmatrix} 1 \\ 2 \\ -1 \end{pmatrix}$ the result is a column vector

$$A\mathbf{x} = \begin{pmatrix} \mathbf{A}_1\mathbf{x} \\ \mathbf{A}_2\mathbf{x} \\ \mathbf{A}_3\mathbf{x} \\ \mathbf{A}_4\mathbf{x} \end{pmatrix} = \begin{pmatrix} 9 \\ 0 \\ 6 \\ 3 \end{pmatrix}$$

As another example,

$$\begin{pmatrix} 2 & 4 & 5 & -2 \\ 1 & 0 & -2 & 1 \end{pmatrix} \begin{pmatrix} 1 \\ 2 \\ 0 \\ 2 \end{pmatrix} = \begin{pmatrix} 6 \\ 3 \end{pmatrix}$$

Thus a system of m linear equations in n unknowns can always be rewritten as one matrix equation. For the matrix A just given, the corresponding equations for the matrix equation

$$A\mathbf{x} = \begin{pmatrix} 9 \\ 0 \\ 6 \\ 3 \end{pmatrix}$$

are

$$\begin{aligned} 2x_1 + 3x_2 - x_3 &= 9 \\ x_1 \qquad\quad + x_3 &= 0 \\ 3x_1 + \ x_2 - x_3 &= 6 \\ 2x_2 + x_3 &= 3 \end{aligned}$$

and our computation shows that a solution is given by $\mathbf{x} = \begin{pmatrix} 1 \\ 2 \\ -1 \end{pmatrix}$.

DEFINITION 1.2.3. The n by n matrix that has the ijth entry equal to 0 if $i \neq j$ and equal to 1 if $i = j$ is called the **identity matrix** of size n. It is denoted by I (if n is clear) or by I_n if we wish to make clear what its size is.

For example, the matrix I_3 is

$$\begin{pmatrix} 1 & 0 & 0 \\ 0 & 1 & 0 \\ 0 & 0 & 1 \end{pmatrix}$$

Note that multiplication by I sends $\mathbf{x}$ to itself, $I\mathbf{x} = \mathbf{x}$. Thus the solution of $I\mathbf{x} = \mathbf{b}$ is $\mathbf{x} = \mathbf{b}$.

The operation of multiplying a matrix by a column vector (which just comes from dot product) satisfies a property called **linearity,** which the dot product has. Given an m by n matrix A and two n by 1 column vectors $\mathbf{v}, \mathbf{w}$, then

$$A(\mathbf{v} + \mathbf{w}) = A\mathbf{v} + A\mathbf{w}$$

Also, if c is any scalar, then $A(c\mathbf{v}) = cA\mathbf{v}$. Later we will abstract these two properties into the definition of a linear transformation, and say that matrix multiplication by A is a linear transformation from $\mathbb{R}^n$ to $\mathbb{R}^m$. For the moment, we just describe their implications toward solving $A\mathbf{x} = \mathbf{b}$. Suppose $\mathbf{v}, \mathbf{w}$ are two solutions. Then $A(\mathbf{v} - \mathbf{w}) = A\mathbf{v} - A\mathbf{w} = \mathbf{b} - \mathbf{b} = \mathbf{0}$, where $\mathbf{0}$ represents the **zero vector,** that is, the column vector all of whose entries are 0.

Thus if $\mathbf{v}$ is one solution of $A\mathbf{x} = \mathbf{b}$ and $\mathbf{w}$ is any other solution, $\mathbf{v} = \mathbf{w} + \mathbf{z}$, where $\mathbf{z} = \mathbf{v} - \mathbf{w}$ is a solution of $A\mathbf{x} = \mathbf{0}$. See Figure 1.3.

DEFINITION 1.2.4. This equation $A\mathbf{x} = \mathbf{0}$ associated with $A\mathbf{x} = \mathbf{b}$ is called the **associated homogeneous equation.**

The problem of finding the general solution to $A\mathbf{x} = \mathbf{b}$ is equivalent to finding one solution of $A\mathbf{x} = \mathbf{b}$ and the general solution of $A\mathbf{x} = \mathbf{0}$. Now let us see what linearity implies about the solutions of $A\mathbf{x} = \mathbf{0}$. If $\mathbf{v}, \mathbf{w}$ are solutions and if c, d are any scalars, then

$$A(c\mathbf{v} + d\mathbf{w}) = A(c\mathbf{v}) + A(d\mathbf{w}) = cA\mathbf{v} + dA\mathbf{w} = \mathbf{0} + \mathbf{0} = \mathbf{0}$$

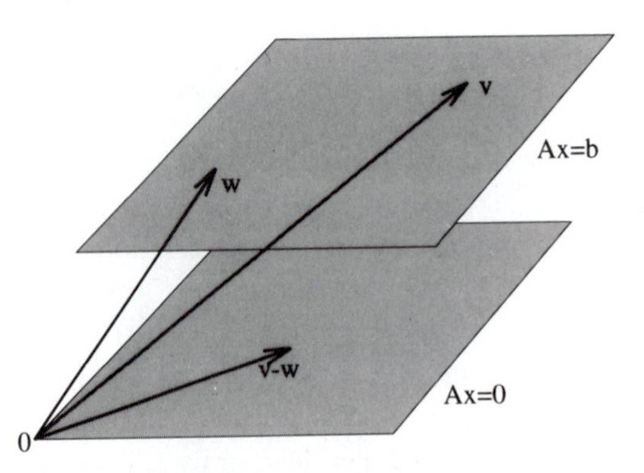

Figure 1.3. Homogeneous solution and general solution

DEFINITION 1.2.5 The expression $c\mathbf{v} + d\mathbf{w}$ is called a **linear combination** of the two vectors $\mathbf{v}$ and $\mathbf{w}$. By a general linear combination of k vectors $\mathbf{v}_1, \ldots, \mathbf{v}_k$ we mean something of the form

$$c_1\mathbf{v}_1 + \cdots + c_k\mathbf{v}_k$$

Here the c_i are scalars.

Thus linear combinations of solutions of the homogeneous equation are also solutions of the homogeneous equation. The general solution of $A\mathbf{x} = \mathbf{b}$ will turn out to be the sum of a particular solution and a general linear combination of some solutions of the homogeneous equation $A\mathbf{x} = \mathbf{0}$. Thus the general solution will be of the form

$$\mathbf{v}_0 + c_1\mathbf{v}_1 + \cdots + c_k\mathbf{v}_k$$

where $\mathbf{v}_0$ is a particular solution and each $\mathbf{v}_i$ is a solution of the associated homogeneous equation for $i = 1, \ldots, k$.

Exercise 1.2.1.

Multiply the row vector $\mathbf{a} = (2 \quad 8 \quad -1)$ by the column vector $\mathbf{b} = \begin{pmatrix} 4 \\ 6 \\ -1 \end{pmatrix}$.

Exercise 1.2.2.

Multiply the matrix $A = \begin{pmatrix} 2 & 8 & -1 \\ 3 & 0 & 1 \\ 0 & 2 & 1 \end{pmatrix}$ by the vector $\mathbf{b}$.

Exercise 1.2.3.

Rewrite the system of equations

$$\begin{aligned} 3x + 9y - 2z &= 2 \\ 2x - 2y \phantom{{}+3z} &= 3 \\ -x + y + 3z &= 0 \end{aligned}$$

as a matrix equation.

Exercise 1.2.4.

Rewrite the matrix equation $A\mathbf{x} = \mathbf{b}$, where

$$A = \begin{pmatrix} 3 & 4 & 5 \\ 1 & 2 & 1 \end{pmatrix}, \quad \mathbf{b} = \begin{pmatrix} 1 \\ 0 \end{pmatrix}$$

as two equations in three unknowns.

Exercise 1.2.5.

Suppose

$$A = \begin{pmatrix} 1 & 2 & -1 \\ 2 & 1 & 1 \\ -1 & 1 & 2 \end{pmatrix}, \quad \mathbf{b} = \begin{pmatrix} 1 \\ 3 \\ -1 \end{pmatrix}, \quad \mathbf{c} = \begin{pmatrix} 3 \\ 1 \\ 1 \end{pmatrix}$$

Compute $A\mathbf{c}, A(\mathbf{b} + \mathbf{c}), A\mathbf{b} + A\mathbf{c}$. Compute $5\mathbf{b} - 3\mathbf{c}$ and show $A(5\mathbf{b} - 3\mathbf{c}) = 5A\mathbf{b} - 3A\mathbf{c}$.

1.3. SOLVING SYSTEMS IN REDUCED FORM

To understand better the reason for the procedure of Gaussian elimination to solve systems of equations, we will first look at some systems that are easy to solve in terms of our methods of solving one equation.

■ ***Example 1.3.1.*** The first system we look at is

$$\begin{aligned} x + y + \phantom{z - {}} 3w &= 2 \\ z - 2w &= 1 \end{aligned}$$

In matrix terms this corresponds to the matrix equation $A\mathbf{v} = \mathbf{b}$, where

$$A = \begin{pmatrix} 1 & 1 & 0 & 3 \\ 0 & 0 & 1 & -2 \end{pmatrix}, \quad \mathbf{v} = \begin{pmatrix} x \\ y \\ z \\ w \end{pmatrix}, \quad \mathbf{b} = \begin{pmatrix} 2 \\ 1 \end{pmatrix}$$ ■

Note that the first nonzero coefficient of each equation is 1 and the variable involved does not appear in the other equation. We will call these variables, which are x and z in this case, the basic variables, and the remaining variables, which are y and w here, the free variables. Then each equation just has one basic variable, and it may be solved as was done earlier in terms of the free variables. We just move all the free variables to the right-hand side, and this will have solved for the basic variables in terms of the free variables. We can then substitute these expressions in terms of the free variables for the basic variables and then write the general solution in terms of a particular solution (when the free variables are all set to zero) plus a linear combination of solutions to the associated homogeneous system (where the right-hand side is set to 0). Those solutions will again come by setting one of the free variables equal to 1 and the remaining free variables equal to 0. In the preceding example, we get

$$\begin{aligned} x &= 2 - y - 3w \\ z &= 1 + 2w \end{aligned}$$

and so the general solution is

$$(x, y, z, w) = (2 - y - 3w, y, 1 + 2w, w) = (2, 0, 1, 0) + y(-1, 1, 0, 0) + w(-3, 0, 2, 1)$$

■ ***Example 1.3.2.*** As another example, consider the equations $A\mathbf{x} = \mathbf{b}$, where $A = \begin{pmatrix} 1 & 2 & 0 & 2 & 0 \\ 0 & 0 & 1 & 3 & 0 \\ 0 & 0 & 0 & 0 & 1 \end{pmatrix}$ and $\mathbf{b} = \begin{pmatrix} 2 \\ 1 \\ 3 \end{pmatrix}$. This corresponds to the system

$$\begin{aligned} x_1 + 2x_2 \quad + 2x_4 \quad &= 2 \\ x_3 + 3x_4 \quad &= 1 \\ x_5 &= 3 \end{aligned}$$

Now the basic variables are the variables corresponding to the first 1's in each row of A (here they are x_1, x_3, x_5). The remaining variables, which are x_2 and x_4, are called the free variables. Then we can solve for the basic variables in terms of the free variables by moving the terms involving the free variables to the right-hand side.

$$\begin{aligned} x_1 &= 2 - 2x_2 - 2x_4 \\ x_3 &= 1 \qquad\quad - 3x_4 \\ x_5 &= 3 \end{aligned}$$

The general solution can then be written as

$$\mathbf{x} = \begin{pmatrix} x_1 \\ x_2 \\ x_3 \\ x_4 \\ x_5 \end{pmatrix} = \begin{pmatrix} 2 - 2x_2 - 2x_4 \\ x_2 \\ 1 - 3x_4 \\ x_4 \\ 3 \end{pmatrix} = \begin{pmatrix} 2 \\ 0 \\ 1 \\ 0 \\ 3 \end{pmatrix} + x_2 \begin{pmatrix} -2 \\ 1 \\ 0 \\ 0 \\ 0 \end{pmatrix} + x_4 \begin{pmatrix} -2 \\ 0 \\ -3 \\ 1 \\ 0 \end{pmatrix} \qquad ■$$

■ ***Example 1.3.3.*** Here is a simple extension of the last example. We first add a row of zeros to A at the bottom and a 0 to $\mathbf{b}$; that is,

$$A = \begin{pmatrix} 1 & 2 & 0 & 2 & 0 \\ 0 & 0 & 1 & 3 & 0 \\ 0 & 0 & 0 & 0 & 1 \\ 0 & 0 & 0 & 0 & 0 \end{pmatrix}, \qquad \mathbf{b} = \begin{pmatrix} 2 \\ 1 \\ 3 \\ 0 \end{pmatrix}$$

Now the solutions to these four equations are just the solutions to the first three which also satisfy the fourth equation. Since all $\mathbf{x}$ satisfy the fourth equation, we see that adding this equation doesn't change the solution set at all. Similarly, we can add any number of zero rows to A as long as we add the same number of zero rows to $\mathbf{b}$ and we will not change the solution set. On the other hand, if we add a zero row to A but add a corresponding nonzero entry to $\mathbf{b}$ at the bottom, then this last equation will have no solutions and so the system $A\mathbf{x} = \mathbf{b}$ will have no solutions. For example, if we use the 4 by 5 A and let $\mathbf{b} = \begin{pmatrix} 2 \\ 1 \\ 3 \\ 4 \end{pmatrix}$, then there are no solutions to $A\mathbf{x} = \mathbf{b}$ since there are no solutions to the last equation. ■

A good question at this point is why we would even consider these last two cases with a zero row for A. The reason is that they arise through the process of Gaussian elimination, which we will use to simplify a general system to one of the forms already discussed.

DEFINITION 1.3.1. We will call an m by n matrix in **reduced normal form** if it has the following properties:

1. The first nonzero entry in each row is a 1.
2. The other entries in the columns containing these "first 1's" are equal to 0.
3. The first 1's move to the right as we move down the rows; that is, if the first nonzero entry in row i is in column $c(i)$, then $i < j$ implies $c(i) < c(j)$.
4. Any zero rows occur at the bottom of the matrix.

A matrix which is in reduced normal form may have some rows of zeros, but they will occur at the bottom of the matrix if there are any by condition 4.

DEFINITION 1.3.2. The variables corresponding to the first 1's, which are indexed by the column numbers $c(i)$, will be called the **basic variables.** The other variables will be called the **free variables.**

Each nonzero row will correspond to an equation where exactly one of the basic variables occurs. It will occur with coefficient 1 in this equation and will not occur in any other equation with nonzero coefficient.

It is easy to solve a system $R\mathbf{x} = \mathbf{b}$ if R is in reduced normal form. We record the method.

Algorithm for Solving $R\mathbf{x} = \mathbf{b}$

1. We first look at the zero rows at the bottom of R, if there are any. If there are any nonzero entries in $\mathbf{b}$ in the corresponding rows, then there is no solution of the system.
2. Suppose that all the entries of $\mathbf{b}$ corresponding to the zero rows are zero. Then we can truncate R and $\mathbf{b}$ by discarding the zero rows (which are imposing no conditions on $\mathbf{x}$) and get an equivalent system of r equations in n unknowns (where r is the number of nonzero rows of R). By moving the free variables to the right-hand side, we can then solve these equations one at a time for the basic variables in terms of the free variables. The general solution will be of the form $\mathbf{x} = \mathbf{v}_0 + c_1\mathbf{v}_1 + \cdots + c_k\mathbf{v}_k$, where the c_k are arbitrary scalars. The number k represents the number of free variables; it is $n - r$, where r is the number of nonzero rows. The vector $\mathbf{v}_0$ is the solution obtained by setting the free variables equal to 0 and solving for the basic variables (which can just be read off in this case). The vectors $\mathbf{v}_i$ are found by replacing $\mathbf{b}$ by the zero vector, setting the ith free variable equal to 1 and the other free variables equal to 0, and solving for the basic variables.

We now apply the algorithm to solve some systems that are in reduced normal form.

■ ***Example 1.3.4.*** $A = \begin{pmatrix} 1 & 8 & 0 & 0 & 2 \\ 0 & 0 & 1 & 0 & 5 \\ 0 & 0 & 0 & 1 & 5 \end{pmatrix}$, $\mathbf{b} = \begin{pmatrix} 1 \\ 2 \\ 1 \end{pmatrix}$

The basic variables are x_1, x_3, and x_4, and the free variables are x_2 and x_5. Setting the free variables equal to 0, we read off that the basic variables are then $x_1 = 1$, $x_3 = 2$, and $x_4 = 1$ and so $\mathbf{v}_0 = \begin{pmatrix} 1 \\ 0 \\ 2 \\ 1 \\ 0 \end{pmatrix}$. To find the solutions $\mathbf{v}_i, i = 1, 2$, we replace $\mathbf{b}$ by $\mathbf{0}$, set $x_2 = 1, x_5 = 0$, and solve for the basic variables $x_1 = -8, x_3 = 0, x_4 = 0$. Next set $x_2 = 0, x_5 = 1$, and solve for $x_1 = -2$, $x_3 = -5$, and $x_4 = -5$. Note that all these solutions can be read off from the columns of A corresponding to the free variables. For $\mathbf{v}_0$, form the vector by putting the free variables equal to zero and filling in the basic variables with the entries of $\mathbf{b}$. For $\mathbf{v}_i$, make the ith free variable equal to 1 and the other free variables equal to 0 and fill in the basic variables with the negatives of the entries of the ith column. Thus the general solution of $A\mathbf{x} = \mathbf{b}$ is

$$\mathbf{x} = \mathbf{v}_0 + c_1\mathbf{v}_1 + c_2\mathbf{v}_2 = \begin{pmatrix} 1 \\ 0 \\ 2 \\ 1 \\ 0 \end{pmatrix} + c_1 \begin{pmatrix} -8 \\ 1 \\ 0 \\ 0 \\ 0 \end{pmatrix} + c_2 \begin{pmatrix} -2 \\ 0 \\ -5 \\ -5 \\ 1 \end{pmatrix}$$ ■

■ ***Example 1.3.5.*** Let B be the matrix with the same first three rows as A but with a zero row at the end. If we make the fourth entry of $\mathbf{b}$ zero, then we get the exact same solutions as earlier. If, on the other hand, we make it to be nonzero, then there will be no solution to the system. ■

■ ***Example 1.3.6.*** $A = \begin{pmatrix} 1 & 0 & 0 & 0 \\ 0 & 1 & 0 & 0 \\ 0 & 0 & 1 & 0 \\ 0 & 0 & 0 & 1 \end{pmatrix}$, $\mathbf{b} = \begin{pmatrix} 1 \\ 2 \\ 3 \\ 4 \end{pmatrix}$

Then it is easy to read off the solution $\mathbf{x} = \mathbf{b}$, which is unique. If we add zero rows to A and $\mathbf{b}$, then we will still get the same unique solution. In the situation of matrices in reduced normal form, the only way to get a unique solution is to have no free variables, which can occur only if the matrix looks like A as just given with some zero rows attached. ■

■ ***Example 1.3.7.*** $A = \begin{pmatrix} 0 & 1 & 0 & 2 & 0 \\ 0 & 0 & 1 & 2 & 0 \\ 0 & 0 & 0 & 0 & 1 \end{pmatrix}$, $\mathbf{b} = \begin{pmatrix} 2 \\ 9 \\ 1 \end{pmatrix}$

We use the method explained in Example 3.4 to read off the general solution to $A\mathbf{x} = \mathbf{b}$ as

$$x = \begin{pmatrix} 0 \\ 2 \\ 9 \\ 0 \\ 1 \end{pmatrix} + c_1 \begin{pmatrix} 1 \\ 0 \\ 0 \\ 0 \\ 0 \end{pmatrix} + c_2 \begin{pmatrix} 0 \\ -2 \\ -2 \\ 1 \\ 0 \end{pmatrix} \qquad \blacksquare$$

In Exercises 1.3.1–1.3.8, find the general solution of $A\mathbf{x} = \mathbf{b}$ for the given A and $\mathbf{b}$. Be sure to check your answers; that is, compute $A\mathbf{x}$ and verify that it is $\mathbf{b}$.

Exercise 1.3.1.

$$A = \begin{pmatrix} 1 & 0 & 3 & 0 \\ 0 & 1 & 2 & 0 \end{pmatrix}, \qquad \mathbf{b} = \begin{pmatrix} 1 \\ 2 \end{pmatrix}$$

Exercise 1.3.2.

$$A = \begin{pmatrix} 0 & 1 & 1 \\ 0 & 0 & 0 \end{pmatrix}, \qquad \mathbf{b} = \begin{pmatrix} 2 \\ 1 \end{pmatrix}$$

Exercise 1.3.3.

$$A = \begin{pmatrix} 0 & 1 & 1 \\ 0 & 0 & 0 \end{pmatrix}, \qquad \mathbf{b} = \begin{pmatrix} 2 \\ 0 \end{pmatrix}$$

Exercise 1.3.4.

$$A = \begin{pmatrix} 1 & 0 & 1 & 0 & 1 \\ 0 & 1 & 1 & 0 & 1 \\ 0 & 0 & 0 & 1 & 1 \end{pmatrix}, \qquad \mathbf{b} = \begin{pmatrix} -1 \\ 7 \\ 2 \end{pmatrix}$$

Exercise 1.3.5.

$$A = \begin{pmatrix} 1 & 0 & 0 & 1 \\ 0 & 0 & 1 & 2 \\ 0 & 0 & 0 & 0 \end{pmatrix}, \qquad \mathbf{b} = \begin{pmatrix} 2 \\ 1 \\ 0 \end{pmatrix}$$

Exercise 1.3.6.

$$A = \begin{pmatrix} 1 & 2 & 3 & 0 \\ 0 & 0 & 0 & 1 \end{pmatrix}, \qquad \mathbf{b} = \begin{pmatrix} 1 \\ 2 \end{pmatrix}$$

Exercise 1.3.7.

$$A = \begin{pmatrix} 1 & 2 & 0 & 0 & 0 & 1 \\ 0 & 0 & 1 & 0 & 0 & 2 \\ 0 & 0 & 0 & 1 & 0 & 1 \\ 0 & 0 & 0 & 0 & 1 & 2 \end{pmatrix}, \qquad \mathbf{b} = \begin{pmatrix} 1 \\ 2 \\ 1 \\ 1 \end{pmatrix}$$

Exercise 1.3.8.

$$A = \begin{pmatrix} 1 & -1 & 0 & 0 & 0 & 0 & 0 & 1 \\ 0 & 0 & 1 & 0 & 0 & 2 & 0 & 1 \\ 0 & 0 & 0 & 1 & 0 & 3 & 0 & 0 \\ 0 & 0 & 0 & 0 & 1 & 1 & 0 & 2 \\ 0 & 0 & 0 & 0 & 0 & 0 & 1 & 0 \end{pmatrix}, \qquad \mathbf{b} = \begin{pmatrix} 1 \\ 2 \\ 3 \\ 1 \\ 1 \end{pmatrix}$$

1.4. GAUSSIAN ELIMINATION AND SOLVING GENERAL SYSTEMS

Now that we know how to find the general solution of a matrix equation $A\mathbf{x} = \mathbf{b}$, where A is in reduced normal form, the next step is to show that we can always change any given equation to an equation in reduced normal form without changing the solution set. The algorithm that we are going to describe for achieving this is called **Gaussian elimination;** it is based on simplifying a set of equations by eliminating one variable at a time from succeeding equations. We will first look at a couple of examples from the equation point of view, keeping track of how the corresponding matrix equation changes. We will than give a formulation entirely in terms of matrices. This will be the one which we will actually implement.

■ ***Example 1.4.1.*** Consider the system of equations

$$\begin{aligned} 2x - 3y &= 1 \\ 4x + y &= 9 \end{aligned}$$

The matrix form is

$$\begin{pmatrix} 2 & -3 \\ 4 & 1 \end{pmatrix} \begin{pmatrix} x \\ y \end{pmatrix} = \begin{pmatrix} 1 \\ 9 \end{pmatrix}$$

We will form a new matrix, which we will call the **augmented matrix** for the equation, by adding $\mathbf{b}$ as the last column. We will separate A from $\mathbf{b}$ by a vertical slash to make clear the fact that we are dealing with an augmented matrix. The augmented matrix here is

$$\left(\begin{array}{cc|c} 2 & -3 & 1 \\ 4 & 1 & 9 \end{array}\right)$$

Note that the augmented matrix serves as a bookkeeping device to keep track of the coefficients and the right-hand side of the equations at the same time. Now we will simplify these equations by eliminating the variable x from the second equation. We first compute the ratio $4/2 = 2$ of coefficients of x in the two equations, multiply the first equation by this ratio 2, and subtract 2 times the first equation from the second. We then get a new system of equations with the old first equation and a new second equation, which is the old second equation minus 2 times the first equation. Here the equations are

$$\begin{aligned} 2x - 3y &= 1 \\ 7y &= 7 \end{aligned}$$

The corresponding augmented matrix is

$$\left(\begin{array}{cc|c} 2 & -3 & 1 \\ 0 & 7 & 7 \end{array}\right)$$

■

The crucial fact about this new system is that it has the same solutions as the old system. It is easy to see that any solution of the original system is a solution to the new system since the operations of multiplying or adding equal quantities don't change the equality. To see that the solutions of the new equations are also solutions of the old ones depends on the fact that the process is reversible. That is, we can recover the old equations by doing similar operations. We get our old second equation from the new second equation by adding two times the first equation to the new second equation.

DEFINITION 1.4.1. The entries in a matrix where the row index equals the column index form what is called the **diagonal** of a matrix. The entries where the row index is greater than the column index are said to lie below the diagonal, and those entries where the row index is less than the column index are said to lie above the diagonal. If all the entries that lie below the diagonal are zero, then the matrix is called **upper triangular.** Similarly, a **lower triangular** matrix is one where all of the entries above the diagonal are zero. A **diagonal matrix** is one where all of the nondiagonal entries are zero.

The augmented matrix $\left(\begin{array}{cc|c} 2 & -3 & 1 \\ 0 & 7 & 7 \end{array}\right)$ as well as the coefficient part $\begin{pmatrix} 2 & -3 \\ 0 & 7 \end{pmatrix}$ are upper triangular. The diagonal entries in each matrix are 2 and 7. An example of a lower triangular matrix is $\begin{pmatrix} 1 & 0 \\ 2 & 1 \end{pmatrix}$.

Returning to our example, we can describe the operation on the augmented matrix as replacing the second row of the matrix by the old second row plus -2 times the first row.

DEFINITION 1.4.2. The operation on a matrix of replacing the ith row by the ith row plus r times the jth row and leaving all the other rows unchanged is denoted $O(i, j; r)$.

Note that the operation is coded in the notation by reading from right to left; that is, one multiplies r times the jth row and adds it to the ith row to form the new ith row. Another reason for the notation is that we will show in the next section that this operation can be achieved by multiplying on the left by a matrix $E(i, j; r)$, which is identical to the identity matrix except that the ij-entry is r.

In our example the new augmented matrix comes from the old one by performing the operation $O(2, 1; -2)$ on it. Note that this operation also describes what has happened to the coefficient part on the augmented matrix as well as what has happened to the right-hand side. Using the augmented matrix is a convenient way of dealing with both of these at once. Continuing with our example, we now want to solve for the variable y. We do this by dividing by the coefficient 7. Equivalently, we are replacing the second equation by $1/7$ times the second equation and leaving the first equation unchanged. Note that this process is also reversible (multiply, instead of divide, by 7). The new system is

$$\begin{aligned} 2x - 3y &= 1 \\ y &= 1 \end{aligned}$$

or in matrix form

$$\left(\begin{array}{cc|c} 2 & -3 & 1 \\ 0 & 1 & 1 \end{array}\right)$$

DEFINITION 1.4.3. We denote the operation of multiplying the ith row of a matrix by the nonzero scalar d and leaving the other rows unchanged by $Om(i; d)$.

We have performed $Om(2; 1/7)$ to find our new augmented matrix. From the equation viewpoint we now want to eliminate y from the first equation by substituting $y = 1$ into it and solving for x. The process of eliminating y from the matrix point of view is to make its coefficient 0 by subtracting the appropriate multiple of the second equation from the first; here that multiple is just the coefficient of y, which is -3. Thus on the matrix side we are performing the operation $O(1, 2; 3)$. The new equations and matrix are

$$\begin{aligned} 2x &= y \\ y &= 1 \end{aligned}$$

and

$$\left(\begin{array}{cc|c} 2 & 0 & 4 \\ 0 & 1 & 1 \end{array}\right)$$

The final step is to solve for x, which is done by dividing the first equation by 2; this is achieved on the matrix level by performing the operation $Om(1; 1/2)$. We get

$$\begin{aligned} x &= 2 \\ y &= 1 \end{aligned}$$

which has matrix equivalent

$$\left(\begin{array}{cc|c} 1 & 0 & 2 \\ 0 & 1 & 1 \end{array}\right)$$

■ ***Example 1.4.2.*** Consider the system of equations

$$\begin{aligned} x - y + z &= 2 \\ 3x + y - z &= 2 \\ x + 3y \phantom{{}-z} &= 4 \end{aligned}$$

which has augmented matrix

$$\left(\begin{array}{ccc|c} 1 & -1 & 1 & 2 \\ 3 & 1 & -1 & 2 \\ 1 & 3 & 0 & 4 \end{array}\right)$$

We first eliminate the variable x from the second and third equations. This involves first replacing the second equation by the second equation minus 3 times the first equation and then replacing the third equation by the third equation minus the first equation. In terms of the augmented matrix, we will first perform operation $O(2, 1; -3)$ and then perform operation $O(3, 1; -1)$. The two steps change the equations and corresponding matrix as follows.

$$\begin{aligned} x - y + z &= 2 \\ 4y - 4z &= -4 \\ 4y - z &= 2 \end{aligned}$$

$$\left(\begin{array}{ccc|c} 1 & -1 & 1 & 2 \\ 0 & 4 & -4 & -4 \\ 0 & 4 & -1 & 2 \end{array}\right)$$

We then ignore the first equation and concentrate on the last two equations. We eliminate the variable y in the third equation. We do this by subtracting the second equation from the third equation. For the matrix we will be applying the operation $O(3, 2; -1)$. The result is

$$\begin{aligned} x - y + z &= 2 \\ 4y - 4z &= -4 \\ 3z &= 6 \end{aligned}$$

$$\left(\begin{array}{ccc|c} 1 & -1 & 1 & 2 \\ 0 & 4 & -4 & -4 \\ 0 & 0 & 3 & 6 \end{array}\right)$$

The system is now in upper triangular form, as represented by the fact that the augmented matrix is an upper triangular matrix. The strategy is now to work from the bottom up, solving for the variables one at a time. This is called the back substitution part of the algorithm. This back substitution part of Gaussian elimination requires much less work in general than the forward elimination step of the algorithm we have just completed. Although it is not necessary from the viewpoint of solving the equation, we will continue to work on the whole augmented matrix as it will lead to a matrix decomposition, which we will use later. The only extra work in this approach is identifying each operation we use.

We first solve for z in the last equation and then eliminate z from the two preceding equations. For the matrix, this will consist of the operations

$$Om(3; 1/3), O(2, 3; 4), O(1, 3; -1)$$

The equations after each of these steps are

$$\begin{aligned} x - y + z &= 2 \\ 4y - 4z &= -4 \\ z &= 2 \end{aligned}$$

$$\begin{aligned} x - y + z &= 2 \\ 4y &= 4 \\ z &= 2 \end{aligned}$$

$$\begin{aligned} x - y &= 0 \\ 4y &= 4 \\ z &= 2 \end{aligned}$$

The corresponding forms of the augmented matrix are

$$\left(\begin{array}{ccc|c} 1 & -1 & 1 & 2 \\ 0 & 4 & -4 & -4 \\ 0 & 0 & 1 & 2 \end{array}\right), \quad \left(\begin{array}{ccc|c} 1 & -1 & 1 & 2 \\ 0 & 4 & 0 & 4 \\ 0 & 0 & 1 & 2 \end{array}\right), \quad \left(\begin{array}{ccc|c} 1 & -1 & 0 & 0 \\ 0 & 4 & 0 & 4 \\ 0 & 0 & 1 & 2 \end{array}\right)$$

We next solve for y and then eliminate y from the first equation. For the matrix this will involve $Om(2; 1/4)$ followed by $O(1, 2; 1)$. This will give us

$$\begin{aligned} x - y \quad &= 0 \\ y \quad &= 1 \\ z &= 2 \end{aligned}$$

and then

$$\begin{aligned} x &= 1 \\ y &= 1 \\ z &= 2 \end{aligned}$$

The matrix changes to

$$\left(\begin{array}{ccc|c} 1 & -1 & 0 & 0 \\ 0 & 1 & 0 & 1 \\ 0 & 0 & 1 & 2 \end{array}\right), \quad \left(\begin{array}{ccc|c} 1 & 0 & 0 & 1 \\ 0 & 1 & 0 & 1 \\ 0 & 0 & 1 & 2 \end{array}\right) \qquad \blacksquare$$

In the next examples we will no longer carry through each operation both on the equations and the augmented matrix. It should be clear from the two examples so far that the augmented matrix carries all the information on how the equations are changing. We will describe in words what we are doing to the equations and symbolically indicate the operation we are using on the matrix.

■ ***Example 1.4.3.***

$$\begin{aligned} x + y - 2z &= 0 \\ 2x + 2y - 3z &= 2 \\ 3x - y + 2z &= 12 \end{aligned}$$

which has augmented matrix

$$\left(\begin{array}{ccc|c} 1 & 1 & -2 & 0 \\ 2 & 2 & -3 & 2 \\ 3 & -1 & 2 & 12 \end{array}\right)$$

Eliminate x from the second and third equation: $O(2,1;-2), O(3,1;-3)$

$$\left(\begin{array}{ccc|c} 1 & 1 & -2 & 0 \\ 0 & 0 & 1 & 2 \\ 3 & -1 & 2 & 12 \end{array}\right), \quad \left(\begin{array}{ccc|c} 1 & 1 & -2 & 0 \\ 0 & 0 & 1 & 2 \\ 0 & -4 & 8 & 12 \end{array}\right)$$

We now work on the variables y and z in the last two equations. Note that y is not involved in the second equation, and so we have to do nothing to eliminate it. However, from the matrix point of view we wish to get to an upper triangular matrix, so we will perform an operation which corresponds to reordering the equations by switching the last two equations.

DEFINITION 1.4.4. We denote by $Op(i, j)$ the operation on a matrix of permuting (i.e., interchanging) the ith and jth rows of a matrix and leaving the other rows unchanged.

Thus applying $Op(2,3)$, we get a matrix which is upper triangular:

$$\left(\begin{array}{ccc|c} 1 & 1 & -2 & 0 \\ 0 & -4 & 8 & 12 \\ 0 & 0 & 1 & 2 \end{array}\right)$$

We now work from the bottom up, solving for z in the third equation and eliminating z from the first two equations. Since the coefficient of z is already 1, the first step requires no work.

Eliminate z in the second equation, then eliminate z in the first equation. We use $O(2,3;-8)$ and $O(1,3;2)$.

$$\left(\begin{array}{ccc|c} 1 & 1 & -2 & 0 \\ 0 & -4 & 0 & -4 \\ 0 & 0 & 1 & 2 \end{array}\right), \quad \left(\begin{array}{ccc|c} 1 & 1 & 0 & 4 \\ 0 & -4 & 0 & -4 \\ 0 & 0 & 1 & 2 \end{array}\right)$$

Next we solve for y and eliminate y in the first equation. We use $Om(2; -1/4)$ and $O(1, 2; -1)$.

$$\left(\begin{array}{ccc|c} 1 & 1 & 0 & 4 \\ 0 & 1 & 0 & 1 \\ 0 & 0 & 1 & 2 \end{array}\right), \quad \left(\begin{array}{ccc|c} 1 & 0 & 0 & 3 \\ 0 & 1 & 0 & 1 \\ 0 & 0 & 1 & 2 \end{array}\right).$$

Thus the solution to this system is $x = 3, y = 1, z = 2$. ■

These last three examples have been somewhat deceptive in that we always ended up with the identity matrix as the coefficient matrix and a unique solution to our system. We know from our earlier discussion that this will not always be possible for a general system, but what is possible is to get an equivalent system where the coefficient matrix is in reduced normal form. Once it is in this form we know how to find all the solutions to this equation. We describe our strategy briefly. We first perform a forward elimination so that the following three properties of reduced normal form are satisfied.

1. If the first nonzero entry in row i is in column $c(i)$, then $i < j$ implies $c(i) < c(j)$.
2. The entries in the column beneath the first nonzero entry in each row are all zero.
3. Any zero rows occur beneath all the nonzero rows.

We may have to permute some rows along the way as in Example 1.4.3 so that the columns containing the first nonzero entries move to the right. There may also be a step, which will be illustrated in Example 1.4.4, where all the lower entries in a column are 0, and so we just move over one column to the right and continue. The matrix at this stage will in particular be upper triangular. So far we have executed the **forward elimination** part of the algorithm. We will call the columns with these first nonzero entries the **basic columns.** They will correspond to the basic variables which we discussed earlier. We then enter the **backward elimination** portion of the algorithm. Starting with the last basic column and working backward, we normalize it to make the first nonzero coefficient 1 and then add multiples of the row containing it to make all of the other entries in the column 0. At the end of this procedure the coefficient part of the augmented matrix will be in reduced normal form so that we can solve the system.

We will work a few more examples to illustrate this algorithm, which is called the **Gaussian elimination algorithm.** In each example we will just give the initial augmented matrix for the system of equations.

■ ***Example 1.4.4.***

$$\left(\begin{array}{ccc|c} 1 & 1 & -2 & 0 \\ 2 & 2 & -3 & 1 \\ 3 & 3 & 1 & 7 \end{array}\right)$$

We first do forward elimination. Using $O(2, 1; -2)$ and $O(3, 1; -3)$ gives

$$\left(\begin{array}{ccc|c} 1 & 1 & -2 & 0 \\ 0 & 0 & 1 & 1 \\ 3 & 3 & 1 & 7 \end{array}\right), \quad \left(\begin{array}{ccc|c} 1 & 1 & -2 & 0 \\ 0 & 0 & 1 & 1 \\ 0 & 0 & 7 & 7 \end{array}\right)$$

Now note that when we disregard the first row and look at the remainder of the second column, both of the entries are zero. Thus we cannot get a nonzero entry in this column by permuting rows, as in the last example. We just move over to the third column and continue the forward elimination algorithm. Using $O(3, 2; -7)$ gives

$$\left(\begin{array}{ccc|c} 1 & 1 & -2 & 0 \\ 0 & 0 & 1 & 1 \\ 0 & 0 & 0 & 0 \end{array}\right)$$

We have completed the forward elimination part of the algorithm. The coefficient part of the matrix now satisfies the property that the entries beneath the first nonzero entry in each row are zero, as well as having the columns containing these first nonzero entries moving over to the right as we move down the rows. We will denote by $c(i)$ the index of the column $\mathbf{A}^{c(i)}$ containing the first nonzero entry for the ith nonzero row. Here $c(1) = 1$, $c(2) = 3$. Note that the last row becomes zero here, so we have essentially reduced our original system of three equations to an equivalent one with two equations. Also note that if we had started off with a different $\mathbf{b}$, we may well have had no solution. For example, if the original augmented matrix was

$$\left(\begin{array}{ccc|c} 1 & 1 & -2 & 0 \\ 2 & 2 & -3 & 1 \\ 3 & 3 & 1 & 8 \end{array}\right)$$

we could reduce with the same operations to

$$\left(\begin{array}{ccc|c} 1 & 1 & -2 & 0 \\ 0 & 0 & 1 & 1 \\ 0 & 0 & 0 & 1 \end{array}\right)$$

and from the last row we see that there are no solutions. Returning to our original system, we now perform the backward elimination part of the algorithm. We will be working on $\mathbf{A}^{c(2)}$, then on $\mathbf{A}^{c(1)}$. For each column $\mathbf{A}^{c(i)}$, we first normalize to make the first nonzero entry of the ith row equal to 1 and then perform row operations on the preceding rows to make all of the other entries in $\mathbf{A}^{c(i)}$ equal to 0. In the system at hand, the two rows are already normalized, and so we only have to perform one operation $O(1, 2; 2)$ to complete the algorithm.

$$\left(\begin{array}{ccc|c} 1 & 1 & 0 & 2 \\ 0 & 0 & 1 & 1 \\ 0 & 0 & 0 & 0 \end{array}\right)$$

Now we have completed the Gaussian elimination algorithm and the coefficient part of the matrix is in normal form. To complete the solution of the problem, we note that the basic variables correspond to the basic columns, $c(1) = 1$ and $c(2) = 3$; thus the basic variables are x_1 and x_3, and the free variable is x_2. Then the general solution can be read off as

$$x = \begin{pmatrix} 2 \\ 0 \\ 1 \end{pmatrix} + c_1 \begin{pmatrix} -1 \\ 1 \\ 0 \end{pmatrix}$$ ■

■ ***Example 1.4.5.***

$$\left(\begin{array}{ccccccc|c} 2 & 4 & 1 & -1 & 0 & 1 & -2 & 0 \\ 4 & 8 & 2 & -2 & 1 & 5 & -4 & 0 \\ 0 & 0 & 0 & 2 & 0 & 4 & 4 & 0 \end{array}\right)$$

Note that the right-hand side will remain 0 throughout all of the operations, so we will only keep the coefficient part of the matrix in our notation. We perform the forward elimination part of the algorithm. We first use $O(2, 1; -2)$ and $Op(2, 3)$.

$$\begin{pmatrix} 2 & 4 & 1 & -1 & 0 & 1 & -2 \\ 0 & 0 & 0 & 0 & 1 & 3 & 0 \\ 0 & 0 & 0 & 2 & 0 & 4 & 4 \end{pmatrix}, \quad \begin{pmatrix} 2 & 4 & 1 & -1 & 0 & 1 & -2 \\ 0 & 0 & 0 & 2 & 0 & 4 & 4 \\ 0 & 0 & 0 & 0 & 1 & 3 & 0 \end{pmatrix}$$

We have $c(1) = 1$, $c(2) = 4$, and $c(3) = 5$. We start the backward elimination part of the algorithm. Column 5 is already in normal form so we begin with column 4. We use $Om(2; 1/2)$ and $O(1, 2; 1)$.

$$\begin{pmatrix} 2 & 4 & 1 & -1 & 0 & 1 & -2 \\ 0 & 0 & 0 & 1 & 0 & 2 & 2 \\ 0 & 0 & 0 & 0 & 1 & 3 & 0 \end{pmatrix}, \quad \begin{pmatrix} 2 & 4 & 1 & 0 & 0 & 3 & 0 \\ 0 & 0 & 0 & 1 & 0 & 2 & 2 \\ 0 & 0 & 0 & 0 & 1 & 3 & 0 \end{pmatrix}$$

Finally, we normalize row 1 by dividing by 2, using $Om(1; 1/2)$

$$\begin{pmatrix} 1 & 2 & .5 & 0 & 0 & 1.5 & 0 \\ 0 & 0 & 0 & 1 & 0 & 2 & 2 \\ 0 & 0 & 0 & 0 & 1 & 3 & 0 \end{pmatrix}$$

The matrix is now in reduced normal form and we can read off the general solution.

The basic variables are x_1, x_4, and x_5 and the free variables are x_2, x_3, x_6, and x_7. Note that the zero vector is the particular solution and the general solution is

$$x = c_1 \begin{pmatrix} -2 \\ 1 \\ 0 \\ 0 \\ 0 \\ 0 \\ 0 \end{pmatrix} + c_2 \begin{pmatrix} -.5 \\ 0 \\ 1 \\ 0 \\ 0 \\ 0 \\ 0 \end{pmatrix} + c_3 \begin{pmatrix} -1.5 \\ 0 \\ 0 \\ -2 \\ -3 \\ 1 \\ 0 \end{pmatrix} + c_4 \begin{pmatrix} 0 \\ 0 \\ 0 \\ -2 \\ 0 \\ 0 \\ 1 \end{pmatrix}$$ ■

■ ***Example 1.4.6.*** In this example we will keep the right-hand side as variable to illustrate what is happening to it during the elimination procedure. We will also use it to determine exactly for which **b** there is a solution.

$$\left(\begin{array}{rrrrrrr|c} 3 & 0 & 1 & 4 & -2 & 0 & 1 & b_1 \\ 6 & 0 & 4 & 8 & -5 & 1 & 2 & b_2 \\ 0 & 0 & 0 & 5 & 4 & 0 & 1 & b_3 \\ -3 & 0 & -1 & 16 & 18 & 0 & 3 & b_4 \\ 0 & 0 & 6 & -10 & -11 & 3 & -2 & b_5 \end{array}\right)$$

We first do the forward elimination step. We work with one column at a time. For column 1 we use $O(2,1;-2)$ and $O(4,1;1)$. Then we use $O(5,2;-3)$ for column 3.

$$\left(\begin{array}{rrrrrrr|c} 3 & 0 & 1 & 4 & -2 & 0 & 1 & b_1 \\ 0 & 0 & 2 & 0 & -1 & 1 & 0 & b_2 - 2b_1 \\ 0 & 0 & 0 & 5 & 4 & 0 & 1 & b_3 \\ 0 & 0 & 0 & 20 & 16 & 0 & 4 & b_4 + b_1 \\ 0 & 0 & 6 & -10 & -11 & 3 & -2 & b_5 \end{array}\right)$$

$$\left(\begin{array}{rrrrrrr|c} 3 & 0 & 1 & 4 & -2 & 0 & 1 & b_1 \\ 0 & 0 & 2 & 0 & -1 & 1 & 0 & b_2 - 2b_1 \\ 0 & 0 & 0 & 5 & 4 & 0 & 1 & b_3 \\ 0 & 0 & 0 & 20 & 16 & 0 & 4 & b_4 + b_1 \\ 0 & 0 & 0 & -10 & -8 & 0 & -2 & 6b_1 + b_5 - 3b_2 \end{array}\right)$$

We then use $O(4,3;-4)$ and $O(5,3;2)$ on column 4.

$$\left(\begin{array}{rrrrrrr|c} 3 & 0 & 1 & 4 & -2 & 0 & 1 & b_1 \\ 0 & 0 & 2 & 0 & -1 & 1 & 0 & -2b_1 + b_2 \\ 0 & 0 & 0 & 5 & 4 & 0 & 1 & b_3 \\ 0 & 0 & 0 & 0 & 0 & 0 & 0 & b_1 - 4b_3 + b_4 \\ 0 & 0 & 0 & 0 & 0 & 0 & 0 & 6b_1 - 3b_2 + 2b_3 + b_5 \end{array}\right)$$

From this stage we can note that a necessary and sufficient condition for a solution is that $b_1 - 4b_3 + b_4 = 0$ and $6b_1 - 3b_2 + 2b_3 + b_5 = 0$. Note also that the right-hand side, which is the result of applying each of our operations to the original **b**, is also the result of multiplying **b** by the matrix

$$O_1 = \begin{pmatrix} 1 & 0 & 0 & 0 & 0 \\ -2 & 1 & 0 & 0 & 0 \\ 0 & 0 & 1 & 0 & 0 \\ 1 & 0 & -4 & 1 & 0 \\ 6 & -3 & 2 & 0 & 1 \end{pmatrix}$$

We will pursue this point further after this example. We now assume that the last two equations are satisfied, and so we will delete these last two rows and proceed.

We do the backward elimination on the matrix which remains after deleting the last two zero rows. We use $Om(3;1/5)$ and $O(1,3;-4)$ on column 4 and $Om(2;1/2)$ and $O(1,2;-1)$ on column 3.

$$\left(\begin{array}{ccccccc|c} 3 & 0 & 1 & 0 & -5.2 & 0 & .2 & b_1 - .8b_3 \\ 0 & 0 & 2 & 0 & -1 & 1 & 0 & -2b_1 + b_2 \\ 0 & 0 & 0 & 1 & .8 & 0 & .2 & .2b_3 \end{array}\right)$$

$$\left(\begin{array}{ccccccc|c} 3 & 0 & 0 & 0 & -4.7 & -.5 & .2 & 2b_1 - .5b_2 - .8b_3 \\ 0 & 0 & 1 & 0 & -.5 & .5 & 0 & -b_1 + .5b_2 \\ 0 & 0 & 0 & 1 & .8 & 0 & .2 & .2b_3 \end{array}\right)$$

We then use $Om(1; 1/3)$ for column 1.

$$\left(\begin{array}{ccccccc|c} 1 & 0 & 0 & 0 & -47/30 & -1/6 & 1/15 & 2/3\,b_1 - 1/6\,b_2 - 4/15\,b_3 \\ 0 & 0 & 1 & 0 & -1/2 & 1/2 & 0 & -b_1 + 1/2\,b_2 \\ 0 & 0 & 0 & 1 & 4/5 & 0 & 1/5 & 1/5\,b_3 \end{array}\right)$$

Since we now have the coefficient matrix in reduced normal form we can now read off the general solution. The basic variables are x_1, x_3, and x_4. The free variables are x_2, x_5, x_6, and x_7. The general solution is

$$\mathbf{x} = \begin{pmatrix} 2/3\,b_1 - 1/6\,b_2 - 4/15\,b_3 \\ 0 \\ -b_1 + 1/2\,b_2 \\ 1/5\,b_3 \\ 0 \\ 0 \\ 0 \end{pmatrix} + c_1 \begin{pmatrix} 0 \\ 1 \\ 0 \\ 0 \\ 0 \\ 0 \\ 0 \end{pmatrix} + c_2 \begin{pmatrix} 47/30 \\ 0 \\ 1/2 \\ -4/5 \\ 1 \\ 0 \\ 0 \end{pmatrix} + c_3 \begin{pmatrix} 1/6 \\ 0 \\ -1/2 \\ 0 \\ 0 \\ 1 \\ 0 \end{pmatrix} + c_4 \begin{pmatrix} -1/15 \\ 0 \\ 0 \\ -1/5 \\ 0 \\ 0 \\ 1 \end{pmatrix}$$

■

We conclude this section with a statement of the Gaussian elimination algorithm to reduce the coefficient matrix to reduced normal form. The form we give is recursive; that is, it calls the algorithm itself for a smaller matrix.

Gaussian Elimination Algorithm

Forward Elimination Step. With the first column, check to see if all coefficients are zero. If so, leave the first column alone, and look at the submatrix with first column omitted and apply the forward elimination step to it. If not, find the first row with a nonzero entry in that row and the first column and permute that row with the first row so this nonzero entry (which is then called the **pivot** for this step) is in the first row. Then add multiples of the first row to lower rows so that the lower entries of the first column are all 0. We are then finished with the first column and first row. Form the submatrix arising from deleting the first row and column and apply the forward elimination step to it.

At the end of the forward elimination step, our matrix satisfies the three properties:

1. If the first nonzero entry in row i is in column $c(i)$, then $i < j$ implies $c(i) < c(j)$.

2. The entries in the column beneath the first nonzero entry in each row are all zero.
3. Any zero rows occur beneath all the nonzero rows.

Backward Elimination Step. This procedure is also inductive, where we work with the basic columns, starting with the last one and working backward. Starting with the last nonzero row, call its first nonzero entry the pivot. Then normalize the row by dividing by the pivot so that the new pivot is 1. Then add multiples of this row to previous rows so that all the entries in the column containing the pivot are 0. Then look at the submatrix obtained by deleting this row and apply the backward elimination step to it.

In the first three exercises that follow, use Gaussian elimination to solve the system given by the augmented matrix. Identify each operation used to reduce the coefficient matrix to reduced normal form.

Exercise 1.4.1.

$$\left(\begin{array}{cc|c} 1 & -1 & 2 \\ 2 & 0 & 2 \end{array}\right)$$

Exercise 1.4.2.

$$\left(\begin{array}{ccc|c} 2 & 4 & 2 & 4 \\ -2 & -4 & 1 & -8 \end{array}\right)$$

Exercise 1.4.3.

$$\left(\begin{array}{ccc|c} 1 & 2 & 3 & 3 \\ 2 & 4 & 8 & 6 \\ 0 & 2 & 2 & 2 \end{array}\right)$$

Note that once we know the operations used to reduce the left-hand side to reduced normal form, we can solve equations with the same left-hand side and different right-hand sides by applying the same operations on the new right-hand side.

Exercise 1.4.4.

Solve $A\mathbf{x} = \begin{pmatrix} 4 \\ 10 \\ 2 \end{pmatrix}$ where A is the matrix from the previous exercise.

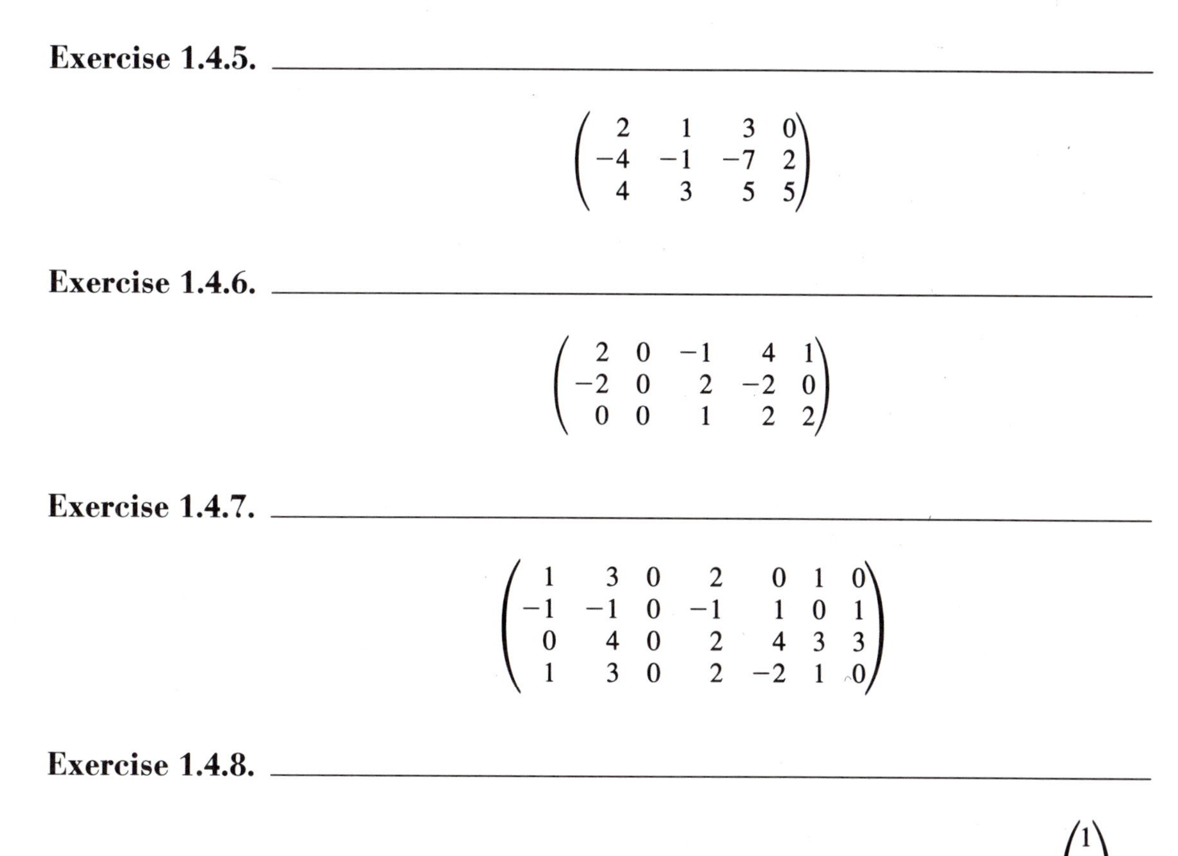

In Exercises 1.4.5–1.4.7, reduce the given matrix to reduced normal form. Identify each row operation used.

Exercise 1.4.5.

$$\begin{pmatrix} 2 & 1 & 3 & 0 \\ -4 & -1 & -7 & 2 \\ 4 & 3 & 5 & 5 \end{pmatrix}$$

Exercise 1.4.6.

$$\begin{pmatrix} 2 & 0 & -1 & 4 & 1 \\ -2 & 0 & 2 & -2 & 0 \\ 0 & 0 & 1 & 2 & 2 \end{pmatrix}$$

Exercise 1.4.7.

$$\begin{pmatrix} 1 & 3 & 0 & 2 & 0 & 1 & 0 \\ -1 & -1 & 0 & -1 & 1 & 0 & 1 \\ 0 & 4 & 0 & 2 & 4 & 3 & 3 \\ 1 & 3 & 0 & 2 & -2 & 1 & 0 \end{pmatrix}$$

Exercise 1.4.8.

With A the matrix in the previous exercise, find the general solution to $A\mathbf{x} = \begin{pmatrix} 1 \\ 1 \\ 7 \\ 1 \end{pmatrix}$.

Exercise 1.4.9.

Determine for which $\mathbf{b}$ (in terms of equations which $\mathbf{b}$ has to satisfy) there is a solution to $A\mathbf{x} = \mathbf{b}$, where

$$A = \begin{pmatrix} 2 & 1 & 1 & 0 & 1 \\ 0 & 1 & 2 & 1 & 3 \\ 6 & 2 & 1 & -1 & 0 \end{pmatrix}$$

For those $\mathbf{b}$ where there is a solution, find the general solution in terms of the coordinates of $\mathbf{b}$.

1.5. ELEMENTARY MATRICES AND ROW OPERATIONS

We now want to explore the Gaussian elimination algorithm from the viewpoint of matrix multiplication. We first expand the definition of the product of a matrix with a column vector to the product of two matrices.

DEFINITION 1.5.1. If A is an m by n matrix and B is an n by p matrix, then we are able to multiply any row vector from A by any column vector in B. We denote the row vectors in A by $\mathbf{A}_1, \ldots, \mathbf{A}_m$ and the column vectors in B by $\mathbf{B}^1, \ldots, \mathbf{B}^p$. We form a new m by p matrix AB (called the product of A and B) whose ij-entry (that is, the entry in the ith row and jth column) is given by the product $\mathbf{A}_i\mathbf{B}^j$ of the ith row of A with the jth column of B. Using summation notation, the formula is

$$(AB)_{ij} = \mathbf{A}_i\mathbf{B}^j = a_{i1}b_{1j} + \cdots + a_{in}b_{nj} = \sum_{k=1}^{n} a_{ik}b_{kj}$$

Note that the number of columns of A must correspond to the number of rows of B for this definition to make sense.

Here are a couple of examples.

$$\begin{pmatrix} 1 & 2 & 3 \\ 2 & 6 & 1 \end{pmatrix}\begin{pmatrix} 2 & 1 \\ 1 & 0 \\ 5 & 1 \end{pmatrix} = \begin{pmatrix} 19 & 4 \\ 15 & 3 \end{pmatrix}, \quad \begin{pmatrix} 2 & 1 \\ 1 & 0 \\ 5 & 1 \end{pmatrix}\begin{pmatrix} 1 & 2 & 3 \\ 2 & 6 & 1 \end{pmatrix} = \begin{pmatrix} 4 & 10 & 7 \\ 1 & 2 & 3 \\ 7 & 16 & 16 \end{pmatrix}$$

These examples illustrate that in general $AB \neq BA$. We express this lack of equality by saying that matrix multiplication is not commutative. In fact, here we are a 2 by 2 matrix and the other is a 3 by 3 matrix. Sometimes AB is defined but BA is not even defined. For example, if A is a 2 by 3 matrix and B is a 3 by 4 matrix, then AB is defined but BA is not defined. Even when A and B are both square matrices so that AB and BA are defined and are of the same size, it is rarely the case that $AB = BA$.

Despite being noncommutative, matrix multiplication does satisfy many other properties that we are familiar with when working with real numbers. For example, if A is an m by n matrix and B, C are n by p matrices, then we can form the sum $B + C$ and $A(B + C) = AB + AC$. Also, $A(rB) = rAB$ for a real scalar r. These two facts just boil down to the fact that the dot product is linear in each term and matrix multiplication is defined in terms of the dot product. We can also verify by direct (tedious) calculation that matrix multiplication is associative.

When we multiply A times B, the ith row $(AB)_i$ of the product just arises from multiplying the ith row of A by B, that is,

$$(AB)_i = \mathbf{A}_iB$$

Moreover, as a row vector it is a linear combination of the row vectors of B with coefficients given by the ith row of A:

$$(AB)_i = a_{i1}\mathbf{B}_1 + \cdots + a_{in}\mathbf{B}_n = \sum_{k=1}^{n} a_{ik}\mathbf{B}_k$$

Similarly, the jth column of AB is a linear combination of the columns of A, with coefficients coming from the jth column of B:

$$(AB)^j = \mathbf{AB}^j = b_{1j}\mathbf{A}^1 + \cdots + b_{nj}\mathbf{A}^n = \sum_{k=1}^{n} b_{kj}\mathbf{A}^k$$

As an example of these statements note that when we multiply

$$\begin{pmatrix} 1 & 2 & 3 \\ -1 & 0 & 2 \\ 1 & 2 & 4 \end{pmatrix}\begin{pmatrix} 1 & 2 \\ 3 & 4 \\ 5 & 6 \end{pmatrix} = \begin{pmatrix} 22 & 28 \\ 9 & 10 \\ 27 & 34 \end{pmatrix}$$

the second row is a linear combination

$$(-1)(1 \quad 2) + 0(3 \quad 4) + 2(5 \quad 6)$$

of the rows of B, and the first column is the linear combination

$$1\begin{pmatrix} 1 \\ -1 \\ 1 \end{pmatrix} + 3\begin{pmatrix} 2 \\ 0 \\ 2 \end{pmatrix} + 5\begin{pmatrix} 3 \\ 2 \\ 4 \end{pmatrix}$$

of the columns of A.

Matrix multiplication can also be interpreted in terms of the composition of functions. Given an m by n matrix A, then there is an associated function L_A from $\mathbb{R}^n$ to $\mathbb{R}^m$ given by identifying column vectors with the corresponding vectors in $\mathbb{R}^n$ and $\mathbb{R}^m$ and setting $L_A(\mathbf{x}) = A\mathbf{x}$. This function has the properties that

$$L_A(\mathbf{x} + \mathbf{y}) = L_A(\mathbf{x}) + L_A(\mathbf{y}) \quad \text{and} \quad L_A(c\mathbf{x}) = cL_A(\mathbf{x})$$

These follow directly from the properties of multiplying a matrix by a vector:

$$A(\mathbf{x} + \mathbf{y}) = A\mathbf{x} + A\mathbf{y} \quad \text{and} \quad A(c\mathbf{x}) = cA\mathbf{x}$$

These properties are called linearity properties of the function L_A, and since it satisfies them, L_A is called a **linear transformation.**

DEFINITION 1.5.2. A linear transformation $L : \mathbb{R}^n \to \mathbb{R}^m$ is a function that satisfies the two properties

$$L(\mathbf{v} + \mathbf{w}) = L(\mathbf{v}) + L(\mathbf{w}) \quad \text{and} \quad L(c\mathbf{v}) = cL(\mathbf{v})$$

Thus each m by n matrix A determines a linear transformation L_A. If A is an m by n matrix and B is an n by p matrix, then we can form the product AB. Since L_B is a linear transformation from $\mathbb{R}^p$ to $\mathbb{R}^n$ and L_A is a linear transformation from $\mathbb{R}^n$ to $\mathbb{R}^m$, we can also form the composition $L_A L_B$ of these linear transformations. Then

$$L_{AB}(\mathbf{x}) = (AB)\mathbf{x} = A(B\mathbf{x}) = L_A L_B(\mathbf{x}); \text{ that is, } L_A L_B = L_{AB}$$

Thus matrix multiplication just corresponds to the composition of functions when we interpret the matrix in terms of the linear transformation which it determines.

We now show that each linear transformation $L : \mathbb{R}^n \to \mathbb{R}^m$ comes from a matrix A; that is, $L = L_A$ for a unique matrix A. Let us denote the **standard basis vectors** in $\mathbb{R}^n$ by $\mathbf{e}_1 = (1, 0, \ldots, 0), \mathbf{e}_2 = (0, 1, 0, \ldots, 0), \ldots, \mathbf{e}_n = (0, \ldots, 0, 1)$, and denote the standard basis vectors in $\mathbb{R}^m$ by $\mathbf{f}_1 = (1, 0, \ldots, 0), \ldots, \mathbf{f}_m = (0, 0, \ldots, 1)$. Then

$$L(\mathbf{e}_j) = \mathbf{A}^j = \sum_{i=1}^{m} a_{ij}\mathbf{f}_i$$

Here we identify the vectors $\mathbf{f}_i, \mathbf{A}^j$ with column vectors in $\mathbb{R}^m$. Then for a general $\mathbf{x}$, we can write $\mathbf{x} = x_1\mathbf{e}_1 + \cdots + x_n\mathbf{e}_n$ and apply linearity of L to write

$$\begin{aligned} L(\mathbf{x}) &= L(x_1\mathbf{e}_1 + \cdots + x_n\mathbf{e}_n) = x_1L(\mathbf{e}_1) + \cdots + x_nL(\mathbf{e}_n) \\ &= x_1\mathbf{A}^1 + \cdots + x_n\mathbf{A}^n = A\mathbf{x} \end{aligned}$$

Here the matrix A is the matrix with columns the vectors $\mathbf{A}^j$ from above. Thus given L, we can determine the matrix from which it arises by just applying L to $\mathbf{e}_j$ and putting the answer $\mathbf{A}^j$ as the jth column vector of A. This is consistent with the fact that $A\mathbf{e}_j = \mathbf{A}^j$, where we are regarding $\mathbf{e}_j$ as a column vector in this equation. We will pursue this relationship in the next chapter when we discuss vector spaces and linear transformations more thoroughly.

Let us return now to see how Gaussian elimination is related to matrix multiplication. We start with a problem $A\mathbf{x} = \mathbf{b}$. After performing the forward elimination step we have an equivalent equation $U\mathbf{x} = \mathbf{c}$. Here U is in a special form; in particular it is upper triangular. We then perform backward elimination to get $R\mathbf{x} = \mathbf{d}$, and we can then read off the general solution from this form of the equation. Now each step in the elimination process is one of the following three. They may be thought of as being applied to the augmented matrix or as being applied to both the coefficient matrix and the right hand side. We will just express what is happening to A, but we will mean that we are performing the same operations on $\mathbf{b}$.

1. Add a multiple of one row to another row; that is, leave all of the rows except the ith one unchanged and replace the $\mathbf{A}_i$ by $\mathbf{A}_i + r\mathbf{A}_j$.
2. Permute two rows; that is, leave all rows except the ith and jth rows unchanged and replace $\mathbf{A}_i$ with $\mathbf{A}_j$ and $\mathbf{A}_j$ with $\mathbf{A}_i$.
3. Multiply a row by a nonzero scalar r; that is, leave all rows except the ith one unchanged and replace $\mathbf{A}_i$ by $r\mathbf{A}_i$.

We gave the names $O(i, j; r)$, $Op(i, j)$, and $Om(i; r)$ to these three operations. Now each operation forms a new matrix from A by replacing one (or two) of the rows of A by a linear combination of other rows. But this also happens when we multiply A by an appropriate matrix O on the left, that is, form OA. The property of matrix multiplication which we are using is that $(OA)_k = O_kA = \sum o_{kl}\mathbf{A}_l$: the kth row of the product is a linear combination of the rows $\mathbf{A}_l$ of A with coefficients coming from the kth row of O.

For operation 1 the matrix O should have the standard basis vectors $\mathbf{e}_1, \ldots, \mathbf{e}_n$ in all rows except the ith one, since each of these rows are supposed to remain unchanged. The ith row should have a 1 in the ii-position and an r in the ij-position to indicate that we are multiplying the ith row by 1 and the jth row by r and summing them to get the new ith row. We denote the corresponding matrix by $E(i, j; r)$. It agrees with the identity matrix except in the ij-position and its value there is r. For example, to perform operation $O(3, 2; -5)$, we multiply A by $E(3, 2; -5)$:

$$\begin{pmatrix} 1 & 0 & 0 \\ 0 & 1 & 0 \\ 0 & -5 & 1 \end{pmatrix}\begin{pmatrix} 1 & 3 & 1 & 4 \\ 0 & 2 & 1 & 1 \\ 0 & 10 & 7 & 2 \end{pmatrix} = \begin{pmatrix} 1 & 3 & 1 & 4 \\ 0 & 2 & 1 & 1 \\ 0 & 0 & 2 & -3 \end{pmatrix}$$

For operation 2 the matrix should have all rows except the ith and jth ones agreeing with the identity matrix to reflect the fact that they are left alone, and the ith row should be $\mathbf{e}_j$ and the jth row should be $\mathbf{e}_i$ to achieve the permutation of these two rows. That is, the matrix $P(i, j)$ which we multiply by to achieve $Op(i, j)$ should have $\mathbf{P}_k = \mathbf{e}_k$, $k \neq i, j$, $\mathbf{P}_i = \mathbf{e}_j$, and $\mathbf{P}_j = \mathbf{e}_i$. For example, to perform $Op(2, 3)$, we multiply by $P(2, 3)$:

$$\begin{pmatrix} 1 & 0 & 0 \\ 0 & 0 & 1 \\ 0 & 1 & 0 \end{pmatrix}\begin{pmatrix} 1 & 2 & 5 & 2 \\ 0 & 0 & 2 & 1 \\ 0 & 2 & 1 & 4 \end{pmatrix} = \begin{pmatrix} 1 & 2 & 5 & 2 \\ 0 & 2 & 1 & 4 \\ 0 & 0 & 2 & 1 \end{pmatrix}$$

For operation 3 the matrix should agree with the identity matrix in all rows except the ith one, and there it should have an r on the diagonal instead of a 1 since we are multiplying the ith row by r. Thus to achieve $Om(i; r)$ we multiply by $D(i; r)$, where $D(i; r)$ agrees with the identity except in the iith position and it is r there. For example, to achieve $Om(2; .5)$, we multiply by $D(2; .5)$:

$$\begin{pmatrix} 1 & 0 \\ 0 & .5 \end{pmatrix}\begin{pmatrix} 2 & 1 & 3 \\ 0 & 2 & 4 \end{pmatrix} = \begin{pmatrix} 2 & 1 & 3 \\ 0 & 1 & 2 \end{pmatrix}$$

DEFINITION 1.5.3. The three types of matrices $E(i, j; r)$, $P(i, j)$, and $D(i; r)$ are called **elementary matrices.** $P(i, j)$ is an example of a special type of matrix called a permutation matrix. A **permutation matrix** P (of size m) is one which has as its rows the standard basis vectors $\mathbf{e}_1, \dots, \mathbf{e}_m$ but in a possibly different order; that is, $\mathbf{P}_i = \mathbf{e}_{s(i)}$, where s is a function from $\{1, \dots, m\}$ to itself which is a **bijection.** This means that s is **onto**—for each k in $\{1, \dots, m\}$, there is an i in $\{1, \dots, m\}$ so that $s(i) = k$—and s is **1–1**—for $i \neq j, s(i) \neq s(j)$. Such a function s is called a **permutation** of the set $\{1, \dots, m\}$.

One of the permutations is the identity permutation, and so the identity matrix is a permutation matrix. The permutation matrices of size m correspond exactly to the permutations of $\{1, \dots, m\}$. There are $m!$ permutations of $\{1, \dots, m\}$, so there are a lot of permutation matrices for m large. However, any nonidentity permutation matrix always will be the product of at most $m - 1$ of the permutation matrices $P(i, j)$. The six permutation matrices of size 3 are

$$I = \begin{pmatrix} 1 & 0 & 0 \\ 0 & 1 & 0 \\ 0 & 0 & 1 \end{pmatrix}, \qquad P(1, 2) = \begin{pmatrix} 0 & 1 & 0 \\ 1 & 0 & 0 \\ 0 & 0 & 1 \end{pmatrix}$$

$$P(1,3) = \begin{pmatrix} 0 & 0 & 1 \\ 0 & 1 & 0 \\ 1 & 0 & 0 \end{pmatrix}, \qquad P(2,3) = \begin{pmatrix} 1 & 0 & 0 \\ 0 & 0 & 1 \\ 0 & 1 & 0 \end{pmatrix}$$

$$P(1,2)P(1,3) = \begin{pmatrix} 0 & 1 & 0 \\ 0 & 0 & 1 \\ 1 & 0 & 0 \end{pmatrix}, \qquad P(1,2)P(2,3) = \begin{pmatrix} 0 & 0 & 1 \\ 1 & 0 & 0 \\ 0 & 1 & 0 \end{pmatrix}$$

Finally, note that the matrix $D(i;r)$ is an example of a diagonal matrix since all its nonzero entries are on the diagonal.

Now let us look at an example of Gaussian elimination and see how it can be interpreted in terms of multiplying by elementary matrices. We look at $A\mathbf{x} = \mathbf{b}$. We will indicate the corresponding equation after i steps as $A(i)\mathbf{x} = \mathbf{b}(i)$ and call $A = A(0), \mathbf{b} = \mathbf{b}(0)$. We start with

$$A = A(0) = \begin{pmatrix} 5 & 1 & 3 \\ 10 & 5 & 8 \\ 15 & 6 & 16 \end{pmatrix}, \qquad \mathbf{b} = \mathbf{b}(0) = \begin{pmatrix} 3 \\ 7 \\ 5 \end{pmatrix}$$

Our first operation is $O(2,1;-2)$, so we multiply by $E(2,1;-2)$ to get

$$A(1) = E(2,1;-2)A(0) = \begin{pmatrix} 5 & 1 & 3 \\ 0 & 3 & 2 \\ 15 & 6 & 16 \end{pmatrix}, \qquad \mathbf{b}(1) = E(2,1;-2)\mathbf{b}(0) = \begin{pmatrix} 3 \\ 1 \\ 5 \end{pmatrix}$$

We then perform $O(3,1;-3)$, so we multiply by $E(3,1;-3)$ to get

$$A(2) = E(3,1;-3)A(1) = \begin{pmatrix} 5 & 1 & 3 \\ 0 & 3 & 2 \\ 0 & 3 & 7 \end{pmatrix}, \qquad \mathbf{b}(2) = E(3,1;-3)\mathbf{b}(1) = \begin{pmatrix} 3 \\ 1 \\ -4 \end{pmatrix}$$

Note that to get from $A = A(0)$ to $A(2)$, we just multiply by the product

$$O(2) = E(3,1;-3)E(2,1;-2) = \begin{pmatrix} 1 & 0 & 0 \\ -2 & 1 & 0 \\ -3 & 0 & 1 \end{pmatrix}$$

We will also be multiplying $\mathbf{b}$ by $O(2)$. Our new equation is related to our original one in that it is achieved by multiplying both sides by $O(2)$; that is, it is $O(2)A\mathbf{x} = O(2)\mathbf{b}$. We will also keep track of how the matrix by which we are multiplying our original equation to get the ith equation is changing, that is, keep track of $O(k)$ so that $A(k) = O(k)A$ and $\mathbf{b}(k) = O(k)\mathbf{b}$. Note that $O(k)$ will just be a product of some elementary matrices.

To achieve the next operation $O(3,2;-1)$, we multiply by $E(3,2;-1)$ to get

$$A(3) = E(3,2;-1)A(2) = \begin{pmatrix} 5 & 1 & 3 \\ 0 & 3 & 2 \\ 0 & 0 & 5 \end{pmatrix}, \qquad \mathbf{b}(3) = E(3,2,-1)\mathbf{b}(2) = \begin{pmatrix} 3 \\ 1 \\ -5 \end{pmatrix}$$

and note

$$O(3) = E(3,2;-1)O(2) = \begin{pmatrix} 1 & 0 & 0 \\ -2 & 1 & 0 \\ -1 & -1 & 1 \end{pmatrix}$$

We are now at the end of the forward elimination part of the algorithm and our new equation has the form $U\mathbf{x} = \mathbf{c}$ where $U = A(3)$ is upper triangular. Moreover, it has arisen from the original equation by multiplying both sides of $A\mathbf{x} = \mathbf{b}$ by $O_1 = O(3)$. Note that here O_1 is lower triangular. That is a special circumstance due to the fact that we never had to permute any rows. It follows from the fact that we are working from the top down, and so the elementary matrices $E(i, j; r)$ we are using will always be lower triangular and the product of lower triangular matrices is lower triangular. We will return to this point later when we discuss further some computational points associated with actual implementation of this algorithm on a computer.

We now perform the backward elimination step. Note that this part of the algorithm is equivalent to doing back substitution to solve for the variables from the bottom up. We first do $Om(3; 1/5)$ by multiplying by $D(3; 1/5)$:

$$A(4) = D(3;1/5)A(3) = \begin{pmatrix} 5 & 1 & 3 \\ 0 & 3 & 2 \\ 0 & 0 & 1 \end{pmatrix}, \qquad \mathbf{b}(4) = D(3;1/5)\mathbf{b}(3) = \begin{pmatrix} 3 \\ 1 \\ -1 \end{pmatrix},$$

$$O(4) = D(3;1/5)O(3) = \begin{pmatrix} 1 & 0 & 0 \\ -2 & 1 & 0 \\ -1/5 & -1/5 & 1/5 \end{pmatrix}$$

We next do $O(2,3;-2)$ by multiplying by $E(2,3;-2)$:

$$A(5) = E(2,3;-2)A(4) = \begin{pmatrix} 5 & 1 & 3 \\ 0 & 3 & 0 \\ 0 & 0 & 1 \end{pmatrix}, \qquad \mathbf{b}(5) = E(2,3;-2)\mathbf{b}(4) = \begin{pmatrix} 3 \\ 3 \\ -1 \end{pmatrix},$$

$$O(5) = E(2,3;-2)O(4) = \begin{pmatrix} 1 & 0 & 0 \\ -8/5 & 7/5 & -2/5 \\ -1/5 & -1/5 & 1/5 \end{pmatrix}$$

We then do $O(1,3;-3)$ by multiplying by $E(1,3;-3)$:

$$A(6) = E(1,3;-3)A(4) = \begin{pmatrix} 5 & 1 & 0 \\ 0 & 3 & 0 \\ 0 & 0 & 1 \end{pmatrix}, \qquad \mathbf{b}(6) = E(1,3;-3)\mathbf{b}(5) = \begin{pmatrix} 6 \\ 3 \\ -1 \end{pmatrix},$$

$$O(6) = E(1,3;-3)O(5) = \begin{pmatrix} 8/5 & 3/5 & -3/5 \\ -8/5 & 7/5 & -2/5 \\ -1/5 & -1/5 & 1/5 \end{pmatrix}$$

We then do $Om(2; 1/3)$ by multiplying by $D(2; 1/3)$:

$$A(7) = D(2; 1/3)A(6) = \begin{pmatrix} 5 & 1 & 0 \\ 0 & 1 & 0 \\ 0 & 0 & 1 \end{pmatrix}, \qquad \mathbf{b}(7) = D(2; 1/3)\mathbf{b}(6) = \begin{pmatrix} 6 \\ 1 \\ -1 \end{pmatrix},$$

$$O(7) = D(2; 1/3)O(6) = \begin{pmatrix} 24/15 & 9/15 & -9/15 \\ -8/15 & 7/15 & -2/15 \\ -3/15 & -3/15 & 3/15 \end{pmatrix}$$

We next do $O(1, 2; -1)$ by multiplying by $E(1, 2; -1)$:

$$A(8) = E(1, 2; -1)A(7) = \begin{pmatrix} 5 & 0 & 0 \\ 0 & 1 & 0 \\ 0 & 0 & 1 \end{pmatrix}, \qquad \mathbf{b}(8) = E(1, 2; -1)\mathbf{b}(7) = \begin{pmatrix} 5 \\ 1 \\ -1 \end{pmatrix}$$

$$O(8) = E(1, 2; -1)O(7) = \begin{pmatrix} 32/15 & 2/15 & -7/15 \\ -8/15 & 7/15 & -2/15 \\ -3/15 & -3/15 & 3/15 \end{pmatrix}$$

The final step is $Om(1; 1/5)$, which we achieve by multiplying by $D(1; 1/5)$:

$$A(9) = D(1; 1/5)A(8) = \begin{pmatrix} 1 & 0 & 0 \\ 0 & 1 & 0 \\ 0 & 0 & 1 \end{pmatrix}, \qquad \mathbf{b}(9) = D(1; 1/5)\mathbf{b}(8) = \begin{pmatrix} 1 \\ 1 \\ -1 \end{pmatrix},$$

$$O(9) = D(1; 1/5)O(9) = 1/75 \begin{pmatrix} 32 & 2 & -7 \\ -40 & 35 & -10 \\ -15 & -15 & 15 \end{pmatrix}$$

Thus if we call the final matrix $O(9) = O$, then multiplying by O has changed our original equation $A\mathbf{x} = \mathbf{b}$ into $I\mathbf{x} = OA\mathbf{x} = O\mathbf{b} = \mathbf{d}$ so that we can solve and get $\mathbf{x} = \mathbf{d}$, which in our example gives $x_1 = 1, x_2 = 1, x_3 = -1$. Now this example is special in the sense that the final row reduced matrix turned out to be the identity. In general, there will be a square m by m matrix O, which will be the product of all the elementary matrices used in each operation, so that $OA = R$, where R is in reduced normal form, and the equation $A\mathbf{x} = \mathbf{b}$ will be equivalent to the equation $R\mathbf{x} = O\mathbf{b}$.

If we stop after the forward elimination step, then we would have a matrix O_1 which is a product of elementary matrices, so that multiplication by O_1 changes $A\mathbf{x} = \mathbf{b}$ into $U\mathbf{x} = O_1\mathbf{b}$. In particular, this much of a simplification is sufficient to answer the question of the existence of a solution to $A\mathbf{x} = \mathbf{b}$. There will exist a solution as long as the entries at the bottom of $O_1\mathbf{b}$ corresponding to any zero rows of U are also zero. The condition that U has no zero rows is necessary to always have a solution for every $\mathbf{b}$. For there is always some $\mathbf{c}$ for which there is no solution of $U\mathbf{x} = \mathbf{c}$ when U has any zero rows (just make the bottom entries of $\mathbf{c}$ nonzero), and so reversing our steps in the Gaussian elimination process we can get back to a matrix equation $A\mathbf{x} = \mathbf{b}$, with the same set of solutions, that is, none. Once the matrix is in the form U at the end of the forward elimination step, we can also solve for the basic variables in terms of the free variables by back substitution if we wish: this process is equivalent to our backward elimination algorithm.

In Exercises 1.5.1–1.5.6, use the matrices

$$A = \begin{pmatrix} 2 & -1 & 1 \\ 1 & -1 & 1 \\ 4 & 4 & -1 \end{pmatrix}, \quad B = \begin{pmatrix} 1 & 2 & 3 \\ 1 & 2 & 1 \end{pmatrix}, \quad C = \begin{pmatrix} 1 & 1 & 1 & 1 \\ 1 & 2 & 3 & 4 \\ 2 & 3 & 4 & 5 \end{pmatrix},$$

$$D = \begin{pmatrix} 1 & 1 & 2 & 1 \\ 2 & 2 & 2 & 2 \\ 3 & 5 & 6 & 5 \end{pmatrix}, \quad \mathbf{a} = \begin{pmatrix} 2 \\ -1 \\ -3 \end{pmatrix}, \quad \mathbf{b} = \begin{pmatrix} 1 \\ 1 \\ 2 \end{pmatrix}, \quad \mathbf{c} = \begin{pmatrix} 1 \\ 1 \\ 0 \end{pmatrix}, \quad \mathbf{d} = \begin{pmatrix} 0 \\ 2 \\ 2 \end{pmatrix}$$

Exercise 1.5.1.

Find the products BA, BC, and AC.

Exercise 1.5.2.

(a) Express the second row of the product BC as a linear combination of the rows of C.

(b) Express the second column of the product BC as linear combination of the columns of B.

Exercise 1.5.3.

(a) Consider the linear transformation $L : \mathbb{R}^2 \to \mathbb{R}^2$ which reflects each vector through the x-axis. In particular, $L(\mathbf{e}_1) = \mathbf{e}_1$ and $L(\mathbf{e}_2) = -\mathbf{e}_2$. Find the matrix A so $L = L_A$.

(b) Answer the same question for the linear transformation which rotates each vector by 45 degrees.

Exercise 1.5.4.

Give each of the elementary matrices which occur in performing Gaussian elimination to reduce A to reduced normal form.

Exercise 1.5.5.

Find the matrix O_1 which occurs at the end of the forward elimination part of the algorithm in solving $C\mathbf{x} = \mathbf{b}$ and $C\mathbf{x} = \mathbf{c}$. Give the equivalent equations $U\mathbf{x} = O_1\mathbf{b}$ and $U\mathbf{x} = O_1\mathbf{c}$ in each case. Which equation has a solution?

Exercise 1.5.6.

Find the matrix O by which you multiply to change the equation $D\mathbf{x} = \mathbf{d}$ into normal form $R\mathbf{x} = O\mathbf{d}$. Solve this equation.

Exercise 1.5.7.

Show that matrix multiplication is associative. You should compute the ij-entry of both $(AB)C$ and $A(BC)$ and verify that they are the same. It is useful to use summation notation.

Exercise 1.5.8.

Show that the product of two lower triangular matrices is lower triangular. Moreover, show that if the matrices also have all 1's on the diagonal such as the matrices $E(i, j; r)$, then the product also has all 1's on the diagonal.

1.6. INVERSES, TRANSPOSES, AND GAUSSIAN ELIMINATION

We now reinterpret the results of the Gaussian elimination algorithm. For each step of the algorithm, we are multiplying both sides of the equation by an elementary matrix E, which is $E(i, j; r)$, $P(i, j)$, or $D(i; s), s \neq 0$. Thus the equation $A\mathbf{x} = \mathbf{b}$ gets changed to a new equation $EA\mathbf{x} = E\mathbf{b}$. Now the reason the solutions to the new system remain the same as for the old system is that the process is reversible. In terms of matrices, this means that there is another elementary matrix F so that $FE = I$, the identity matrix (of size m). This brings up the concept of the inverse of a matrix.

DEFINITION 1.6.1. A matrix B is called an **inverse** for a matrix A if $BA = AB = I$. Here A, B are supposed to be square matrices of the same size, and by I we mean the identity matrix of this size. A matrix A with an inverse B is called **invertible.** The inverse B is denoted by A^{-1}. An invertible matrix is also called **nonsingular,** and one that is not invertible is called **singular.**

For the matrix $E(i, j; r)$, the matrix $E(i, j; -r)$ is its inverse. For $P(i, j)$, its inverse is the same matrix $P(i, j)$. The inverse of $D(i; s), s \neq 0$, is $D(i; 1/s)$.

$$\begin{pmatrix} 1 & 0 & 0 \\ 0 & 1 & 0 \\ r & 0 & 1 \end{pmatrix}\begin{pmatrix} 1 & 0 & 0 \\ 0 & 1 & 0 \\ -r & 0 & 1 \end{pmatrix} = I, \quad \begin{pmatrix} 0 & 1 & 0 \\ 1 & 0 & 0 \\ 0 & 0 & 1 \end{pmatrix}\begin{pmatrix} 0 & 1 & 0 \\ 1 & 0 & 0 \\ 0 & 0 & 1 \end{pmatrix} = I,$$

$$\begin{pmatrix} 1 & 0 & 0 \\ 0 & r & 0 \\ 0 & 0 & 1 \end{pmatrix}\begin{pmatrix} 1 & 0 & 0 \\ 0 & 1/r & 0 \\ 0 & 0 & 1 \end{pmatrix} = I$$

If C, D are invertible matrices of size m, then the product CD will also be invertible, and its inverse will be the product of the inverses of C and D, *but in the opposite order;* that is, $(CD)^{-1} = D^{-1}C^{-1}$. For

$$CD(D^{-1}C^{-1}) = C(DD^{-1})C^{-1} = CIC^{-1} = CC^{-1} = I$$

Similarly, $(D^{-1}C^{-1})CD = I$. Thus $D^{-1}C^{-1}$ satisfies the defining property of the inverse.

Note that this property completely determines the inverse; that is, *inverses are unique.* Suppose D, E both satisfy the property to be the inverse of C. Then

$$D = DI = D(CE) = (DC)E = IE = E$$

Returning to the elementary matrices involved in Gaussian elimination, each of them is invertible. We know that the product of two invertible matrices is invertible. We claim that the product of any number of invertible matrices will be invertible. The concept necessary to prove this is mathematical induction. We will use this for many different arguments (in fact we have already alluded to it before) so we will now state it precisely and apply it to this problem.

Principle of Mathematical Induction.

Let $P(n)$ be a statement involving the natural number n. Then $P(n)$ is true for all natural numbers n if

1. $P(1)$ is true.
2. The implication—$P(k)$ is true implies $P(k+1)$ is true—holds.

The second part is sometimes replaced by the statement

2′. The implication—$P(j)$ is true for all $j \leq k$ implies $P(k+1)$ is true—holds.

The basis for this principle is that all natural numbers arise from the number 1 by taking successors. Thus the idea is that once we know $P(1)$ is true, then

$$P(1) \text{ is true} \Rightarrow P(2) \text{ is true} \Rightarrow P(3) \text{ is true} \Rightarrow \cdots \Rightarrow P(n) \text{ is true}$$

The second form is sometimes more convenient to apply and is equivalent to the first.

Let us apply this principle to prove our claim.

PROPOSITION 1.6.1. If $A_1, \ldots, A_n$ are invertible, then so is the product $A_1 \cdots A_n$. Moreover, the inverse is the product $A_n^{-1} \cdots A_1^{-1}$ of the inverses in the opposite order.

Proof. We first verify this is true for $n = 1$. In that case it just says that if A_1 is invertible, then A_1 is invertible with inverse A_1^{-1}, so it is true. We now verify the inductive step. We assume that the theorem is true when $n = k$ and try to prove it for $n = k + 1$. Thus we suppose that $A_1, \ldots, A_{k+1}$ are invertible. Our induction hypothesis tells us that $A_1 \cdots A_k$ is invertible with inverse $A_k^{-1} \cdots A_1^{-1}$. We saw earlier that the product AB of two invertible matrices is invertible with inverse $B^{-1}A^{-1}$. We then write the product $A_1 \cdots A_{k+1} = (A_1 \cdots A_k)A_{k+1}$ and apply the last statement to say that this is invertible with inverse

$$A_{k+1}^{-1}(A_1 \cdots A_k)^{-1} = A_{k+1}^{-1}A_k^{-1} \cdots A_1^{-1} \qquad \blacksquare$$

For a general product $E_k \ldots E_1$ of k elementary matrices, the inverse will be the product of the inverses $E_1^{-1} \ldots E_k^{-1}$ in the opposite order. Thus at the end of the Gaussian elimination algorithm we will have replaced the equation $A\mathbf{x} = \mathbf{b}$ by a new equation $OA\mathbf{x} = O\mathbf{b}$, where $OA = R$ will be in reduced normal form and O will be invertible. O will be the product of all the elementary matrices used in the algorithm, and its inverse will be the product of elementary matrices which are the inverses of these in the opposite order.

We now restrict our attention to the case where A is a square matrix of size n. Then the matrix R in reduced normal form which can occur at the end of the Gaussian elimination algorithm either will be the identity matrix I or will have at least one row of zeros at the bottom. This follows from the fact that the columns containing the basic variables will be the first r columns of the identity matrix. If $r < n$ there will be a row of zeros at the bottom of the matrix, and if $r = n$ then R is the identity matrix.

If $R = I$, then we will have $OA = R = I$. Now start with the equation $AX = I$ and try to solve for X. If we multiply by O, we get $OAX = (OA)X = IX = X$ on the left-hand side and $OI = O$ on the right-hand side. Thus the solution is $X = O$, and $AO = I$ as well. Thus A is invertible and $O = A^{-1}$. Note that we can tell whether there will be n nonzero rows by the end of the forward elimination part of the algorithm since the number of zero rows is unchanged during the backward elimination step. Suppose that R does have a zero row. Then if we try to solve $Ax = O^{-1}\mathbf{e}_n$, we multiply by O to get the equivalent equation $R\mathbf{x} = \mathbf{e}_n$, which has no solution since the last row of R is zero. But if A had an inverse A^{-1}, then multiplying $A\mathbf{x} = \mathbf{b}$ by A^{-1} would give $\mathbf{x} = A^{-1}\mathbf{b}$ as a solution. Thus we can conclude

PROPOSITION 1.6.2. The square matrix A is invertible iff its reduced normal form R is equal to I. This is equivalent to knowing there are no zero rows in R and so the number r of nonzero rows of R is n.

When $r = n$, we can solve for the inverse of A by solving the matrix equation $AX = I$ for X. This can be done by forming an augmented matrix with the matrix A giving the first n columns and the identity matrix giving the last n columns, and using Gaussian elimination to reduce A to the identity matrix. The matrix appearing on the right-hand side will be the solution to $AX = I$ and so will be A^{-1}. This method of finding the inverse is called the **Gauss–Jordan method.** The same method can be used even if we don't already know that A is invertible. However, we should stop and conclude that it is not invertible whenever $c(i) \neq i$, since this will imply that there will be a zero row in R. This will be somewhat inefficient in that we will be doing a lot of unnecessary calculations on the right side before we stop. A somewhat better method is to perform Gaussian elimination on A and keep track of the operations used. If at any point we get $c(i) \neq i$, we stop and conclude that A is not invertible. If we can reduce A to the identity, then we just repeat the operations on the identity matrix to find A^{-1}. Note that this method is consistent with the fact that O is the inverse of A when A is invertible.

Here are three examples. In the first one we will perform the Gauss–Jordan algorithm to find the inverse of A. In the second and third we will try to reduce A to the identity by Gaussian elimination, recording our steps along the way and applying them to I if A is invertible.

■ ***Example 1.6.1.*** $A = \begin{pmatrix} 2 & 2 & 1 \\ 2 & 1 & -1 \\ 3 & 2 & 1 \end{pmatrix}$.

We form the augmented matrix

$$(A \mid I) = \left(\begin{array}{ccc|ccc} 2 & 2 & 1 & 1 & 0 & 0 \\ 2 & 1 & -1 & 0 & 1 & 0 \\ 3 & 2 & 1 & 0 & 0 & 1 \end{array}\right)$$

We then perform the Gaussian elimination algorithm to the augmented matrix. We record the operations used and the new augmented matrix.

$$\left(\begin{array}{ccc|ccc} 2 & 2 & 1 & 1 & 0 & 0 \\ 0 & -1 & -2 & -1 & 1 & 0 \\ 3 & 2 & 1 & 0 & 0 & 1 \end{array}\right), \qquad O(2,1;-1)$$

$$\left(\begin{array}{ccc|ccc} 2 & 2 & 1 & 1 & 0 & 0 \\ 0 & -1 & -2 & -1 & 1 & 0 \\ 0 & -1 & -.5 & -1.5 & 0 & 1 \end{array}\right), \qquad O(3,1;-1.5)$$

$$\left(\begin{array}{ccc|ccc} 2 & 2 & 1 & 1 & 0 & 0 \\ 0 & -1 & -2 & -1 & 1 & 0 \\ 0 & 0 & 1.5 & -.5 & -1 & 1 \end{array}\right), \qquad O(3,2;-1)$$

$$\left(\begin{array}{ccc|ccc} 2 & 2 & 1 & 1 & 0 & 0 \\ 0 & -1 & -2 & -1 & 1 & 0 \\ 0 & 0 & 1 & -1/3 & -2/3 & 2/3 \end{array}\right), \qquad Om(3;2/3)$$

$$\left(\begin{array}{ccc|ccc} 2 & 2 & 1 & 1 & 0 & 0 \\ 0 & -1 & 0 & -5/3 & -1/3 & 5/3 \\ 0 & 0 & 1 & -1/3 & -2/3 & 2/3 \end{array}\right), \qquad O(2,3;2)$$

$$\left(\begin{array}{ccc|ccc} 2 & 2 & 0 & 5/3 & 2/3 & -2/3 \\ 0 & -1 & 0 & -5/3 & -1/3 & 5/3 \\ 0 & 0 & 1 & -1/3 & -2/3 & 2/3 \end{array}\right), \qquad O(1,3;-1)$$

$$\left(\begin{array}{ccc|ccc} 2 & 2 & 0 & 5/3 & 2/3 & -2/3 \\ 0 & 1 & 0 & 5/3 & 1/3 & -5/3 \\ 0 & 0 & 1 & -1/3 & -2/3 & 2/3 \end{array}\right), \qquad Om(2;-1)$$

$$\left(\begin{array}{ccc|ccc} 2 & 0 & 0 & -2 & 0 & 2 \\ 0 & 1 & 0 & 5/3 & 1/3 & -5/3 \\ 0 & 0 & 1 & -1/3 & -2/3 & 2/3 \end{array}\right), \qquad O(1,2;-2)$$

$$\left(\begin{array}{ccc|ccc} 1 & 0 & 0 & -1 & 0 & 1 \\ 0 & 1 & 0 & 5/3 & 1/3 & -5/3 \\ 0 & 0 & 1 & -1/3 & -2/3 & 2/3 \end{array}\right), \qquad Om(1;1/2)$$

Thus the inverse of A is $\begin{pmatrix} -1 & 0 & 1 \\ 5/3 & 1/3 & -4/3 \\ -1/3 & -2/3 & 2/3 \end{pmatrix}$. ∎

∎ ***Example 1.6.2.*** We try to find the inverse of $B = \begin{pmatrix} 1 & 2 & 3 \\ 4 & 5 & 6 \\ 7 & 8 & 9 \end{pmatrix}$ if it exists.

We use $O(2,1;-4), O(3,1;-7)$, and $O(3,2;-2)$ to do the forward elimination step.

$$\begin{pmatrix} 1 & 2 & 3 \\ 0 & -3 & -6 \\ 7 & 8 & 9 \end{pmatrix}, \quad \begin{pmatrix} 1 & 2 & 3 \\ 0 & -3 & -6 \\ 0 & -6 & -12 \end{pmatrix}, \quad \begin{pmatrix} 1 & 2 & 3 \\ 0 & -3 & -6 \\ 0 & 0 & 0 \end{pmatrix}$$

Hence B is not invertible. ∎

∎ ***Example 1.6.3.*** We try to invert

$$C = \begin{pmatrix} 1 & 0 & 1 & 0 \\ 1 & 1 & 0 & 1 \\ 0 & 0 & 1 & 1 \\ 1 & 1 & 1 & 1 \end{pmatrix}$$

We first apply $O(2,1;-1), O(4,1;-1), O(4,2;-1)$.

$$\begin{pmatrix} 1 & 0 & 1 & 0 \\ 0 & 1 & -1 & 1 \\ 0 & 0 & 1 & 1 \\ 1 & 1 & 1 & 1 \end{pmatrix}, \quad \begin{pmatrix} 1 & 0 & 1 & 0 \\ 0 & 1 & -1 & 1 \\ 0 & 0 & 1 & 1 \\ 0 & 1 & 0 & 1 \end{pmatrix}, \quad \begin{pmatrix} 1 & 0 & 1 & 0 \\ 0 & 1 & -1 & 1 \\ 0 & 0 & 1 & 1 \\ 0 & 0 & 1 & 0 \end{pmatrix}$$

We next apply $O(4,3;-1), Om(4;-1), O(3,4;-1)$.

$$\begin{pmatrix} 1 & 0 & 1 & 0 \\ 0 & 1 & -1 & 1 \\ 0 & 0 & 1 & 1 \\ 0 & 0 & 0 & -1 \end{pmatrix}, \quad \begin{pmatrix} 1 & 0 & 1 & 0 \\ 0 & 1 & -1 & 1 \\ 0 & 0 & 1 & 1 \\ 0 & 0 & 0 & 1 \end{pmatrix}, \quad \begin{pmatrix} 1 & 0 & 1 & 0 \\ 0 & 1 & -1 & 1 \\ 0 & 0 & 1 & 0 \\ 0 & 0 & 0 & 1 \end{pmatrix}$$

We then apply $O(2,4;-1), O(2,3;1), O(1,3;-1)$.

$$\begin{pmatrix} 1 & 0 & 1 & 0 \\ 0 & 1 & -1 & 0 \\ 0 & 0 & 1 & 0 \\ 0 & 0 & 0 & 1 \end{pmatrix}, \quad \begin{pmatrix} 1 & 0 & 1 & 0 \\ 0 & 1 & 0 & 0 \\ 0 & 0 & 1 & 0 \\ 0 & 0 & 0 & 1 \end{pmatrix}, \quad \begin{pmatrix} 1 & 0 & 0 & 0 \\ 0 & 1 & 0 & 0 \\ 0 & 0 & 1 & 0 \\ 0 & 0 & 0 & 1 \end{pmatrix}$$

Having arrived at the identity, we now find C^{-1} by performing the same operations on the identity matrix.

$$\begin{pmatrix} 1 & 0 & 0 & 0 \\ -1 & 1 & 0 & 0 \\ 0 & 0 & 1 & 0 \\ 0 & 0 & 0 & 1 \end{pmatrix}, \quad \begin{pmatrix} 1 & 0 & 0 & 0 \\ -1 & 1 & 0 & 0 \\ 0 & 0 & 1 & 0 \\ -1 & 0 & 0 & 1 \end{pmatrix}, \quad \begin{pmatrix} 1 & 0 & 0 & 0 \\ -1 & 1 & 0 & 0 \\ 0 & 0 & 1 & 0 \\ 0 & -1 & 0 & 1 \end{pmatrix},$$

$$\begin{pmatrix} 1 & 0 & 0 & 0 \\ -1 & 1 & 0 & 0 \\ 0 & 0 & 1 & 0 \\ 0 & -1 & -1 & 1 \end{pmatrix}, \quad \begin{pmatrix} 1 & 0 & 0 & 0 \\ -1 & 1 & 0 & 0 \\ 0 & 0 & 1 & 0 \\ 0 & 1 & 1 & -1 \end{pmatrix}, \quad \begin{pmatrix} 1 & 0 & 0 & 0 \\ -1 & 1 & 0 & 0 \\ 0 & -1 & 0 & 1 \\ 0 & 1 & 1 & -1 \end{pmatrix},$$

$$\begin{pmatrix} 1 & 0 & 0 & 0 \\ -1 & 0 & -1 & 1 \\ 0 & -1 & 0 & 1 \\ 0 & 1 & 1 & -1 \end{pmatrix}, \quad \begin{pmatrix} 1 & 0 & 0 & 0 \\ -1 & -1 & -1 & 2 \\ 0 & -1 & 0 & 1 \\ 0 & 1 & 1 & -1 \end{pmatrix}, \quad \begin{pmatrix} 1 & 1 & 0 & -1 \\ -1 & -1 & -1 & 2 \\ 0 & -1 & 0 & 1 \\ 0 & 1 & 1 & -1 \end{pmatrix} = C^{-1}$$

■

Exercise 1.6.1.

Give inverses of each of the following matrices.

$$\text{(a)} \begin{pmatrix} 1 & 0 & 0 \\ 2 & 1 & 0 \\ 0 & 0 & 1 \end{pmatrix}, \quad \text{(b)} \begin{pmatrix} 1 & 0 & 0 \\ 0 & 3 & 0 \\ 0 & 0 & 1 \end{pmatrix}, \quad \text{(c)} \begin{pmatrix} 0 & 1 & 0 \\ 1 & 0 & 0 \\ 0 & 0 & 1 \end{pmatrix}$$

Exercise 1.6.2.

Let A, B, C be the matrices in parts (a), (b), and (c).

(a) Find the inverse of AB.

(b) Find the inverse of ABC.

Exercise 1.6.3.

Use the Gauss–Jordan algorithm to find the inverse of $A = \begin{pmatrix} 1 & 2 & -1 \\ 0 & 2 & 1 \\ 1 & 4 & 1 \end{pmatrix}$. Use the inverse A^{-1} to solve $A\mathbf{x} = \begin{pmatrix} 2 \\ 1 \\ 3 \end{pmatrix}$.

Exercise 1.6.4.

Decide whether the following matrices are invertible. Do not find the inverse.

$$A = \begin{pmatrix} 1 & 1 & 1 \\ 0 & 3 & 1 \\ 0 & 0 & 5 \end{pmatrix}, B = \begin{pmatrix} 1 & 3 & 1 \\ 0 & 2 & 1 \\ 3 & 11 & 4 \end{pmatrix}, C = \begin{pmatrix} 1 & 3 & 1 \\ 0 & 2 & 1 \\ 3 & 11 & 5 \end{pmatrix}, D = AB, E = AC.$$

Exercise 1.6.5.

For each matrix A, B, C which is invertible in the previous exercise, reduce it to the identity and record each elementary operation used. Find the inverse by applying the same operations to the identity.

We now consider another operation on matrices that will turn out to be closely related to solving $A\mathbf{x} = \mathbf{b}$ as well. Given an m by n matrix A, there is a related n by m matrix which is called the **transpose** of A and is written A^t (this is read as A-transpose).

DEFINITION 1.6.2. For an m by n matrix A the transpose B of A, which is denoted A^t, is the n by m matrix whose ij-entry is the ji-entry of A; that is, $b_{ij} = a_{ji}$.

The transpose is found by interchanging the rows and columns of A; that is, the ith row of A becomes the ith column of A^t and the jth column of A becomes the jth row of A^t. Note that $(A^t)^t = A$. Here is an example:

$$\text{If } A = \begin{pmatrix} 1 & 2 & 3 & 4 \\ 5 & 6 & 7 & 8 \\ 9 & 10 & 11 & 12 \end{pmatrix}, \text{ then } A^t = \begin{pmatrix} 1 & 5 & 9 \\ 2 & 6 & 10 \\ 3 & 7 & 11 \\ 4 & 8 & 12 \end{pmatrix}.$$

A great deal of the importance of the operation of taking the transpose is that it is closely connected with the dot product. If we identify vectors in $\mathbb{R}^n$ with n by 1 column vectors and denote the dot product of two such column vectors by $\langle \mathbf{v}, \mathbf{w} \rangle$, then $\langle \mathbf{v}, \mathbf{w} \rangle = \mathbf{v}^t\mathbf{w}$, the product of the row vector $\mathbf{v}^t$ and the column vector $\mathbf{w}$. This is not surprising since we used the dot product to define the product of a row vector with a column vector.

When we take the product of an m by n matrix A and an n by p matrix B, then B^t is a p by n matrix and A^t is an n by m matrix so we can also form the product of B^t and A^t. We now show

$$(AB)^t = B^t A^t$$

by computing the ij-entry of each side.

$$((AB)^t)_{ij} = (AB)_{ji} = \mathbf{A}_j\mathbf{B}^i = (a_{j1} \quad \dots \quad a_{jn})\begin{pmatrix} b_{1i} \\ \vdots \\ b_{nj} \end{pmatrix}$$

$$(B^tA^t)_{ij} = (B^t)_i(A^t)^j = (\mathbf{B}^i)^t(\mathbf{A}_j)^t = (b_{1i} \quad \dots \quad b_{ni})\begin{pmatrix} a_{j1} \\ \vdots \\ a_{jn} \end{pmatrix}$$

Equality follows since each side gives the dot product $\langle \mathbf{A}_j, \mathbf{B}^i \rangle$.

If we look at the dot product of $A\mathbf{x}$ with $\mathbf{c}$, we get the following important equation:

$$\langle A\mathbf{x}, \mathbf{c} \rangle = (A\mathbf{x})^t\mathbf{c} = (\mathbf{x}^tA^t)\mathbf{c} = \mathbf{x}^t(A^t\mathbf{c}) = \langle \mathbf{x}, A^t\mathbf{c} \rangle$$

Suppose we are interested in knowing for which $\mathbf{b}$ we can solve $A\mathbf{x} = \mathbf{b}$. We have seen earlier that this is characterized by some equations involving the coordinates of $\mathbf{b}$ which arise when we reduce A to reduced normal form R and there are some zero rows at the bottom of R. In fact, if there are r nonzero rows and $k = m - r$ zero rows, then we have a solution of $R\mathbf{x} = O\mathbf{b}$ exactly when the last k rows of $O\mathbf{b}$ are zero. If we denote the transposes of the last k rows of O (we take the transpose in order to get a column vector) by $\mathbf{c}_1, \dots, \mathbf{c}_k$, then multiplying these rows by $\mathbf{b}$ is just taking the dot product $\langle \mathbf{c}_i, \mathbf{b} \rangle$ for $i = 1, \dots, k$. Thus there is a solution $\mathbf{x}$ of $A\mathbf{x} = \mathbf{b}$ iff $\langle \mathbf{c}_i, \mathbf{b} \rangle = 0$ for $i = 1, \dots, k$, where $\mathbf{c}_1, \dots, \mathbf{c}_k$ are the transposes of the last $k = m - r$ rows of O.

Note that $OA = R$ implies that $R^t = A^tO^t$. Since the last k rows of R are zero, restricting to the last k columns gives $A^t\mathbf{c}_i = \mathbf{0}, i = 1, \dots, k$.

$$\begin{pmatrix} \dots \\ \vdots \\ \dots \\ \mathbf{c}_1^t \\ \vdots \\ \mathbf{c}_k^t \end{pmatrix} A = \begin{pmatrix} \mathbf{R}_1 \\ \vdots \\ \mathbf{R}_r \\ \mathbf{0} \\ \vdots \\ \mathbf{0} \end{pmatrix}, \qquad A^t\begin{pmatrix} \vdots & \dots & \vdots & \mathbf{c}_1 & \dots & \mathbf{c}_k \end{pmatrix} = (\mathbf{R}_1^t \quad \dots \quad \mathbf{R}_r^t \quad \mathbf{0} \quad \dots \quad \mathbf{0})$$

This leads to the following criterion.

Criterion for the Solvability of $A\mathbf{x} = \mathbf{b}$

There is a solution of $A\mathbf{x} = \mathbf{b}$ iff $\langle \mathbf{b}, \mathbf{c} \rangle = 0$ for every $\mathbf{c}$ satisfying $A^t\mathbf{c} = \mathbf{0}$.

If the condition on the right is satisfied, then we will have $\langle \mathbf{b}, \mathbf{c}_i \rangle = 0$ since the $\mathbf{c}_i$ satisfy $A^t\mathbf{c}_i = \mathbf{0}$. We know that means that there is a solution of $A\mathbf{x} = \mathbf{b}$. For the converse, we note that if $\mathbf{x}$ is a solution of $A\mathbf{x} = \mathbf{b}$ and $\mathbf{c}$ satisfies $A^t\mathbf{c} = \mathbf{0}$, then

$$\langle \mathbf{b}, \mathbf{c} \rangle = \langle A\mathbf{x}, \mathbf{c} \rangle = \langle \mathbf{x}, A^t\mathbf{c} \rangle = \langle \mathbf{x}, \mathbf{0} \rangle = 0$$

It is also true that the solutions of $A\mathbf{x} = \mathbf{0}$ are precisely those vectors $\mathbf{x}$ so that $\langle \mathbf{x}, A^t\mathbf{c} \rangle = 0$ for all $\mathbf{c}$ in $\mathbb{R}^m$. For $A\mathbf{x} = \mathbf{0}$ implies that

$$0 = \langle \mathbf{0}, \mathbf{c} \rangle = \langle A\mathbf{x}, \mathbf{c} \rangle = \langle \mathbf{x}, A^t\mathbf{c} \rangle \text{ so } \langle \mathbf{x}, A^t\mathbf{c} \rangle = 0 \text{ for all } \mathbf{c}$$

Conversely, if $\langle \mathbf{x}, A^t\mathbf{c} \rangle = 0$ for all $\mathbf{c}$, then $\langle A\mathbf{x}, \mathbf{c} \rangle = 0$ for all $\mathbf{c}$. Choosing $\mathbf{c} = A\mathbf{x}$ implies $\langle A\mathbf{x}, A\mathbf{x} \rangle = \mathbf{0}$, but this can happen only for $A\mathbf{x} = \mathbf{0}$.

We thus conclude:

Characterization of Solutions of $A\mathbf{x} = \mathbf{0}$

The solutions of $A\mathbf{x} = \mathbf{0}$ are precisely those vectors $\mathbf{x}$ which satisfy

$$\langle \mathbf{x}, A^t\mathbf{c} \rangle = 0 \text{ for all } \mathbf{c} \text{ in } \mathbb{R}^m$$

These two statements show that there is an intimate connection between the equations $A\mathbf{x} = \mathbf{b}$ and $A^t\mathbf{c} = \mathbf{y}$. We will pursue this further in the next two chapters.

Exercise 1.6.6.

Find A^t when $A = \begin{pmatrix} 1 & 3 & 5 \\ 2 & 4 & 6 \\ 2 & 1 & 3 \end{pmatrix}$ and B^t for $B = \begin{pmatrix} 3 & 2 \\ 2 & 1 \\ 1 & 4 \end{pmatrix}$. Verify that $(AB)^t = B^tA^t$.

Exercise 1.6.7.

Let P be a permutation matrix. Show that $P^t = P^{-1}$; that is, verify that $P^tP = PP^t = I$.

Exercise 1.6.8.

Show that for a square matrix A,

$$A^t = A^{-1} \text{ iff } \langle A\mathbf{x}, A\mathbf{y} \rangle = \langle \mathbf{x}, \mathbf{y} \rangle$$

for all $\mathbf{x}, \mathbf{y}$.

(Hint: If $\mathbf{x} = \mathbf{e}_i, \mathbf{y} = \mathbf{e}_j$, then $\langle \mathbf{x}, \mathbf{y} \rangle = 0$ if $i \neq j$ and 1 if $i = j$ but $\mathbf{x}^tB\mathbf{y} = b_{ij}$, the ij-entry of B.)

Exercise 1.6.9.

Consider the matrix $A = \begin{pmatrix} 1 & 2 & 2 \\ 1 & 3 & 1 \\ 2 & 5 & 3 \end{pmatrix}$. Use the matrix O with $OA = R$ to find a solution of $A^t\mathbf{c} = \mathbf{0}$. Similarly, use the matrix O_1 arising at the end of the forward elimination part of the algorithm (so that $O_1A = U$) to find a solution of $A^t\mathbf{c} = \mathbf{0}$. Find all solutions of $A^t\mathbf{c} = \mathbf{0}$.

Exercise 1.6.10.

For the matrix A of the preceding exercise, use the criterion for a solution of $A\mathbf{x} = \mathbf{b}$ to find all solutions of $A^t\mathbf{c} = \mathbf{0}$ as follows. Note that for the columns of A, there are the equations

$$A\begin{pmatrix}1\\0\\0\end{pmatrix} = \begin{pmatrix}1\\1\\2\end{pmatrix}, \quad A\begin{pmatrix}0\\1\\0\end{pmatrix} = \begin{pmatrix}2\\3\\5\end{pmatrix}, \quad A\begin{pmatrix}0\\0\\1\end{pmatrix} = \begin{pmatrix}2\\1\\3\end{pmatrix}$$

Find all vectors $\mathbf{c}$ which are perpendicular to these three vectors. Rewrite the condition for doing this in terms of solving $A^t\mathbf{c} = \mathbf{0}$.

1.7. DETERMINANTS

We now discuss the determinant of a square matrix. Our primary uses of determinants will be to solve the eigenvalue–eigenvector problem for a matrix as well as a criterion for linear independence. However, we will also discuss a few other applications of the determinant briefly.

Many students will have encountered the determinant in the context of small matrices before, such as in a several variables calculus course. To begin our discussion we will first look at the determinant for a 2 by 2 matrix and a 3 by 3 matrix. For the 2 by 2 matrix $A = \begin{pmatrix}a & b\\ c & d\end{pmatrix}$, $\det A = ad - bc$. For a 3 by 3 matrix

$$A = \begin{pmatrix}a_{11} & a_{12} & a_{13}\\ a_{21} & a_{22} & a_{23}\\ a_{31} & a_{32} & a_{33}\end{pmatrix}$$

$\det A$ is given by the more complicated formula

$$\det A = a_{11}(a_{22}a_{33} - a_{23}a_{32}) - a_{12}(a_{21}a_{33} - a_{23}a_{31}) + a_{13}(a_{21}a_{32} - a_{22}a_{31})$$

When these six terms are written with the positive terms first, there is a simple device to remember them by repeating the first two columns of the matrix and then multiplying along the positive diagonals with a plus sign and multiplying along the negative diagonals with a minus sign.

$$\det A = (a_{11}a_{22}a_{33} + a_{12}a_{23}a_{31} + a_{13}a_{21}a_{32}) - (a_{31}a_{22}a_{13} + a_{32}a_{23}a_{11} + a_{33}a_{21}a_{12})$$

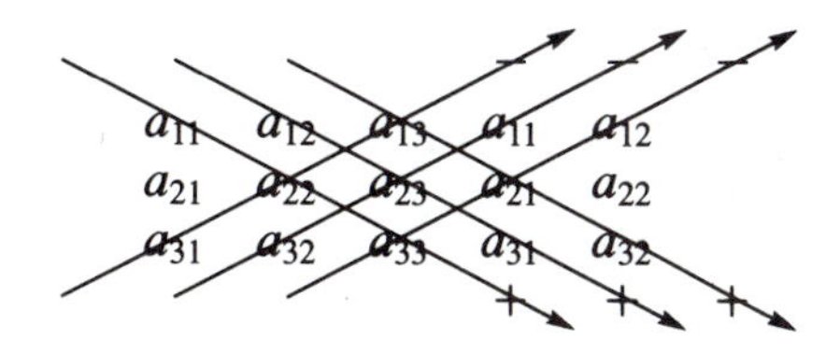

We will use the standard notation of denoting the determinant of a matrix by replacing the curved parentheses around the matrix by vertical lines. For example,

$$\det\begin{pmatrix} 1 & 4 & 0 \\ 2 & 2 & -1 \\ 3 & -2 & 4 \end{pmatrix} = \begin{vmatrix} 1 & 4 & 0 \\ 2 & 2 & -1 \\ 3 & -2 & 4 \end{vmatrix} = -38, \quad \det\begin{pmatrix} a & b \\ c & d \end{pmatrix} = \begin{vmatrix} a & b \\ c & d \end{vmatrix} = ad - bc$$

When we are trying to emphasize certain properties of the determinant, however, we will continue to use the symbol det before the matrix.

Unfortunately, formulas for $\det A$ get increasingly more complicated as the size of A grows. As we will see, $\det A$ will turn out to involve a sum of $n!$ terms, each of which is a product of n of the entries of the matrix. This grows quite quickly with n, and so even evaluating the determinant using a computer directly from definitions like those above becomes infeasible for n large. Thus we will not approach the determinant from the point of view of a definition like the two just given but rather from a set of properties which characterize the determinant. The actual computation of determinants will involve using Gaussian elimination to find a matrix which is simpler but which has the same (or closely related) determinant. We first give the three properties which characterize the determinant of an n by n matrix A. Later we will justify that there is in fact a definition of determinant which satisfies the three properties.

We derive the definition from the properties themselves, thus showing that there is a unique determinant function satisfying the properties. The determinant is a function from the n by n matrices to the real numbers. From the point of view of our properties, we wish to identify an n by n matrix with the n rows of the matrix. Thus we want to think of the determinant as a function defined on n-tuples of row vectors in $\mathbb{R}^n$ with value in the real numbers; that is, $\det A = \det(\mathbf{A}_1, \ldots, \mathbf{A}_n)$. Now if $n - 1$ of the rows are held fixed and the other row is allowed to vary, this determines a function from $\mathbb{R}^n$ to $\mathbb{R}$. The first property that $\det A$ is supposed to satisfy is that this function should be linear. Since this is supposed to hold no matter which row one varies, this property is called **multilinearity.** The second property, which is called the **alternating property,** is that when two rows of a matrix are interchanged, the determinant of the new matrix should be the negative of the determinant of the original matrix. The final property is that the determinant of the identity matrix should be 1. Thus the three characterizing properties of the determinant are

1. Multilinearity

$$\begin{aligned} &\det(\mathbf{A}_1, \ldots, \mathbf{A}_{i-1}, b\mathbf{B}_i + c\mathbf{C}_i, \mathbf{A}_{i+1}, \ldots, \mathbf{A}_n) \\ &= b\det(\mathbf{A}_1, \ldots, \mathbf{A}_{i-1}, \mathbf{B}_i, \mathbf{A}_{i+1}, \ldots, \mathbf{A}_n) + c\det(\mathbf{A}_1, \ldots, \mathbf{A}_{i-1}, \mathbf{C}_i, \mathbf{A}_{i+1}, \ldots, \mathbf{A}_n) \end{aligned}$$

2. Alternating property

$$\det(\ldots, \mathbf{A}_i, \ldots, \mathbf{A}_j, \ldots) = -\det(\ldots, \mathbf{A}_j, \ldots, \mathbf{A}_i, \ldots)$$

3. Normalization

$$\det I = \det(\mathbf{e}_1, \ldots, \mathbf{e}_n) = 1$$

Note that a consequence of multilinearity (1) is

4. If any row is the zero row, then $\det A = 0$.

This follows from (1) in that we can think of the zero row as zero times any row vector and then use (1) to pull the scalar 0 out.

A consequence of (2) is

5. If a matrix has two equal rows then $\det A = 0$.

The alternating property says that when we interchange two rows, the determinant changes sign. But interchanging two equal rows doesn't change the matrix. Hence we get $\det A = -\det A$, which implies that $\det A = 0$.

Before continuing with the other properties of the determinant, let us see how these three properties lead to the definition just given for the determinant of a 2 by 2 matrix.

$$\det\begin{pmatrix} a & b \\ c & d \end{pmatrix} = \det(a\mathbf{e}_1 + b\mathbf{e}_2, c\mathbf{e}_1, +d\mathbf{e}_2) = \quad \text{(by (1))}$$

$$a\det(\mathbf{e}_1, c\mathbf{e}_1 + d\mathbf{e}_2) + b\det(\mathbf{e}_2, c\mathbf{e}_1 + d\mathbf{e}_2) = \quad \text{(by (1))}$$

$$ac\det(\mathbf{e}_1, \mathbf{e}_1) + ad\det(\mathbf{e}_1, \mathbf{e}_2) + bc\det(\mathbf{e}_2, \mathbf{e}_1) + bd\det(\mathbf{e}_2, \mathbf{e}_2)$$

The first and last terms are 0 by (5), and the second term is ad by (3). We can apply (2) and (3) to show that the third term is $-bc$. Thus the determinant is given by the formula $ad - bc$. A similar but more complicated calculation would also verify that these properties lead to the definition of the determinant of a 3 by 3 matrix given earlier.

Let us now note some other properties that (1) and (3) imply. Suppose we apply an elementary row operation to A to add a multiple of the jth row to the ith row. Then the determinant is unchanged.

6. $\det(\mathbf{A}_1, \dots, \mathbf{A}_j, \dots, \mathbf{A}_i + r\mathbf{A}_j, \dots \mathbf{A}_n) = \det(\mathbf{A}_1, \dots, \mathbf{A}_j, \dots, \mathbf{A}_i, \dots, \mathbf{A}_n)$.

Use (1) to rewrite the first term as $\det A + r\det(\mathbf{A}_1, \dots, \mathbf{A}_j, \dots, \mathbf{A}_j, \dots, \mathbf{A}_n)$ and then note that the second term is 0 by (5) since there are two equal rows.

Note that this means that if we perform Gaussian elimination on A to get $O_1 A = U$, then $\det A = \pm\det U$, where the sign is +1 if there are an even number of interchanges of rows involved and is -1 if there are an odd number of interchanges involved. If A is singular, then U has a zero row and so the determinant is zero. If A is not singular, then U is an upper triangular matrix with nonzero entries on the diagonal. Working backward as in the backward elimination step of Gaussian elimination, we can use elementary row operations to make U into a diagonal matrix D with the same determinant. Now applying (1) to the rows of D one row at a time gives that $\det D$ is the product of the diagonal entries of D with $\det I$, and (3) implies this is just the product of the diagonal entries of U. The argument expressing $\det U$ as the product of the diagonal entries also applies to a lower triangular matrix with nonzero diagonal entries, except we use forward elimination to reduce to a diagonal matrix. An upper (lower) triangular matrix with a zero diagonal entry can be reduced using elementary row operations to a matrix with a zero row by backward (forward) elimination, and hence has zero determinant. We summarize this discussion with the following properties:

7. The determinant of a triangular matrix is the product of the diagonal entries.
8. A square matrix A is singular iff $\det A = 0$.
9. If $A = O_1 U$ from the forward elimination step of Gaussian elimination, then $\det A = \pm \det U$. Then (7) implies that the determinant of A is $\pm$ product of the pivots.

Note that this says that no matter how we perform Gaussian elimination to reduce A to a triangular matrix, the product of the pivots will be the same up to sign (although the pivots themselves will be different). It also says that the matrix will be invertible iff $\det A \neq 0$, since A is invertible iff U has no zero rows.

We next give two more properties of $\det A$ which are very useful for theoretical reasons.

10. $\det AB = (\det A)(\det B)$.

We first outline an elegant proof, whose details are left as an exercise. Note that if B is singular (so there is a nonzero solution of $B\mathbf{x} = \mathbf{0}$), then AB will also be singular (with the same solution). In this case both sides will be 0, and so the equality holds. Thus let us suppose that B is nonsingular, and so $\det B \neq 0$. We then consider the function $D(A) = \det(AB)/\det(B)$. In a later exercise you will verify that $D(A)$ satisfies the three properties which characterize the determinant of A; hence $D(A) = \det A$. But this means that $\det AB/\det B = \det A$, which gives (10).

Here is a more computational proof. We note that the properties (1)–(3) of the determinant and (7) can be interpreted as saying that

$$\det EB = \det(E)\det(B)$$

whenever E is an elementary matrix. Then this implies that

$$\det(OB) = \det(O)\det(B)$$

where O is a product of elementary matrices. A formal argument would be by induction using $(EO)B = E(OB)$ to give

$$\det(EO)B = \det E(OB) = \det E \det(OB) = \det E \det O \det B = \det(EO)\det B$$

Now if A is invertible, it is a product of elementary matrices so the result follows in this case. If A is not invertible, then $\det A = 0$. There will be an O which is invertible so that OA has a zero row, and hence OAB also has a zero row. Hence $\det OAB = 0 = \det O \det(AB)$. Since $\det O \neq 0$, we have $\det AB = 0$, and so the formula holds in this case as well.

11. $\det A^t = \det A$.

By multiplying $O_1 A = U$ by $M = O_1^{-1}$ we write $A = MU$. M is a product of permutation matrices which interchange a pair of rows (which have det -1) and matrices $E(i, j, r)$ (which have det 1) and so $\det M = \pm 1$. M^t is the product of the transposes of these matrices in the opposite order. For the elementary matrix $E(i, j, r)$, its transpose is the elementary matrix $E(j, i, r)$, which also has determinant 1. For a permutation matrix which interchanges two rows, its transpose is just the matrix itself, and so it has the same determinant. Thus $\det M^t = \det M$. Since U and U^t are triangular matrices with the same diagonal entries, $\det U = \det U^t$. Thus

$$\det A^t = \det U^t M^t = (\det U^t)(\det M^t) = (\det U)(\det M) = \det MU = \det A$$

Since the operation of taking the transpose interchanges rows and columns, this means that all the properties of the determinant which we have discussed in terms of rows will also be true when we consider them as properties of the columns of the matrix. Thus elementary column operations don't change the determinant of a matrix, interchanging two columns of a matrix changes the determinant by multiplying by -1, and the determinant is a multilinear function of the columns of the matrix. For example,

$$\begin{vmatrix} 1 & 0 & 3 & 2 \\ -2 & 4 & 0 & 1 \\ -1 & 0 & 9 & 2 \\ 0 & 0 & 3 & 1 \end{vmatrix} = \begin{vmatrix} 1 & 0 & 3 & 2 \\ 0 & 4 & 0 & 0 \\ -1 & 0 & 9 & 2 \\ 0 & 0 & 3 & 1 \end{vmatrix}$$

(using column operations with the second column)

$$= \begin{vmatrix} 1 & 0 & -3 & 0 \\ 0 & 4 & 0 & 0 \\ -1 & 0 & 3 & 0 \\ 0 & 0 & 3 & 1 \end{vmatrix} \text{(using } O(4,3;-2)) = 0$$

since the third row is the negative of the first row.

We now do some examples which illustrate the foregoing properties.

We first use Gaussian elimination to compute the determinants of some matrices.

■ ***Example 1.7.1.*** (a) If $A = \begin{pmatrix} 4 & 2 & 1 \\ 3 & 0 & 1 \\ 0 & 2 & 3 \end{pmatrix}$, then A reduces to $U = \begin{pmatrix} 4 & 2 & 1 \\ 0 & -3/2 & 1/2 \\ 0 & 0 & 10/3 \end{pmatrix}$ with no row exchanges, so $\det A = \det U = -20$.

(b) If $A = \begin{pmatrix} 1 & 3 & -1 & 2 \\ 0 & 0 & 2 & 1 \\ 1 & 2 & 3 & 1 \\ 0 & 2 & 1 & 2 \end{pmatrix}$, then A reduces to $U = \begin{pmatrix} 1 & 3 & -1 & 2 \\ 0 & -1 & 4 & -1 \\ 0 & 0 & 2 & 1 \\ 0 & 0 & 0 & -9/2 \end{pmatrix}$ with one row exchange, so $\det A = -\det U = -9$.

(c) If $A = \begin{pmatrix} 1 & 2 & 1 & 2 & 3 \\ 2 & 5 & 3 & 1 & 2 \\ 3 & 1 & 0 & 1 & 1 \\ 2 & 1 & 1 & 4 & 2 \\ 1 & 2 & 1 & 2 & 1 \end{pmatrix}$, then A reduces to $U = \begin{pmatrix} 1 & 2 & 1 & 2 & 3 \\ 0 & 1 & 1 & -3 & -4 \\ 0 & 0 & 2 & -20 & -28 \\ 0 & 0 & 0 & 11 & 12 \\ 0 & 0 & 0 & 0 & -2 \end{pmatrix}$ with no row exchanges, so $\det A = \det U = -44$. ■

■ ***Example 1.7.2.*** (a) $A = \begin{pmatrix} 0 & 1 & 0 & 0 \\ 0 & 0 & 1 & 0 \\ 0 & 0 & 0 & 1 \\ 1 & 0 & 0 & 0 \end{pmatrix}$ is a permutation matrix. It can be changed to the identity matrix with three interchanges (first flip rows 1 and 4, then 2 and 4, then 3 and 4) and so $\det A = -\det I = -1$.

(b) $B = \begin{pmatrix} 0 & 1 & 0 & 0 \\ 0 & 1 & 1 & 0 \\ 0 & 1 & 1 & 1 \\ 1 & 1 & 1 & 1 \end{pmatrix}$ is reduced to the matrix A in 1.7.2 by row operations (subtract row 3 from row 4, then subtract row 2 from row 3, then subtract row 1 from row 2) and so $\det B = \det A = -1$. ■

■ ***Example 1.7.3.*** Suppose that A is the 4 by 4 matrix $\begin{pmatrix} \mathbf{u} \\ \mathbf{v} \\ \mathbf{w} \\ \mathbf{x} \end{pmatrix}$, where $\mathbf{u}, \mathbf{v}, \mathbf{w}, \mathbf{x}$ denote the rows of A. Then we can find the relationship between $\det A$ and the determinants of related matrices by using the properties of the determinant. For example, if B is given by $B = \begin{pmatrix} \mathbf{u} \\ 2\mathbf{u} - 3\mathbf{v} \\ \mathbf{u} + \mathbf{v} + 3\mathbf{w} - \mathbf{x} \\ \mathbf{u} - \mathbf{x} \end{pmatrix}$, then by performing row operations on B we can reduce it to $C = \begin{pmatrix} \mathbf{u} \\ -3\mathbf{v} \\ 3\mathbf{w} \\ -\mathbf{x} \end{pmatrix}$. Using the multilinearity property allows us to see that $\det B = \det C = 9 \det A$. Alternatively, $B = DA$, where $D = \begin{pmatrix} 1 & 0 & 0 & 0 \\ 2 & -3 & 0 & 0 \\ 1 & 1 & 3 & -1 \\ 1 & 0 & 0 & -1 \end{pmatrix}$ and $\det D = 9$ implies $\det B = 9 \det A$. ■

■ ***Example 1.7.4.*** We can frequently combine many of the techniques just outlined to compute the determinant. Here is an example. If we start with $A = \begin{pmatrix} 1 & 0 & 1 & 1 \\ 2 & 1 & 0 & 1 \\ 1 & 0 & 0 & 2 \\ 1 & 1 & 1 & 1 \end{pmatrix}$, then subtracting row 4 from row 2 changes this to $B = \begin{pmatrix} 1 & 0 & 1 & 1 \\ 1 & 0 & -1 & 0 \\ 1 & 0 & 0 & 2 \\ 1 & 1 & 1 & 1 \end{pmatrix}$.

Adding the first column to the third changes this to $C = \begin{pmatrix} 1 & 0 & 2 & 1 \\ 1 & 0 & 0 & 0 \\ 1 & 0 & 1 & 2 \\ 1 & 1 & 2 & 1 \end{pmatrix}$. Interchanging the first and second rows gives $D = \begin{pmatrix} 1 & 0 & 0 & 0 \\ 1 & 0 & 2 & 1 \\ 1 & 0 & 1 & 2 \\ 1 & 1 & 2 & 1 \end{pmatrix}$ and $\det D = -\det A$. Now row operations with the first row changes D to $\begin{pmatrix} 1 & 0 \\ 0 & E \end{pmatrix}$ with $E = \begin{pmatrix} 0 & 2 & 1 \\ 0 & 1 & 2 \\ 1 & 2 & 1 \end{pmatrix}$ and $\det D = \det E$ since the pivots for D will be 1 and the pivots for E. But $\det E = -\begin{vmatrix} 1 & 0 & 0 \\ 0 & 1 & 2 \\ 0 & 2 & 1 \end{vmatrix} = 3$ (we subtract the first row from the third, then interchange the first and third rows). Thus $\det A = -3$. Note that we have essentially used the properties of the determinant to reduce the problem of computing $\det A$ to a problem of computing a 2 by 2 determinant. ■

Exercise 1.7.1.

Use Gaussian elimination to compute the determinants of the following matrices.

$$\text{(a)} \begin{pmatrix} 3 & 2 & 4 \\ 1 & 2 & 1 \\ 4 & 2 & 1 \end{pmatrix}, \quad \text{(b)} \begin{pmatrix} 3 & 1 & 4 \\ 2 & 2 & 2 \\ 4 & 1 & 1 \end{pmatrix}, \quad \text{(c)} \begin{pmatrix} 5 & 1 & 0 & 1 \\ 2 & 1 & 1 & 0 \\ 2 & 1 & 1 & 1 \\ 1 & 2 & 0 & 0 \end{pmatrix},$$

$$\text{(d)} \begin{pmatrix} 1 & -1 & 2 & 1 & 0 \\ 2 & 1 & 1 & 1 & 0 \\ 1 & 0 & 2 & 1 & 1 \\ 1 & 0 & 1 & 1 & 1 \\ -1 & -1 & 1 & -1 & 0 \end{pmatrix}$$

Exercise 1.7.2.

Compute the determinants of the following matrices.

$$\text{(a)} \begin{pmatrix} 0 & 1 & 0 & 0 \\ 1 & 0 & 0 & 0 \\ 0 & 0 & 0 & 1 \\ 0 & 0 & 1 & 0 \end{pmatrix}, \quad \text{(b)} \begin{pmatrix} 0 & 0 & 0 & 1 \\ 0 & 0 & 1 & 1 \\ 0 & 1 & 1 & 1 \\ 1 & 1 & 1 & 1 \end{pmatrix}, \quad \text{(c)} \begin{pmatrix} 1 & 2 & 1 & 3 \\ 1 & 2 & 0 & 1 \\ 1 & 2 & 1 & 1 \\ 1 & 1 & 1 & 1 \end{pmatrix}, \quad \text{(d)} \begin{pmatrix} 2 & 1 & 0 & 1 \\ 1 & 9 & 1 & 3 \\ 1 & 0 & 0 & 2 \\ 1 & 8 & 0 & 2 \end{pmatrix}$$

Exercise 1.7.3.

Suppose $A = \begin{pmatrix} \mathbf{u} \\ \mathbf{v} \\ \mathbf{w} \\ \mathbf{x} \end{pmatrix}$ and $\det A = -3$.

(a) Find $\det B$, if $B = \begin{pmatrix} \mathbf{u} - \mathbf{v} \\ \mathbf{v} \\ \mathbf{u} + \mathbf{v} - 3\mathbf{x} \\ \mathbf{w} + \mathbf{u} \end{pmatrix}$.

(b) Find $\det C$, if $C = \begin{pmatrix} \mathbf{u} + \mathbf{v} \\ \mathbf{u} - 2\mathbf{v} - \mathbf{x} \\ \mathbf{u} + 2\mathbf{v} - 4\mathbf{w} + 2\mathbf{x} \\ \mathbf{u} - \mathbf{v} + \mathbf{w} - \mathbf{x} \end{pmatrix}$. (Hint: Write $C = DA$.)

Exercise 1.7.4.

For a fixed matrix B with $\det B \neq 0$, verify that the function $D(A) = \det(AB)/\det B$ satisfies the three properties which characterize the determinant.

We now discuss some other techniques for the computation of determinants. One should keep in mind that although these techniques may be quite useful for the computation of determinants of special or small matrices, the use of Gaussian elimination as discussed earlier will in general be much more efficient for computing determinants. The first technique we will discuss is expansion by cofactors. We first introduce some notation.

DEFINITION 1.7.1. Given an n by n matrix A, for each pair of integers i, j with $1 \leq i, j \leq n$, there is a related $(n-1)$ by $(n-1)$ matrix $A(i, j)$, the ij-**minor,** which is obtained from A by deleting the ith row and jth column of A. The signed determinant $A_{ij} = (-1)^{i+j} \det A(i, j)$ is called the ij-**cofactor.**

For example, if $A = \begin{pmatrix} 1 & 2 & -1 & 0 \\ 2 & 1 & 4 & 2 \\ 3 & 2 & -1 & 1 \\ 2 & 1 & 3 & 2 \end{pmatrix}$, then $A(2,3) = \begin{pmatrix} 1 & 2 & 0 \\ 3 & 2 & 1 \\ 2 & 1 & 2 \end{pmatrix}$, $A_{23} = 5$ and

$A(4,2) = \begin{pmatrix} 1 & -1 & 0 \\ 2 & 4 & 2 \\ 3 & -1 & 1 \end{pmatrix}$, $A_{42} = 2$.

Note that the signs $(-1)^{i+j}$ form a checkerboard pattern of $+1$'s and -1's starting with a $+1$ in the upper left-hand corner.

$$\begin{pmatrix} + & - & + & - & + \\ - & + & - & + & - \\ + & - & + & - & + \\ - & + & - & + & - \\ + & - & + & - & + \end{pmatrix}$$

Then the basic formula for expansion by cofactors using the ith row is

$$\det A = \sum_{k=1}^{n} a_{ik}A_{ik} = a_{i1}A_{i1} + \cdots + a_{in}A_{in}$$

There is also a similar formula for expansion using the jth column. Then

$$\det A = \sum_{k=1}^{n} a_{kj}A_{kj} = a_{1j}A_{1j} + \cdots + a_{nj}A_{nj}$$

What expansion by cofactors is doing is expressing the determinant of an n by n matrix as a linear combination of the determinants of n submatrices of size $n-1$ by $n-1$ (the minors). For example, a 4 by 4 determinant is reduced to computing four 3 by 3 determinants. Of course, each of those can be reduced to computing three 2 by 2 determinants. This idea could be used to give the inductive definition of the determinant, using as a starting point the determinant of a 2 by 2 matrix. The determinant of an n by n matrix requires the addition of $n!$ terms, each of which is a product of n factors. Computationally, this is very inefficient for large (or even not so large) n.

If $A = \begin{pmatrix} 1 & 2 & 1 & 0 \\ 3 & 1 & 3 & 2 \\ 2 & 1 & 1 & 0 \\ 1 & 2 & 1 & 2 \end{pmatrix}$, the determinant can be computed by expanding by any row or column. Since the fourth column contains two zeros, it is easiest to expand by it. This gives $\det A = 2\begin{vmatrix} 1 & 2 & 1 \\ 2 & 1 & 1 \\ 1 & 2 & 1 \end{vmatrix} + 2\begin{vmatrix} 1 & 2 & 1 \\ 3 & 1 & 3 \\ 2 & 1 & 1 \end{vmatrix}$. The first determinant is 0, and the second is $1(-2) - 2(-3) + 1(1) = 5$ (using expansion by the first row).

This method can be combined with the earlier ones for an alternate computational method. First, simplify A by subtracting the first row from the last to get $\det A = \det B$ with

$$B = \begin{pmatrix} 1 & 2 & 1 & 0 \\ 3 & 1 & 3 & 2 \\ 2 & 1 & 1 & 0 \\ 0 & 0 & 0 & 2 \end{pmatrix}$$

Expansion of $\det B$ by the last row reduces it to $2 \det B(4,4)$, which is the determinant we computed earlier when we expanded by the first row. Alternatively, some Gaussian elimination could have been used in conjunction with this expansion.

To justify this formula, we first consider a special case. For a matrix of the form $A = \begin{pmatrix} 1 & 0 \\ 0 & B \end{pmatrix}$, we first note $\det A = \det B$. This could be verified using the property of the determinant as being $\pm$ product of pivots or just the fact that $f(B) = \det \begin{pmatrix} 1 & 0 \\ 0 & B \end{pmatrix}$ satisfies (1)–(3) and so must be the determinant. In the foregoing notation, the zeros denote the zero vectors of appropriate sizes. Now suppose that the ith row of A is the vector $\mathbf{e}_k$. Then by adding multiples of this row to other rows, we can get a matrix with the same determinant and the kth column has all zeros except that the ik-entry is 1. Then by permuting the rows we can put this nonzero entry in the $1k$-position and shift all the earlier rows down by one. This will involve $(i-1)$ flips. We can then permute columns to put this in the 11-position by $(k-1)$ more interchanges of columns and shift all the earlier columns to the right by one. We then will get the matrix $\begin{pmatrix} 1 & 0 \\ 0 & A(i,k) \end{pmatrix}$ which has determinant $\det A(i,k)$. Since we did $i+k-2$ interchanges, we get that the determinant is $A_{ik} = (-1)^{i+k} \det A(i,k)$ in this case. We are using here that $(-1)^{i+k} = (-1)^{i+k-2}$.

To get the formula for expansion by the ith row, write

$$\det A = \det(\mathbf{A}_1, \ldots, \mathbf{A}_n)$$

Now write the ith row as a linear combination $a_{i1}\mathbf{e}_1 + \ldots + a_{in}\mathbf{e}_n$, and then use multilinearity on the ith row to rewrite

$$\begin{aligned} \det A &= \det(\mathbf{A}_1, \ldots, a_{i1}\mathbf{e}_1 + \cdots + a_{in}\mathbf{e}_n, \ldots, \mathbf{A}_n) \\ &= a_{i1} \det(\mathbf{A}_1, \ldots, \mathbf{e}_1, \ldots, \mathbf{A}_n) + \cdots + a_{in} \det(\mathbf{A}_1, \ldots, \mathbf{e}_n, \ldots, \mathbf{A}_n) \\ &= a_{i1}A_{i1} + \cdots + a_{in}A_{in} \end{aligned}$$

A similar argument gives the formula for expansion by the jth column.

This formula can be used to give a formula for the inverse of a matrix. Again note that this formula is computationally inefficient except for very small matrices.

DEFINITION 1.7.2. The matrix $\operatorname{adj} A$, called the **adjunct matrix** of A, is the matrix whose ij-entry is the cofactor A_{ji}—*note* the switch of indices.

Note that classically this matrix $\operatorname{adj} A$ was called the adjoint, but we will use the term adjunct since the term adjoint is used in a different manner today.

When we multiply the ith row of A by the ith column of $\operatorname{adj} A$, we get $a_{i1}A_{i1} + \cdots + a_{in}A_{in} = \det A$. On the other hand, when we multiply the ith row of A by the jth column of $\operatorname{adj} A$, where $i \neq j$, we get $a_{i1}A_{j1} + \cdots + a_{in}A_{jn}$. This is the determinant of the matrix formed by replacing the jth row of A by the ith row, leaving all other rows alone, and then expanding the determinant by the new jth row. Since this new matrix has two equal rows (the ith and jth rows), its determinant will be zero. These two computations can be put together to see that $A(\operatorname{adj} A) = (\det A)I$. If $\det A \neq 0$, then we saw earlier that

A is invertible. We can then use the preceding formula to give

$$A^{-1} = (1/\det A)(\operatorname{adj} A)$$

As an example, suppose $A = \begin{pmatrix} 1 & 2 & 1 \\ 2 & 1 & 3 \\ 0 & 1 & 2 \end{pmatrix}$. Then

$$\operatorname{adj} A = \begin{pmatrix} A_{11} & A_{21} & A_{31} \\ A_{12} & A_{22} & A_{32} \\ A_{13} & A_{23} & A_{33} \end{pmatrix} = \begin{pmatrix} -1 & -3 & 5 \\ -4 & 2 & -1 \\ 2 & -1 & -3 \end{pmatrix}, \qquad \det A = -7$$

and so $A^{-1} = (-1/7)\begin{pmatrix} -1 & -3 & 5 \\ -4 & 2 & -1 \\ 2 & -1 & -3 \end{pmatrix}$.

Exercise 1.7.5.

Compute the determinant of $A = \begin{pmatrix} 2 & 1 & 4 & 2 \\ 0 & 2 & 1 & 2 \\ 0 & 2 & 1 & 1 \\ 2 & 0 & 1 & 0 \end{pmatrix}$

(a) by expansion by the first column.

(b) by expansion by the fourth row.

Exercise 1.7.6.

Compute the adjunct matrix for matrix A and use it to compute A^{-1}.

Exercise 1.7.7.

Consider the equation $A\mathbf{x} = \mathbf{b}$, where A is invertible. Then $\mathbf{x} = A^{-1}\mathbf{b}$ is the unique solution. Write $A^{-1} = (1/\det A)(\operatorname{adj} A)$ and show $x_i = \det B(i)/\det A$, where $B(i)$ is the matrix obtained from A by replacing the ith column of A by the vector $\mathbf{b}$. Hint: x_i comes from multiplying $1/\det(A)$ times the result of multiplying the ith row of the adjunct matrix times the vector $\mathbf{b}$. This last multiplication needs to be identified with computing $\det B(i)$ by expanding by the ith column.

This formula is known as **Cramer's rule.** It is computationally inefficient except for small matrices. Use it to solve $A\mathbf{x} = \mathbf{b}$ for A, where $\mathbf{b} = (1, 2, 4, 3)$.

We now look at a means of defining the determinant so that it satisfies properties (1)–(3). We verify that this is the only possible definition which satisfies these three properties by deriving it from the properties.

$$\begin{aligned}\det A &= \det(\mathbf{A}_1, \ldots, \mathbf{A}_n) = \det(a_{11}\mathbf{e}_1 + \cdots + a_{1n}\mathbf{e}_n, \ldots, a_{n1}\mathbf{e}_1 + \cdots + a_{nn}\mathbf{e}_n) \\ &= \sum_\sigma a_{1\sigma(1)} \cdots a_{n\sigma(n)} \det(\mathbf{e}_{\sigma(1)}, \ldots, \mathbf{e}_{\sigma(n)})\end{aligned}$$

Here the summation is obtained by applying multilinearity to each row and we are summing over all functions from $\{1, \ldots, n\}$ to $\{1, \ldots, n\}$. However, whenever $\sigma(i) = \sigma(j)$ for different i and j, the matrix whose determinant is being computed will have two equal rows and so the determinant will be zero. Thus we only need to sum over the functions which send $\{1, \ldots, n\}$ to distinct values. Such functions are the permutations of $\{1, \ldots, n\}$—all they are doing are reordering the numbers $1, \ldots, n$. The corresponding matrix is then the permutation matrix which is determined by the permutation. If this permutation is expressible as the product of k flips of two of the numbers (so that it takes k flips of the corresponding rows to get back to the identity) then property (2) implies that $\det(\mathbf{e}_{\sigma(1)}, \ldots, \mathbf{e}_{\sigma(n)}) = (-1)^k$. We call $\epsilon(\sigma) = \det(\mathbf{e}_{\sigma(1)}, \ldots, \mathbf{e}_{\sigma(n)})$ the **sign of the permutation,** and we get the formula

$$\det A = \sum_\sigma \epsilon(\sigma) a_{1\sigma(1)} \ldots a_{n\sigma(n)}$$

We call a permutation σ **odd** if $\epsilon(\sigma) = -1$ and **even** if $\epsilon(\sigma) = 1$. We need to verify that an odd permutation is the product of an odd number of flips and an even permutation is the product of an even number of flips to justify our computation of $\epsilon(\sigma)$ from property (2). To see this we give another way to decide whether a permutation should be even or odd without writing it as a product of flips. For a permutation σ, write $[\sigma(1) \ldots \sigma(n)]$ as the images of $1, \ldots, n$. Let $\mu(\sigma)$ be the number of pairs of numbers which are out of their natural order. For example, for the permutation [3 2 5 1 4], the pairs $(3, 2), (3, 1), (2, 1), (5, 1), (5, 4)$ are out of their natural order, so $\mu(\sigma) = 5$. We claim that $\mu(\sigma)$ is congruent mod 2 to the number of flips used to bring σ back to the identity, no matter how we do this. To see this, we look at one flip and see how it changes the number $\mu(\sigma)$. If we flip adjacent numbers, such as flipping 5 and 1, then all other comparisons between two numbers will be unaffected by this flip, but there will be a change in the order of these two. Thus one flip of adjacent numbers will change the parity of $\mu(\sigma)$ from odd to even or from even to odd. We then check that any interchange of two numbers can be achieved by an odd number of flips of adjacent numbers—if they are separated by p numbers, then it takes $p + 1$ flips of adjacent numbers to move the right one to the position of the left one and p flips of adjacent numbers to get the left one to the old position of the right one. Thus the number $\mu(\sigma)$ will change parity for one flip, even if the numbers are not adjacent. Since $\mu(I)$ is 0, this means that if it takes an odd number of flips to get σ to the identity, $\mu(\sigma)$ will be odd. Thus the parity of $\mu(\sigma)$ determines whether it takes an odd number or an even number of flips to bring σ back to the identity, independent of how we do this. Thus we can *define*

$$\det A = \sum_\sigma \epsilon(\sigma) a_{1\sigma(1)} \cdots a_{n\sigma(n)}$$

where $\epsilon(\sigma)$ is *defined* to be $(-1)^{\mu(\sigma)}$ and the argument shows that this is consistent with properties (2) and (3) and that the three properties determine this formula uniquely. Here we are summing over all permutations of $\{1, \ldots n\}$. With this definition property (1) is also satisfied. We note that although this definition is useful for theoretical reasons to justify that there is in fact some function which satisfies (1)–(3)—note that we have been using this during our whole discussion—it is not an efficient means to compute the determinant except in very special circumstances.

Exercise 1.7.8.

Use the permutation definition of the determinant to compute $\det A$, where $A = \begin{pmatrix} 2 & 1 & 3 \\ 2 & 1 & 4 \\ 0 & 1 & 3 \end{pmatrix}$. Indicate the six terms that are being added as well as the corresponding six permutations of $\{1, 2, 3\}$.

Exercise 1.7.9.

For each of the following permutations find the number $\mu(\sigma)$ and indicate how many flips it takes to bring each back to the identity.
(a) [3 1 4 2 5], (b) [5 2 6 3 1 4]

We complete this discussion of determinants by discussing one of the oldest applications of the determinant. We will restrict our discussion to $\mathbb{R}^2$ and $\mathbb{R}^3$, where the geometry involved is easiest to visualize. The same ideas can be used in higher dimensions as well. In $\mathbb{R}^2$ consider a parallelogram with edges given by vectors $\mathbf{v}, \mathbf{w}$. If $\mathbf{v}$ and $\mathbf{w}$ are orthogonal to one another, then the parallelogram is a rectangle and the area of the parallelogram is just given by the product of the lengths of the edges. Since the square of the length is given by the dot product of the vector with itself,

$$\begin{aligned} \text{area}^2 &= l(\mathbf{v})^2 l(\mathbf{w})^2 = \begin{vmatrix} l(\mathbf{v})^2 & 0 \\ 0 & l(\mathbf{w})^2 \end{vmatrix} = \begin{vmatrix} \mathbf{v}^t \\ \mathbf{w}^t \end{vmatrix} \begin{vmatrix} \mathbf{v} & \mathbf{w} \end{vmatrix} \\ &= \det A^t \det A = (\det A)^2 \end{aligned}$$

where $A = (\mathbf{v} \quad \mathbf{w})$. Thus the area is given by

$$\text{area} = |\det A|$$

When $\mathbf{w}$ is not perpendicular to $\mathbf{v}$, write $\mathbf{w} = a\mathbf{v} + \mathbf{z}$, where $\mathbf{z}$ is perpendicular to $\mathbf{v}$ as in Figure 1.4. Then the area of the parallelogram is the same as the area of the corresponding rectangle with sides given by vectors $\mathbf{v}, \mathbf{z}$ (as shown by an easy geometric argument or by using the formula that the area is the product of the length of the base

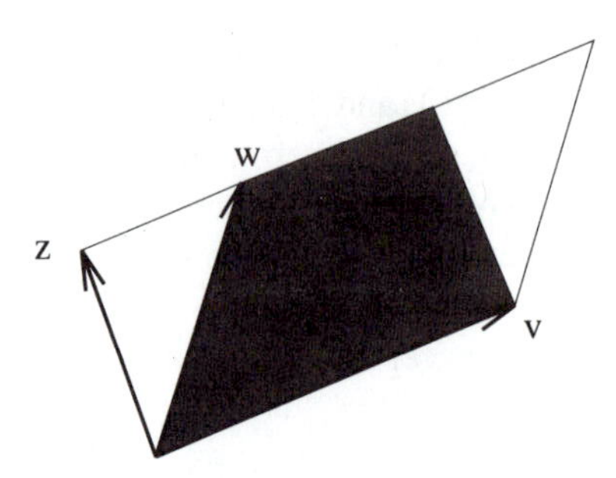

Figure 1.4. Area of parallelogram

times the altitude, which is the length of $\mathbf{z}$). On the other hand, the properties of the determinant say that $\det(\mathbf{v} \quad \mathbf{w}) = \det(\mathbf{v} \quad a\mathbf{v} + \mathbf{z}) = \det(\mathbf{v} \quad \mathbf{z})$. Thus the area is still given by the formula area $= |\det(\mathbf{v} \quad \mathbf{w})|$.

For example, if a parallelogram has adjacent edges given by the vectors $\mathbf{v} = (2, 1)$ and $\mathbf{w} = (-3, 2)$, then its area is given by the formula area $= \left\|\begin{matrix} 2 & -3 \\ 1 & 2 \end{matrix}\right\| = 7$. The area of the corresponding triangle has these two vectors as two adjacent edges at a common vertex and is thus given by 1/2 of the area of the parallelogram, or 7/2.

If we are given the parallelogram (or triangle) in terms of its vertices, then we must first determine the vectors along the adjacent edges. For example, if the triangle has vertices $\mathbf{x} = (1, 2)$, $\mathbf{y} = (3, 3)$, $\mathbf{z} = (-1, 5)$, then if we look at the vertex $\mathbf{x}$, the vectors involved are $\mathbf{v} = \mathbf{y} - \mathbf{x} = (2, 1)$ and $\mathbf{w} = \mathbf{z} - \mathbf{x} = (-2, 3)$ and so the area of the triangle is given by area $= (1/2)\left\|\begin{matrix} 2 & -2 \\ 1 & 3 \end{matrix}\right\| = 4$. The area of the corresponding parallelogram would thus be 8. There is an interesting formula related to this calculation which builds in the computation of $\mathbf{v}$ and $\mathbf{w}$. For what we have shown is that for the parallelogram, the area is given by $\|\mathbf{y} - \mathbf{x} \quad \mathbf{z} - \mathbf{x}\|$. But this can be rewritten as

$$\left\|\begin{matrix} \mathbf{x} & \mathbf{y} & \mathbf{z} \\ 1 & 1 & 1 \end{matrix}\right\|$$

(by subtracting the first column from each of the following columns and then expanding by the third row). Thus we get the following formula for the area of a triangle with vertices $(x_1, x_2), (y_1, y_2), (z_1, z_2)$,

$$\text{area (triangle)} = (1/2)\left\|\begin{matrix} x_1 & y_1 & z_1 \\ x_2 & y_2 & z_2 \\ 1 & 1 & 1 \end{matrix}\right\|$$

In dimension three similar reasoning will show that the volume of a parallelepiped, which has as three edges vectors $\mathbf{u}, \mathbf{v}, \mathbf{w}$ emanating from a common vertex, will be given by the formula

$$\text{volume} = \|\mathbf{u} \quad \mathbf{v} \quad \mathbf{w}\|$$

We now look at an example for a parallelepiped in $\mathbb{R}^3$. If the adjacent edges are given by $\mathbf{u} = (2, 1, 3), \mathbf{v} = (1, 2, 1), \mathbf{w} = (-1, 2, 3)$, then the volume is given by

$$\text{volume} = \left\| \begin{matrix} 2 & 1 & -1 \\ 1 & 2 & 2 \\ 3 & 1 & 3 \end{matrix} \right\| = 16$$

One place where these formulas are applied is in the change of variables formula for integrals over regions in $\mathbb{R}^n$. We will just restrict our attention to integrals over regions in the plane. Suppose there are two regions R, S and a differentiable function g which sends R to S and has a differentiable inverse. Then small pieces of R will be mapped to corresponding pieces of S under g. The integral over R of a continuous function h is defined by subdividing R into small regions (usually rectangles), choosing a point in each region, taking the product of h evaluated at the chosen point times the area of the small region, and then summing these products over all the small regions in the subdivision. The theory says that as the area of the small regions involved goes to zero, the sums (called Riemann sums) tend to a limit which is the integral. There is a similar statement in the region S. The function g allows us to relate integrals over S to integrals over R. If we start with a function f defined on S, then there is a corresponding function given by the composition $f \circ g$ of f with g. If we subdivide R into small regions, say, rectangles, then there will be a corresponding subdivision of S into the images of these rectangles. We depict this in Figure 1.5.

Using these image regions we can get an approximation of the integral of f over S by $\int_S f(y)\,dA \sim \sum f(y_i)A(g(R_i))$, where y_i is a point in the region $g(R_i)$ and $A(g(R_i))$ denotes the area of the region $g(R_i)$. We can choose $y_i = g(x_i)$, with $x_i \in R_i$. Here R_i denotes one of the regions into which we have subdivided R. Now $A(g(R_i))$ is closely related to $A(R_i)$. A good approximation is found by approximating g by its linear approximation at the point x_i, which is given by multiplying by the Jacobian matrix

$$\begin{pmatrix} \partial g^1/\partial x^1 & \partial g^1/\partial x^2 \\ \partial g^2/\partial x^1 & \partial g^2/\partial x^2 \end{pmatrix}(x_i)$$

The determinant of this matrix is sometimes called the Jacobian determinant of the function and is denoted $J(g)(x_i)$. When we multiply by a matrix with columns $\mathbf{v}, \mathbf{w}$, this means that the standard basis vectors are sent to $\mathbf{v}$ and $\mathbf{w}$ and so a rectangle with sides which are multiples of $\mathbf{e}_1, \mathbf{e}_2$ will be sent to a parallelogram. The area of the image

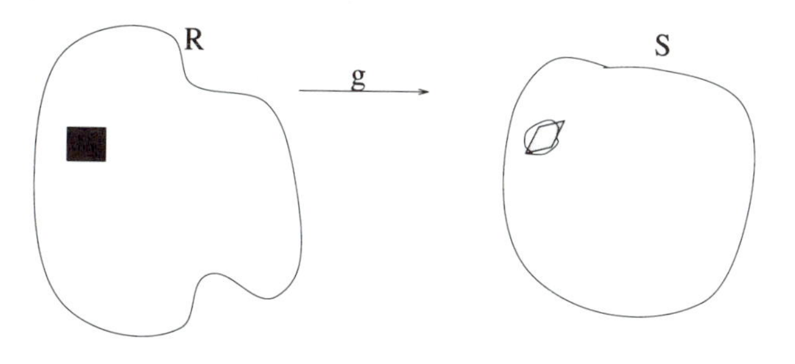

Figure 1.5. Change of variables

parallelogram will then be the product of the original area (i.e. product of the lengths of the sides) times the area of the parallelogram with sides $\mathbf{v}, \mathbf{w}$. This latter area is then given by $\left\| \mathbf{v} \quad \mathbf{w} \right\|$. Here the notation indicates the absolute value of the determinant. Applying this to our situation we see that the sum approximating the integral is itself approximated by

$$\sum_i f(g(x_i))|J(g)(x_i)|A(R_i)$$

which is a Riemann sum for

$$\int_R f(g(x))|J(g)(x)|\, dA$$

This then gives the motivation (as well as a rough sketch of the proof) for the theorem on the change of variables, which says that

$$\int_R f(g(x))|J(g)(x)|\, dA = \int_S f(y)\, dA$$

The analogous result holds in higher dimensions as well, where the Jacobian matrix has entries which are the various partial derivatives of the component functions of g with respect to the variables in R.

An example of the use of this in $\mathbb{R}^2$ occurs when we change to use polar coordinates and make the substitution $x = r\cos(\theta), y = r\sin(\theta)$. Then the Jacobian matrix involved is $\begin{pmatrix} \cos(\theta) & \sin(\theta) \\ -r\sin(\theta) & r\cos(\theta) \end{pmatrix}$, which has $J(g) = r$. The change of variables formula then specializes to give the formula

$$\int_S f(x, y)dx\, dy = \int_R f(r\cos(\theta), r\sin(\theta))r\, dr\, d\theta$$

Exercise 1.7.10.

Give the area of the triangle with vertices $(2, 3), (1, 4), (-1, 5)$.

Exercise 1.7.11.

Find the volume of the parallelepiped whose edges adjacent to a vertex are the vectors $(1, 1, 0), (0, 1, 1), (1, -1, 1)$.

Exercise 1.7.12.

Spherical coordinates (r, θ, ϕ) for a point in 3-space are related to rectangular coordinates by $x = r\cos(\theta)\sin(\phi), y = r\sin(\theta)\sin(\phi), z = r\cos(\phi)$. Here r measures the distance to the origin, θ measures the angle with the x-axis of the projection of the point onto the xy-plane, and ϕ measures the angle with the positive z-axis made in the half plane determined by the z-axis and the vector from the origin to the point. Find the Jacobian determinant for the change of variables from rectangular to spherical coordinates just given. Use this to find the volume enclosed by the sphere of radius 2.

1.8. COMPUTATIONAL NOTES

We have discussed the Gaussian elimination algorithm and used it to reduce a system of equations to a normal form from which we can read off the solution. Solving linear equations, as well as solving other problems in linear algebra which we will encounter later, is an everyday activity in all areas of science and business. There are two main reasons for this. First, linear phenomena tend to arise in many different sorts of applications. Second, there are lots of phenomena which are not really linear but have reasonable linear models. This leads to a linear problem which can be solved and gives useful information about the original problem. When such problems arise, they frequently involve very large matrices which would make it difficult to do the calculations by hand. Fortunately, there exist efficient algorithms such as Gaussian elimination which can be implemented on the computer.

A new problem enters at this point, however. Because most computer languages handle accurately only an approximation of a real number (i.e., they only handle a finite number of digits in an expansion of the number), there are errors introduced every time an operation such as multiplication is performed. Thus the final result may be inaccurate because all these small errors can add up to a rather large error.

Another source of inaccuracy in computations is the initial equations we are dealing with. First, the model we are using to get the equations may be a poor one, which is chosen primarily because it leads to a problem that can be solved. Even if the model is good, the initial data which is fed into the model (such as the coefficients in the equations) may have errors. Thus it is important to understand how small changes in the initial data can lead to changes in the solution. We would like to not only design an algorithm which will solve the problem in the abstract, such as the Gaussian elimination algorithm we have been discussing, but will do it in a way which minimizes the effects of such errors.

The analysis of such errors belongs to the subject of numerical analysis, and we will not really discuss this here. See [4, 3] for a thorough treatment of these issues. However, we want to present a couple of examples which illustrate some of the ideas involved. For simplicity in our discussion we will restrict ourselves to square matrices. When we perform the process of Gaussian elimination, the resulting matrix R in reduced normal form is either the identity, and there is a unique solution, or there is a zero row, and there is either no solution or multiple solutions. Moreover, very small changes in the initial data can change from one of these cases to the other. As an example, consider the following four systems of equations:

$$\begin{aligned} x + y &= 1 \\ x + 1.0001y &= 1 \end{aligned} \tag{1}$$

$$\begin{aligned} x + y &= 1 \\ x + 1.0001y &= 1.0001 \end{aligned} \tag{2}$$

$$\begin{aligned} x + y &= 1 \\ x + y &= 1 \end{aligned} \tag{3}$$

$$\begin{aligned} x + y &= 1 \\ x + y &= 1.0001 \end{aligned} \tag{4}$$

Now changing from one system to another involves changing a coefficient by a small amount (here .0001, but the same principle will apply even if it is a much smaller change). The effect on the solution is dramatic, however. Equations (1) and (2) both have a unique solution, but for (1) it is $x = 1, y = 0$ and for (2) it is $x = 0, y = 1$. This is a rather large change for such a small change in the initial equations. Equation (3) has not only each of these as solutions but has an infinite number of solutions. Finally, equation (4) has no solutions at all.

The problems that are presented when we encounter equations of this sort are impossible to avoid in a situation where there are inaccuracies either from the initial data or from round-off errors in solving the equations themselves. What we can do, however, is devise tests on the initial equations which measure how sensitive they are to small changes in the data and so be forewarned about inaccuracies in the results that may arise. For these equations, what is happening in (3) and (4) is that there will be a row of zeros in the reduced normal form (we say the equations are singular) and in (1) and (2) a small change will make them singular even though they are not singular. These nearly singular equations are sometimes called "computationally singular" because of the large errors in the solutions which can result from errors introduced by the computational process.

Another type of problem which can occur is one which can be helped by devising a better algorithm. For simplicity let us assume we are dealing with a computer which can only keep three significant digits in a number. For example, when we add 1.00 and .00139, the computer gets 1.00, and the last three digits are lost. Now consider the following system, which was introduced by Forsythe and Moler in [4] to illustrate the need for pivoting:

$$\begin{aligned} .0001x + y &= 1 \\ x + y &= 2 \end{aligned}$$

This equation has exact solution $x = 10000/9999, y = 9998/9999$. When we perform Gaussian elimination (taking rounding to three digits into account), we first eliminate x from the second equation to get

$$\begin{aligned} .0001x + y &= 1 \\ -10000y &= -10000 \end{aligned}$$

which gives $y = 1$ and then $x = 0$. Thus the rounding has produced a big error in the x value of the "solution." On the other hand, if we first change the equations by permuting them, we get the system

$$\begin{aligned} x + y &= 2 \\ .0001x + y &= 1 \end{aligned}$$

Gaussian elimination (with rounding) then gives

$$\begin{aligned} x + y &= 2 \\ y &= 1 \end{aligned}$$

which has solution $y = 1$ and $x = 1$, which is much closer to the correct solution. This leads to the computational device which is called **pivoting**. There are actually many forms of pivoting which are used, and there frequently is an interplay between the accuracy achieved and the cost in computation time in achieving the result. We will only discuss a simple form of pivoting which is sometimes called partial pivoting and is essentially what we used in the example. Note that the Gaussian elimination algorithm works on one column at a time. We interchanged rows before only when we had to; that is, the first entry in a column on which we were working in the forward elimination part of the algorithm was a zero. Then we moved down the column to find the first nonzero entry (if there was one) and permuted rows to bring it to the first position (in the remainder of the matrix from some row on). In partial pivoting, we first find the maximal entry (in terms of absolute value) in a column and then permute rows so that it is the first entry. It is then called the **pivot** for the next operations of making the entries beneath it in that column 0. Here is an example of Gaussian elimination with partial pivoting.

$$\begin{pmatrix} 1 & 2 & 0 \\ 2 & 1 & 4 \\ 4 & 0 & 2 \end{pmatrix}$$

The maximal entry in the first column is 4. We perform $Op(1,3)$ to get the 4 to the (1,1) position. It then becomes the pivot for working with the first column. Doing $O(2,1;-1/2)$ and then $O(3,1;-1/4)$ finishes the algorithm on the first column.

$$\begin{pmatrix} 4 & 0 & 2 \\ 2 & 1 & 4 \\ 1 & 2 & 0 \end{pmatrix}, \quad \begin{pmatrix} 4 & 0 & 2 \\ 0 & 1 & 3 \\ 0 & 2 & -1/2 \end{pmatrix}$$

We then look at the part of the matrix consisting of the last two rows and columns and look at the second column. We interchange the last two rows to get the new pivot 2 in the (2,2) position and then make the entry beneath it 0. We use the operations $Op(2,3)$ and $O(3,2;-1/2)$.

$$\begin{pmatrix} 4 & 0 & 2 \\ 0 & 2 & -1/2 \\ 0 & 1 & 3 \end{pmatrix}, \quad \begin{pmatrix} 4 & 0 & 2 \\ 0 & 2 & -1/2 \\ 0 & 0 & 13/4 \end{pmatrix}$$

This ends the forward elimination part of the algorithm; the rest of the algorithm is unchanged from before.

Most commercial algorithms will use some sort of pivoting to improve the accuracy of their results. For the remainder of the discussion we will assume we are doing the partial pivoting described earlier. We consider now only the forward elimination part of the algorithm. We also assume that the algorithm stops when the matrix is determined to be nearly singular, and so we will deal with just the nonsingular case. At the end of the forward elimination step we will have $O_1 A = U$. Now for each column we will do an interchange (we can store this interchange information in a vector $IPVT$, where $IPVT(k)$ gives the row which is interchanged with the kth one) and some row operations. When we do the operation of adding r times the jth row to the ith one, we can store that information in a lower triangular matrix by making the (i, j) entry equal to r. Thus for the preceding Gaussian elimination, with partial pivoting we can store all the information in the row vector $IPVT = (3 \quad 3)$, which tells us that for the first column we first interchange the first and third rows, and then for the second column we interchange the second and third rows. Note that we always leave the last row and column alone so $IPVT$ only has $n-1$ entries for an n by n matrix. The lower triangular matrix with the numbers r is

$$\begin{pmatrix} 0 & 0 & 0 \\ -1/2 & 0 & 0 \\ -1/4 & -1/2 & 0 \end{pmatrix}$$

Now if we fill in the upper half of the matrix with the matrix U, which occurs at the end of the algorithm, we can form a matrix

$$\begin{pmatrix} 4 & 0 & 2 \\ -1/2 & 2 & -1/2 \\ -1/4 & -1/2 & 13/4 \end{pmatrix}$$

which together with $IPVT$ not only tells us what we get for U at the end of the forward elimination step but also contains in coded form the list of operations which you have performed on A to get to U. In particular, if we wish to solve $A\mathbf{x} = \mathbf{b}$ for a variety of right-hand sides $\mathbf{b}$, we can use this information to do this efficiently. What we first have to do is figure out what the corresponding right-hand side in the equivalent equation $U\mathbf{x} = \mathbf{c}$ is. To do this we perform the same operations on $\mathbf{b}$ as we did on A. As an example, let $\mathbf{b} = (1 \quad 2 \quad 0)^t$. $IPVT(1) = 3$ tells us to first do $Op(1,3)$ to get $(0 \quad 2 \quad 1)^t$, then $O(2,1;-1/2)$ to get $(0 \quad 2 \quad 1)^t$, then $O(3,1;-1/4)$ to get $(0 \quad 2 \quad 1)^t$. We then look at $IPVT(2) = 3$ to tell us to perform $Op(2,3)$ to get $(0 \quad 1 \quad 2)^t$ and then $O(3,2;-1/2)$ to get $(0 \quad 1 \quad 3/2)^t$. Thus the equation which is equivalent to

$$\left(\begin{array}{ccc|c} 1 & 2 & 0 & 1 \\ 2 & 1 & 4 & 2 \\ 4 & 0 & 2 & 0 \end{array}\right) \quad \text{is} \quad \left(\begin{array}{ccc|c} 4 & 0 & 2 & 0 \\ 0 & 2 & -1/2 & 1 \\ 0 & 0 & 13/4 & 3/2 \end{array}\right)$$

From this it is easy to solve to get the solution $(-3/13, 8/13, 6/13)$. The matrix

$$\begin{pmatrix} 4 & 0 & 2 \\ -1/2 & 2 & -1/2 \\ -1/4 & -1/2 & 13/4 \end{pmatrix}$$

which allows one (together with $IPVT$) to construct the equivalent system $U\mathbf{x} = \mathbf{c}$ to $A\mathbf{x} = \mathbf{b}$ that occurs at the end of forward elimination is the output of a routine called the LU decomposition of A. The reason for the notation is that there is a matrix L (which is just O_1^{-1}, where $O_1 A = U$) so that $A = LU$ and that in a special case L can be read off from the preceding matrix. The special case involved is when there is *no* pivoting done (i.e., there are no row exchanges, so $IPVT = (1 \ \ 2 \ \ 3 \ \ \dots \ \ n-1)$). Fortunately, this special case does arise for useful classes of matrices such as the positive definite symmetric matrices which we will study later. Then the matrix L will just be the part of the output matrix that lies below the diagonal *with the signs changed* and with ones added on the diagonal. Let us do a simple example without any row interchanges to illustrate this.

$$\begin{pmatrix} 2 & -1 & 0 \\ -1 & 2 & -1 \\ 0 & -1 & 2 \end{pmatrix} \xrightarrow{O(2,1;1/2)} \begin{pmatrix} 2 & -1 & 0 \\ 0 & 3/2 & -1 \\ 0 & -1 & 2 \end{pmatrix} \xrightarrow{O(3,2;2/3)} \begin{pmatrix} 2 & -1 & 0 \\ 0 & 3/2 & -1 \\ 0 & 0 & 4/3 \end{pmatrix}$$

Then we record the information in the matrix

$$\begin{pmatrix} 2 & -1 & 0 \\ 1/2 & 3/2 & -1 \\ 0 & 2/3 & 4/3 \end{pmatrix}$$

and get $L = \begin{pmatrix} 1 & 0 & 0 \\ -1/2 & 1 & 0 \\ 0 & -2/3 & 1 \end{pmatrix}$ and $U = \begin{pmatrix} 2 & -1 & 0 \\ 0 & 3/2 & -1 \\ 0 & 0 & 4/3 \end{pmatrix}$ with $LU = A$. The reason this works is that $L = O_1^{-1}$ and $O_1 = O(3,2;2/3)O(2,1;1/2)$. Thus $L = O_1^{-1} = O(2,1;-1/2)O(3,2;-2/3)$ since we just take the products of the inverses in the opposite order. When we take this product, we can think of multiplying the right matrix by I, then the next one by the result, and so on. But they are now in the correct order so that the row operations are working from the bottom up and at each step we just change one number. In this example $O(3,2;-2/3)$ changes the (3,2) entry to $-2/3$. Finally, $O(2,1;-1/2)$ changes the $(2,1)$ entry to $-1/2$ and so we end up with L as we did earlier. When there are row exchanges, then $A = LU$ with L a permuted lower triangular matrix. There will be a permutation matrix P related to $IPVT$ so that $PA = L'U$ with L' lower triangular. The general case is developed in the Exercises and allows us to apply the analysis given for no exchanges by replacing A by PA.

Once we have an LU decomposition for a matrix, it is quite easy to solve the system $A\mathbf{x} = \mathbf{b}$. We first solve $L\mathbf{c} = \mathbf{b}$ for $\mathbf{c}$ and then solve $U\mathbf{x} = \mathbf{c}$ for $\mathbf{x}$. Since L is invertible (its inverse is just O_1 with $O_1 A = U$), there will be a unique solution for $\mathbf{c}$. When L is in fact lower triangular (with ones on the diagonal) as happens when there are no permutations involved, then it is easy to solve $L\mathbf{c} = \mathbf{b}$ from the top down and solve $U\mathbf{x} = \mathbf{c}$ from the bottom up.

As an example, consider $A\mathbf{x} = \mathbf{b}$ where we know

$$L = \begin{pmatrix} 1 & 0 & 0 \\ 1 & 1 & 0 \\ 3 & 0 & 1 \end{pmatrix}, \qquad U = \begin{pmatrix} 1 & 3 & 2 \\ 0 & 2 & 2 \\ 0 & 0 & 3 \end{pmatrix}, \qquad \mathbf{b} = (1,0,1)$$

Then solving $L\mathbf{c} = \mathbf{b}$ for $\mathbf{c}$ gives $c_1 = 1, c_2 = -1, c_3 = -2$, and solving $U\mathbf{x} = \mathbf{c}$ for $\mathbf{x}$ gives $x_3 = -2/3, x_2 = 1/6, x_1 = 11/6$. Basically, the process of solving for $\mathbf{c}$ and then solving for $\mathbf{x}$ involves the back elimination part of the algorithm (done in a forward manner for L).

There is a lack of symmetry involved in the LU decomposition when we can achieve it with a lower triangular L as was done here. For the entries on the diagonal of L are all 1 and they need not be for U. We will assume that all the entries on the diagonal of U are nonzero for the moment. Then these diagonal entries can be factored out of each row and U rewritten as $U = DV$, where D is a diagonal matrix and V is an upper triangular matrix with 1's on the diagonal. For example, if

$$U = \begin{pmatrix} 2 & -1 & 0 \\ 0 & 3/2 & -1 \\ 0 & 0 & 4/3 \end{pmatrix}$$

then we can factor

$$U = \begin{pmatrix} 2 & 0 & 0 \\ 0 & 3/2 & 0 \\ 0 & 0 & 4/3 \end{pmatrix} \begin{pmatrix} 1 & -1/2 & 0 \\ 0 & 1 & -2/3 \\ 0 & 0 & 1 \end{pmatrix} = DV$$

Thus starting with

$$A = \begin{pmatrix} 2 & -1 & 0 \\ -1 & 2 & -1 \\ 0 & -1 & 2 \end{pmatrix}$$

we can factor it as LDV, where

$$L = \begin{pmatrix} 1 & 0 & 0 \\ -1/2 & 1 & 0 \\ 0 & -2/3 & 1 \end{pmatrix}$$

Notice the relationship between L and V here, $V = L^t$. This depends on two facts. First, the original matrix A was **symmetric,** which means that $A = A^t$. But then $A = LDV$ implies $A^t = V^t D^t L^t$. But $D^t = D$ and V^t is a lower triangular matrix with 1's on the diagonal and L^t is an upper triangular matrix with 1's on the diagonal. Thus $A = A^t$ means that $LDV = V^t D L^t$. The second fact that one needs is that in this type of LDV decomposition the factors are unique, which then implies that $V = L^t$. To see this, suppose $LDV = L'D'V'$, where L, L' are lower triangular with 1's on the diagonal, D, D' are diagonal matrices, and V, V' are upper triangular with 1's on the diagonal. Multiply the equation by L^{-1} on the left and $V'^{-1}D'^{-1}$ on the right to get $(DV)(D'V')^{-1} = L^{-1}L'$. The left-hand side will be an upper triangular matrix (since the product of upper triangular matrices is upper triangular). The right-hand side will be a lower triangular matrix with 1's on the diagonal since the product of two such matrices still has this property. We are also using the fact that the inverse of a lower triangular matrix is lower triangular and it has 1's on the diagonal if the original matrix did (and similarly for upper triangular matrices). But the only matrix which is both upper triangular and lower triangular with 1's on the diagonal is the identity matrix. From this we can conclude that $L = L'$ and thus get the equation $DV = D'V'$. Since the diagonal entries of the product are just given by those of D and D', we get $D = D'$ and then $V = V'$. We are using here that the diagonal entries of D and D' are all nonzero to cancel D and D' from the foregoing equation (by multiplying by D^{-1}). The entries in D will just be the pivots from the forward elimination algorithm.

A symmetric matrix A is called **positive definite** if these pivots are all positive. Thus for a positive definite matrix we will get a decomposition $A = LDL^t$ with the entries of D all positive. By taking the square roots of each of those entries we can form a diagonal matrix $\sqrt{D}$ with $\sqrt{D}\sqrt{D} = D$. But then $L\sqrt{D} = M$ will be a lower triangular matrix with positive entries on the diagonal and $A = MM^t$. This decomposition of a positive definite symmetric matrix is called the **Cholesky decomposition** of A. For the matrix $A = \begin{pmatrix} 2 & -1 & 0 \\ -1 & 2 & -1 \\ 0 & -1 & 2 \end{pmatrix}$ the Cholesky decomposition is MM^t with

$$M = \begin{pmatrix} \sqrt{2} & 0 & 0 \\ -1/2 & \sqrt{3}/2 & 0 \\ 0 & -2/3 & 4/\sqrt{3} \end{pmatrix}$$

Use partial pivoting in Exercises 1.8.1–1.8.3.

Exercise 1.8.1.

Give the factors L and U in the LU decomposition for the matrix

$$A = \begin{pmatrix} 2 & 1 & 1 \\ 1 & 3 & 0 \\ 1 & 0 & 4 \end{pmatrix}$$

Use the LU decomposition to solve $A\mathbf{x} = \mathbf{b}$ for $\mathbf{b} = (4, 5, -2)$ by solving $L\mathbf{c} = \mathbf{b}$ for $\mathbf{c}$ and $U\mathbf{x} = \mathbf{c}$ for $\mathbf{x}$.

Exercise 1.8.2.

Give the Cholesky decomposition for the matrix in Exercise 1.8.1.

Exercise 1.8.3.

Give the factors L and U in the LU decomposition for the matrix

$$A = \begin{pmatrix} 1 & 2 & 1 \\ 2 & 2 & 4 \\ 4 & 2 & 0 \end{pmatrix}$$

Use the LU decomposition to solve $A\mathbf{x} = (0 \quad -8 \quad 6)^t$.

1.9. TWO BASIC APPLICATIONS OF GAUSSIAN ELIMINATION

Linear equations arise in a wide variety of areas. In this section we will discuss two basic examples which lead to linear equations. We will return to look at these examples later where we will use additional techniques we have developed for their analysis. The exercises at the end of the chapter contain a few more applications whose solutions are given using methods of this chapter.

Our first application focuses on the use of the inverse. The model we look at is called an open Leontief input–output economic model. It is named after the economist Wassily Leontief, who was awarded the Nobel prize for his work in 1973. It is based on a matrix C, called the consumption matrix, which measures how much input goes into a unit (equal to some number of dollars of product) of output. We assume there are n industries. The entry c_{ij} gives the units of input from industry i required in producing a unit of output of industry j. If $\mathbf{p} = (x_1, \ldots, x_n)$ is the production vector giving the units produced by each industry, then $C\mathbf{p}$ gives the amount consumed internally in this production, and $(I - C)\mathbf{p}$ gives the amount available for external demand. We would like to see if we can meet a given external demand $\mathbf{d}$ for each industry's output. That means that we want to solve $(I - C)\mathbf{p} = \mathbf{d}$. If $I - C$ is invertible, the solution is then given by the vector $\mathbf{p} = (I - C)^{-1}\mathbf{d}$. Of course, this only makes sense when the answer is nonnegative. Since demand will be nonnegative, the condition we need to be able to meet any demand is that $(I - C)^{-1}$ is itself nonnegative. We will return to look at this condition after we have introduced eigenvalues in Chapter 4, but for the moment we look at an example where it is satisfied.

■ ***Example 1.9.1.*** Suppose there are three industries and the consumption matrix is given by

$$C = \begin{pmatrix} .4 & .3 & .2 \\ .1 & .5 & .1 \\ .4 & .2 & .2 \end{pmatrix}$$

Then

$$(I - C)^{-1} = \begin{pmatrix} 2.5676 & 1.8919 & .8784 \\ .8108 & 2.7027 & .5405 \\ 1.4865 & 1.6216 & 1.8243 \end{pmatrix}$$

and thus any demand can be met. For example, if the demand vector is given as $\mathbf{d} = (100, 50, 25)$, then the production from each of the industries should be given by $\mathbf{p} = (I - C)^{-1}\mathbf{d} = (373.31, 229.73, 275.34)$. ■

Exercise 1.9.1.

Consider an economy with three industries and consumption matrix $C = \begin{pmatrix} .5 & .3 & .1 \\ .2 & .4 & .2 \\ .2 & .1 & .2 \end{pmatrix}$.

Find the matrix $(I - C)^{-1}$ and find the production vector $\mathbf{p}$ necessary to meet a demand vector $\mathbf{d} = (100, 80, 59)$.

Our second example concerns an electric circuit. This circuit will be a simple one with only voltage sources and resistors along branches, and no external current sources. There are two basic laws from circuits which will lead to a system of equations for current flow in the circuit. They are:

Kirchhoff's Laws

1. The sum of the signed currents flowing into a node must equal zero.
2. The net voltage drops around a loop of a circuit are zero.

Ohm's Law

At any resistor the voltage drop is proportional to the current flowing through the resistor.

$$V = RI$$

■ ***Example 1.9.2.*** We apply these laws to the circuit in Figure 1.6 to illustrate how they lead to a system of equations which we can solve via Gaussian elimination to determine the current flowing through each segment of the circuit. The arrows on each branch indicate the direction for the current. Each branch has a resistor with positive resistance, and we have a voltage source on the first branch. Units of current are amperes, units of voltage are volts, and units of resistance are ohms.

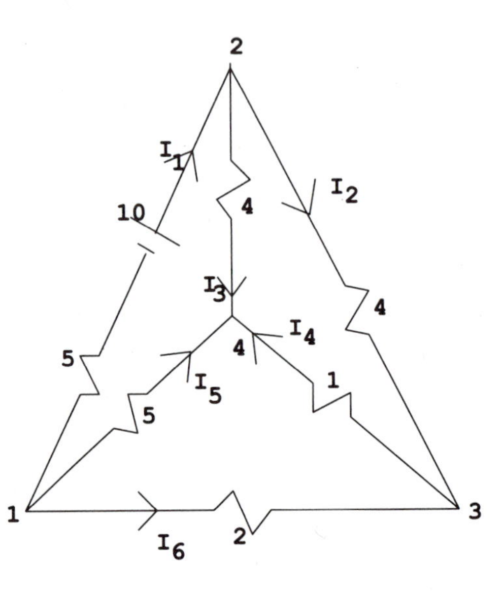

Figure 1.6. Circuit for Example 1.9.2

We first apply Kirchhoff's first law at the first three nodes. The equation coming from the fourth node is just the sum of these three equations.

$$\begin{aligned} -I_1 - I_5 - I_6 &= 0 \\ I_1 - I_2 - I_3 &= 0 \\ I_2 - I_4 + I_6 &= 0 \end{aligned}$$

Here there are three basic small loops. We can form other loops (e.g., a large outer loop), but these won't lead to new information, just a combination of the three equations we get from these loops. Applying Ohm's law and the second Kirchhoff's law gives:

$$\begin{aligned} 5I_1 + 4I_3 - 5I_5 &= 10 \\ 4I_2 - 4I_3 + I_4 &= 0 \\ -I_4 + 5I_5 - 2I_6 &= 0 \end{aligned}$$

Putting these six equations together gives one matrix equation.

$$\begin{pmatrix} -1 & 0 & 0 & 0 & -1 & -1 \\ 1 & -1 & -1 & 0 & 0 & 0 \\ 0 & 1 & 0 & -1 & 0 & 1 \\ 5 & 0 & 4 & 0 & -5 & 0 \\ 0 & 4 & -4 & 1 & 0 & 0 \\ 0 & 0 & 0 & -1 & 5 & -2 \end{pmatrix} \begin{pmatrix} I_1 \\ I_2 \\ I_3 \\ I_4 \\ I_5 \\ I_6 \end{pmatrix} = \begin{pmatrix} 0 \\ 0 \\ 0 \\ 10 \\ 0 \\ 0 \end{pmatrix}$$

The solution is $(1.1814, .6156, .5657, -.1997, -.3661, -.8153)$. The negative signs just indicate that the current is flowing in the opposite direction of the arrow. ■

Exercise 1.9.2.

(a) Show that the equation coming from the fourth node is implied by the equations from the other nodes.

(b) Show that the equation coming from the outer loop is implied by the other equations from the three small loops.

Exercise 1.9.3.

Find the current flowing in an analogous circuit where we add a voltage source of 5 volts on edge 3.

Exercise 1.9.4.

Set up and solve linear equations which give the current flowing in each branch in Figure 1.7.

Exercise 1.9.5.

Set up and solve linear equations which give the current flowing in each branch in Figure 1.8. Explain the form of the solution obtained in terms of the nature of the circuit.

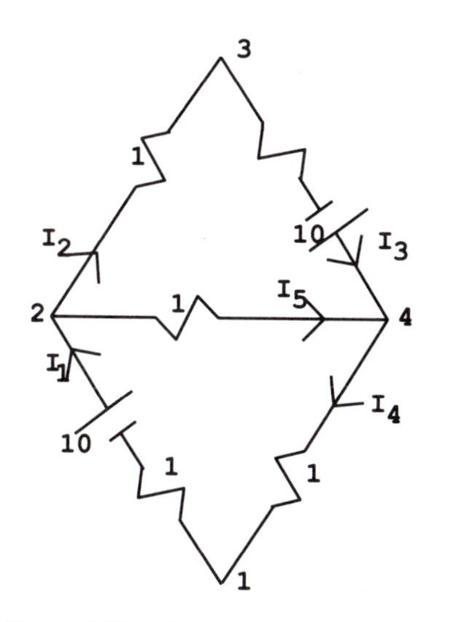

Figure 1.7. Circuit for Exercise 1.9.4

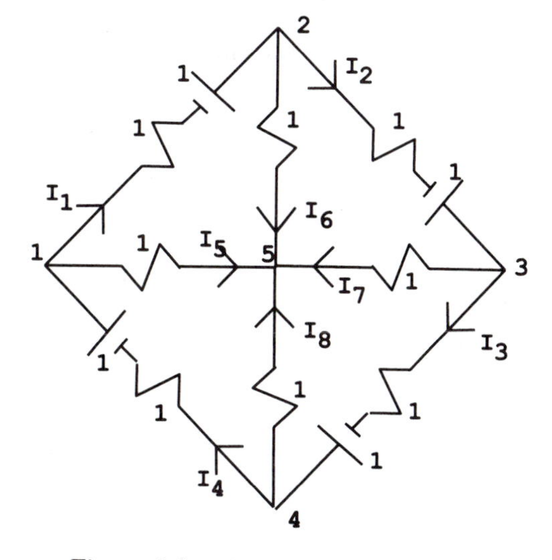

Figure 1.8. Circuit for Exercise 1.9.5

1.10. CHAPTER 1 EXERCISES

1.10.1.

$$\begin{aligned} x_1 + x_3 &= -2 \\ x_2 + 4x_3 + x_5 &= 3 \\ x_4 - 2x_5 &= 0 \end{aligned}$$

(a) Rewrite this system in the form $A\mathbf{x} = \mathbf{b}$.
(b) Give the general solution of the system.

1.10.2.

$$\begin{aligned} x_1 + x_2 - x_3 &= 5 \\ 2x_1 + x_3 - x_4 &= 2 \\ 2x_2 - 3x_3 + x_4 &= 8 \end{aligned}$$

(a) Rewrite this equation in the form $A\mathbf{x} = \mathbf{b}$.
(b) Give the general solution to the system.

1.10.3.

$$\begin{aligned} x_1 - 4x_2 + 3x_3 - 3x_4 + x_5 &= 0 \\ 2x_1 - 8x_2 + 7x_3 - 6x_4 &= 1 \\ 2x_1 - 7x_2 + 6x_3 - 6x_4 + 2x_5 &= 2 \\ x_2 - 2x_5 &= 2 \end{aligned}$$

(a) Rewrite this equation in the form $A\mathbf{x} = \mathbf{b}$.
(b) Give the general solution to the system.

1.10.4. Consider the system $A\mathbf{x} = \mathbf{b}$, with

$$A = \begin{pmatrix} 2 & 1 & 1 & 0 & 0 \\ 0 & 1 & 3 & 2 & 1 \\ 2 & 2 & 4 & 2 & 1 \end{pmatrix}, \quad \mathbf{b} = \begin{pmatrix} 0 \\ 1 \\ 1 \end{pmatrix}$$

(a) Write down the linear equations in the variables $x_1, \ldots, x_5$ corresponding to this system.
(b) Give the general solution of this system.
(c) Show that there is no solution of the corresponding system when the right hand side is changed to $\mathbf{b} = \begin{pmatrix} 1 \\ 1 \\ 1 \end{pmatrix}$.

1.10.5. Consider the system $A\mathbf{x} = \mathbf{b}$, with

$$A = \begin{pmatrix} 0 & 1 & 2 & -1 & 0 & 1 \\ 1 & 1 & 2 & 2 & 1 & 0 \\ 1 & 2 & 2 & 2 & 1 & 1 \\ 2 & 4 & 6 & 3 & 2 & 2 \\ 1 & 0 & 2 & 2 & 1 & -1 \end{pmatrix}, \quad \mathbf{b} = \begin{pmatrix} 5 \\ 2 \\ 7 \\ 16 \\ 1 \end{pmatrix}$$

(a) Write down the linear equations in the variables $x_1, \ldots, x_6$ corresponding to this system.
(b) Give the general solution of this system.

(c) Give conditions in the coefficients of **b** which must be satisfied for $A\mathbf{x} = \mathbf{b}$ to have a solution.

In Exercises 1.10.6–1.10.13 perform the forward elimination part of the Gaussian elimination algorithm. In each case tell whether $A\mathbf{x} = \mathbf{b}$ has a unique solution or not. If not, give conditions on **b** for which it has a solution. Also, give the matrix O_1 so that $O_1A = U$, where U is the matrix at the end of the forward elimination step.

1.10.6.

$$\begin{pmatrix} 1 & 2 \\ 3 & 4 \end{pmatrix}$$

1.10.7.

$$\begin{pmatrix} 1 & 2 & -1 \\ 3 & 1 & 3 \\ 6 & 7 & 0 \end{pmatrix}$$

1.10.8.

$$\begin{pmatrix} -3 & 2 & 1 \\ 7 & -4 & -3 \\ 1 & 0 & -1 \end{pmatrix}$$

1.10.9.

$$\begin{pmatrix} 4 & 1 & 3 \\ 5 & 1 & 1 \\ -1 & -3 & 1 \end{pmatrix}$$

1.10.10.

$$\begin{pmatrix} 2 & 0 & 3 & 2 & -1 \\ 6 & 2 & 13 & 1 & 1 \\ 4 & -1 & 4 & -4 & -6 \end{pmatrix}$$

1.10.11.

$$\begin{pmatrix} 1 & 0 & -1 \\ 2 & 1 & 3 \\ 4 & 1 & 1 \end{pmatrix}$$

1.10.12.

$$\begin{pmatrix} 1 & 2 & 3 & 4 & 5 \\ 6 & 7 & 8 & 9 & 10 \\ 11 & 12 & 13 & 14 & 15 \end{pmatrix}$$

1.10.13.

$$\begin{pmatrix} 2 & 1 & 0 & -1 & 2 \\ 1 & 2 & -1 & 0 & -2 \\ 4 & -1 & -1 & -1 & -1 \\ 5 & 2 & -8 & 3 & -2 \\ 1 & 9 & -4 & -2 & -4 \end{pmatrix}$$

1.10.14. Give the matrix M so that $MA = B$ in each case.

$$A = \begin{pmatrix} 2 & 1 & 4 \\ 4 & 2 & 1 \\ 4 & 5 & 3 \end{pmatrix}$$

(a) $B = \begin{pmatrix} 2 & 1 & 4 \\ 0 & 0 & -7 \\ 4 & 5 & 3 \end{pmatrix}$, (b) $B = \begin{pmatrix} 2 & 1 & 4 \\ 0 & 0 & -7 \\ 0 & 3 & -5 \end{pmatrix}$, (c) $B = \begin{pmatrix} 2 & 1 & 0 \\ 0 & 3 & 0 \\ 0 & 0 & 1 \end{pmatrix}$

1.10.15. Give the matrix M so that $MA = B$ in each case.

$$A = \begin{pmatrix} 1 & 2 & -3 & 4 \\ 2 & 4 & -5 & 9 \\ 2 & 2 & -3 & 4 \\ 5 & 8 & -11 & 17 \end{pmatrix}$$

(a) $B = \begin{pmatrix} 1 & 2 & -3 & 4 \\ 0 & 0 & 1 & 1 \\ 2 & 2 & -3 & 4 \\ 5 & 8 & -11 & 17 \end{pmatrix}$, (b) $B = \begin{pmatrix} 1 & 2 & -3 & 4 \\ 0 & 0 & 1 & 1 \\ 0 & -2 & 3 & -4 \\ 0 & -2 & 4 & 3 \end{pmatrix}$,

(c) $B = \begin{pmatrix} 1 & 2 & -3 & 4 \\ 0 & -2 & 3 & -4 \\ 0 & 0 & 1 & 1 \\ 0 & 0 & 0 & 0 \end{pmatrix}$

In Exercises 1.10.16–1.10.23 find the general solution of the equations given in terms of their augmented matrix $(A|\mathbf{b})$. Indicate which row operations are used in reducing the matrix to reduced normal form.

1.10.16.
$$\left(\begin{array}{ccc|c} 1 & 1 & 1 & 2 \\ 2 & 1 & 2 & 1 \\ 1 & 1 & 1 & 3 \end{array}\right)$$

1.10.17.
$$\left(\begin{array}{ccc|c} 2 & 4 & 4 & 12 \\ 1 & 2 & 0 & 8 \\ -1 & -2 & -8 & 0 \end{array}\right)$$

1.10.18.
$$\left(\begin{array}{ccccc|c} 1 & 1 & 0 & 2 & 1 & 4 \\ 2 & 1 & -1 & 0 & 2 & 5 \\ 4 & 3 & -1 & 4 & 4 & 13 \end{array}\right)$$

1.10.19.
$$\left(\begin{array}{ccccc|c} 1 & 1 & -1 & 3 & 1 & 3 \\ 2 & 0 & 1 & 1 & -2 & 1 \\ 1 & -1 & 0 & 2 & -3 & -2 \\ 0 & 6 & -1 & -1 & 12 & 15 \end{array}\right)$$

1.10.20.
$$\left(\begin{array}{ccccc|c} 1 & 1 & -1 & 3 & 1 & 2 \\ 2 & 0 & 1 & 1 & -2 & 1 \\ 1 & -1 & 0 & 2 & -3 & 1 \\ 0 & 6 & -1 & -1 & 12 & -2 \end{array}\right)$$

1.10.21.
$$\left(\begin{array}{ccccc|c} 1 & 0 & 2 & 0 & 4 & 1 \\ 1 & 1 & 2 & 4 & 3 & 2 \\ 0 & 0 & 1 & 1 & 0 & 4 \end{array}\right)$$

1.10.22.

$$\left(\begin{array}{ccc|c} 3 & 1 & 4 & 8 \\ 3 & 1 & 5 & 9 \\ 6 & 2 & 9 & 17 \\ 9 & 3 & 14 & 26 \end{array}\right)$$

1.10.23.

$$\left(\begin{array}{cccccc|c} 1 & 2 & 3 & -1 & -2 & -3 & 0 \\ 2 & 3 & 4 & -2 & -3 & -4 & 1 \\ 3 & 4 & 5 & -3 & -4 & -5 & 2 \\ 4 & 5 & 6 & -4 & -5 & -6 & 3 \\ 5 & 6 & 7 & -5 & -6 & -7 & 4 \end{array}\right)$$

1.10.24. Consider the matrix equation $A\mathbf{x} = \mathbf{b}$ where $A = \begin{pmatrix} 2 & 1 & 3 \\ 4 & 2 & 1 \\ 8 & 4 & 7 \end{pmatrix}$.

(a) Give the vector $\mathbf{c}$ so that $A\mathbf{x} = \mathbf{b}$ has a solution iff the dot product $\langle \mathbf{c}, \mathbf{b} \rangle = 0$.

(b) Describe geometrically what the set of all $\mathbf{b}$ so that $A\mathbf{x} = \mathbf{b}$ has a solution "looks like" in $\mathbb{R}^3$ (i.e., is it a point, a line, a plane, or all of $\mathbb{R}^3$?). Sketch this set as well as the line through the vector $\mathbf{c}$.

(c) Write every such $\mathbf{b}$ (so that there is a solution of $A\mathbf{x} = \mathbf{b}$) in the form $a_1\mathbf{v}_1 + \ldots + a_k\mathbf{v}_k$ for some k. (Hint: Solve the equation from part (a).)

1.10.25. Consider the matrix equation $A\mathbf{x} = \mathbf{b}$, where

$$A = \begin{pmatrix} 1 & 2 & -4 & -1 \\ 3 & -1 & 9 & -17 \\ 4 & -4 & 20 & -28 \\ -2 & 0 & -4 & 10 \end{pmatrix}$$

(a) Give two vectors $\mathbf{c}_1, \mathbf{c}_2$ so that $A\mathbf{x} = \mathbf{b}$ has a solution iff

$$\langle \mathbf{c}_1, \mathbf{b} \rangle = 0, \langle \mathbf{c}_2, \mathbf{b} \rangle = 0$$

(b) Write every such $\mathbf{b}$ (so that there is a solution of $A\mathbf{x} = \mathbf{b}$) in the form $a_1\mathbf{v}_1 + \cdots + a_k\mathbf{v}_k$ for some k. (Hint: Solve the equations from part (a).)

1.10.26. Consider the matrix equation $A\mathbf{x} = \mathbf{b}$, where

$$A = \begin{pmatrix} 1 & -3 & 2 & 0 & 1 & 3 & 1 & 9 \\ 3 & -2 & 1 & 0 & 3 & -2 & 1 & 8 \\ 3 & -2 & -1 & 9 & -3 & -1 & 0 & 1 \\ 9 & -20 & 11 & 9 & 3 & 17 & 6 & 55 \\ 0 & -7 & 3 & 9 & -6 & 12 & 1 & 12 \\ 17 & -9 & 0 & 18 & 5 & -13 & 3 & 25 \end{pmatrix}$$

(a) Give a matrix C so that $A\mathbf{x} = \mathbf{b}$ has a solution iff $C\mathbf{b} = \mathbf{0}$. (Hint: C can be recovered from the matrix O_1 with $O_1A = U$.)

(b) Give an example of a vector $\mathbf{b}$ for which one can solve $A\mathbf{x} = \mathbf{b}$ and one for which one can't solve the equation.

(c) For the vector given in part (b) for which there is a solution, find the general solution to the equation.

1.10.27. Show that a square matrix is invertible iff it is a product of elementary matrices.

In Exercises 1.10.28–1.10.32 find the inverse of the given matrix if it exists and write the inverse as a product of elementary matrices.

1.10.28.
$$\begin{pmatrix} 5 & 3 & 1 \\ 4 & 2 & 3 \\ 9 & 5 & 4 \end{pmatrix}$$

1.10.29.
$$\begin{pmatrix} 1 & 1 & 1 \\ 1 & 2 & 1 \\ 1 & 3 & 2 \end{pmatrix}$$

1.10.30.
$$\begin{pmatrix} 1 & 0 & 0 & 0 \\ -3 & 1 & 0 & 0 \\ 4 & 0 & 1 & 0 \\ 8 & -3 & 0 & 1 \end{pmatrix}$$

1.10.31.
$$\begin{pmatrix} 1 & 4 & 0 & 0 \\ -2 & 0 & 1 & 0 \\ 0 & 1 & 0 & 0 \\ 1 & 0 & 0 & 1 \end{pmatrix}$$

1.10.32.
$$\begin{pmatrix} 1 & 3 & 0 & 2 & -1 \\ 0 & 1 & -1 & 0 & 0 \\ 0 & 0 & -1 & -3 & 1 \\ 0 & 0 & 0 & 1 & 2 \\ 0 & 0 & 0 & 0 & 1 \end{pmatrix}$$

In Exercises 1.10.33–1.10.41 find the inverse of each matrix if it exists.

1.10.33.
$$\begin{pmatrix} 2 & 1 \\ 1 & 2 \end{pmatrix}$$

1.10.34.
$$\begin{pmatrix} 1 & 3 \\ -1 & 2 \end{pmatrix}$$

1.10.35.
$$\begin{pmatrix} 1 & 2 & 3 \\ 2 & 5 & 8 \\ -3 & -4 & -4 \end{pmatrix}$$

1.10.36.
$$\begin{pmatrix} 1 & -1 & 2 \\ 2 & 1 & 0 \\ 3 & 0 & 2 \end{pmatrix}$$

1.10.37.
$$\begin{pmatrix} 1 & -1 & 2 \\ 2 & 1 & 0 \\ 3 & 0 & 1 \end{pmatrix}$$

1.10.38.

$$\begin{pmatrix} 2 & 0 & 0 & 0 \\ -2 & 1 & 0 & 0 \\ 4 & 2 & 4 & 0 \\ 0 & 2 & 1 & 1 \end{pmatrix}$$

1.10.39.

$$\begin{pmatrix} 4 & -4 & 8 & 0 \\ -4 & 5 & -6 & 2 \\ 8 & -6 & 36 & 8 \\ 0 & 2 & 8 & 6 \end{pmatrix}$$

1.10.40.

$$\begin{pmatrix} 1 & 2 & 3 & 4 & 5 \\ 6 & 7 & 8 & 9 & 10 \\ 11 & 12 & 13 & 14 & 15 \\ 16 & 17 & 18 & 19 & 20 \\ 21 & 22 & 23 & 24 & 25 \end{pmatrix}$$

1.10.41.

$$\begin{pmatrix} 1 & 2 & 3 & 4 & 5 \\ 6 & 7 & 8 & 9 & 10 \\ 11 & 12 & 14 & 15 & 16 \\ 17 & 19 & 19 & 20 & 20 \\ 21 & 25 & 30 & 40 & 50 \end{pmatrix}$$

1.10.42. Let

$$E = \begin{pmatrix} 1 & 0 & 0 \\ 0 & 1 & 0 \\ r & 0 & 1 \end{pmatrix}, \qquad D = \begin{pmatrix} 1 & 0 & 0 \\ 0 & r & 0 \\ 0 & 0 & 1 \end{pmatrix}, \qquad P = \begin{pmatrix} 0 & 1 & 0 \\ 1 & 0 & 0 \\ 0 & 0 & 1 \end{pmatrix}$$

(a) Find formulas for $E^{100}, D^{100}, P^{100}$.

(b) Prove each of your formulas by induction.

1.10.43. Find the inverse of the following product of 3 by 3 elementary matrices:

$$A = E(3, 2; -1)E(3, 1; -4)E(2, 1; 2)$$

1.10.44. Find the inverse of the following product of the elementary 5 by 5 matrices.

$$A = E(4, 2; -3)D(3; 4)P(4, 5)E(3, 5; 1)$$

1.10.45. Find the inverse of the following product of elementary 10 by 10 matrices:

$$A = E(5, 3; -4)P(7, 9)E(9, 7; 5)$$

1.10.46. Use Gaussian elimination to explain why an invertible lower triangular matrix is the product of lower triangular elementary matrices. Use this to show that the inverse of a lower triangular matrix is lower triangular.

1.10.47. Let $A = \begin{pmatrix} 2 & 1 \\ 1 & 2 \end{pmatrix}$. Compute A^2. Show that

$$A^2 - 4A + 3I = 0$$

1.10.48. Let $A = \begin{pmatrix} 1 & 2 & 3 \\ 2 & -2 & 1 \\ 3 & 2 & 1 \end{pmatrix}$. Compute A^3 and show that

$$A^3 = 18A + 28I$$

1.10.49. Let $A = \begin{pmatrix} 2 & 1 & 0 \\ 0 & -1 & 2 \\ 1 & 3 & -1 \end{pmatrix}$.

(a) Compute A^3 and show that $A^3 = 9A - 8I$.
(b) Solve for a, b, c so that $A^6 = aA^2 + bA + cI$.

1.10.50. Denote by $M(i, j)$ the matrix with 1 in the ijth position and 0 elsewhere.
(a) Show that $M(i, j)A$ is a matrix that has zeros except in the ith row and whose ith row is the jth row of $A : (M(i, j)A)_i = A_j$.
(b) Show that $AM(i, j)$ is a matrix that has zeros except in the jth column and whose jth column is the ith column of $A : (AM(i, j))^j = A^i$.

1.10.51. Suppose that $A = \begin{pmatrix} a & b \\ c & d \end{pmatrix}$ commutes with every 2 by 2 matrix B; that is, $AB = BA$. Show that $a = d$ and $b = c = 0$. (Hint: Use Exercise 1.10.50.)

1.10.52. Suppose that an n by n matrix commutes A with every other n by n matrix $B, AB = BA$.
(a) By choosing $B = M(i, i)$ (see Exercise 1.10.50) show that all the off-diagonal entries must be 0.
(b) By choosing $B = M(i, j)$ show that all the diagonal entries must be equal. Conclude that A must be a multiple of the identity.

1.10.53. Give examples of matrices A, B, C so that
(a) $A\mathbf{x} = \mathbf{b}$ has a unique solution for any $\mathbf{b}$.
(b) $B\mathbf{x} = \mathbf{b}$ has infinitely many solutions for any $\mathbf{b}$.
(c) $C\mathbf{x} = \mathbf{b}$ has no solution for $\mathbf{b} = (1 \quad 1 \quad 0)^t$ and C is a 3 by 3 matrix.
(d) Does there exist a 3 by 3 matrix D so that $D\mathbf{x} = \mathbf{b}$ has infinitely many solutions for every $\mathbf{b}$ in $\mathbb{R}^3$? Either give an example or explain why this can't happen.

1.10.54. Let $A = \begin{pmatrix} 1 & -1 & -1 \\ 1 & 0 & a \\ 1 & a & 0 \end{pmatrix}$. Give conditions on a so that

$$A\mathbf{x} = \begin{pmatrix} 2 \\ 1 \\ 1 \end{pmatrix}$$

has
(a) a unique solution
(b) more than one solution
(c) no solution

1.10.55. Let $A = \begin{pmatrix} 1 & 0 & 1 \\ 1 & 1 & a \\ 0 & -a & -2 \end{pmatrix}$, $\mathbf{b} = \begin{pmatrix} 1 \\ 0 \\ -1 \end{pmatrix}$. Give conditions on a so that $A\mathbf{x} = \mathbf{b}$ has

(a) a unique solution
(b) more than one solution
(c) no solution

1.10.56. Suppose $A\mathbf{x} = \mathbf{b}$ (for a 3 by 4 matrix A) has a solution exactly when $b_1 + b_2 + b_3 = 0$. Give a vector $\mathbf{c}$ so that $A'\mathbf{c} = 0$. Give an example of a 3 by 4 matrix A with this property.

1.10.57. Give an example of a 4 by 5 matrix so that $A\mathbf{x} = \mathbf{b}$ has a solution exactly when $b_1 + b_2 + b_3 + b_4 = 0$.

1.10.58. Find conditions on a, b, c so that

$$\begin{pmatrix} 2 & 4 & 1 \\ -4 & -7 & 0 \\ 0 & -1 & -2 \end{pmatrix} \mathbf{x} = \begin{pmatrix} a \\ b \\ c \end{pmatrix}$$

has a solution.

1.10.59. Give conditions on a, b, c, d so that

$$\begin{pmatrix} 2 & 3 & -1 & 4 \\ 3 & 1 & -1 & 0 \\ 0 & 2 & 1 & 1 \\ -1 & 0 & -1 & 3 \end{pmatrix} \mathbf{x} = \begin{pmatrix} a \\ b \\ c \\ d \end{pmatrix}$$

has a solution.

1.10.60. Suppose A is a 3 by 3 matrix.

(a) Show that $(1, 1, 1)$ is a solution of $A'\mathbf{x} = \mathbf{0}$ iff the sum of the rows of A is the zero row.
(b) Show that $(1, 1, 1)$ is a solution of $A\mathbf{x} = \mathbf{0}$ iff the sum of the columns of A is the zero column.
(c) Generalize this result to n by n matrices.

1.10.61. Suppose that $A = \begin{pmatrix} A_{11} & A_{12} \\ A_{21} & A_{22} \end{pmatrix}$, $B = \begin{pmatrix} B_{11} & B_{12} \\ B_{21} & B_{22} \end{pmatrix}$ decompose into blocks where A_{ij} is of size m_i by n_j and B_{jk} is of size n_j by p_k. Show that the product AB can be expressed in terms of the multiplication of the blocks as

$$AB = \begin{pmatrix} A_{11}B_{11} + A_{12}B_{21} & A_{11}B_{12} + A_{12}B_{22} \\ A_{21}B_{11} + A_{22}B_{21} & A_{21}B_{12} + A_{22}B_{22} \end{pmatrix}$$

1.10.62. With the notation of the preceding exercise show that if A and B are block upper triangular in the sense that $A_{21} = B_{21}$ are zero matrices, then the product will also be block upper triangular.

1.10.63. Multiply the following matrices using the block multiplication formula of Exercise 1.10.61.

(a) $\begin{pmatrix} 2 & 1 & 0 \\ 1 & 2 & 0 \\ 0 & 0 & 1 \end{pmatrix}\begin{pmatrix} 3 & 2 & 0 \\ 1 & 1 & 0 \\ 0 & 0 & 7 \end{pmatrix}$

(b) $\begin{pmatrix} 2 & 1 & 2 & 1 \\ 4 & 1 & 5 & 4 \\ 0 & 0 & 0 & 5 \\ 0 & 0 & 0 & 1 \end{pmatrix}\begin{pmatrix} 1 & 2 & 1 & 1 & -1 \\ 0 & 1 & 0 & 2 & 2 \\ -1 & 1 & 2 & 0 & 1 \\ 0 & 0 & 0 & 1 & 1 \end{pmatrix}$

1.10.64. With the notation of Exercise 1.10.61, suppose the off-diagonal blocks of A and B are zero and the diagonal blocks are square and invertible. Show that AB is square and invertible.

1.10.65. Show that if A, B are symmetric matrices, then AB is symmetric iff $AB = BA$. Give an example of symmetric matrices with

(a) AB not symmetric.

(b) AB symmetric.

1.10.66. Determine when a product of elementary matrices is again an elementary matrix.

1.10.67. Suppose that A is an m by n matrix which row reduces to a matrix U with r nonzero rows at the end of the forward elimination step. An n by m matrix B is called a **left inverse** for A if $BA = I_n$ and is called a **right inverse** for A if $AB = I_m$.

(a) Show that A has a right inverse iff one can solve $A\mathbf{x} = \mathbf{b}$ for all $\mathbf{b} \in \mathbb{R}^m$.

(b) Show that A has a right inverse iff $r = m$.

(c) Show that A has a left inverse iff $A\mathbf{x} = \mathbf{0}$ has a unique solution $\mathbf{x} = \mathbf{0}$.

(d) Show that A has a left inverse iff $r = n$.

(e) Show that only a square matrix can be invertible (has both left and right inverses) and this requires $r = m = n$.

1.10.68. Show that a square matrix A has an inverse iff A^t has an inverse.

1.10.69. Suppose $L : \mathbb{R}^2 \to \mathbb{R}^2$ is a linear transformation with $L(1,0) = (\sqrt{3}/2, 1/2)$ and $L(0,1) = (-1/2, \sqrt{3}/2)$. Find A so that $L(\mathbf{x}) = A\mathbf{x}$ for all $\mathbf{x}$. Describe geometrically where L sends a vector in $\mathbb{R}^2$.

1.10.70. Suppose $L : \mathbb{R}^2 \to \mathbb{R}^2$ is a linear transformation with $L(1,0) = (0,1)$, $L(0,1) = (1,0)$. Find A so that $L(\mathbf{x}) = A\mathbf{x}$ for all $\mathbf{x}$. Describe geometrically where L sends a vector in $\mathbb{R}^2$.

1.10.71. Suppose $L : \mathbb{R}^3 \to \mathbb{R}^2$ is a linear transformation with

$$L(1,0,0) = (2,3), L(0,1,0) = (-1,3), L(0,0,1) = (1,0)$$

Find the matrix A so that $L(\mathbf{x}) = A\mathbf{x}$ for all $\mathbf{x}$.

1.10.72. Suppose that $L : \mathbb{R}^4 \to \mathbb{R}^4$ is the linear transformation so that $L(\mathbf{e}_1) = \mathbf{0}$, $L(\mathbf{e}_i) = \mathbf{e}_{i-1}$ for $i \geq 2$.

(a) Find the matrix A so that $L(\mathbf{x}) = A\mathbf{x}$.

(b) Show that if we take the composition of L with itself four times ($L \circ L \circ L \circ L$, also denoted L^4), then this composition sends each of the vectors $\mathbf{e}_1, \dots, \mathbf{e}_4$ to $\mathbf{0}$.

(c) Without computing A^4, use the preceding exercise to show that it must be the zero matrix.

1.10.73. Generalize the preceding exercise to linear transformations $L : \mathbb{R}^n \to \mathbb{R}^n$.

1.10.74. Compute determinants of the following matrices.

$$A = \begin{pmatrix} 3 & 2 & 1 \\ 1 & 0 & 1 \\ -1 & 1 & 0 \end{pmatrix}, B = \begin{pmatrix} 1 & 2 & 3 \\ 0 & 3 & 1 \\ 0 & 0 & 4 \end{pmatrix}, C = BA, D = B^{-1}, E = \begin{pmatrix} 1 & 2 & 3 & 4 \\ 5 & 6 & 7 & 8 \\ 9 & 10 & 11 & 12 \\ 13 & 14 & 15 & 16 \end{pmatrix}$$

1.10.75. Which of the matrices in the previous exercise are invertible? Justify your answer in terms of determinants. Use the adjunct matrix to find the inverse of B.

1.10.76. Compute determinants of the following matrices, using properties of determinants.

$$\text{(a)} \begin{pmatrix} 1 & 0 & -1 & 1 & 2 \\ 0 & 0 & 0 & 2 & 1 \\ 0 & 3 & 9 & 119 & -1 \\ 0 & 0 & 0 & 0 & -2 \\ 0 & 0 & 4 & -3 & -1 \end{pmatrix}, \quad \text{(b)} \begin{pmatrix} 1 & 1 & 1 & 1 & 1 & 1 \\ 1 & 1 & 1 & 1 & 1 & 1 \\ 3 & -2 & 83 & -5 & 12 & 11 \\ 23 & 32 & 21 & 12 & 1 & 9 \\ -4 & 2 & 1 & 1 & -3 & 5 \\ 9 & 2 & 10 & 11 & 15 & 99 \end{pmatrix}$$

1.10.77. Compute determinants of the following matrices, using properties of determinants.

$$\text{(a)} \begin{pmatrix} 0 & 0 & 0 & 4 & 0 & 0 & 0 & 0 & 0 \\ 0 & 0 & 3 & 0 & 0 & 0 & 0 & 0 & 0 \\ 0 & 0 & 0 & 0 & 0 & 0 & 7 & 0 & 0 \\ 1 & 0 & 0 & 0 & 0 & 0 & 0 & 0 & 0 \\ 0 & 0 & 0 & 0 & 0 & 0 & 0 & 0 & 9 \\ 0 & 0 & 0 & 0 & 5 & 0 & 0 & 0 & 0 \\ 0 & 0 & 0 & 0 & 0 & 0 & 0 & 8 & 0 \\ 0 & 2 & 0 & 0 & 0 & 0 & 0 & 0 & 0 \\ 0 & 0 & 0 & 0 & 0 & 6 & 0 & 0 & 0 \end{pmatrix}, \quad \text{(b)} \begin{pmatrix} 1 & 1 & 1 & 1 & 1 & 1 & 1 & 1 & 1 \\ 1 & 2 & 2 & 2 & 2 & 2 & 2 & 2 & 2 \\ 2 & 2 & 3 & 3 & 3 & 3 & 3 & 3 & 3 \\ 3 & 3 & 3 & 4 & 4 & 4 & 4 & 4 & 4 \\ 4 & 4 & 4 & 4 & 5 & 5 & 5 & 5 & 5 \\ 5 & 5 & 5 & 5 & 5 & 6 & 6 & 6 & 6 \\ 6 & 6 & 6 & 6 & 6 & 6 & 7 & 7 & 7 \\ 7 & 7 & 7 & 7 & 7 & 7 & 7 & 8 & 8 \\ 8 & 8 & 8 & 8 & 8 & 8 & 8 & 8 & 9 \end{pmatrix},$$

$$\text{(c)} \begin{pmatrix} 1 & 1 & 1 & 1 & 1 & 1 & 1 & 1 & 1 \\ 1 & 2 & 2 & 2 & 2 & 2 & 2 & 2 & 2 \\ 2 & 2 & 3 & 3 & 3 & 3 & 3 & 3 & 3 \\ 3 & 3 & 3 & 4 & 4 & 4 & 4 & 4 & 4 \\ 4 & 4 & 4 & 4 & 5 & 5 & 5 & 5 & 5 \\ 5 & 5 & 5 & 5 & 5 & 6 & 6 & 6 & 6 \\ 6 & 6 & 6 & 6 & 6 & 6 & 7 & 7 & 7 \\ 7 & 7 & 7 & 7 & 7 & 7 & 7 & 8 & 8 \\ 8 & 8 & 8 & 8 & 8 & 8 & 8 & 8 & 8 \end{pmatrix}$$

1.10.78. Suppose A has a block decomposition as in Exercise 1.10.61 where the block A_{21} is zero and the diagonal blocks are square. Show that $\det(A) = \det(A_{11})\det(A_{22})$.

1.10.79. Use the preceding exercise to compute the determinants of the following matrices.

$$\text{(a) } A = \begin{pmatrix} 2 & 3 & 4 & 7 \\ 4 & 1 & 2 & 5 \\ 0 & 0 & 3 & 1 \\ 0 & 0 & 1 & 3 \end{pmatrix}, \quad \text{(b) } B = \begin{pmatrix} 3 & 4 & 5 & -1 & 2 \\ 0 & 0 & 3 & 2 & 4 \\ 0 & 1 & 3 & 6 & 2 \\ 0 & 0 & 0 & 2 & 1 \\ 0 & 0 & 0 & 1 & -1 \end{pmatrix}$$

1.10.80. Use Exercise 1.10.78 to compute the determinants of the following matrices.

$$\text{(a) } A = \begin{pmatrix} 1 & 4 & 3 & 8 & 7 & -2 \\ 6 & -3 & -4 & 3 & 2 & 2 \\ 0 & 0 & 4 & 1 & 4 & 4 \\ 0 & 0 & -3 & 2 & 2 & 1 \\ 0 & 0 & 0 & 0 & 3 & 2 \\ 0 & 0 & 0 & 0 & 2 & 1 \end{pmatrix}, \quad \text{(b) } B = \begin{pmatrix} 0 & 0 & 0 & 0 & 3 & 2 \\ 0 & 0 & 0 & 0 & 2 & 1 \\ 1 & 4 & 3 & 8 & 7 & -2 \\ 6 & -3 & -4 & 3 & 2 & 2 \\ 0 & 0 & 4 & 1 & 4 & 4 \\ 0 & 0 & -3 & 2 & 2 & 1 \end{pmatrix}$$

1.10.81. Use Exercise 1.10.78 to compute the determinant of the following matrix.

$$A = \begin{pmatrix} 0 & 0 & 5 & 4 & 0 & 0 & 0 & 0 & 0 \\ 0 & 0 & 2 & 1 & 0 & 0 & 0 & 0 & 0 \\ 4 & 1 & 0 & 0 & 0 & 0 & 0 & 0 & 0 \\ -3 & 1 & 0 & 0 & 0 & 0 & 0 & 0 & 0 \\ 0 & 0 & 0 & 0 & 4 & 3 & 1 & 0 & 0 \\ 0 & 0 & 0 & 0 & 2 & 0 & 1 & 0 & 0 \\ 0 & 0 & 0 & 0 & 3 & 0 & 2 & 0 & 0 \\ 0 & 0 & 0 & 0 & 0 & 0 & 0 & 2 & -1 \\ 0 & 0 & 0 & 0 & 0 & 0 & 0 & 1 & 3 \end{pmatrix}$$

1.10.82. (a) Find the area of the triangle with vertices at

$$(2,3), (4,1), (3,0)$$

(b) Find the volume of the parallelepiped with edges at a common vertex given by the vectors

$$(1,2,-1), (2,1,0), (3,1,1)$$

1.10.83. Show that if T is a parallelogram in $\mathbb{R}^3$ with adjacent edges at a vertex given by the vectors $\mathbf{v}, \mathbf{w}$, then the area of the parallelogram is given by the length of the cross product $\mathbf{v} \times \mathbf{w}$, which is defined by

$$\det \begin{pmatrix} \mathbf{e}_1 & \mathbf{e}_2 & \mathbf{e}_3 \\ v_1 & v_2 & v_3 \\ w_1 & w_2 & w_3 \end{pmatrix}$$

(Hint: First note that the cross product is perpendicular to each of $\mathbf{v}, \mathbf{w}$ and the product of the area of the parallelogram and the length of the cross product is given by the volume of the parallelepiped with base the given parallelogram and altitude given by the cross product. This latter volume can be computed via determinants.)

1.10.84. Use determinants to find all values a so that $A - aI$ is *not* invertible, where $A = \begin{pmatrix} 3 & 1 & -2 \\ -2 & 2 & 2 \\ 0 & 1 & 1 \end{pmatrix}$. Such a number a is called an **eigenvalue** for the matrix A. For each eigenvalue, find all solutions $\mathbf{v}$ to $(A - aI)\mathbf{v} = \mathbf{0}$. These vectors $\mathbf{v}$ are called **eigenvectors** for the eigenvalue a. (Hint: A condition for a to be an eigenvalue is that $\det(A - aI) = 0$. The determinant on the left is a polynomial of degree 3 in the variable a and one is to find its roots.)

1.10.85. With reference to the Exercise 1.10.84, find all eigenvalues and eigenvectors for the matrix $A = \begin{pmatrix} -7 & 6 & 4 \\ -6 & 6 & 3 \\ -12 & 8 & 7 \end{pmatrix}$. (Hint: One of the eigenvalues is 1.)

1.10.86. With reference to Exercise 1.10.84 show that a matrix is invertible iff 0 is *not* an eigenvalue.

1.10.87. How many different determinants are there for 10 by 10 matrices which have all entries either 0 or 1 and at most 11 nonzero entries? Justify your answer in terms of the permutation definition of the determinant.

1.10.88. Show that the determinant as defined by the permutation definition satisfies the multilinearity property.

1.10.89. Suppose $A = LU$, where

$$L = \begin{pmatrix} 1 & 0 & 0 \\ 2 & 1 & 0 \\ 4 & 1 & 1 \end{pmatrix}, \quad U = \begin{pmatrix} 2 & 4 & 1 & 3 & 5 \\ 0 & 0 & 2 & 1 & 4 \\ 0 & 0 & 0 & 2 & 1 \end{pmatrix}$$

(a) Solve $A\mathbf{x} = \begin{pmatrix} 1 \\ 2 \\ 0 \end{pmatrix}$ by first solving $L\mathbf{c} = \begin{pmatrix} 1 \\ 2 \\ 0 \end{pmatrix}$ for $\mathbf{c}$ and then solving $U\mathbf{x} = \mathbf{c}$ for $\mathbf{x}$.

(b) Give the row operations which are used to reduce A to U with no row exchanges. (Hint: This is coded in L.)

(c) Give the matrix O_1 so that $O_1 A = U$.

1.10.90. Repeat the previous exercise for

$$L = \begin{pmatrix} 1 & 0 & 0 \\ 4 & 1 & 0 \\ 1 & 0 & 1 \end{pmatrix}, \quad U = \begin{pmatrix} 1 & 1 & 1 & 2 & 4 \\ 0 & 2 & 1 & 2 & 1 \\ 0 & 0 & 0 & 0 & 3 \end{pmatrix}$$

For Exercises 1.10.91–1.10.93 give the LU decomposition that is found by using partial pivoting and the LU decomposition that is found without partial pivoting. If any

L appears which is not lower triangular, find a permutation matrix P so that $L = PL'$ where L' is lower triangular. Show that $P^tA = L'U$ in this case.

1.10.91.
$$\begin{pmatrix} 1 & 2 & 3 \\ 4 & 4 & 6 \\ 7 & 8 & 9 \end{pmatrix}$$

1.10.92.
$$\begin{pmatrix} 1 & 2 & 3 & 4 & 5 \\ -1 & 3 & 2 & 1 & 6 \\ 5 & 4 & 3 & 2 & 1 \end{pmatrix}$$

1.10.93.
$$\begin{pmatrix} 1 & 2 & 3 & 4 \\ 2 & 1 & 3 & 4 \\ 3 & 3 & 1 & 2 \\ 4 & 4 & 2 & 1 \end{pmatrix}$$

In Exercises 1.10.94–1.10.96 give the Cholesky decomposition.

1.10.94.
$$\begin{pmatrix} 1 & 2 \\ 2 & 5 \end{pmatrix}$$

1.10.95.
$$\begin{pmatrix} 5 & 1 & 2 \\ 1 & 5 & 2 \\ 2 & 2 & 5 \end{pmatrix}$$

1.10.96.
$$\begin{pmatrix} 10 & 5 & 2 & 1 \\ 5 & 10 & -3 & 2 \\ 2 & -3 & 10 & 1 \\ 1 & 2 & 1 & 10 \end{pmatrix}$$

In Exercises 1.10.97–1.10.98 give a permutation matrix P so that when we execute the Gaussian elimination algorithm with partial pivoting on PA, there are no row interchanges required.

1.10.97.
$$\begin{pmatrix} 1 & 3 & -1 \\ 3 & 1 & 2 \\ 5 & 2 & 1 \end{pmatrix}$$

1.10.98.
$$\begin{pmatrix} 3 & -2 & 4 & 1 \\ 2 & 1 & -4 & 1 \\ 5 & 0 & 2 & 4 \\ 3 & 4 & 2 & 1 \end{pmatrix}$$

1.10.99. Show that for any matrix A there is a permutation matrix P so that we can execute the Gaussian elimination algorithm with partial pivoting on PA with no row interchanges required.

1.10.100. A dieter wants to combine three foods, say, tuna salad, bread, and a vegetable mixture, to make a meal with specified contributions of protein, carbohydrates, and fat. Each ounce of tuna salad has .8 ounces of protein and .2 ounces of fat. Each ounce of bread has .25 ounces of protein, .7 ounces of carbohydrates and .05 ounces of fat. Each ounce of the vegetable mixture has .6 ounces of carbohydrates, .3 ounces of protein, and .1 ounces of fat.

We can organize the information we are given with a table.

	Tuna	*Bread*	*Veggies*	*Goal*
protein	.8	.25	.3	2.95
carbohydrates	0	.7	.6	3.3
fat	.2	.05	.1	.75

How many ounces each of tuna salad, bread, and vegetable mixture should the dieter consume if he desires to end up with 2.95 ounces of protein, 3.3 ounces of carbohydrates, and .75 ounces of fat?

1.10.101. With respect to the previous problem, show that it is in fact impossible to choose a diet which gives 3 ounces of protein, 3 ounces of carbohydrates, and 1 ounce of fat.

1.10.102. With the same data as in the previous problem, how many ounces of each of the foods should be eaten to give 4 ounces of protein, 5 ounces of carbohydrates, and 1 ounce of fat.

1.10.103. An exercise program is to consist of 140 minutes of exercise a week, which is taken from three types of exercise for variety: bicycling, swimming, and exercise machines. We will denote the hours per week from each activity by b, s, m. The calories burned in one hour of exercise (in thousands) in each type of exercise are given in the accompanying table. Each exercise is also given a pleasure factor, which is also listed in the table. Formulate an exercise program (in terms of how many hours per week of each exercise) to burn 1800 calories but still have a total pleasure factor of 2.

	Bicycling	*Swimming*	*Machines*
1000 calories/hour	.6	.45	1
pleasure/hour	.9	1.2	.6

1.10.104. A company makes four different types of furniture: a sofa, a chair, a dresser, and a bed. The company's work can be broken into carpentry, assembly, and finishing. The table gives the division of hours needed to produce each item. The total amount of time available each week is 150 hours for carpentry, 150 hours for assembly, and 100 hours for finishing.

	Sofa	*Chair*	*Dresser*	*Bed*	*Hours Available*
carpentry	2	1	1.5	1.5	150
assembly	2	1.5	1	1	150
finishing	1	.7	.5	1	100

Suppose that there is a profit margin of \$80 for a sofa, \$50 for a chair, \$60 for a dresser, and \$75 for a bed. Formulate a production schedule for each type of furniture which maximizes the profits.

1.10.105. In the same situation as the last exercise, suppose that the table describing the hours of labor is changed to

	Sofa	*Chair*	*Dresser*	*Bed*	*Hours Available*
carpentry	1.8	1.2	1.4	1.4	180
assembly	1.5	1.4	1	1	160
finishing	.8	.6	.4	.9	80

Find a production schedule which maximizes the profit.

1.10.106. Suppose in the Leontief input–output model the consumption matrix is $\begin{pmatrix} .1 & .6 & .3 \\ .2 & 0 & .6 \\ .1 & .4 & .1 \end{pmatrix}$. What is the production vector necessary to meet the demand of 100 units from each industry?

1.10.107. Using the Leontief input-output model, suppose that the output of one unit from each of three industries requires equal amounts of input from each of the industries. Thus the consumption matrix will have the form $C = \begin{pmatrix} c & c & c \\ c & c & c \\ c & c & c \end{pmatrix}$. For what values of c will the economy be able to meet any demand; that is, when will $(I - C)^{-1}$ be a nonnegative matrix?

1.10.108. Consider an economy with three industries and a consumption matrix $C = \begin{pmatrix} .4 & .2 & .1 \\ .3 & .3 & .4 \\ .2 & .5 & .1 \end{pmatrix}$. Find the matrix $(I - C)^{-1}$ and the production vector $\mathbf{p}$ necessary to meet a demand vector $\mathbf{d} = (1000, 800, 400)$.

1.10.109. Consider an economy with five industries and a consumption matrix

$$C = \begin{pmatrix} .15 & .15 & .15 & .15 & .15 \\ .15 & .15 & .15 & .15 & .15 \\ .15 & .15 & .15 & .15 & .15 \\ .15 & .15 & .15 & .15 & .15 \\ .15 & .15 & .15 & .15 & .15 \end{pmatrix}.$$

Find the matrix $(I - C)^{-1}$ and the production vector $\mathbf{p}$ necessary to meet a demand vector $\mathbf{d} = (1000, 1000, 1000, 1000, 1000)$.

In Exercises 1.10.110–1.10.111 find the equations for the currents in the given circuits and solve them.

1.10.110.

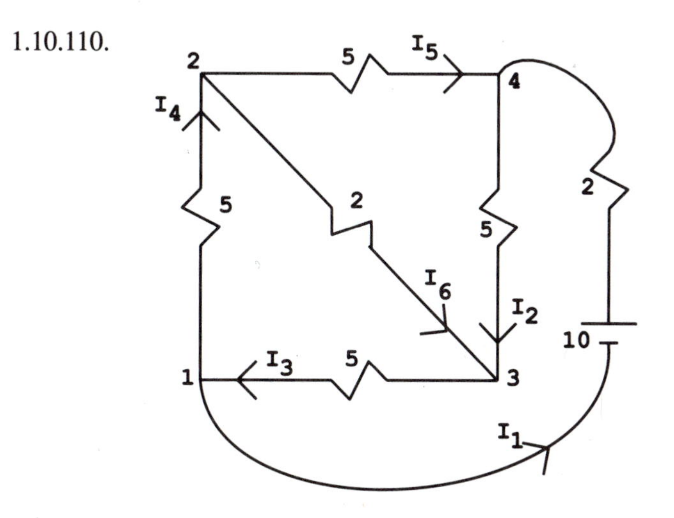

Figure 1.9. Circuit 10.110

1.10.111.

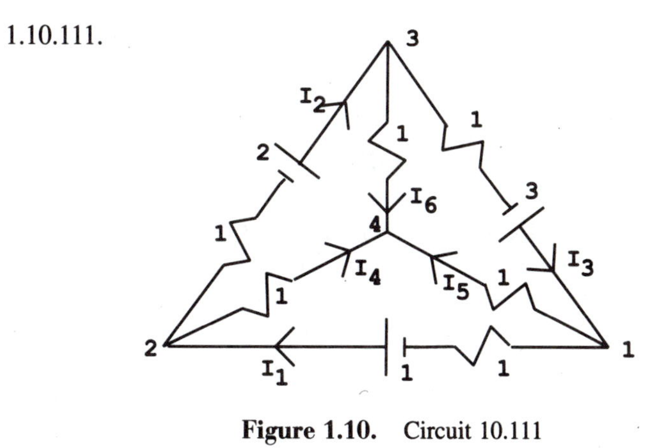

Figure 1.10. Circuit 10.111

CHAPTER 2

VECTOR SPACES AND LINEAR TRANSFORMATIONS

2.1. BASIC DEFINITIONS AND EXAMPLES

In this chapter we apply the methods of Chapter 1 to study general vector spaces and linear transformations. We will also use the ideas of vector space theory to find efficient approaches to problems in matrix algebra. To motivate the definitions, we recall some of the basic properties of vector operations in $\mathbb{R}^n$ as well as the effect of multiplying by a matrix. Consider the operation of adding two vectors in $\mathbb{R}^n$. First, note that the result is again a vector in $\mathbb{R}^n$. This is sometimes called the **closure property** of addition.

$$\text{For } \mathbf{v}, \mathbf{w} \in \mathbb{R}^n, \mathbf{v} + \mathbf{w} \in \mathbb{R}^n \qquad \textbf{closure under addition} \tag{1.1}$$

The operation is both associative and commutative; that is, the formulas

$$(\mathbf{v} + \mathbf{w}) + \mathbf{z} = \mathbf{v} + (\mathbf{w} + \mathbf{z}) \qquad \textbf{associative property} \tag{1.2}$$

$$\mathbf{v} + \mathbf{w} = \mathbf{w} + \mathbf{v} \qquad \textbf{commutative property} \tag{1.3}$$

hold. Also, there is a vector $\mathbf{0} = (0, 0, \ldots, 0)$ with the property that

$$\mathbf{v} + \mathbf{0} = \mathbf{0} + \mathbf{v} = \mathbf{v} \qquad \mathbf{0} \text{ is called an } \textbf{additive identity} \tag{1.4}$$

Furthermore, given any vector $\mathbf{v} \in \mathbb{R}^n$, there is another vector $-\mathbf{v}$, so that

$$\mathbf{v} + (-\mathbf{v}) = (-\mathbf{v}) + \mathbf{v} = \mathbf{0} \qquad -\mathbf{v} \text{ is called an } \textbf{additive inverse} \text{ for } \mathbf{v} \tag{1.5}$$

Any set V with an operation + satisfying the five properties given (closure, associativity, commutativity, existence of identity, existence of inverses) is called an **abelian group**. Thus $\mathbb{R}^n$ is an abelian group under the operation of vector addition. However, there is another operation in $\mathbb{R}^n$, and that is the operation of multiplying a real number (frequently referred to as a **scalar** to distinguish it from a vector) times a vector to get a new vector.

$$\text{Given } a \in \mathbb{R}, \mathbf{v} \in \mathbb{R}^n \text{ then } a\mathbf{v} \in \mathbb{R}^n \quad \textbf{closure under scalar multiplication} \tag{1.6}$$

This fact is sometimes referred to as the **closure property of scalar multiplication.** Moreover, this scalar multiplication obeys some properties (here we denote scalars by italics $a, b, c, \ldots$ and vectors by boldface $\mathbf{v}, \mathbf{w}, \mathbf{x}, \ldots$):

$$a(b\mathbf{v}) = (ab)\mathbf{v} \qquad \textbf{associative property} \tag{1.7}$$

$$a(\mathbf{v} + \mathbf{w}) = a\mathbf{v} + a\mathbf{w} \qquad \textbf{distributive property 1} \tag{1.8}$$

$$(a + b)\mathbf{v} = a\mathbf{v} + b\mathbf{v} \qquad \textbf{distributive property 2} \tag{1.9}$$

$$1\mathbf{v} = \mathbf{v} \text{ for all } \mathbf{v} \qquad \textbf{1 is the identity for scalar multiplication} \tag{1.10}$$

DEFINITION 2.1.1. A set V that has an operation $+$ so that the set forms an abelian group under $+$ and an operation of multiplying a real number a by an element $\mathbf{v} \in V$ to get another element $a\mathbf{v} \in V$ satisfying the foregoing properties is called a **vector space** (over the real numbers).

The definition is modeled on the properties of vectors in $\mathbb{R}^n$. However, it turns out to apply to a large class of examples which arise in mathematics as well as various applications of mathematics to other areas. There is a more general definition which allows scalars other than the real numbers. We will discuss this in Chapter 4 in connection with the eigenvalue–eigenvector problem when we look at complex vector spaces, but until then, we will discuss only vector spaces over the real numbers. We will generally just denote the operation of scalar multiplication by juxtaposition $a\mathbf{v}$ of the real number and the vector as was done earlier. Sometimes, however, it will be convenient to put a dot between the scalar and the vector to emphasize this operation and we will use the notation $a \cdot \mathbf{v}$ for scalar multiplication of the scalar a times the vector $\mathbf{v}$.

We start off by looking at a few examples and general constructions of vector spaces.

■ ***Example 2.1.1.*** Consider all functions from a fixed set X to the real numbers. Then we can define $f + g$ in the usual way,

$$(f + g)(x) = f(x) + g(x)$$

To verify the group properties we just use the corresponding properties of addition of real numbers. For example, to see the commutativity of addition we note $(f+g)(x) = f(x)+g(x) = g(x)+f(x) = (g+f)(x)$. Thus we are just using the definition of addition in $\mathbb{R}$ and the fact that addition of real numbers is commutative. Associativity is similarly verified. The identity element is just the zero function: $f(x) = 0$ for all x. The inverse of f is the function $(-f)$ given by $(-f)(x) = -f(x)$. Thus the verification that the functions form an abelian group rests entirely on the corresponding properties of the real numbers. Next consider scalar multiplication via $(af)(x) = af(x)$; that is, we use the multiplication in the real numbers to define it. Then we can easily verify all the multiplicative properties which a vector space is supposed to satisfy. For example,

$$(a(f+g))(x) = a(f+g)(x) = a(f(x)+g(x)) = af(x)+ag(x) = (af+ag)(x)$$

implies that $a(f+g) = af+ag$. Thus functions from X to the real numbers form a vector space, which we will denote by $\mathcal{F}(X,\mathbb{R})$.

As a special case, consider the set $X = \{1,2,3\}$. Then the set of functions $\mathcal{F}(\{1,2,3\},\mathbb{R})$ is easily identified with $\mathbb{R}^3$ by identifying the function $f:\{1,2,3\}\to\mathbb{R}$ with the vector $(f(1), f(2), f(3))$ in $\mathbb{R}^3$. The operation of function addition in the vector space $\mathcal{F}(\{1,2,3\},\mathbb{R})$ corresponds to usual vector addition under this identification:

$$f+g \leftrightarrow ((f+g)(1),(f+g)(2),(f+g)(3)) = (f(1),f(2),f(3)) + (g(1),g(2),g(3))$$

Similarly the operation of multiplying a function by a scalar corresponds to the usual scalar multiplication in $\mathbb{R}^3$:

$$(cf) \leftrightarrow ((cf)(1),(cf)(2),(cf)(3)) = c(f(1),f(2),f(3))$$

Thus these function spaces can be thought of as a type of generalization of the vector spaces $\mathbb{R}^n$.

If we check through each step in the verification process that $\mathcal{F}(X,\mathbb{R})$ is a vector space, we see that the important property that $\mathbb{R}$ satisfies is just that it forms a vector space under addition and scalar multiplication. Thus we could check that if V was any vector space, then the functions from any set X into V, denoted $\mathcal{F}(X,V)$, is also a vector space with the operations

$$(f+g)(x) = f(x)+g(x)$$
$$(af)(x) = af(x)$$

In these definitions, the addition and scalar multiplication are the ones defined in the vector space V. The verification will just use the definition and the corresponding properties in V. If we choose $V = \mathbb{R}^3$, for example, we would see that the functions from X to $\mathbb{R}^3$ form a vector space. As a special case, when $X = \mathbb{R}$, $\mathcal{F}(\mathbb{R},\mathbb{R}^3)$ is the vector space of curves in 3-space. ■

■ ***Example 2.1.2.*** We are usually not just interested in any functions, but rather ones with some special property. For example, we may be interested in functions from the real numbers which are infinitely differentiable. We choose infinitely differentiable instead of just once differentiable for our convenience so that we may take as many derivatives as we wish without worrying about their existence. This is a subset $\mathfrak{D}(\mathbb{R}, \mathbb{R})$ of $\mathfrak{F}(\mathbb{R}, \mathbb{R})$. Since the elements of $\mathfrak{D}(\mathbb{R}, \mathbb{R})$ are in particular elements of $\mathfrak{F}(\mathbb{R}, \mathbb{R})$, we can define the operations of addition and scalar multiplication for them. That the result again is differentiable is shown in the first course in calculus: the derivative of the sum is the sum of the derivatives and the derivative of a constant times a function is the constant times the derivative of the function.

$$(f + g)'(t) = f'(t) + g'(t), (cf)'(t) = cf'(t)$$

This means that the same operations serve to define operations of addition and scalar multiplication on the subset $\mathfrak{D}(\mathbb{R}, \mathbb{R})$. We say that the subset $\mathfrak{D}(\mathbb{R}, \mathbb{R})$ is **closed under the operations of addition and scalar multiplication**. We also check that the zero function is differentiable and the additive inverse $-f$ of a differentiable function is differentiable. ■

The fact that the zero function and inverses of differentiable functions are differentiable also follows from some general facts about subsets of vector spaces which are closed under addition and scalar multiplication, which we now verify.

LEMMA 2.1.1. Let V be a vector space.

(a) The additive identity $\mathbf{0}$ is uniquely determined by its properties; it satisfies $0\mathbf{v} = \mathbf{0}, a\mathbf{0} = \mathbf{0}$. Moreover, $a\mathbf{v} = \mathbf{0}$ iff $a = 0$ or $\mathbf{v} = \mathbf{0}$.

(b) The inverse $-\mathbf{v}$ of a vector $\mathbf{v} \in V$ is uniquely determined by its properties; it satisfies $(-1)\mathbf{v} = -\mathbf{v}$.

Proof. (a) If $\mathbf{u}, \mathbf{v}$ both satisfy the property of being an identity, then

$$\mathbf{u} = \mathbf{u} + \mathbf{v} = \mathbf{v}$$

the first equality arising since $\mathbf{v}$ is the identity and the second from $\mathbf{u}$ being the identity. Denote the identity by $\mathbf{0}$. Now to see that $0\mathbf{v} = \mathbf{0}$, note that

$$\mathbf{v} = (1 + 0)\mathbf{v} = \mathbf{v} + 0\mathbf{v}$$

But there is an inverse $(-\mathbf{v})$ of $\mathbf{v}$ so that $(-\mathbf{v}) + \mathbf{v} = \mathbf{0}$. Adding it to the preceding equation and using associativity gives $\mathbf{0} = 0\mathbf{v}$. Next consider $a\mathbf{0}$ and write

$$a\mathbf{0} = a(\mathbf{0} + \mathbf{0}) = a\mathbf{0} + a\mathbf{0}$$

The same device of adding the inverse of $a\mathbf{0}$ to both sides of the equation gives $\mathbf{0} = a\mathbf{0}$. If $a\mathbf{v} = \mathbf{0}$ and $a \neq 0$, then multiplying by $(1/a)$ gives (using the associative property of multiplying a scalar by a vector)

$$\mathbf{v} = (1/a)\mathbf{0} = \mathbf{0}$$

(b) For uniqueness of the inverse, note that if $\mathbf{x}, \mathbf{y}$ are two vectors satisfying the property of being an inverse for $\mathbf{v}$, then

$$\mathbf{x} = \mathbf{x} + \mathbf{0} = \mathbf{x} + (\mathbf{v} + \mathbf{y}) = (\mathbf{x} + \mathbf{v}) + \mathbf{y} = \mathbf{0} + \mathbf{y} = \mathbf{y}$$

Note that this is exactly the same proof (with only a change in notation) that was used to show that the inverse of a matrix is unique. What is going on here is just that the invertible matrices form a (nonabelian) group under the operation of matrix multiplication. Returning to our claim, we next check that $(-1)\mathbf{v}$ satisfies the property for being the inverse of $\mathbf{v}$:

$$\mathbf{0} = 0\mathbf{v} = (1 + (-1))\mathbf{v} = \mathbf{v} + (-1)\mathbf{v}$$

Similarly,

$$(-1)\mathbf{v} + \mathbf{v} = \mathbf{0}$$ ■

Since $0\mathbf{v} = \mathbf{0}$ and $(-1)\mathbf{v} = -\mathbf{v}$ are the results of multiplying $\mathbf{v}$ by the scalars 0 and -1, this implies that any subset of a vector space which satisfies the closure property under the same operations of addition and scalar multiplication will also satisfy the properties of existence of an identity and inverses. Next note that all the other properties of a vector space such as associativity of the multiplication are inherited by a subset from the original vector space (e.g., $(f + g) + h = f + (g + h)$ for differentiable functions because the equality holds for all functions). Thus this shows that $\mathcal{D}(\mathbb{R}, \mathbb{R})$ is a vector space using the same operations as $\mathcal{F}(\mathbb{R}, \mathbb{R})$. It is called a subspace of $\mathcal{F}(\mathbb{R}, \mathbb{R})$.

DEFINITION 2.1.2. A **subspace** of a vector space is a nonempty subset that, when endowed with the operations of addition and scalar multiplication of the vector space, itself satisfies the properties of a vector space.

What we have seen so far is that if a subset satisfies the closure properties under addition and scalar multiplication, then this is enough to imply that it is a subspace. Let us write this as a proposition as this is the natural way we will encounter most vector spaces.

PROPOSITION 2.1.2. If V is a vector space and S is a subset which is closed under the operations of addition and scalar multiplication in V (i.e., if $\mathbf{v}, \mathbf{w} \in S$ and $a \in \mathbb{R}$, then $\mathbf{v} + \mathbf{w} \in S$ and $a\mathbf{v} \in S$), then S is a subspace of V (and, in particular, S is a vector space).

Note that a subset S which is a subspace of V must contain the additive identity $\mathbf{0}$ of V since $\mathbf{0} = 0\mathbf{v}$ for any $\mathbf{v} \in S$. This fact gives one of the easiest ways to recognize that a particular subset is not a subspace and can be used as a first check on a subset to see if it might be a subspace. If it satisfies this check, then we can check to see if it is closed under scalar multiplication and addition to see if it is a subspace. We record this first check as a proposition.

PROPOSITION 2.1.3. Any subspace S of a vector space V must contain the zero vector $\mathbf{0}$ of V. Thus a subset S which doesn't contain the zero vector $\mathbf{0}$ is not a subspace.

We now look at some more vector spaces which are either subspaces of $\mathcal{F}(X, V)$ for some vector space V or are subspaces of $\mathbb{R}^n$.

■ ***Example 2.1.3.*** The polynomial functions $\mathcal{P}(\mathbb{R}, \mathbb{R})$ form a subspace of $\mathcal{F}(\mathbb{R}, \mathbb{R})$. Recall that a polynomial is a function of the form

$$p(x) = a_0 + a_1 x + \cdots + a_{n-1}x^{n-1} + a_n x^n$$

Here the constants a_i are taken to be real numbers. It is said to be of **degree** n if the coefficient $a_n \neq 0$. $\mathcal{P}(\mathbb{R}, \mathbb{R})$ is a subspace since the sum of two polynomials is a polynomial and the product of a real number with a polynomial gives a polynomial. ■

■ ***Example 2.1.4.*** The polynomials of degree less than or equal to a fixed number n, denoted $\mathcal{P}^n(\mathbb{R}, \mathbb{R})$, also form a subspace since adding a polynomial of degree n and one of degree m gives a polynomial whose degree is at most the maximum of m and n (it may be lower if they are of the same degree and the top terms cancel) and multiplying by a constant either preserves the degree or gives the zero polynomial. Note that any such polynomial $p(x) = a_0 + a_1 x + \cdots + a_n x^n$ is completely determined by the $(n+1)$ real numbers $(a_0, a_1, \ldots, a_n)$. Moreover, the operations of addition of two polynomials and scalar multiplication of a polynomial by a real number are completely reflected in the corresponding operations in $\mathbb{R}^{n+1}$ on the coefficients. Later we will formally give the definition of two vector spaces being isomorphic and show that this means that $\mathcal{P}^n(\mathbb{R}, \mathbb{R})$ is isomorphic to $\mathbb{R}^{n+1}$. From the point of view of vector space theory, this will mean that it satisfies all the algebraic properties tied to the vector space structure that $\mathbb{R}^{n+1}$ does and will allow us to use techniques from $\mathbb{R}^{n+1}$ such as matrix algebra calculations to solve problems about polynomials. ■

■ ***Example 2.1.5.*** Another example of a subspace is all polynomials which vanish at 0. If we have two such polynomials p, q, then the sum $p+q$ and the scalar multiple ap still will vanish at 0 since

$$(p+q)(0) = p(0) + q(0) = 0 + 0 = 0, (ap)(0) = ap(0) = a0 = 0 \quad ■$$

We next give some examples of subsets which are not subspaces.

■ ***Example 2.1.6.*** Consider the polynomials which have value 1 at 0. This is not a subspace since it does not contain the zero polynomial (which of course has value 0 at 0). This illustrates that you can frequently tell subsets which aren't subspaces by checking whether the zero vector lies in the subset. Of course there are also examples of subsets which aren't subspaces which contain the zero vector. For example, the subset of all polynomials which are nonnegative at 0 contains the zero polynomial, but it is not a subspace because it is not closed under scalar multiplication: the constant function 1 satisfies this property but $(-1)1 = -1$ does not. Note that this subset is closed under addition. ■

■ ***Example 2.1.7.*** Another important operation from analysis is that of taking a definite integral. Consider the vector space of all functions from the interval $[0, 1]$ to the real numbers, $\mathcal{F}([0, 1], \mathbb{R})$. Then the subset $\mathcal{I}([0, 1], \mathbb{R})$ of all functions which are integrable on this interval, that is, so the integral $\int_0^1 f(t)\,dt$ exists, is a subspace. This follows from the fact that the sum of integrable functions is integrable and the product of a real number by an integrable function is integrable. Contained in the class of integrable functions are the continuous functions $\mathcal{C}([0, 1], \mathbb{R})$. This is a subspace since the sum of continuous functions is continuous and the product of a real number with a continuous function gives a continuous function. ■

■ ***Example 2.1.8.*** Consider the m by n matrices $\mathcal{M}(m, n)$. Then we have defined the operations of addition of two such matrices and multiplication of a matrix by a scalar. It is easy to verify that this gives a vector space since all the operations are taking place coordinatewise. The verification boils down to the fact that the real numbers form a vector space. Alternatively, an element of $\mathcal{M}(m, n)$ can be identified with a vector in $\mathbb{R}^{mn}$ (in terms of its mn entries), and this identification respects the vector space operations in $\mathcal{M}(m, n)$ and $\mathbb{R}^{mn}$. Thus the verification that $\mathcal{M}(m, n)$ is a vector space becomes essentially identical to verifying that $\mathbb{R}^{mn}$ is a vector space. Later we will show that $\mathcal{M}(m, n)$ and $\mathbb{R}^{mn}$ are isomorphic vector spaces. ■

■ ***Example 2.1.9.*** Consider an m by n matrix A. Associated with A are two important vector spaces which are subspaces, one of $\mathbb{R}^n$ and the other of $\mathbb{R}^m$. First, consider the subset $\mathcal{N}(A)$ of solutions of $A\mathbf{x} = \mathbf{0}$. This is a subspace of $\mathbb{R}^n$. For if $\mathbf{x}, \mathbf{y}$ are in $\mathcal{N}(A)$ and a is a real number, then

$$A(\mathbf{x} + \mathbf{y}) = A\mathbf{x} + A\mathbf{y} = \mathbf{0} + \mathbf{0} = \mathbf{0}$$

and

$$A(a\mathbf{x}) = aA(\mathbf{x}) = a\mathbf{0} = \mathbf{0}$$

DEFINITION 2.1.3. $\mathcal{N}(A) = \{\mathbf{x} \in \mathbb{R}^n : A\mathbf{x} = \mathbf{0}\}$ is called the **null space** of A.

While we are discussing this, note that the set of solutions of $A\mathbf{x} = \mathbf{w}$ for $\mathbf{w} \neq \mathbf{0}$ is *not* a subspace since it does not contain the zero vector.

A second subspace associated to A is the set $\mathcal{R}(A)$ of all vectors in $\mathbb{R}^m$ of the form $A\mathbf{x}$ for some $\mathbf{x} \in \mathbb{R}^n$; alternatively, we could define $\mathcal{R}(A)$ as the set of all $\mathbf{w}$ for which $A\mathbf{x} = \mathbf{w}$ has a solution.

If $\mathbf{w}_1, \mathbf{w}_2$ are in $\mathcal{R}(A)$, this means that there are vectors $\mathbf{x}_1, \mathbf{x}_2$ so that $A\mathbf{x}_i = \mathbf{w}_i$, $i = 1, 2$. Then $A(\mathbf{x}_1 + \mathbf{x}_2) = A\mathbf{x}_1 + A\mathbf{x}_2 = \mathbf{w}_1 + \mathbf{w}_2$, so $\mathbf{w}_1 + \mathbf{w}_2$ is in $\mathcal{R}(A)$. Also, $A\mathbf{x} = \mathbf{w}$ implies $A(a\mathbf{x}) = a\mathbf{w}$, so that $\mathbf{w}$ being in $\mathcal{R}(A)$ implies $a\mathbf{w}$ is in $\mathcal{R}(A)$ for any real number a.

DEFINITION 2.1.4. $\mathcal{R}(A) = \{\mathbf{w} \in \mathbb{R}^m : \mathbf{w} = A\mathbf{x} \text{ for some } \mathbf{x} \in \mathbb{R}^n\}$ is called the **range** of A.

The reason for calling this subspace the range of A is that we are thinking of multiplication by A sending $\mathbf{x}$ to $A\mathbf{x}$ as being a function from $\mathbb{R}^n$ to $\mathbb{R}^m$ and $\mathcal{R}(A)$ is just the range of this function. There are also subspaces $\mathcal{N}(A^t)$ and $\mathcal{R}(A^t)$ associated with the transpose of A. The interaction of these four subspaces will be a major theme in the study of the vector spaces related to solving $A\mathbf{x} = \mathbf{w}$. ■

■ ***Example 2.1.10.*** Suppose that V is a vector space and $\mathbf{v}_1, \ldots, \mathbf{v}_k$ are vectors in V. Then the set S of all linear combinations $c_1\mathbf{v}_1 + \cdots + c_k\mathbf{v}_k$ is a subspace of V. If $\mathbf{v} = c_1\mathbf{v}_1 + \cdots + c_k\mathbf{v}_k$ and $\mathbf{w} = d_1\mathbf{v}_1 + \cdots + d_k\mathbf{v}_k$, then the sum

$$\mathbf{v} + \mathbf{w} = (c_1 + d_1)\mathbf{v}_1 + \cdots + (c_k + d_k)\mathbf{v}_k$$

is also a linear combination of these vectors, so the subset is closed under addition. Note that we are using both the associativity and commutativity of addition as well as the distributive property of scalar multiplication to rewrite the sum. Also,

$$a\mathbf{v} = (ac_1)\mathbf{v}_1 + \cdots + (ac_k)\mathbf{v}_k$$

so the subset is closed under scalar multiplication. S is called the **subspace spanned by** $\mathbf{v}_1, \ldots, \mathbf{v}_k$. The vectors $\mathbf{v}_1, \ldots, \mathbf{v}_k$ are said to **span** this subspace. We will denote this subspace by $\text{span}(\mathbf{v}_1, \ldots, \mathbf{v}_k)$. ■

DEFINITION 2.1.5. The subspace spanned by $\mathbf{v}_1, \ldots, \mathbf{v}_k$, which is denoted by $\text{span}(\mathbf{v}_1, \ldots, \mathbf{v}_k)$, is the set of linear combinations of these vectors:

$$\text{span}(\mathbf{v}_1, \ldots, \mathbf{v}_k) = \{\mathbf{v} \in V : \mathbf{v} = c_1\mathbf{v}_1 + \cdots + c_k\mathbf{v}_k \text{ for some } c_i \in \mathbb{R}\}$$

We now restrict our attention to $V = \mathbb{R}^3$. Let $\mathbf{v}_1 = (1, 1, 3)$. Then $\text{span}(\mathbf{v}_1)$ is just the line through the origin and $\mathbf{v}_1$, which is all vectors which are constant multiples of $(1, 1, 3)$, that is, $(a, a, 3a)$ for some a. Now consider a second vector $\mathbf{v}_2$. If $\mathbf{v}_2$ already lies in $\text{span}(\mathbf{v}_1)$, then $\text{span}(\mathbf{v}_1, \mathbf{v}_2) = \text{span}(\mathbf{v}_1)$. For $\mathbf{v}_2 = c\mathbf{v}_1$ and so if $\mathbf{v} = c_1\mathbf{v}_1 + c_2\mathbf{v}_2$, then $\mathbf{v} = (c_1 + cc_2)\mathbf{v}_1$ and thus lies in $\text{span}(\mathbf{v}_1)$. There always is an inclusion $\text{span}(\mathbf{v}_1) \subset \text{span}(\mathbf{v}_1, \mathbf{v}_2)$ since a multiple of $\mathbf{v}_1$ is a linear combination of $\mathbf{v}_1$ and $\mathbf{v}_2$ with the coefficient of $\mathbf{v}_2$ equal to 0. In general $\text{span}(\mathbf{v}_1, \ldots, \mathbf{v}_k) \subset \text{span}(\mathbf{v}_1, \ldots, \mathbf{v}_k, \ldots, \mathbf{v}_m)$ by the same argument. Suppose $\mathbf{v}_2$ is not a multiple of $\mathbf{v}_1$, say, $\mathbf{v}_2 = (0, 1, 2)$. Then $\text{span}(\mathbf{v}_1, \mathbf{v}_2)$ will now be a larger subspace. In fact, it will be a plane through the origin which contains the vectors $\mathbf{v}_1$ and $\mathbf{v}_2$. To write this plane in a more usual form we can find its normal vector by taking the cross product of $(1, 1, 3)$ and $(0, 1, 2)$, which is $(-1, -2, 1)$, so its equation is $-x - 2y + z = 0$. If we take a third vector $\mathbf{v}_3$, then $\text{span}(\mathbf{v}_1, \mathbf{v}_2, \mathbf{v}_3)$ will be equal to $\text{span}(\mathbf{v}_1, \mathbf{v}_2)$ if $\mathbf{v}_3$ already lies in the plane $\text{span}(\mathbf{v}_1, \mathbf{v}_2)$ (see Exercise 2.1.7). If $\mathbf{v}_3$ does not lie in that plane (for example, $\mathbf{v}_3 = (1, 1, 1)$), then it can be shown that every vector in $\mathbb{R}^3$ will lie in $\text{span}(\mathbf{v}_1, \mathbf{v}_2, \mathbf{v}_3)$ and so $\mathbb{R}^3 = \text{span}(\mathbf{v}_1, \mathbf{v}_2, \mathbf{v}_3)$. What is happening here is a part of a general fact about subspaces of $\mathbb{R}^3$. The only subspaces of $\mathbb{R}^3$ are the zero vector $\mathbf{0}$, a line through the origin, a plane through the origin, or $\mathbb{R}^3$

itself. Moreover, each such subspace arises as the subspace spanned by at most three vectors, with a line being span($\mathbf{v}_1$) and a plane span($\mathbf{v}_1, \mathbf{v}_2$) when $\mathbf{v}_2$ does not lie in the line through $\mathbf{v}_1$.

Exercise 2.1.1.

Determine which of the following subsets of $\mathbb{R}^3$ is a subspace.

(a) the solutions of $x - 3y + 4z = 0$

(b) the solutions of $x^2 - y^2 = 0$

(c) the solutions of $x^2 - y^2 + z^2 = 0$

(d) the solutions of

$$\begin{aligned} x - 4y + z &= 0 \\ x + 4y - z &= 1 \end{aligned}$$

(e) the points with $x \neq 0$

Exercise 2.1.2.

Show that each of the following subsets is a subspace by identifying it with either $\mathcal{N}(A)$ or $\mathcal{R}(A)$ for some matrix A.

(a) the solutions of

$$\begin{aligned} 3x_1 + 2x_2 - x_3 + x_4 &= 0 \\ 2x_1 - 3x_2 + x_3 - 5x_4 &= 0 \\ x_1 + x_2 + 4x_3 - 7x_4 &= 0 \end{aligned}$$

(b) the set of all vectors $\mathbf{b} = (b_1, b_2, b_3)$ for which there is a solution of

$$\begin{aligned} 3x_1 + 2x_2 - x_3 + x_4 &= b_1 \\ 2x_1 - 3x_2 + x_3 - 5x_4 &= b_2 \\ x_1 + x_2 + 4x_3 - 7x_4 &= b_3 \end{aligned}$$

Exercise 2.1.3.

(a) For part (a) of Exercise 2.1.2 solve the system of equations to write the general solution in the form of $c_1\mathbf{v}_1 + \cdots + c_k\mathbf{v}_k$ and so show that the subspace is of the form span($\mathbf{v}_1, \ldots, \mathbf{v}_k$).

(b) For part (b) of Exercise 2.1.2 use the fact that multiplication by a matrix A expresses vectors in the range of A as linear combinations of the columns of A

$$A\mathbf{x} = x_1\mathbf{A}^1 + \cdots + x_n\mathbf{A}^n$$

to express the subspace from part (b) as span($\mathbf{w}_1, \ldots, \mathbf{w}_l$) for some $\mathbf{w}_1, \ldots, \mathbf{w}_l$.

Exercise 2.1.4.

Determine whether the following subsets of $\mathscr{F}(\mathbb{R}, \mathbb{R})$ are subspaces.

(a) the polynomials of degree greater than 3

(b) the polynomials which have value 0 at $x = 1$ and $x = 4$

(c) the functions so that $\int_0^1 f(t)\,dt = 0$

(d) the solutions of the differential equation $y' - y = 0$

(e) the solutions of the differential equation $y' - y = e^t$

(f) the functions which fail to be continuous at the point $t = 0$

Exercise 2.1.5.

Determine which of the following subsets of the vector space of 4 by 4 matrices are subspaces.

(a) the **symmetric matrices,** that is, those which satisfy $A = A^t$

(b) the **skew symmetric matrices,** that is, those which satisfy $A = -A^t$

(c) the **orthogonal matrices,** that is, those that satisfy $A^{-1} = A^t$

(d) the permutation matrices

Exercise 2.1.6.

For each of the four classes of matrices in Exercise 2.1.5, consider that set with the operation of matrix multiplication as addition and the usual scalar multiplication. Determine which of the properties for a vector space the set with those operations satisfies.

Exercise 2.1.7.

Suppose that $\mathbf{v}_1, \ldots, \mathbf{v}_k, \mathbf{v}_{k+1}, \ldots, \mathbf{v}_m$ is a collection of vectors in a vector space V. Show that $\text{span}(\mathbf{v}_1, \ldots, \mathbf{v}_k)$ is contained in $\text{span}(\mathbf{v}_1, \ldots, \mathbf{v}_m)$. Give one example in $\mathbb{R}^3$ where they are equal and one example where they are not equal. Show that $\text{span}(\mathbf{v}_1, \ldots, \mathbf{v}_k) = \text{span}(\mathbf{v}_1, \ldots, \mathbf{v}_k, \mathbf{v}_{k+1})$ iff $\mathbf{v}_{k+1}$ is in the subspace $\text{span}(\mathbf{v}_1, \ldots, \mathbf{v}_k)$.

2.2. SPAN, LINEAR INDEPENDENCE, BASIS, AND DIMENSION

We now wish to define the concepts of span, linear independence, basis, and dimension. These are fundamental concepts which are central to the rest of the book. We first recall and extend the definition of the span which was introduced in Section 2.1.

DEFINITION 2.2.1 The vectors $\mathbf{v}_1, \ldots, \mathbf{v}_k$ in a vector space V span a subspace S if every vector $\mathbf{v}$ in S is a linear combination $c_1\mathbf{v}_1 + \cdots + c_k\mathbf{v}_k$ of the vectors. We say that S is spanned by $\mathbf{v}_1, \ldots, \mathbf{v}_k$. The subspace spanned by $\mathbf{v}_1, \ldots, \mathbf{v}_n$ is denoted by $\text{span}(\mathbf{v}_1, \ldots, \mathbf{v}_n)$. More generally, if K is a subset of a vector space V, then the subspace $\text{span}(K)$ spanned by K is the set of all *finite* linear combinations $c_1\mathbf{v}_1 + \cdots + c_p\mathbf{v}_p$, where $\mathbf{v}_i$ is in K and p can be any nonnegative integer.

■ ***Example 2.2.1.*** In $\mathbb{R}^3$ the subspace spanned by $\mathbf{e}_1 = (1, 0, 0)$ and $\mathbf{e}_2 = (0, 1, 0)$ is the xy-plane, since each vector in the xy-plane is

$$(x, y, 0) = x\mathbf{e}_1 + y\mathbf{e}_2$$

The subspace spanned by $\mathbf{e}_1, \mathbf{e}_2, \mathbf{e}_3 = (0, 0, 1)$ is all of $\mathbb{R}^3$, since any vector $(x, y, z) \in \mathbb{R}^3$ may be written as

$$(x, y, z) = x\mathbf{e}_1 + y\mathbf{e}_2 + z\mathbf{e}_3$$

Note that if we add another vector, say $\mathbf{v}_4 = (3, 1, 4)$, the subspace spanned by $\mathbf{e}_1, \mathbf{e}_2, \mathbf{e}_3, \mathbf{v}_4$ is still $\mathbb{R}^3$. By Exercise 2.1.7, adding more vectors can only increase the subspace which is being spanned, but since it is already all of $\mathbb{R}^3$ any collection of vectors which contains $\mathbf{e}_1, \mathbf{e}_2, \mathbf{e}_3$ must span $\mathbb{R}^3$. ■

■ ***Example 2.2.2.*** The vector space $\mathcal{M}(2, 2)$ of all 2 by 2 matrices is spanned by the matrices $\begin{pmatrix} 1 & 0 \\ 0 & 0 \end{pmatrix}, \begin{pmatrix} 0 & 1 \\ 0 & 0 \end{pmatrix}, \begin{pmatrix} 0 & 0 \\ 1 & 0 \end{pmatrix}$, and $\begin{pmatrix} 0 & 0 \\ 0 & 1 \end{pmatrix}$ since the general 2 by 2 matrix $\begin{pmatrix} a & b \\ c & d \end{pmatrix}$ is the linear combination

$$\begin{pmatrix} a & b \\ c & d \end{pmatrix} = a\begin{pmatrix} 1 & 0 \\ 0 & 0 \end{pmatrix} + b\begin{pmatrix} 0 & 1 \\ 0 & 0 \end{pmatrix} + c\begin{pmatrix} 0 & 0 \\ 1 & 0 \end{pmatrix} + d\begin{pmatrix} 0 & 0 \\ 0 & 1 \end{pmatrix}$$

of these four matrices. ■

■ ***Example 2.2.3.*** The vector space $\mathcal{P}^n(\mathbb{R}, \mathbb{R})$ of all polynomials of degree less than or equal to n is spanned by the polynomials $1, x, x^2, \ldots, x^n$ since the general polynomial $a_0 + a_1x + \cdots + a_nx^n$ of degree less than or equal to n is a linear combination

$$a_0 \cdot 1 + a_1 \cdot x + \cdots + a_n \cdot x^n$$ ■

■ ***Example 2.2.4.*** Let K be the subset of $\mathcal{P}(\mathbb{R}, \mathbb{R})$ consisting of $1, x^2, \ldots, x^{2i}, \ldots$, that is, all even monomials x^{2i}. Then the subspace spanned by K is the subspace of all even polynomials

$$\text{span}(K) = \{a_0 + a_1x^2 + \cdots + a_nx^{2n}\}$$

where n can be any nonnegative integer. ■

Before giving our next example, we define what we mean by the row space and the column space of a matrix.

DEFINITION 2.2.2. For an m by n matrix A the **column space** of A is the subspace spanned by the column vectors of A; that is, it is the subspace of linear combinations of the columns of A.

When the column vectors are identified with vectors in $\mathbb{R}^m$, this becomes a subspace of $\mathbb{R}^m$. Because of the formula

$$A\mathbf{c} = c_1\mathbf{A}^1 + \cdots + c_n\mathbf{A}^n$$

we can identify the column space with the range of A. For this reason we will use the notation $\mathcal{R}(A)$ for the column space.

$$\mathcal{R}(A) = \{A\mathbf{c} : \mathbf{c} \in \mathbb{R}^n\} \leftrightarrow \text{span}(\mathbf{A}^1, \ldots, \mathbf{A}^n) = \text{column space of } A$$

DEFINITION 2.2.3 The **row space** of A is the subspace spanned by the row vectors of A; that is, it is the subspace of linear combinations of the rows of A.

When the row vectors are identified with vectors in $\mathbb{R}^n$, this becomes a subspace of $\mathbb{R}^n$. Because of the formula

$$\mathbf{c}^t A = c_1\mathbf{A}_1 + \cdots + c_m\mathbf{A}_m$$

and $(\mathbf{c}^t A) = (A^t\mathbf{c})^t$, we can identify the row space with the range of A^t. For this reason we will use the notation $\mathcal{R}(A^t)$ for the row space.

$$\mathcal{R}(A^t) = \{A^t\mathbf{c} : \mathbf{c} \in \mathbb{R}^m\} \leftrightarrow \text{span}(\mathbf{A}_1, \ldots, \mathbf{A}_m) = \text{row space of } A$$

■ ***Example 2.2.5.*** Let $A = \begin{pmatrix} 2 & 4 & 1 \\ 3 & 1 & 4 \\ 5 & 5 & 5 \end{pmatrix}$. The row space is just all linear combinations $c_1\,(2 \quad 4 \quad 1) + c_2\,(3 \quad 1 \quad 4) + c_3\,(5 \quad 5 \quad 5)$. But note that $(5 \quad 5 \quad 5) = (2 \quad 4 \quad 1) + (3 \quad 1 \quad 4)$ and so this can be rewritten as

$$(c_1 + c_3)(2 \quad 4 \quad 1) + (c_2 + c_3)(3 \quad 1 \quad 4) = d_1\,(2 \quad 4 \quad 1) + d_2\,(3 \quad 1 \quad 4)$$

Thus we only need two of the row vectors to describe the row space. Geometrically, it is a plane in $\mathbb{R}^3$ with normal vector the cross product $(15, -5, -10)$ of these two vectors and so is given by the equation $15x - 5y - 10z = 0$. The column space is just all linear combinations $c_1\begin{pmatrix} 2 \\ 3 \\ 5 \end{pmatrix} + c_2\begin{pmatrix} 4 \\ 1 \\ 5 \end{pmatrix} + c_3\begin{pmatrix} 1 \\ 4 \\ 5 \end{pmatrix}$. We have an equality

$$\begin{pmatrix} 4 \\ 1 \\ 5 \end{pmatrix} = 3\begin{pmatrix} 2 \\ 3 \\ 5 \end{pmatrix} - 2\begin{pmatrix} 1 \\ 4 \\ 5 \end{pmatrix}$$

Thus every vector in the column space can be rewritten as a linear combination

$$d_1\begin{pmatrix} 2 \\ 3 \\ 5 \end{pmatrix} + d_2\begin{pmatrix} 1 \\ 4 \\ 5 \end{pmatrix}$$

Thus we again only need two vectors to span this subspace. It is the plane

$$-x - y + z = 0$$

■

A basic question regarding the concept of span is whether a given vector $\mathbf{v}$ lies in the subspace spanned by $\mathbf{v}_1, \ldots, \mathbf{v}_n$. In the case of vectors in $\mathbb{R}^m$, this is just a problem of solving linear equations that can be done with Gaussian elimination. We look at an example.

■ ***Example 2.2.6.*** Determine whether the vector $(1, 2, 0, 1)$ lies in the subspace spanned by $(1, 2, 1, 0), (-1, 2, 1, 4), (1, 2, 1, 3)$.

To solve this problem, form a matrix which has the three vectors as its columns:

$$A = \begin{pmatrix} 1 & -1 & 1 \\ 2 & 2 & 2 \\ 1 & 1 & 1 \\ 0 & 4 & 3 \end{pmatrix}$$

Note that

$$c_1(1, 2, 1, 0) + c_2(-1, 2, 1, 4) + c_3(1, 2, 1, 3) = A\mathbf{c}$$

Thus there is a solution of

$$c_1(1, 2, 1, 0) + c_2(-1, 2, 1, 4) + c_3(1, 2, 1, 3) = (1, 2, 0, 1)$$

exactly when there is a solution $\mathbf{c}$ of the matrix equation

$$\begin{pmatrix} 1 & -1 & 1 \\ 2 & 2 & 2 \\ 1 & 1 & 1 \\ 0 & 4 & 3 \end{pmatrix} \mathbf{c} = \begin{pmatrix} 1 \\ 2 \\ 0 \\ 1 \end{pmatrix}$$

Using Gaussian elimination the augmented matrix

$$\left(\begin{array}{ccc|c} 1 & -1 & 1 & 1 \\ 2 & 2 & 2 & 2 \\ 1 & 1 & 1 & 0 \\ 0 & 4 & 3 & 1 \end{array}\right) \text{ reduces to } \left(\begin{array}{ccc|c} 1 & -1 & 1 & 1 \\ 0 & 4 & 0 & 0 \\ 0 & 0 & 3 & 1 \\ 0 & 0 & 0 & -1 \end{array}\right)$$

at the end of the forward elimination step, and so there is no solution. Hence $(1, 2, 0, 1)$ is not in the subspace spanned by $(1, 2, 1, 0), (-1, 2, 1, 4), (1, 2, 1, 3)$. However, the same method would show that $(1, 2, 1, 1)$ is in this subspace since

$$\begin{pmatrix} 1 & -1 & 1 \\ 2 & 2 & 2 \\ 1 & 1 & 1 \\ 0 & 4 & 3 \end{pmatrix} \mathbf{c} = \begin{pmatrix} 1 \\ 2 \\ 1 \\ 1 \end{pmatrix}$$

has the solution $\mathbf{c} = (2/3, 0, 1/3)$. ■

We can use the method from the last example to give an algorithm for determining when $\mathbf{v}$ lies in the subspace spanned by $\mathbf{v}_1, \ldots, \mathbf{v}_n$.

Algorithm for Determining Span

A vector $\mathbf{v}$ lies the the subspace spanned by vectors $\mathbf{v}_1, \ldots, \mathbf{v}_n$ in $\mathbb{R}^m$ exactly when the matrix equation $A\mathbf{c} = \mathbf{v}$ has a solution, where A is the matrix which has the vectors $\mathbf{v}_1, \ldots, \mathbf{v}_n$ as its columns.

Example 2.2.4 illustrates the point that we can frequently reduce the number of vectors needed from a particular spanning set. To see how to find the minimal possible number of vectors in a spanning set for a subspace, we introduce the concept of linear independence.

DEFINITION 2.2.4. The vectors $\mathbf{v}_1, \ldots, \mathbf{v}_k$ in a vector space V are **linearly independent** if

$$c_1\mathbf{v}_1 + \cdots + c_k\mathbf{v}_k = \mathbf{0} \text{ iff all } c_i = 0$$

If this does not hold, the vectors are called **linearly dependent**.

One way of expressing the fact that k vectors are linearly dependent is to say that one of them is a linear combination of the others. Suppose there is a relation $c_1\mathbf{v}_1 + \cdots + c_k\mathbf{v}_k = \mathbf{0}$ with not all the $c_i = 0$. Assume c_j is one of the coefficients which is not zero. Then we can solve for $\mathbf{v}_j$ as a linear combination of the other vectors. Just rewrite the equation as

$$c_j\mathbf{v}_j = -(c_1\mathbf{v}_1 + \cdots + c_{j-1}\mathbf{v}_{j-1} + c_{j+1}\mathbf{v}_{j+1} + \cdots + c_k\mathbf{v}_k)$$

and then solve for $\mathbf{v}_j$ by dividing by c_j. Thus linear dependence is sometimes expressed by saying that we can write one of the vectors in terms of the others and linear independence is expressed by saying that this is not possible. We will frequently just use the words *independent* and *dependent* to mean linearly independent and linearly dependent, respectively. To verify whether a collection of vectors is independent or not, write out the equation

$$c_1\mathbf{v}_1 + \cdots + c_k\mathbf{v}_k = \mathbf{0}$$

and solve for the c_i. Note that there is always the solution $\mathbf{c} = \mathbf{0}$, which is called the **trivial solution.** Any other solution is called a **nontrivial solution.** If there is a nontrivial solution, then the vectors are dependent. A nontrivial solution is called a **dependency relation** of the vectors. If, on the other hand, there is only the trivial solution, then the vectors are independent.

We will almost always be interested in finite sets of vectors when discussing independence. However, to be complete, we now extend the definition to take care of infinite sets of vectors.

DEFINITION 2.2.5. A set K of vectors is called independent if any finite subset is independent; that is, if $\mathbf{v}_1, \cdots, \mathbf{v}_n \in K$ and $c_1\mathbf{v}_1 + \cdots + c_n\mathbf{v}_n = \mathbf{0}$, then $c_1 = \cdots = c_n = 0$.

As an example, the set $\{1, x, x^2, \ldots\}$ is an independent set in $\mathcal{P}(\mathbb{R}, \mathbb{R})$ since any finite subset will be independent.

■ ***Example 2.2.7.*** The vectors $\mathbf{e}_1, \mathbf{e}_2, \ldots, \mathbf{e}_n$ in $\mathbb{R}^n$ are independent. For

$$c_1\mathbf{e}_1 + c_2\mathbf{e}_2 + \cdots + c_n\mathbf{e}_n = \mathbf{0}$$

implies $(c_1, \ldots, c_n) = \mathbf{0}$ and so $c_1 = \cdots = c_n = 0$. Note that these vectors also span $\mathbb{R}^n$.

Let A be an m by n matrix and consider the column vectors, which we identify with n vectors in $\mathbb{R}^m$. Then $c_1A^1 + \cdots + c_nA^n = A\mathbf{c}$ so we are just looking for solutions of $A\mathbf{c} = 0$. If $\mathbf{c} = \mathbf{0}$ is the only solution, then the vectors are independent; if there is a nontrivial solution, the vectors are dependent. But there will be a nontrivial solution iff there are some free variables when we perform Gaussian elimination. This gives a method to determine whether a collection of n vectors in $\mathbb{R}^m$ is independent or not.

Let $\mathbf{v}_1, \ldots, \mathbf{v}_n$ be n vectors in $\mathbb{R}^m$. Form the matrix A with the jth column vector $\mathbf{A}^j = \mathbf{v}_j$. Perform Gaussian elimination on A and find the free variables. The vectors $\mathbf{v}_1, \ldots, \mathbf{v}_n$ will be independent iff there are no free variables. Note that if $n > m$ there always will be at least one free variable, so we see that $n > m$ implies that any n vectors in $\mathbb{R}^m$ are dependent. Note that the determinant provides a check on whether there will be free variables when $m = n$. In this case the vectors will be independent iff $\det A \neq 0$. ■

■ ***Example 2.2.8.*** Let us use this criterion for the vectors $(2,3,5), (1,4,2), (1,-1,3)$ in $\mathbb{R}^3$. We form the matrix $A = \begin{pmatrix} 2 & 1 & 1 \\ 3 & 4 & -1 \\ 5 & 2 & 3 \end{pmatrix}$. We solve $A\mathbf{c} = \mathbf{0}$ for $\mathbf{c}$ by Gaussian elimination. A reduces to $U = \begin{pmatrix} 2 & 1 & 1 \\ 0 & 5/2 & -5/2 \\ 0 & 0 & 0 \end{pmatrix}$, so there is a free variable and the vectors are dependent. The dependency relation is found by solving for $\mathbf{c}$ to get $\mathbf{c} = (-1, 1, 1)$, which means that $-(2,3,5) + (1,4,2) + (1,-1,3) = (0,0,0)$. ■

We now record the algorithm just given for deciding whether a collection of vectors in $\mathbb{R}^m$ is independent and its consequence.

Algorithm for Independence

If $\mathbf{v}_1, \ldots, \mathbf{v}_n$ is a collection of vectors in $\mathbb{R}^m$, then they are independent iff the matrix equation $A\mathbf{c} = \mathbf{0}$ has only the trivial solution, where A is the matrix whose jth column vector is $\mathbf{v}_j$ (regarded as a column vector). If $n > m$, the vectors are dependent. If $n = m$, they are independent iff $\det A \neq 0$.

■ ***Example 2.2.9.*** Consider the polynomials $p_1 = 2 - 3x^2, p_2 = 1 + 2x - x^2, p_3 = 1 + x + x^2$. To check whether they are independent, we write $c_1p_1 + c_2p_2 + c_3p_3 = \mathbf{0}$ and solve for $\mathbf{c}$. This becomes $c_1(2-3x^2)+c_2(1+2x-x^2)+c_3(1+x+x^2) = \mathbf{0}$. Collecting terms gives $(2c_1 + c_2 + c_3) + (2c_2 + c_3)x + (-3c_1 - c_2 + c_3)x^2 = \mathbf{0}$.

But the zero polynomial is characterized by having all the coefficients equal to zero, so this gives a matrix equation $A\mathbf{c} = \mathbf{0}$, where $A = \begin{pmatrix} 2 & 1 & 1 \\ 0 & 2 & 1 \\ -3 & -1 & 1 \end{pmatrix}$. This reduces to $\begin{pmatrix} 2 & 1 & 1 \\ 0 & 2 & 1 \\ 0 & 0 & 9/4 \end{pmatrix}$, and so the only solution is $\mathbf{c} = \mathbf{0}$. Thus the three polynomials are independent.

Suppose we now add a fourth polynomial $4-3x+2x^2$. Then these four polynomials can be seen to be dependent with no work at all, for they will lead to a system $B\mathbf{c} = \mathbf{0}$ with B obtained from A by adding a fourth column $(4 \quad -3 \quad 2)^t$. Since this system has more columns than rows, there will have to be a free variable and thus a nontrivial solution. ■

Note that in this example we essentially end up showing independence by replacing $p(x) = a+bx+cx^2$ by the vector $(a, b, c) \in \mathbb{R}^3$ and showing the corresponding vectors in $\mathbb{R}^3$ are independent. Our simplification of $c_1p_1 + c_2p_2 + c_3p_3 = \mathbf{0}$ led to $c_1\mathbf{A}^1 + c_2\mathbf{A}^2 + c_3\mathbf{A}^3 = A\mathbf{c} = \mathbf{0}$.

Consider a matrix U which arises at the end of the forward elimination step of the Gaussian elimination algorithm. It will have r nonzero rows with the first nonzero entry in row i lying in column $c(i)$. Moreover, $i < j$ implies that $c(i) < c(j)$ and all the entries beneath the $(i, c(i))$-entry in the $c(i)$th column are zero. We will call the $(i, c(i))$-entry the ith pivot. We claim that these r nonzero rows of U are linearly independent. Before proving this generally we look at an example which illustrates the idea of the proof.

■ ***Example 2.2.10.*** Let

$$U = \begin{pmatrix} 2 & 3 & 1 & -1 & 2 \\ 0 & 0 & 4 & 2 & 3 \\ 0 & 0 & 0 & 2 & 1 \\ 0 & 0 & 0 & 0 & 0 \end{pmatrix}$$

To see that the first three rows are independent we look at a general linear combination being $\mathbf{0}$:

$$c_1(2, 3, 1, -1, 2) + c_2(0, 0, 4, 2, 3) + c_3(0, 0, 0, 2, 1) = \mathbf{0}$$

Looking at the first coordinate gives $2c_1 = 0$, so $c_1 = 0$. We then look at the remaining equation

$$c_2(0, 0, 4, 2, 3) + c_3(0, 0, 0, 2, 1) = \mathbf{0}$$

which comes from the second and third rows. Looking at the third coordinate gives $4c_2 = 0$, so $c_2 = 0$. This then gives the equation $c_3(0, 0, 0, 2, 1) = \mathbf{0}$, so looking at the fourth coordinate gives $c_3 = 0$. Another way to look at this calculation is that once we have shown $c_1 = 0$, we have reduced to the case of a matrix formed by the second and third rows, and this allows us to give an inductive proof. This is the idea behind the proof that follows. ■

We now prove that the first r rows $\mathbf{U}_1, \ldots, \mathbf{U}_r$ of the matrix U at the end of the forward elimination algorithm are independent. We use induction on r, using the fact that when we delete the first row we have a new matrix which satisfies the same conditions as U. We first note that if $r = 1$, then one vector is always independent as long as it is nonzero. Suppose that the assertion is true for $r < k$ and that $r = k$. Assume

$$c_1\mathbf{U}_1 + \cdots + c_k\mathbf{U}_k = \mathbf{0}$$

Then examine the $c(1)$-entry in this sum. In all rows except $\mathbf{U}_1$ it is 0, and so in the sum we get the product of c_1 and the first pivot. The pivot is nonzero, so this implies that $c_1 = 0$. We then get the relation

$$c_2\mathbf{U}_2 + \cdots + c_k\mathbf{U}_k = \mathbf{0}$$

which implies by the induction assumption that $c_2 = \cdots = c_k = 0$.

Next consider the r columns in U which contain the pivots. We claim that they are linearly independent as well. We again first look at the matrix in Example 2.2.10.

■ ***Example 2.2.11.***

$$U = \begin{pmatrix} 2 & 3 & 1 & -1 & 2 \\ 0 & 0 & 4 & 2 & 3 \\ 0 & 0 & 0 & 2 & 1 \\ 0 & 0 & 0 & 0 & 0 \end{pmatrix}$$

The columns that contain the pivots are the first, third, and fourth columns. Thus we show they are independent. Let

$$c_1\begin{pmatrix} 2 \\ 0 \\ 0 \\ 0 \end{pmatrix} + c_2\begin{pmatrix} 1 \\ 4 \\ 0 \\ 0 \end{pmatrix} + c_3\begin{pmatrix} -1 \\ 2 \\ 2 \\ 0 \end{pmatrix} = \mathbf{0}$$

Then working from the back, note that the third coefficient is $2c_3 = 0$, which implies $c_3 = 0$. Looking at the second coefficient gives $4c_2 = 0$; hence $c_2 = 0$. Finally, the first coefficient gives $2c_1 = 0$; hence $c_1 = 0$. ■

We now prove the general case by induction on r, using the fact that when we delete the columns beyond $\mathbf{U}^{c(r-1)}$, the matrix is in the same form, and the first $r - 1$ columns containing the pivots of the original matrix become the columns which contain the pivots of the resulting matrix. The result is trivial for $r = 1$. Then assume that it is true for $r < k$ and let $r = k$. We assume

$$c_1\mathbf{U}^{c(1)} + \cdots + c_k\mathbf{U}^{c(k)} = \mathbf{0}$$

and solve for $\mathbf{c}$. Looking at the kth entry, the only column with nonzero entry is $\mathbf{U}^{c(k)}$, and so this linear combination has kth entry c_k times the kth pivot. Since the kth pivot is nonzero, this implies $c_k = 0$. This then implies

$$c_1\mathbf{U}^{c(1)} + \cdots + c_{(k-1)}\mathbf{U}^{c(k-1)} = \mathbf{0}$$

Our induction assumption then implies that $c_1 = \cdots = c_{(k-1)} = 0$.

We record this result as a proposition for future use.

PROPOSITION 2.2.1. If U is the matrix which occurs at the end of the forward elimination step of the Gaussian elimination algorithm on a matrix A, then the rows (respectively, columns) which contain the pivots are linearly independent vectors in $\mathbb{R}^n$ (respectively, $\mathbb{R}^m$).

■ ***Example 2.2.12.*** As a specific example consider the matrix

$$U = \begin{pmatrix} 2 & 3 & 1 & 5 & 3 & 2 \\ 0 & 0 & 2 & 2 & 4 & 1 \\ 0 & 0 & 0 & 0 & 4 & 1 \\ 0 & 0 & 0 & 0 & 0 & 0 \end{pmatrix}$$

Then Proposition 2.2.1 says that the first three rows are independent and the first, third, and fifth columns are independent. ■

A similar argument says that the same result will be true if we use the matrix R at the end of backward elimination step instead of using U. The argument is even easier for R since the columns of R containing the pivots are just the vectors $\mathbf{e}_1, \ldots, \mathbf{e}_r$. Of course, we could just use the fact that R satisfies the same properties which were used about U in the proof.

We now want to combine the two concepts of independence and spanning.

DEFINITION 2.2.6. The vectors $\mathbf{v}_1, \ldots, \mathbf{v}_n$ in a vector space V form a **basis** for V if they are linearly independent *and* they span V. More generally, a set K (possibly infinite) is a basis for a vector space V when K is an independent set and spans V in the sense that every vector in V is a *finite* linear combination of some of these vectors.

We will be interested primarily in vector spaces with finite bases. The prototypical example of a basis is the set of vectors $\mathbf{e}_1, \ldots, \mathbf{e}_n \in \mathbb{R}^n$, which is called the **standard basis** for $\mathbb{R}^n$. An example of an infinite basis for a vector space is the set of monomials $1, x, x^2, \ldots$ which is a basis for $\mathcal{P}(\mathbb{R}, \mathbb{R})$. Note that any collection of k vectors which are linearly independent will be a basis for the subspace which they span (which may or may not be the whole vector space V). In our examples of spanning sets, we find the following bases:

1. The four matrices

$$\begin{pmatrix} 1 & 0 \\ 0 & 0 \end{pmatrix}, \begin{pmatrix} 0 & 1 \\ 0 & 0 \end{pmatrix}, \begin{pmatrix} 0 & 0 \\ 1 & 0 \end{pmatrix}, \begin{pmatrix} 0 & 0 \\ 0 & 1 \end{pmatrix}$$

form a basis for all 2 by 2 matrices.

2. The polynomials $1, x, \ldots, x^n$ form a basis for the vector space $\mathcal{P}^n(\mathbb{R}, \mathbb{R})$ of polynomials of degree less than or equal to n.
3. The infinite collection $1, x^2, x^4, \ldots$ forms a basis for the even polynomials.
4. For the matrix $A = \begin{pmatrix} 2 & 4 & 1 \\ 3 & 1 & 4 \\ 5 & 5 & 5 \end{pmatrix}$ the first two rows form a basis for the row space and the first two columns form a basis for the column space.

There are many bases for a given vector space (in fact, an infinite number). In the last example we could have also found bases by performing the forward elimination step of Gaussian elimination and using the nonzero rows for a basis of the row space. The columns containing the pivots will not even be in the column space for the original matrix. However, the corresponding columns of the original matrix will turn out to give a basis as we will show in Section 2.5.

We want to show that the number of elements in a basis for a vector space is well defined; that is, there cannot be two bases with different numbers of vectors. We prove this as a corollary of the following assertion.

PROPOSITION 2.2.2. If $\mathbf{v}_1, \ldots, \mathbf{v}_m$ span V and if $\mathbf{w}_1, \ldots, \mathbf{w}_n$ are vectors in V, then $n > m$ implies $\mathbf{w}_1, \ldots, \mathbf{w}_n$ are linearly dependent.

COROLLARY 2.2.3. Two bases for a vector space have the same number of elements.

An alternative statement of the conclusion of Proposition 2.2.2 is

If $\mathbf{w}_1, \ldots, \mathbf{w}_n$ are independent, then $n \leq m$.

To see that the corollary follows from Proposition 2.2.2, we start with the assumption that $\mathbf{v}_1, \ldots, \mathbf{v}_m$ and $\mathbf{w}_1, \ldots, \mathbf{w}_n$ form bases for V. Then Proposition 2.2.2 implies $n \leq m$ since the $\mathbf{v}_i$ span V and the $\mathbf{w}_j$ are linearly independent. On the other hand, the fact that the $\mathbf{w}_j$ span V and the $\mathbf{v}_i$ are linearly independent implies that $m \leq n$. Hence $m = n$. This proves the result when both bases have a finite number of elements. There cannot be one basis with a finite number of elements and another with an infinite number since Proposition 2.2.2 would then imply that the infinite collection is dependent.

DEFINITION 2.2.7. The **dimension** of a vector space V is the number of vectors in any basis for V. We call V **finite dimensional** when this number is finite and **infinite dimensional** when it is infinite. We denote the dimension of a vector space V as $\dim V$.

We will primarily be interested in finite dimensional vector spaces here.
We now prove Proposition 2.2.2.

Proof. We assume

$$c_1\mathbf{w}_1 + \cdots + c_n\mathbf{w}_n = \mathbf{0}$$

and try to find a nontrivial solution. We first express $\mathbf{w}_j$ in terms of the $\mathbf{v}_i$ using the assumption that the $\mathbf{v}_i$ span V. Suppose

$$\mathbf{w}_j = a_{1j}\mathbf{v}_1 + \cdots + a_{mj}\mathbf{v}_m$$

We then rewrite $c_1\mathbf{w}_1 + \cdots + c_n\mathbf{w}_n = \mathbf{0}$ as

$$c_1(a_{11}\mathbf{v}_1 + \cdots + a_{m1}\mathbf{v}_m) + \cdots + c_n(a_{1n}\mathbf{v}_1 + \cdots + a_{mn}\mathbf{v}_m) = \mathbf{0}$$

Rearranging terms gives

$$(a_{11}c_1 + \cdots + a_{1n}c_n)\mathbf{v}_1 + \cdots + (a_{m1}c_1 + \cdots + a_{mn}c_n)\mathbf{v}_m = \mathbf{0}$$

A solution of these equations occurs when the coefficients of all the $\mathbf{v}_i$ are 0. This condition gives a matrix equation $A\mathbf{c} = \mathbf{0}$, where A is an m by n matrix and $n > m$. Since there will be a free variable, we have a nontrivial solution $\mathbf{c}$, which then implies that the vectors $\mathbf{w}_i$ are linearly dependent. ■

Let us give the dimensions of some of the examples we have already discussed.

- $\mathbb{R}^n$ has dimension n since it has a basis of the n vectors $\mathbf{e}_1, \ldots, \mathbf{e}_n$. In $\mathbb{R}^3$ a line has dimension 1 and a plane has dimension 2.
- The vector space $\mathcal{M}(2,2)$ of 2 by 2 matrices has dimension 4 since it has as a basis the four matrices

$$\begin{pmatrix} 1 & 0 \\ 0 & 0 \end{pmatrix}, \begin{pmatrix} 0 & 1 \\ 0 & 0 \end{pmatrix}, \begin{pmatrix} 0 & 0 \\ 1 & 0 \end{pmatrix}, \begin{pmatrix} 0 & 0 \\ 0 & 1 \end{pmatrix}$$

  More generally, the vector space $\mathcal{M}(m,n)$ has dimension mn. It has as a basis the matrices $N(i,j), 1 \le i \le m, 1 \le j \le n$, with all entries zero except the ijth one, which is 1. If A is an m by n matrix with ij-entry a_{ij}, then $A = \sum_{i,j=1}^{n} a_{ij}N(i,j)$. This shows that these matrices span $\mathcal{M}(m,n)$. If there is a linear combination $\sum_{i,j=1}^{n} a_{ij}N(i,j) = \mathbf{0}$, then the ij-entry a_{ij} of the linear combination will be 0, and so all $a_{ij} = 0$. Hence the $N(i,j)$ are independent.
- The vector space $\mathcal{P}(\mathbb{R},\mathbb{R})$ is infinite dimensional. This follows from the fact that $1, x, x^2, \ldots$ forms an infinite basis.
- The vector space $\mathcal{P}^n(\mathbb{R},\mathbb{R})$ has dimension $n+1$ since it has basis

$$1, x, x^2, \ldots, x^n$$

We next note a consequence of having a basis $\mathbf{v}_1, \ldots, \mathbf{v}_n$ for V. Since these vectors span V, we know that any vector $\mathbf{v}$ can be represented as a linear combination

$\mathbf{v} = c_1\mathbf{v}_1 + \cdots + c_n\mathbf{v}_n$. We claim that the numbers c_i are unique. Suppose that there is another representation $\mathbf{v} = d_1\mathbf{v}_1 + \cdots + d_n\mathbf{v}_n$. Then taking the difference of $\mathbf{v}$ and itself gives

$$\mathbf{0} = (c_1 - d_1)\mathbf{v}_1 + \cdots + (c_n - d_n)\mathbf{v}_n$$

Since the $\mathbf{v}_i$ are linearly independent this gives $(c_1 - d_1) = \cdots = (c_n - d_n) = 0$, so $c_i = d_i$ for all i. We record this for future reference.

PROPOSITION 2.2.4. If $\mathbf{v}_1, \ldots, \mathbf{v}_n$ is a basis for V, then each vector $\mathbf{v}$ in V can be written uniquely as $\mathbf{v} = c_1\mathbf{v}_1 + \cdots + c_n\mathbf{v}_n$.

Suppose we already know that the dimension of a vector space V is n and we are given n vectors $\mathbf{v}_1, \ldots, \mathbf{v}_n \in V$. To see that it is a basis we would have to know that the vectors are independent and they span V. But we claim that one of these properties implies the other when the dimension is n. First, suppose that $\mathbf{v}_1, \ldots, \mathbf{v}_n$ are independent. Then they must also span V. Assume $\mathbf{v} \in V$ is given. Then $\mathbf{v}_1, \ldots, \mathbf{v}_n, \mathbf{v}$ are $n+1$ vectors in an n dimensional vector space and so they must be dependent. Thus there are $c_1, \ldots, c_{n+1}$ which are not all zero with

$$c_1\mathbf{v}_1 + \cdots + c_n\mathbf{v}_n + c_{n+1}\mathbf{v} = \mathbf{0}$$

We could not have $c_{n+1} = 0$ since this would contradict the independence of $\mathbf{v}_1, \ldots, \mathbf{v}_n$. Thus we can solve for $\mathbf{v}$ as a linear combination of the $\mathbf{v}_i$, showing that they span V. On the other hand, if we know that the $\mathbf{v}_i$ span V, then they have to be independent. If not, one of them could be written as a linear combination of the others. Then we can find $n-1$ vectors which span V, contradicting the fact that the dimension of V is n and so there are n independent vectors in V. Note that the dimension of a vector space is always less than or equal to the number of vectors in any spanning set, since we can always find a basis as a subcollection of the spanning set by throwing out dependent vectors one at a time. This discussion proves the following useful proposition.

PROPOSITION 2.2.5. If the dimension of V is n and $\mathbf{v}_1, \ldots, \mathbf{v}_n$ are n vectors in V, then they are a basis if they are independent *OR* if they span V.

Exercise 2.2.1.

In each part determine whether the vector $\mathbf{v}$ is in the subspace spanned by $\mathbf{v}_1, \mathbf{v}_2, \mathbf{v}_3$. If it is, give constants c_1, c_2, c_3 so that $\mathbf{v} = c_1\mathbf{v}_1 + c_2\mathbf{v}_2 + c_3\mathbf{v}_3$.

(a) $\mathbf{v} = (1, 1, -1, 2), \mathbf{v}_1 = (2, 0, -1, 1), \mathbf{v}_2 = (3, 1, 1, 1), \mathbf{v}_3 = (-3, 1, 2, 1)$

(b) $\mathbf{v} = (11, 1, -1, 2), \mathbf{v}_1 = (2, 0, -1, 1), \mathbf{v}_2 = (3, 1, 1, 1), \mathbf{v}_3 = (-3, 1, 2, 1)$

Exercise 2.2.2.

For each part determine whether the given vectors are linearly independent.

(a) $(1, 2, 3), (4, 5, 6), (7, 8, 9)$

(b) $(2, 1, 4), (3, 9, 2), (3, 6, 2), (4, 7, 2)$

(c) $(1, 1, 1), (1, 1, 0), (1, 0, 0)$

(d) $(1, 5, 2, 7), (3, 2, 1, 5), (3, 2, 1, 1), (2, 6, 2, 1)$

Exercise 2.2.3.

For each part determine whether the given collection of polynomials is independent.

(a) $1, 1 + x, 1 + x + x^2$

(b) $3 + x - x^2, x + x^2, 6 + 3x - x^2$

Exercise 2.2.4.

Consider the vector space of 2 by 2 matrices and the subspace S of symmetric 2 by 2 matrices, that is, those of the form $\begin{pmatrix} a & b \\ b & c \end{pmatrix}$. Show that the three matrices $\begin{pmatrix} 1 & 0 \\ 0 & 0 \end{pmatrix}, \begin{pmatrix} 0 & 1 \\ 1 & 0 \end{pmatrix}$, and $\begin{pmatrix} 0 & 0 \\ 0 & 1 \end{pmatrix}$ span S. Is this a basis for S? Justify your assertion.

Exercise 2.2.5.

Show that the vectors $(1, 1, 1), (1, 1, 0), (1, 0, 0)$ span $\mathbb{R}^3$.

Exercise 2.2.6.

Show that there can't be a set of two vectors that span $\mathbb{R}^3$.

Exercise 2.2.7.

Determine whether span$((1, 2, 1), (3, 1, 2), (2, -1, 1))$ is a line through $\mathbf{0}$, a plane through $\mathbf{0}$, or all of $\mathbb{R}^3$. If it is a line or a plane, write its equation in the usual form.

Exercise 2.2.8.

Find a basis for the set of skew symmetric 3 by 3 matrices

$$\begin{pmatrix} 0 & a & b \\ -a & 0 & c \\ -b & -c & 0 \end{pmatrix}$$

Exercise 2.2.9.

Show that any subspace of $\mathbb{R}^3$ is either $\{\mathbf{0}\}$, a line, a plane, or all of $\mathbb{R}^3$.

2.3. LINEAR TRANSFORMATIONS

We have seen some examples of vector spaces. We next wish to discuss the appropriate types of functions between vector spaces. Since a vector space has two operations on it, we would like to consider functions which are consistent with those operations.

DEFINITION 2.3.1. If V and W are vector spaces, a **linear transformation** $L : V \to W$ is a function satisfying the two properties

$$L(\mathbf{v}_1 + \mathbf{v}_2) = L(\mathbf{v}_1) + L(\mathbf{v}_2) \quad \text{and} \quad L(a\mathbf{v}) = aL(\mathbf{v})$$

where $\mathbf{v}_1, \mathbf{v}_2, \mathbf{v}$ are in V and a is a real number.

Another way of expressing the linearity property which a linear transformation must satisfy is

$$L(a_1\mathbf{v}_1 + a_2\mathbf{v}_2) = a_1L(\mathbf{v}_1) + a_2L(\mathbf{v}_2)$$

If L is a linear transformation, then it will satisfy this property since

$$L(a_1\mathbf{v}_1 + a_2\mathbf{v}_2) = L(a_1\mathbf{v}_1) + L(a_2\mathbf{v}_2) = a_1L(\mathbf{v}_1) + a_2L(\mathbf{v}_2)$$

On the other hand, if L satisfies the linearity property in terms of linear combinations, then choosing $a_2 = 0$ shows that L satisfies the correct property in terms of scalar multiplication, and choosing $a_1 = a_2 = 1$ shows that L satisfies the linearity property with respect to addition. In an upcoming exercise it is shown that a linear transformation satisfies

$$L(a_1\mathbf{v}_1 + \cdots + a_k\mathbf{v}_k) = a_1L(\mathbf{v}_1) + \cdots + a_kL(\mathbf{v}_k)$$

that is, L sends a general linear combination to a linear combination.

A linear transformation $L : V \to W$ satisfies $L(\mathbf{0}) = \mathbf{0}$:

$$L(\mathbf{0}) = L(0 \cdot \mathbf{v}) = 0 \cdot L(\mathbf{v}) = \mathbf{0}$$

Here the first $\mathbf{0}$ is the additive identity in V, and the second one is the additive identity in W.

Recall that we noted earlier that the function $L : \mathbb{R}^n \to \mathbb{R}^m$ given by $L(\mathbf{v}) = A\mathbf{v}$ (where A is an m by n matrix and $\mathbf{v}$ is being considered as a column vector) satisfied this property and so is a linear transformation. Also, we saw that each linear transformation from $\mathbb{R}^n$ to $\mathbb{R}^m$ arises in this way for some matrix A. We will generalize this in Section 2.7 by showing how to associate matrices to linear transformations.

We now look at some examples of linear transformations.

■ ***Example 2.3.1.*** Consider the map $L_A : \mathbb{R}^2 \to \mathbb{R}^2$ given by multiplying by the matrix $A = \begin{pmatrix} 2 & 1 \\ 1 & 2 \end{pmatrix}$. We saw earlier that multiplication by a matrix is a linear transformation. To get a sense of where L_A sends vectors, Figure 2.1 depicts the image of a triangle. The original triangle is given with solid lines and the image with dashed lines.

Figure 2.2 shows where the unit circle is sent by L_A. ■

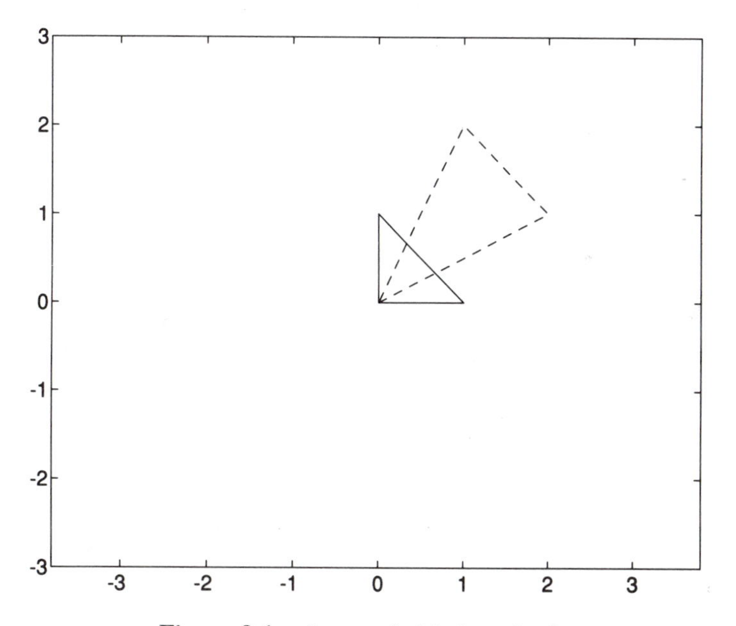

Figure 2.1. Image of triangle under L_A

■ ***Example 2.3.2.*** Consider the function $L : \mathbb{R}^2 \to \mathbb{R}^2$ which reflects each vector through the x-axis. This has a formula $L(x, y) = (x, -y)$. It arises by multiplication by the matrix $\begin{pmatrix} 1 & 0 \\ 0 & -1 \end{pmatrix}$ and so is a linear transformation.

Figure 2.3 shows the image of two triangles under this reflection in the x-axis.

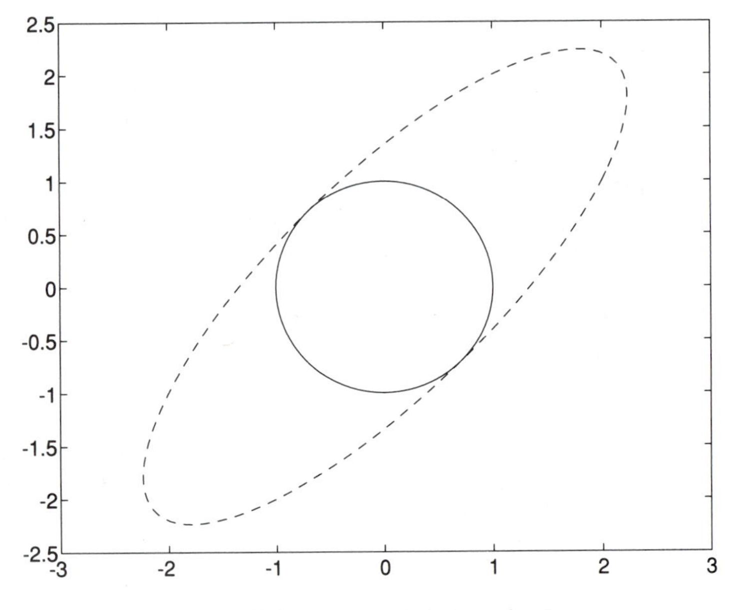

Figure 2.2. Image of circle under L_A

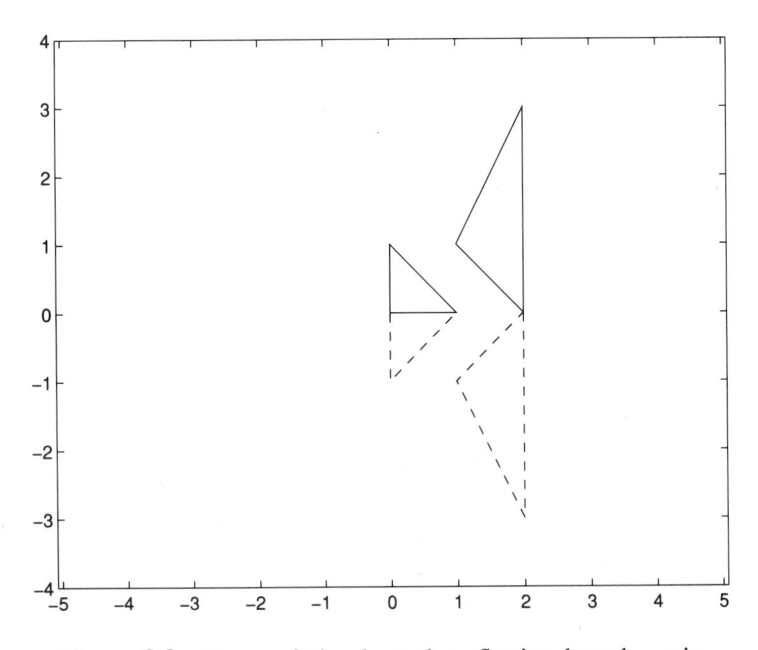

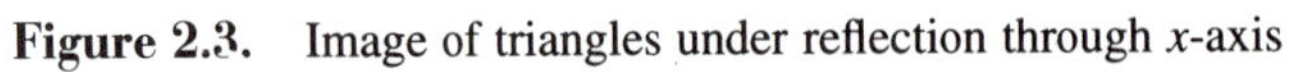

Figure 2.3. Image of triangles under reflection through x-axis

Suppose that we choose another line through which to reflect. For example, let us reflect through the line $x = 2y$. A reflection through a line is characterized by its sending any vector in the line to itself and sending a vector perpendicular to the line to its negative. To see what reflection through a line through the origin does to any arbitrary vector $\mathbf{v}$, we first write $\mathbf{v} = \mathbf{v}_1 + \mathbf{v}_2$, where $\mathbf{v}_1$ lies on the line $x = 2y$ and $\mathbf{v}_2$ lies on the perpendicular line $y = -2x$. That this is possible, and there is a unique way to do this, uses the fact that $(2, 1), (-1, 2)$ is a basis of $\mathbb{R}^2$. We then define

$$L(\mathbf{v}) = L(\mathbf{v}_1 + \mathbf{v}_2) = \mathbf{v}_1 - \mathbf{v}_2$$

If $\mathbf{v} = \mathbf{v}_1 + \mathbf{v}_2$ and $\mathbf{w} = \mathbf{w}_1 + \mathbf{w}_2$, then $\mathbf{v} + \mathbf{w} = (\mathbf{v}_1 + \mathbf{w}_1) + (\mathbf{v}_2 + \mathbf{w}_2)$ and so

$$L(\mathbf{v} + \mathbf{w}) = (\mathbf{v}_1 + \mathbf{w}_1) - (\mathbf{v}_2 + \mathbf{w}_2) = (\mathbf{v}_1 - \mathbf{v}_2) + (\mathbf{w}_1 - \mathbf{w}_2) = L(\mathbf{v}) + L(\mathbf{w})$$

Also,

$$L(c\mathbf{v}) = L(c\mathbf{v}_1 + c\mathbf{v}_2) = c\mathbf{v}_1 - c\mathbf{v}_2 = c(\mathbf{v}_1 - \mathbf{v}_2) = cL(\mathbf{v})$$

Thus L is a linear transformation. Alternatively, this linearity can seen geometrically.

Since the reflection L is a linear transformation from $\mathbb{R}^2$ to $\mathbb{R}^2$, there is some 2 by 2 matrix A so that L is just multiplication by A. One of our tasks will be to find this matrix.

Figure 2.4 depicts the image of the same two triangles under this reflection. ■

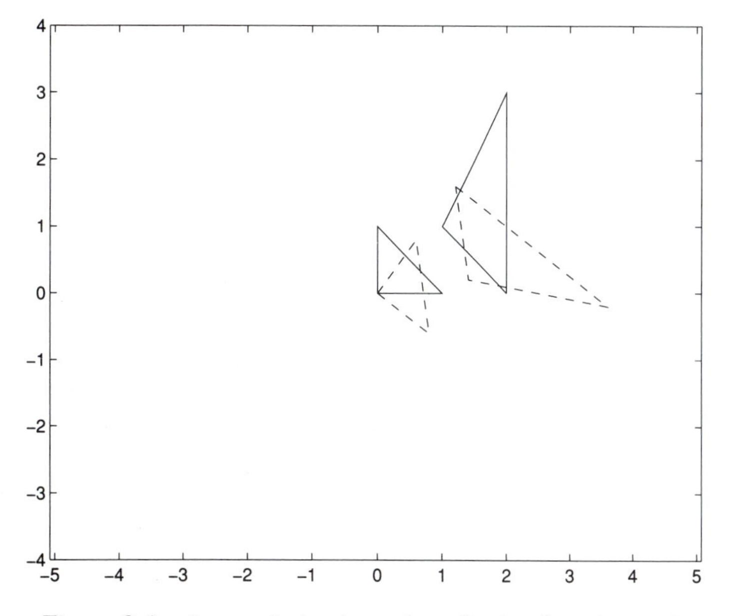

Figure 2.4. Image of triangles under reflection through $x = 2y$

■ ***Example 2.3.3.*** Let $L : \mathbb{R}^2 \to \mathbb{R}^2$ be the rotation of 60 degrees. Then L is a linear transformation since if we add two vectors and rotate the sum, we get the same answer as when we first rotate the two vectors and then add the results. To see this, we think of forming the parallelogram associated with adding the two vectors and then rotating the whole parallelogram to illustrate adding the two rotated vectors. Also, if a vector is multiplied by a constant and then the result is rotated, we get the same result as when we first rotate and then multiply by the constant. To see which matrix corresponds to L, we compute $L(\mathbf{e}_1) = (1/2, \sqrt{3}/2)$ and $L(\mathbf{e}_2) = (-\sqrt{3}/2, 1/2)$. Thus L corresponds to multiplying by the matrix $\begin{pmatrix} 1/2 & -\sqrt{3}/2 \\ \sqrt{3}/2 & 1/2 \end{pmatrix}$.

Figure 2.5 depicts the image of a triangle under this rotation.

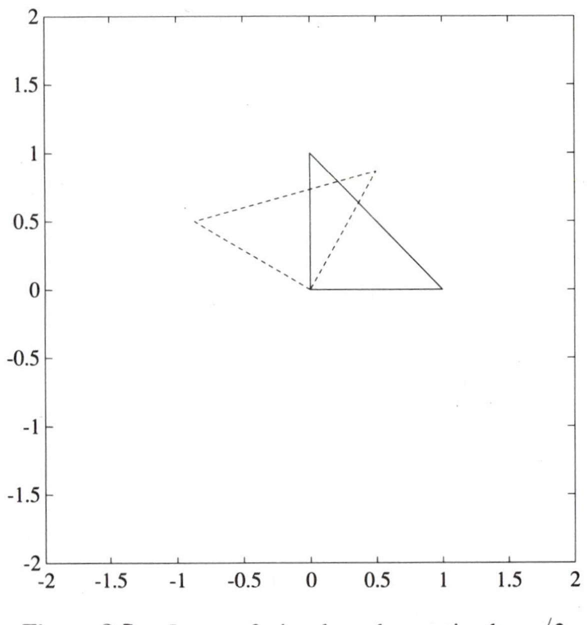

Figure 2.5. Image of triangle under rotation by $\pi/3$

A similar calculation shows that a rotation by angle θ corresponds to multiplication by the matrix $\begin{pmatrix} \cos\theta & -\sin\theta \\ \sin\theta & \cos\theta \end{pmatrix}$. More generally, we can consider a rotation in $\mathbb{R}^3$ where there is some fixed axis. This will also give a linear transformation L, and we will find A with $L = L_A$. ■

■ ***Example 2.3.4.*** Consider the map $L : \mathbb{R}^3 \to \mathbb{R}^3$ which projects each vector onto the xy-plane, $L(x, y, z) = (x, y, 0)$. This corresponds to multiplying by the matrix

$\begin{pmatrix} 1 & 0 & 0 \\ 0 & 1 & 0 \\ 0 & 0 & 0 \end{pmatrix}$ where the columns indicate the images of the vectors $\mathbf{e}_1, \mathbf{e}_2, \mathbf{e}_3$ under L.

Since it comes from matrix multiplication, it will be a linear transformation. A somewhat more complicated projection could come from choosing a different plane P in $\mathbb{R}^3$ and projecting each vector orthogonally onto this plane. What this means is that we would decompose each vector as $\mathbf{v} + \mathbf{w}$, where $\mathbf{v}$ lies in P and $\mathbf{w}$ lies in the line through the origin which is perpendicular to P, and $L(\mathbf{v} + \mathbf{w}) = \mathbf{v}$. This is the generalization of what is happening in the first example since we are sending $(x, y, z) = (x, y, 0) + (0, 0, z)$ to $(x, y, 0)$. This will be a linear transformation since if $\mathbf{z}_1 = \mathbf{v}_1 + \mathbf{w}_1$ and $\mathbf{z}_2 = \mathbf{v}_2 + \mathbf{w}_2$, then $\mathbf{z}_1 + \mathbf{z}_2 = (\mathbf{v}_1 + \mathbf{v}_2) + (\mathbf{w}_1 + \mathbf{w}_2)$ and so

$$L(\mathbf{z}_1 + \mathbf{z}_2) = \mathbf{v}_1 + \mathbf{v}_2 = L(\mathbf{z}_1) + L(\mathbf{z}_2)$$

Also, $c\mathbf{z}_1 = c\mathbf{v}_1 + c\mathbf{w}_1$ implies that

$$L(c\mathbf{z}_1) = c\mathbf{v}_1 = cL(\mathbf{z}_1)$$

We are using here both that P and the perpendicular line are subspaces as well as the fact that each vector in $\mathbb{R}^3$ has a unique decomposition as the sum of a vector in P and a vector in the perpendicular line. We will verify these facts later and discuss this linear transformation, which we will call **projection** onto the plane P, in much more detail. In particular, it must arise as multiplication by some matrix and so one of our tasks will be to find that matrix. ∎

∎ ***Example 2.3.5.*** Let $\mathcal{D}(\mathbb{R}, \mathbb{R})$ be the vector space of differentiable functions with infinitely many derivatives. Define $D : \mathcal{D}(\mathbb{R}, \mathbb{R}) \to \mathcal{D}(\mathbb{R}, \mathbb{R})$ by $D(f) = f'$, where f' denotes the derivative of f. Then

$$D(af + bg) = (af + bg)' = af' + bg' = aD(f) + bD(g)$$

and so D is a linear transformation. ∎

∎ ***Example 2.3.6.*** Let $\mathcal{I}_0 : \mathcal{D}(\mathbb{R}, \mathbb{R}) \to \mathcal{D}(\mathbb{R}, \mathbb{R})$ be the map which takes the antiderivative h of f satisfying $h(0) = 0$. This may also be written as

$$\mathcal{I}_0(f)(x) = \int_0^x f(t)\, dt$$

For example, $\mathcal{I}_0(x^2) = x^3/3$. $\mathcal{I}_0(f)$ is thus characterized by $D\mathcal{I}_0(f) = f$ and $\mathcal{I}_0(f)(0) = 0$. Then

$$\mathcal{I}_0(af + bg) = a\mathcal{I}_0(f) + b\mathcal{I}_0(g)$$

since

$$D(a\mathcal{I}_0(f) + b\mathcal{I}_0(g)) = aD\mathcal{I}_0(f) + bD\mathcal{I}_0(g) = af + bg$$

and

$$(a\mathcal{I}_0(f) + b\mathcal{I}_0(g))(0) = a\mathcal{I}_0(f)(0) + b\mathcal{I}_0(g)(0) = 0 + 0 = 0$$

We could also verify this using the definite integral. ∎

If $L_1, \ldots, L_k$ are linear transformations from V to W, then we can form a linear combination $a_1L_1 + \cdots + a_kL_k$ using the formula

$$(a_1L_1 + \cdots + a_kL_k)(\mathbf{v}) = a_1L_1(\mathbf{v}) + \cdots + a_kL_k(\mathbf{v})$$

Here the vector space operations in W are being used to define addition and scalar multiplication on the right-hand side. It is easy to check that the resulting function is a linear transformation. This is just the verification that the linear transformations form a subspace of the vector space of all functions from V to W. We denote this vector space of all linear transformations from V to W as $\mathcal{L}(V, W)$. We use this idea in Example 2.3.7.

∎ ***Example 2.3.7.*** Consider the differential equation $y'' - 3y' + 2y = 0$. Let $D : \mathcal{D}(\mathbb{R}, \mathbb{R}) \to \mathcal{D}(\mathbb{R}, \mathbb{R})$ be the linear transformation which takes the derivative of a function, $D(y) = y'$. The composition $D \circ D$ of D with itself, which we will denote by D^2, will correspond to taking the second derivative: $D^2(y) = D(D(y)) = D(y') = (y')' = y''$. Then the left-hand side of the equation can be rewritten as the linear combination $(D^2 - 3D + 2I)y$, where $I(y) = y$. Thus the solutions of the differential equation are those functions y which $L = D^2 - 3D + 2I$ sends to the zero function $\mathbf{0}$. Note that the linear transformation L factors as the composition of $(D - 2I)$ and $(D - I)$, $L = (D-2I)(D-I) = (D-I)(D-2I)$. Here we are using juxtaposition of $D-2I$ and $D-I$ to denote the composition of these two linear transformations. The equality can be verified by evaluation on an arbitrary function. It depends on the fact that D commutes with I : $DI = ID = D$. Because of this equality any function which $D - I$ or $D - 2I$ sends to $\mathbf{0}$ is sent to $\mathbf{0}$ by L. Note that $(D - I)y = y' - y, (D - 2I)y = y' - 2y$. Thus any solution of $y' - 2y = 0$ or $y' - y = 0$ is a solution of the preceding second-order equation. Two such solutions are $y = e^{2t}$ and $y = e^t$. Since L is a linear transformation, linear combinations of solutions to $Ly = \mathbf{0}$ are also solutions: if $L(y_1) = 0 = L(y_2)$, then

$$L(c_1y_1 + c_2y_2) = c_1L(y_1) + c_2L(y_2) = c_1 \cdot 0 + c_2 \cdot 0 = 0$$

Thus $c_1e^{2t} + c_2e^t$ is a solution to $Ly = \mathbf{0}$. We will see later that this will imply that the general solution of this equation is $c_1e^{2t} + c_2e^t$. ∎

∎ ***Example 2.3.8.*** Our last example will consist of a number of functions which are not linear transformations.

(a) Let $f : \mathbb{R} \to \mathbb{R}$ be the function $f(x) = x^2$. Then $f(2x) = 4x^2 = 4f(x) \neq 2f(x)$ and so f is not a linear transformation.

(b) Let $F : \mathcal{F}(\mathbb{R}, \mathbb{R}) \to \mathcal{F}(\mathbb{R}, \mathbb{R})$ be given by $F(f) = |f|$; that is, F takes a function to the absolute value of itself. Note that $f + (-f) = \mathbf{0}$ but that $\mathbf{0} =$

$F(\mathbf{0}) = F(f + (-f)) \neq F(f) + F(-f) = |f| + |f| = 2|f|$ unless $f = \mathbf{0}$. Thus F is not a linear transformation.

(c) Let $S : \mathcal{M}(2,2) \to \mathcal{M}(2,2)$ be defined by $S(A) = A^2$. Then $S(2A) = 4A^2 = 4S(A)$, and S is not a linear transformation.

(d) Consider $S : \mathcal{D}(\mathbb{R}, \mathbb{R}) \to \mathcal{D}(\mathbb{R}, \mathbb{R})$ given by $S(y) = y'' - (y')^2$. Then $S(cy) = cy'' - c^2(y')^2 \neq cS(y)$ if $c \neq 1$. Thus S is not a linear transformation. The differential equation $Sy = 0$ is called a nonlinear differential equation. The properties of solutions of linear and nonlinear differential equations are noticeably different. For linear differential equations, the solutions form a vector space (when the right-hand side is 0), whereas they do not for nonlinear equations.

(e) Let $\mathcal{I}_1 : \mathcal{D}(\mathbb{R}, \mathbb{R}) \to \mathcal{D}(\mathbb{R}, \mathbb{R})$ be defined by letting $\mathcal{I}_1(f)$ be the antiderivative of f which satisfies $(\mathcal{I}_1(f))(0) = 1$. Then $\mathcal{I}_1(0) = 1 \neq 0$, so $\mathcal{I}_1$ is not a linear transformation. ■

Exercise 2.3.1.

Show that if $L : V \to W$ is a linear transformation, then

$$L(a_1\mathbf{v}_1 + \cdots + a_k\mathbf{v}_k) = a_1L(\mathbf{v}_1) + \cdots + a_kL(\mathbf{v}_k)$$

(Hint: Use induction on k.)

Exercise 2.3.2.

Verify that the linear transformations form a subspace of the vector space $\mathcal{F}(V, W)$ of all functions from the vector space V to the vector space W.

Exercise 2.3.3.

For each of the following functions, determine whether it is a linear transformation.

(a) $L : \mathcal{P}^n(\mathbb{R}, \mathbb{R}) \to \mathbb{R}^{n+1}$, where $L(a_0 + a_1x + \cdots + a_nx^n) = (a_0, a_1, \ldots, a_n)$.

(b) $M : R^{n+1} \to \mathcal{P}^n(\mathbb{R}, \mathbb{R})$, where $M(a_0, \ldots, a_n) = a_0 + a_1x + \cdots + a_nx^n$. Show that LM is the identity linear transformation of $\mathbb{R}^{n+1}$ and ML is the identity linear transformation of $\mathcal{P}^n(\mathbb{R}, \mathbb{R})$.

(c) $H : \mathbb{R}^2 \to \mathbb{R}^3$, $H(x, y) = (x, y, xy)$.

(d) $F : \mathcal{D}(\mathbb{R}, \mathbb{R}) \to \mathcal{D}(\mathbb{R}, \mathbb{R})$, where $F(y) = y'' - 2y + e^x$.

(e) $T : \mathcal{M}(m, n) \to \mathcal{M}(n, m)$, where $T(A) = A^t$.

(f) $H_f : \mathcal{D}(\mathbb{R}, \mathbb{R}) \to \mathcal{D}(\mathbb{R}, \mathbb{R})$ where $H_f(g) = f \cdot g$, where $(f \cdot g)(x) = f(x)g(x)$.

(g) $T : \mathcal{D}(\mathbb{R}, \mathbb{R}) \to \mathbb{R}$, $T(f) = \int_0^1 f(t)\,dt$.

(h) $L : \mathbb{R}^3 \to \mathbb{R}^3$, $L(\mathbf{v}) = \mathbf{v} \times \mathbf{e}_1$, where $\times$ denotes the cross product.

(i) $L : \mathbb{R}^3 \to \mathbb{R}^2$, $L(x, y, z) = (xy, yz)$.

We close this section by connecting the topics of linear transformations and bases.

PROPOSITION 2.3.1. Let V, W be vector spaces, $\mathcal{V} = \{\mathbf{v}_1, \ldots, \mathbf{v}_n\}$ be a basis for V, and $\mathbf{w}_1, \ldots, \mathbf{w}_n \in W$. Then there exists a unique linear transformation $L : V \to W$ satisfying $L(\mathbf{v}_i) = \mathbf{w}_i$. In particular, this means that a linear transformation $L : V \to W$ is completely determined by its values on the basis $\mathcal{V}$.

Before we give the proof of this proposition, we remark as to how strong a condition its conclusion is. A general function $f : \mathbb{R}^2 \to \mathbb{R}$ is a very complicated object, and its values on two points of $\mathbb{R}^2$ gives very little information about it. Just think of the total graph of the function and the information given by two points on that graph. On the other hand, a linear transformation $L : \mathbb{R}^2 \to \mathbb{R}$ is completely determined by knowing $L(\mathbf{e}_1) = a, L(\mathbf{e}_2) = b$. The linear transformation is then

$$L(x, y) = L(x\mathbf{e}_1 + y\mathbf{e}_2) = xL(e_1) + yL(\mathbf{e}_2) = ax + by$$

Its graph is now the plane which passes through the three points

$$(0, 0, 0), (1, 0, a), (0, 1, b)$$

In general, a linear transformation $L : \mathbb{R}^n \to \mathbb{R}^m$ was shown to be multiplication by a matrix A, where $A^j = L(\mathbf{e}_j)$, and so is completely determined by the values $L(\mathbf{e}_j)$ on the standard basis $\mathbf{e}_1, \ldots, \mathbf{e}_n$. This proposition says a similar phenomenon is true for any linear transformation. We will pursue this further in Section 2.7.

We now give the proof of the proposition.

Proof. For existence we can define L as follows. Any $\mathbf{v} \in V$ is uniquely expressible as

$$\mathbf{v} = c_1\mathbf{v}_1 + \cdots + c_n\mathbf{v}_n$$

Then define

$$L(\mathbf{v}) = c_1\mathbf{w}_1 + \cdots + c_n\mathbf{w}_n$$

By definition, $L(\mathbf{v}_i) = \mathbf{w}_i$ since

$$\mathbf{v}_i = 0 \cdot \mathbf{v}_1 + \cdots + 0 \cdot \mathbf{v}_{i-1} + 1 \cdot \mathbf{v}_i + 0 \cdot \mathbf{v}_{i+1} + \cdots + 0 \cdot \mathbf{v}_n$$

We need to see that L is in fact a linear transformation. Suppose

$$\mathbf{v} = c_1\mathbf{v}_1 + \cdots + c_n\mathbf{v}_n, \mathbf{u} = d_1\mathbf{v}_1 + \cdots + d_n\mathbf{v}_n$$

Then

$$\mathbf{v} + \mathbf{u} = (c_1 + d_1)\mathbf{v}_1 + \cdots + (c_n + d_n)\mathbf{v}_n$$

and

$$\begin{aligned} L(\mathbf{v} + \mathbf{u}) &= (c_1 + d_1)\mathbf{w}_1 + \cdots + (c_n + d_n)\mathbf{w}_n \\ &= (c_1\mathbf{w}_1 + \cdots + c_n\mathbf{w}_n) + (d_1\mathbf{w}_1 + \cdots + d_n\mathbf{w}_n) \\ &= L(\mathbf{v}) + L(\mathbf{u}) \end{aligned}$$

Also,

$$\begin{aligned} L(a\mathbf{v}) &= L(ac_1\mathbf{v}_1 + \cdots + ac_n\mathbf{v}_n) \\ &= ac_1\mathbf{w}_1 + ac_n\mathbf{w}_n \\ &= a(c_1\mathbf{w}_1 + \cdots + c_n\mathbf{w}_n) \\ &= aL(\mathbf{v}) \end{aligned}$$

For uniqueness, suppose that $L : V \to W$ is a linear transformation satisfying $L(\mathbf{v}_i) = \mathbf{w}_i$. Then for any $\mathbf{v} \in V$, there are unique constants $c_1, \ldots, c_n$ with

$$\mathbf{v} = c_1\mathbf{v}_1 + \cdots + c_n\mathbf{v}_n$$

Then L being a linear transformation implies

$$L(\mathbf{v}) = L(c_1\mathbf{v}_1 + \cdots + c_n\mathbf{v}_n) = c_1L(\mathbf{v}_1) + \cdots + c_nL(\mathbf{v}_n) = c_1\mathbf{w}_1 + \cdots + c_n\mathbf{w}_n$$

Thus L is completely determined by its values on this basis. ■

■ ***Example 2.3.9.*** Consider the plane $V \subset \mathbb{R}^3$ given by the equation $x + y + z = 0$. Since the solutions to this equation are of the form $c_1(-1, 1, 0) + c_2(-1, 0, 1)$, the vectors $\mathbf{v}_1 = (-1, 1, 0), \mathbf{v}_2 = (-1, 0, 1)$ span this plane V. They are also independent since $c_1(-1, 1, 0) + c_2(-1, 0, 1) = (0, 0, 0)$ means $(-c_1 - c_2, c_1, c_2) = (0, 0, 0)$ and hence $c_1 = c_2 = 0$. Thus these two vectors form a basis for this plane. Now consider the two vectors in $\mathbb{R}^2$ given by $\mathbf{w}_1 = (2, 3), \mathbf{w}_2 = (-1, 2)$. Then Proposition 2.3.1 says that there is a unique linear transformation $L : V \to \mathbb{R}^2$ with $L(\mathbf{v}_i) = \mathbf{w}_i$. To find where L sends a general vector $\mathbf{v} \in V$, write $\mathbf{v} = c_1\mathbf{v}_1 + c_2\mathbf{v}_2$ and then $L(\mathbf{v}) = c_1(2, 3) + c_2(-1, 2)$. For example, consider $L(-4, 3, 1)$. First, solve for c_1, c_2 with

$$(-4, 3, 1) = c_1(-1, 1, 0) + c_2(-1, 0, 1) = (-c_1 - c_2, c_1, c_2)$$

This is easy to solve to get $c_1 = 3, c_2 = 1$. Then $L(-4, 3, 1) = 3(2, 3) + (-1, 2) = (5, 11)$.

To find a formula for the case where L sends a general vector $(x, y, z) \in V$, we have to first solve for c_1, c_2 with

$$(x, y, z) = c_1(-1, 1, 0) + c_2(-1, 0, 1)$$

The solution is $c_1 = y, c_2 = z$. Then

$$L(x, y, z) = y(2, 3) + z(-1, 2) = (2y - z, 3y + 2z)$$ ■

Exercise 2.3.4.

Suppose $L : V = \text{span}((1, 0, 1), (0, 1, 1)) \to \mathbb{R}^4$ is a linear transformation satisfying

$$L(1, 0, 1) = (1, 2, 3, 4), \ L(0, 1, 1) = (0, 1, 0, 1)$$

(a) Write the vector $(2, 3, 5)$ as a linear combination

$$(2, 3, 5) = c_1(1, 0, 1) + c_2(0, 1, 1)$$

and find $L(2, 3, 5)$.

(b) For a general vector $(x, y, z) \in V$, write

$$(x, y, z) = c_1(1, 0, 1) + c_2(0, 1, 1)$$

Solve for c_1, c_2 in terms of x, y, z and then find a formula for $L(x, y, z)$.

Exercise 2.3.5.

Suppose that

$$L : \text{span}((1, 2, 3), (4, 5, 6)) \to \mathbb{R}^4$$

is a linear transformation satisfying

$$L(1, 2, 3) = (1, 2, 3, 4), L(4, 5, 6) = (0, 1, 0, 1)$$

(a) Write the vector $(10, 11, 12)$ as a linear combination

$$(10, 11, 12) = c_1(1, 2, 3) + c_2(4, 5, 6)$$

and find $L(10, 11, 12)$.

(b) For a general vector $(x, y, z) \in V$, write

$$(x, y, z) = c_1(1, 2, 3) + c_2(4, 5, 6)$$

Solve for c_1, c_2 in terms of x, y, z and then find a formula for $L(x, y, z)$.

2.4. ISOMORPHIC VECTOR SPACES AND DIMENSION

Our discussion in the last section on bases and dimension shows that a linear transformation $L : V \to W$ is completely determined by its values on a basis for V. Some linear transformations have the property that a basis $\{\mathbf{v}_1, \ldots, \mathbf{v}_n\}$ of V is sent to a basis $\{\mathbf{w}_1, \ldots, \mathbf{w}_n\}$ of W. When this happens, we can define a linear transformation $M : W \to V$ so that $M(\mathbf{w}_i) = \mathbf{v}_i, i = 1, \ldots, n$. We have $ML(\mathbf{v}_i) = \mathbf{v}_i, LM(\mathbf{w}_i) = \mathbf{w}_i$. Since the linear transformations ML, LM are determined by the values on these bases, $ML(\mathbf{v}) = \mathbf{v}, LM(\mathbf{w}) = \mathbf{w}$.

DEFINITION 2.4.1. A linear transformation $L : V \to W$ is called an **isomorphism** if there is a linear transformation $M : W \to V$ (called the **inverse** of L) which satisfies $ML(\mathbf{v}) = \mathbf{v}$ for all $\mathbf{v} \in V$ and $LM(\mathbf{w}) = \mathbf{w}$ for all $\mathbf{w} \in W$. The vector spaces V and W are called **isomorphic**.

■ ***Example 2.4.1.*** The linear transformation $L : \mathcal{P}^2 \to \mathbb{R}^3$ given by

$$L(a_0 + a_1 x + a_2 x^2) = (a_0, a_1, a_2)$$

which sends a polynomial to its coefficients satisfies

$$L(1) = \mathbf{e}_1, L(x) = \mathbf{e}_2, L(x^2) = \mathbf{e}_3$$

Define a linear transformation $M : \mathbb{R}^3 \to \mathcal{P}^2(\mathbb{R}, \mathbb{R})$ by sending these basis vectors of $\mathbb{R}^3$ back to the original basis for $\mathcal{P}^2(\mathbb{R}, \mathbb{R})$:

$$M(\mathbf{e}_1) = 1, M(\mathbf{e}_2) = x, M(\mathbf{e}_3) = x^2$$

The full formula for M is

$$M(a, b, c) = a + bx + cx^2$$

Then L and M are isomorphisms, and the vector spaces $\mathcal{P}^2$ and $\mathbb{R}^3$ are isomorphic. ■

Our discussion preceding the definition of isomorphism gives the following basic means of recognizing an isomorphism.

PROPOSITION 2.4.1. Let $L : V \to W$ be a linear transformation. Let $\mathcal{V} = \{\mathbf{v}_1, \ldots, \mathbf{v}_n\}$ be a basis for V, and $\mathbf{w}_i = L(\mathbf{v}_i)$. Then L is an isomorphism iff $\mathbf{w}_1, \ldots, \mathbf{w}_n$ is a basis for W.

Proof. We already saw that if $\mathbf{w}_1, \ldots, \mathbf{w}_n$ is a basis for W, then we can define an inverse M for L by the formula $M(\mathbf{w}_i) = \mathbf{v}_i$. Suppose that we know L is an isomorphism and has inverse M. We need to show that $\mathbf{w}_1 = L(\mathbf{v}_1), \ldots, \mathbf{w}_n = L(\mathbf{v}_n)$ is a basis for W. We first show that they are independent. Suppose

$$c_1\mathbf{w}_1 + \cdots + c_n\mathbf{w}_n = \mathbf{0}$$

Then

$$c_1M(\mathbf{w}_1) + \cdots + c_nM(\mathbf{w}_n) = M(c_1\mathbf{w}_1 + \cdots + c_n\mathbf{w}_n) = M(\mathbf{0}) = \mathbf{0}$$

But $ML(\mathbf{v}) = \mathbf{v}$ implies $M(\mathbf{w}_i) = ML(\mathbf{v}_i) = \mathbf{v}_i$. Hence

$$c_1\mathbf{v}_1 + \cdots + c_n\mathbf{v}_n = c_1M(\mathbf{w}_1) + \cdots + c_nM(\mathbf{w}_n) = \mathbf{0}$$

and the independence of $\mathbf{v}_1, \ldots, \mathbf{v}_n$ implies $c_1 = \cdots = c_n = 0$. We next need to show that $\mathbf{w}_1, \ldots, \mathbf{w}_n$ span W. Let $\mathbf{w} \in W$. Then $M(\mathbf{w}) = c_1\mathbf{v}_1 + \cdots + c_n\mathbf{v}_n$ for some $c_1, \ldots, c_n$ since $\mathbf{v}_1, \ldots, \mathbf{v}_n$ span V. Then

$$\begin{aligned} \mathbf{w} &= LM(\mathbf{w}) = L(c_1\mathbf{v}_1 + \cdots + c_n\mathbf{v}_n) \\ &= c_1L(\mathbf{v}_1) + \cdots + c_nL(\mathbf{v}_n) = c_1\mathbf{w}_1 + \cdots + c_n\mathbf{w}_n \end{aligned}$$ ■

An informal statement of the proposition is that an isomorphism is a linear transformation which sends a basis to a basis.

COROLLARY 2.4.2. If V and W are finite dimensional vector spaces, then they are isomorphic iff they have the same dimension.

Proof. Since an isomorphism $L : V \to W$ sends a basis of V to a basis of W, these two bases will have the same number of vectors, and so V and W will have the same dimension. On the other hand, if V and W have the same dimension, let $\mathbf{v}_1, \ldots, \mathbf{v}_n$ be a basis for V and let $\mathbf{w}_1, \ldots, \mathbf{w}_n$ be a basis for W. Then define a linear transformation $L : V \to W$ by defining $L(\mathbf{v}_i) = \mathbf{w}_i, i = 1, \ldots, n$. Since L will send a basis to a basis, it is an isomorphism and so V and W are isomorphic. ■

As a special case of the corollary, any vector space of dimension n is isomorphic to $\mathbb{R}^n$. There is a special form of isomorphism which will be quite useful to us in studying an n-dimensional vector space. It just comes from sending a given basis of V to the standard basis of $\mathbb{R}^n$ and is what was used in Example 2.4.1. Suppose that $\mathcal{V} = \{\mathbf{v}_1, \ldots, \mathbf{v}_n\}$ is a basis for V. Then the linear transformation $L : V \to \mathbb{R}^n$ defined by $L(\mathbf{v}_i) = \mathbf{e}_i$ is an isomorphism. For a general vector $\mathbf{v} \in V$, we first write $\mathbf{v} = c_1\mathbf{v}_1 + \cdots + c_n\mathbf{v}_n$ and then

$$L(\mathbf{v}) = L(c_1\mathbf{v}_1 + \cdots + c_n\mathbf{v}_n) = c_1\mathbf{e}_1 + \cdots + c_n\mathbf{e}_n = (c_1, \ldots, c_n)$$

The inverse M is given by

$$M(c_1, \ldots, c_n) = c_1\mathbf{v}_1 + \cdots + c_n\mathbf{v}_n$$

We will describe this isomorphism L by saying it sends a vector to its coordinates with respect to the basis $\mathcal{V}$. We will study it further in Section 2.7. It will allow us to transfer problems in a general vector space back to problems in $\mathbb{R}^n$ where we can use matrix techniques.

The relation between vector spaces of being isomorphic is an equivalence relation. This means that it satisfies three properties:

1. *Reflexivity: V is isomorphic to V.* An isomorphism is just given by the identity linear transformation in each direction.
2. *Symmetry: V isomorphic to W implies W is isomorphic to V.* V isomorphic to W means that there exist linear transformations $L : V \to W$ and $M : W \to V$ so that $ML(\mathbf{v}) = \mathbf{v}$ and $LM(\mathbf{w}) = \mathbf{w}$. But M and L then may be used to show that W is isomorphic to V.
3. *Transitivity: V isomorphic to W and W isomorphic to X implies that V is isomorphic to X.* V isomorphic to W implies that there are linear transformations $L : V \to W$ and $M : W \to V$ with $LM(\mathbf{w}) = \mathbf{w}$ and $ML(\mathbf{v}) = \mathbf{v}$. W isomorphic to X implies that there exist linear transformations $N : W \to X$ and $P : X \to W$ with $NP(\mathbf{x}) = \mathbf{x}$ and $PN(\mathbf{w}) = \mathbf{w}$. Then the compositions $NL : V \to X$ and $MP : X \to V$ will be linear transformations so that $(MP)(NL)(\mathbf{v}) = M(PN)(L(\mathbf{v})) = M(L(\mathbf{v})) = ML(\mathbf{v}) = \mathbf{v}$ and $(NL)(MP)(\mathbf{x}) = N(LM)(P(\mathbf{x})) = N(P(\mathbf{x})) = NP(\mathbf{x}) = \mathbf{x}$. Thus V and X are isomorphic.

Let us analyze the definition of an isomorphism some more. One necessary property of $L : V \to W$ is that it is a **bijection**. This means

1. L maps V **onto** W (each vector $\mathbf{w} \in W$ is $L(\mathbf{v})$ for some $\mathbf{v} \in V$). This is true for an isomorphism because $L(M(\mathbf{w})) = \mathbf{w}$.

2. L is $\mathbf{1}-\mathbf{1}$, which means that $L(\mathbf{v}) = L(\mathbf{v}')$ implies that $\mathbf{v} = \mathbf{v}'$. This follows for an isomorphism because

$$L(\mathbf{v}) = L(\mathbf{v}') \text{ implies } \mathbf{v} = ML(\mathbf{v}) = ML(\mathbf{v}') = \mathbf{v}'$$

In general, L being a bijection is equivalent to there being a function $M : W \to V$ which satisfies $ML(\mathbf{v}) = \mathbf{v}$ and $LM(\mathbf{w}) = \mathbf{w}$, which is then called the inverse of L. However, it turns out that M will have to be a linear transformation just because L is. For suppose L is a linear transformation and M satisfies the foregoing properties. Then if $\mathbf{w}, \mathbf{w}' \in W$ and $a, b \in \mathbb{R}$, we write $M(\mathbf{w}) = \mathbf{v}$ and $M(\mathbf{w}') = \mathbf{v}'$. Then $L(\mathbf{v}) = \mathbf{w}$ and $L(\mathbf{v}') = \mathbf{w}'$. But L linear implies that

$$L(a\mathbf{v} + b\mathbf{v}') = aL(\mathbf{v}) + bL(\mathbf{v}') = a\mathbf{w} + b\mathbf{w}'$$

Then $ML(a\mathbf{v} + b\mathbf{v}') = M(a\mathbf{w} + b\mathbf{w}')$ and

$$ML(a\mathbf{v} + b\mathbf{v}') = a\mathbf{v} + b\mathbf{v}' = aM(\mathbf{w}) + bM(\mathbf{w}')$$

and so M is linear. Thus to check that a linear transformation $L : V \to W$ is an isomorphism it is enough to know that L is a bijection.

We now introduce a related concept, that of the null space.

DEFINITION 2.4.2. If $L : V \to W$ is a linear transformation, then the null space $\mathcal{N}(L) = \{\mathbf{v} \in V : L(\mathbf{v}) = \mathbf{0}\}$.

First, note that when L is just multiplication by a matrix A, then $\mathcal{N}(L) = \mathcal{N}(A)$. Next, note that $\mathbf{0} \in \mathcal{N}(L)$ since $L(\mathbf{0}) = \mathbf{0}$. If L is a $1-1$, then $L(\mathbf{v}) = \mathbf{0} = L(\mathbf{0})$ implies $\mathbf{v} = 0$ and so $\mathcal{N}(L) = \{\mathbf{0}\}$.

We next claim that we need only to know that the null space $\mathcal{N}(L) = \{\mathbf{0}\}$ to see that L is $1-1$. For if $L(\mathbf{v}) = L(\mathbf{v}')$ and $\mathcal{N}(L) = \{\mathbf{0}\}$, then $L(\mathbf{v} - \mathbf{v}') = \mathbf{0}$ and so $\mathbf{v} - \mathbf{v}' = \mathbf{0}$ and $\mathbf{v} = \mathbf{v}'$.

Finally, consider a linear transformation $L : V \to W$, where the dimension of V equals the dimension of W. If we start with a basis of V and look at the image vectors, then they will be independent iff they span W. L is $1-1$ implies that they are independent. On the other hand, L is onto implies that they span W. Thus we conclude that L is an isomorphism if either L is $1-1$ or is onto.

We have shown the following facts.

PROPOSITION 2.4.3.

1. If $\dim V = \dim W = n$, then V and W are isomorphic to each other. In particular, each is isomorphic to $\mathbb{R}^n$. Isomorphism is an equivalence relation on the set of vector spaces. Any two finite dimensional vector spaces are isomorphic iff they have the same dimension.
2. If $L : V \to W$ is a linear transformation which is a bijection, then L is an isomorphism.

3. If $L : V \to W$ is a linear transformation with null space $\{\mathbf{0}\}$, then L is $1-1$.
4. A linear transformation $L : V \to W$ is an isomorphism iff it sends a basis of V to a basis of W.
5. If $\dim V = \dim W$ and $L : V \to W$ is a linear transformation, then L is an isomorphism if L is $1-1$ OR L is onto.

■ ***Example 2.4.2.*** Consider the space T of 3 by 3 upper triangular matrices, and the space S of 3 by 3 symmetric matrices. Each is six dimensional, so they are isomorphic. An explicit isomorphism is given by

$$L\begin{pmatrix} a & b & c \\ 0 & d & e \\ 0 & 0 & f \end{pmatrix} = \begin{pmatrix} a & b & c \\ b & d & e \\ c & e & f \end{pmatrix}$$

Note that each space is isomorphic to $\mathbb{R}^6$. ■

■ ***Example 2.4.3.*** Consider the linear transformation

$$L : R^4 \to \mathcal{M}(2,2), L(x_1, x_2, x_3, x_4) = \begin{pmatrix} x_1 & x_1 - x_2 \\ x_1 + 2x_3 & x_1 + x_2 - x_4 \end{pmatrix}$$

To show that this is an isomorphism it suffices to show that $\mathcal{N}(L) = \{\mathbf{0}\}$. But $L(\mathbf{x}) = \mathbf{0}$ means $A\mathbf{x} = \mathbf{0}$ where $A = \begin{pmatrix} 1 & 0 & 0 & 0 \\ 1 & -1 & 0 & 0 \\ 1 & 0 & 2 & 0 \\ 1 & 1 & 0 & -1 \end{pmatrix}$. Since $\det A = 2$, A is invertible, and so the only solution is $\mathbf{x} = \mathbf{0}$. ■

Exercise 2.4.1.

Consider the linear transformation $L : \mathbb{R}^3 \to S$, where S is the vector space of symmetric 2 by 2 matrices. L is defined by $L(x, y, z) = \begin{pmatrix} x & y \\ y & z \end{pmatrix}$. Verify that L is a vector space isomorphism.

Exercise 2.4.2.

Show that if $L : V \to W$ satisfies $L(\mathbf{v}_i) = \mathbf{w}_i, i = 1, \ldots, n$, and the $\mathbf{w}_i$ are independent, then the $\mathbf{v}_i$ are independent.

Exercise 2.4.3.

Can a line in $\mathbb{R}^3$ be isomorphic to a plane in $\mathbb{R}^3$? Justify your answer.

Exercise 2.4.4.

Suppose $\mathbf{v}_1, \ldots, \mathbf{v}_n$ and $\mathbf{w}_1, \ldots, \mathbf{w}_n$ are both bases for a vector space V. Show that there is an n by n matrix A with ij-entry a_{ij} so that $\mathbf{v}_j = \sum_{j=1}^{n} a_{ij}\mathbf{w}_i$ and an n by n matrix B with pm-entry b_{pm} with $\mathbf{w}_m = \sum_{p=1}^{n} b_{pm}\mathbf{v}_p$. Show that A and B are inverses of each other.

Exercise 2.4.5.

Suppose that $L : V \to W$ is a linear transformation. Let $\mathbf{v}_1, \ldots, \mathbf{v}_n$ be a basis for V and $\mathbf{w}_i = L(\mathbf{v}_i)$. Show that if L is $1-1$, then the $\mathbf{w}_i$ are independent and if L is onto, then the $\mathbf{w}_i$ span W.

Exercise 2.4.6.

Show that if S is a subspace of T of the same dimension, they are equal. (Hint: Start with a basis of S and show that it is also a basis for T.)

2.5. LINEAR TRANSFORMATIONS AND SUBSPACES

We now want to apply some of the ideas developed in the previous sections to study certain subspaces associated to linear transformations. By restricting ourselves to those linear transformations from $\mathbb{R}^n$ to $\mathbb{R}^m$ which arise from multiplication by a matrix A, we will also study the relationship between the row space and column spaces of A and the null spaces of A and A^t. Recall that in the last section we defined the null space of a linear transformation. We now define the range of a linear transformation.

DEFINITION 2.5.1. Let $L : V \to W$ be a linear transformation. Associated with L is the **range** of L, $\mathcal{R}(L) = \{\mathbf{w} \in W : L(\mathbf{v}) = \mathbf{w} \text{ for some } \mathbf{v} \in V\}$.

When L arises by multiplying by a matrix A, the range of L is what we called $\mathcal{R}(A)$ previously.

DEFINITION 2.5.2. We define the **nullity** $n(L)$ to be the dimension of $\mathcal{N}(L)$ and the **rank** $r(L)$ to be the dimension of $\mathcal{R}(L)$.

We now consider the case when $\dim V$ is finite. Since $\mathcal{N}(L) \subset V$, $n(L)$ is finite. Since the range of L is spanned by the image under L of basis vectors for V, then $\mathcal{R}(L)$ is finite dimensional, and so $r(L)$ is also finite. In this situation there is the following theorem, which says that the rank and the nullity must sum up to the dimension of the domain space. In the case of the linear transformation L_A, this turns out to be the statement that the total number n of variables is the sum of the number r of basic variables (which correspond to a basis of $\mathcal{R}(A)$) plus the number of free variables (which correspond to a basis of $\mathcal{N}(A)$).

THEOREM 2.5.1. (RANK–NULLITY THEOREM) Assume $L : V \to W$ is a linear transformation and V is finite dimensional. Then

$$r(L) + n(L) = \dim V$$

Proof. Let $\mathbf{v}_1, \ldots, \mathbf{v}_k$ be a basis for $\mathcal{N}(L)$ and $\mathbf{w}_1, \ldots, \mathbf{w}_r$ be a basis for $\mathcal{R}(L)$. Here $k = n(L)$ and $r = r(L)$. To verify the result we need to find a basis for V which has $k + r$ vectors since any basis will have $\dim V$ vectors. Since $\mathbf{w}_i \in \mathcal{R}(L)$, there are vectors $\mathbf{u}_1, \ldots, \mathbf{u}_r$ in V so that $L(\mathbf{u}_i) = \mathbf{w}_i, i = 1, \ldots, r$. We now claim that $\mathbf{v}_1, \ldots, \mathbf{v}_k, \mathbf{u}_1, \ldots, \mathbf{u}_r$ is a basis for V. We first show that these vectors are independent. Suppose $c_1\mathbf{v}_1 + \cdots + c_k\mathbf{v}_k + d_1\mathbf{u}_1 + \cdots + d_r\mathbf{u}_r = \mathbf{0}$. Applying L to both sides and using the facts that $L(\mathbf{v}_i) = \mathbf{0}$ (since $\mathbf{v}_i \in \mathcal{N}(L)$) and $L(\mathbf{u}_i) = \mathbf{w}_i$ gives $d_1\mathbf{w}_1 + \cdots + d_r\mathbf{w}_r = \mathbf{0}$. But the $\mathbf{w}_i$ are independent since they are a basis for $\mathcal{R}(L)$, and so this implies that $d_i = 0, i = 1, \ldots, r$. This then gives $c_1\mathbf{v}_1 + \cdots + c_k\mathbf{v}_k = \mathbf{0}$. The independence of the $\mathbf{v}_i$ (since they are a basis for $\mathcal{N}(L)$) implies that the $c_i = 0, i = 1, \ldots, k$. Thus the $\mathbf{v}_i$ and $\mathbf{u}_i$ are independent.

We now show that the $\mathbf{v}_i$ and $\mathbf{u}_i$ span V. Let $\mathbf{v} \in V$ be given. Then $L(\mathbf{v})$ is in $\mathcal{R}(L)$ and $\{\mathbf{w}_i, i = 1, \ldots, r\}$ form a basis for $\mathcal{R}(L)$, so there are scalars $d_1, \ldots, d_r$ with $L(\mathbf{v}) = d_1\mathbf{w}_1 + \cdots + d_r\mathbf{w}_r$. We can use the fact that $L(\mathbf{u}_i) = \mathbf{w}_i$ to see that $L(d_1\mathbf{u}_1 + \cdots + d_r\mathbf{u}_r) = d_1\mathbf{w}_1 + \cdots + d_r\mathbf{w}_r = L(\mathbf{v})$. Thus $L(\mathbf{v} - (d_1\mathbf{u}_1 + \cdots + d_r\mathbf{u}_r)) = \mathbf{0}$ and so $\mathbf{v} - (d_1\mathbf{u}_1 + \cdots + d_r\mathbf{u}_r)$ is in $\mathcal{N}(L)$. Since the $\mathbf{v}_i$ give a basis for $\mathcal{N}(L)$, we have $\mathbf{v} - (d_1\mathbf{u}_1 + \cdots + d_r\mathbf{u}_r) = c_1\mathbf{v}_1 + \cdots + c_k\mathbf{v}_k$ for some c_i. But this implies that $\mathbf{v} = c_1\mathbf{v}_1 + \cdots + c_k\mathbf{v}_k + d_1\mathbf{u}_1 + \cdots + d_r\mathbf{u}_r$, which shows that the $\mathbf{u}_i$ and $\mathbf{v}_i$ span V. ■

We now apply Theorem 2.5.1 to the linear transformation $L_A : \mathbb{R}^n \to \mathbb{R}^m$, where L_A is the linear transformation which arises by multiplying by an m by n matrix A. Then $\mathcal{N}(L_A) = \mathcal{N}(A)$ and $\mathcal{R}(L_A) = \mathcal{R}(A)$. There is a solution of $A\mathbf{x} = \mathbf{b}$ exactly when $\mathbf{b}$ is in $\mathcal{R}(A)$. Moreover, the solutions of $A\mathbf{x} = \mathbf{b}$ are of the form $\mathbf{v}_0 + c_1\mathbf{v}_1 + \cdots + c_k\mathbf{v}_k$, where $\mathbf{v}_0$ is a particular solution and $c_1\mathbf{v}_1 + \cdots + c_k\mathbf{v}_k$ denotes the general solution of $A\mathbf{x} = \mathbf{0}$; that is, this linear combination denotes the general vector in the null space $\mathcal{N}(A)$. But this just means that we want $\mathbf{v}_1, \ldots, \mathbf{v}_k$ to be a basis for this null space.

Our earlier discussion in Chapter 1 of solving $A\mathbf{x} = \mathbf{0}$ will now allow us to find a basis for $\mathcal{N}(A)$. We just use Gaussian elimination to reduce A to a matrix R in reduced normal form. Then the algorithm for solving $R\mathbf{x} = \mathbf{b}$ in Section 1.3 tells us how to read off the general solution as a linear combination of $n - r$ solutions, where there is one solution corresponding to each free variable. This implies that these $n - r$ solutions span $\mathcal{N}(A)$. That they are a basis then depends on seeing that they are independent. If we call these solutions $\mathbf{v}_1, \ldots, \mathbf{v}_k, k = n - r$, and denote by $f(i)$ the ith free variable, then they satisfy the property that the entry in the $f(i)$ position of $\mathbf{v}_j$ is 0 unless $i = j$, in which case it is 1. Thus if we have $c_1\mathbf{v}_1 + \cdots + c_k\mathbf{v}_k = \mathbf{0}$, then the entry in the $f(i)$ position is c_i, and so $c_i = 0$ for all i. Thus the dimension of $\mathcal{N}(A)$ is $n - r$, where r is the number of basic variables.

■ ***Example 2.5.1.*** Consider the matrix $A = \begin{pmatrix} 1 & 2 & 3 \\ 4 & 5 & 6 \\ 7 & 8 & 9 \end{pmatrix}$. Then the Gaussian elimination algorithm gives

$$O_1 = \begin{pmatrix} 1 & 0 & 0 \\ -4 & 1 & 0 \\ 1 & -2 & 1 \end{pmatrix}, \quad U = \begin{pmatrix} 1 & 2 & 3 \\ 0 & -3 & -6 \\ 0 & 0 & 0 \end{pmatrix}, \quad R = \begin{pmatrix} 1 & 0 & -1 \\ 0 & 1 & 2 \\ 0 & 0 & 0 \end{pmatrix}$$

From the form of R, we then see that a basis for $\mathcal{N}(A)$ is given by $(1, -2, 1)$. ■

The number r which occurs is also the number of nonzero rows in the reduced normal matrix R (or the matrix U, which has the same number of nonzero rows). In particular, this implies that the number r does not depend on the method used to reduce A to normal form (in particular, whether we use partial pivoting or not) since it arises as $n - \dim \mathcal{N}(A)$. The proposition will now imply that $\dim \mathcal{R}(L) = r$. We now want to describe how to find a basis for $\mathcal{R}(L)$. We look at the matrix U at the end of the forward elimination step and find the columns which contain the pivots. We showed that these were independent. Now we look at the corresponding columns in the matrix A. Each of these columns is in the column space of A, which is $\mathcal{R}(A)$. There are r of them since there are r pivots. Suppose there is a relation $a_1A^{c(1)} + \cdots + a_rA^{c(r)} = \mathbf{0}$. Then this gives a solution of $A\mathbf{x} = \mathbf{0}$, where the entry in the $c(i)$th position is a_i and all other entries which don't correspond to the basic columns are 0. This vector then would be a solution to $U\mathbf{x} = \mathbf{0}$ and so would give a similar relation $a_1U^{c(1)} + \cdots + a_rU^{c(r)} = \mathbf{0}$. Since these columns of U are independent, the $a_i = 0$, which then shows that the corresponding columns of A are independent. Since we have r independent vectors in an r-dimensional vector space $\mathcal{R}(A)$, they must be a basis.

Here is an alternative argument which shows the $A^{c(1)}, \ldots, A^{c(r)}$ are independent. Multiplication by O_1 is an isomorphism from $\mathbb{R}^m$ to itself since O_1 is invertible. It sends the column space of A to the column space of U and restricts to an isomorphism between these column spaces. It also sends $A^{c(1)}, \ldots, A^{c(r)}$ to $U^{c(1)}, \ldots, U^{c(r)}$. Since an isomorphism sends a basis to a basis, the fact that $U^{c(1)}, \ldots, U^{c(r)}$ is a basis of the column space of U implies that $A^{c(1)}, \ldots, A^{c(r)}$ is a basis for the column space of A.

■ ***Example 2.5.2.*** We return to the matrix A in the last example. The columns of U which contained the free variables were the first two. Thus our method says that if we take the first two columns of A, we get a basis for $\mathcal{R}(A)$. This gives the basis $(1, 4, 7), (2, 5, 8)$. Another way to look at this is that once we know that the dimension of the range of A is two, all we have to do is produce two independent vectors in the range of A to have a basis. Our algorithm tells us that the first two columns work, but for this matrix any two columns would also work. ■

We now look at the row space, which is also $\mathcal{R}(A^t)$. Here we claim that the nonzero rows of U form a basis. We know that they are independent. Thus we only need to see that they are in the row space and that they span it. To see that they are in the row space, we consider the process of Gaussian elimination. Each step replaces a row by a (special) linear combination of the other rows. Thus each row in the new matrix after a step lies in the row space of the original matrix. Thus the row space for the new matrix is contained in the row space for the old matrix. Since each step is reversible, this implies that the rows of the old matrix are linear combinations of the new rows. Thus the old row space is contained in the new row space. We conclude that the row space never changes throughout the process of Gaussian elimination, even though the rows do. At the end of the forward elimination algorithm, there will be r nonzero rows and $m - r$ zero rows. Thus the linear combinations of these rows will be the same as the linear combinations of the nonzero rows. But this just means that the nonzero rows of U will span the row space. Thus they provide a basis for the row space. Note that this implies that the dimension of the row space is equal to the dimension of the column space.

DEFINITION 2.5.3. The dimension of the row space of a matrix is called the **row rank** of the matrix, and the dimension of the column space is called the **column rank.** What we have just shown is sometimes expressed by saying that the row rank of a matrix is equal to the column rank. The **rank** of the matrix is then defined to be this common number. It is also the same as the rank of the linear transformation given by multiplying by the matrix.

We could have found another basis for the column space by first taking the transpose of the matrix, perform the forward elimination step on it, and then take the nonzero rows. Our computations have thus shown us that we would have gotten the same number r of nonzero rows as we did with A, even though we would get a different basis for the column space this way. Note also that we could use the matrix R in the argument to get a different basis for the row space.

■ ***Example 2.5.3.*** Continuing with our last example, we could get a basis for the row space by using the nonzero rows of U, which would give $(1, 2, 3), (0, -3, -6)$. Alternatively, we could use the nonzero rows of R, giving the basis $(1, 0, -1), (0, 1, 2)$. Another way to get a basis would be to use the fact that the row space is two dimensional, and so if we choose two independent rows in A, these would give a basis. Here we could choose any two rows of A. If we chose the first two rows, we get the basis $(1, 2, 3), (4, 5, 6)$. ■

We now will describe how to find a basis for the null space of A^t. We first note that using the proposition applied to multiplication by A^t as a linear transformation from $\mathbb{R}^m$ to $\mathbb{R}^n$ together with the dimension of $\mathcal{R}(A^t)$ being equal to the number r of nonzero rows, we get that the dimension of $\mathcal{N}(A^t)$ is equal to $m - r$. Thus we need only to find $m - r$ independent vectors in the null space of A^t. But we have an equation $O_1 A = U$, which when taking transposes gives $A^t O_1^t = U^t$. Since the last $m - r$ columns of U^t

are zero, this means that each of the last $m-r$ columns of O_1^t is in the null space of A^t. But O_1^t is invertible (being the product of elementary matrices), and so its columns are independent (since a dependency relation would give a nontrivial solution to $O_1^t\mathbf{c} = \mathbf{0}$ and an invertible matrix has only the trivial solution). Thus the columns of O_1^t are independent, and so the last $m-r$ columns give $m-r$ independent vectors in the $m-r$-dimensional space $\mathcal{N}(A^t)$. Hence these columns are a basis for $\mathcal{N}(A^t)$.

■ ***Example 2.5.4.*** We continue with our example. Since the last row of the matrix O_1 is $(1 \quad -2 \quad 1)$, our discussion gives that a basis for $\mathcal{N}(A^t)$ is $(1, -2, 1)$. ■

Recall that in Chapter 1 we showed that $A\mathbf{x} = \mathbf{b}$ has a solution iff $\mathbf{b}$ is perpendicular to the last $m-r$ columns of O_1^t. Since these form a basis for $\mathcal{N}(A^t)$, this may be restated as

$$\mathbf{b} \in \mathcal{R}(A) \text{ iff } \langle \mathbf{b}, \mathbf{c} \rangle = 0 \text{ for all } \mathbf{c} \in \mathcal{N}(A^t)$$

Recall that we also showed in Chapter 1 the following fact, which we now restate in terms of $\mathcal{N}(A)$:

$$\mathbf{x} \in \mathcal{N}(A) \text{ iff } \langle \mathbf{x}, \mathbf{y} \rangle = 0 \text{ for every } \mathbf{y} \in \mathcal{R}(A^t)$$

We will pursue each of these facts further in Chapter 3 when we discuss orthogonality more deeply. In the terminology introduced there, $\mathcal{N}(A)$ and $\mathcal{R}(A^t)$ are orthogonal complements in $\mathbb{R}^n$, and $\mathcal{N}(A^t)$ and $\mathcal{R}(A)$ are orthogonal complements in $\mathbb{R}^m$.

DEFINITION 2.5.4. The subspaces

$$\mathcal{N}(A), \mathcal{N}(A^t), \mathcal{R}(A), \mathcal{R}(A^t)$$

are called the **four fundamental subspaces associated with the matrix** A.

The terminology *fundamental subspaces* for these four important subspaces is due to Gilbert Strang in his text [7], where he stresses their importance in solving $A\mathbf{x} = \mathbf{b}$.

We now state the algorithms developed earlier for finding bases for each of the fundamental subspaces.

Algorithms for Bases for the Four Fundamental Subspaces

Suppose A is a m by n matrix. Use the Gaussian elimination algorithm to write $O_1 A = U$ and $OA = R$, where U is the matrix at the end of the forward elimination step and R is the matrix in reduced normal form. Let r denote the number of nonzero rows of U (or R; they are the same). We identify all basis vectors with column vectors, which will at times necessitate taking transposes.

1. A basis for $\mathcal{N}(A)$, which is of dimension $n-r$, is given by the $n-r$ vectors which are found corresponding to the $n-r$ free variables by assigning that free variable to 1 and the other free variables to 0, and solving for the basic variables.

This basis can be read off from the matrix R: for the ith vector, assign the ith free variable to 1 and the other free variables to 0, and fill in the basic variables with the negatives of the entries in the nonzero part of the column of R containing the ith free variable.

2. A basis for $\mathcal{R}(A)$, which is of dimension r, is given by the columns of A corresponding to the basic variables.
3. A basis of $\mathcal{R}(A^t)$, which is of dimension r, is given by the transposes of nonzero rows of U (or of R).
4. A basis of $\mathcal{N}(A^t)$, which is of dimension $m - r$, is given by the transposes of the last $m - r$ rows of O_1 (or of O).

■ ***Example 2.5.5.***

$$A = \begin{pmatrix} 1 & 3 & 2 & 1 \\ 4 & 1 & 0 & 2 \\ 1 & 3 & 2 & 2 \\ 0 & 1 & 0 & 0 \end{pmatrix}$$

Then $U = \begin{pmatrix} 1 & 3 & 2 & 1 \\ 0 & -11 & -8 & -2 \\ 0 & 0 & -.73 & -.18 \\ 0 & 0 & 0 & 1 \end{pmatrix}$. This means that A is invertible, and so $\mathcal{R}(A) = \mathbb{R}^4 = \mathcal{R}(A^t)$ and $\mathcal{N}(A) = \mathcal{N}(A^t) = \{\mathbf{0}\}$. We can then take the standard basis of $\mathbb{R}^4$ for $\mathcal{R}(A)$ and $\mathcal{R}(A^t)$. Any time there are no zero rows of U this means that $m = r$, and so the range of A is all of $\mathbb{R}^m$ and $\mathcal{N}(A^t) = \{\mathbf{0}\}$. Any time there are no free variables this means that $n = r$, and so the row space is $\mathbb{R}^n$ and $\mathcal{N}(A) = \{\mathbf{0}\}$. ■

■ ***Example 2.5.6.*** $A = \begin{pmatrix} 1 & 0 & 2 & 1 & 3 \\ 2 & 4 & 3 & 1 & 3 \\ 0 & 0 & 0 & 0 & 1 \\ 3 & 4 & 5 & 2 & 6 \end{pmatrix}$. Then

$$O_1 = \begin{pmatrix} 1 & 0 & 0 & 0 \\ -2 & 1 & 0 & 0 \\ 0 & 0 & 1 & 0 \\ -1 & -1 & 0 & 1 \end{pmatrix}, U = \begin{pmatrix} 1 & 0 & 2 & 1 & 3 \\ 0 & 4 & -1 & -1 & -3 \\ 0 & 0 & 0 & 0 & 1 \\ 0 & 0 & 0 & 0 & 0 \end{pmatrix}$$

and

$$R = \begin{pmatrix} 1 & 0 & 2 & 1 & 0 \\ 0 & 1 & -1/4 & -1/4 & 0 \\ 0 & 0 & 0 & 0 & 1 \\ 0 & 0 & 0 & 0 & 0 \end{pmatrix}$$

Thus a basis for $\mathcal{N}(A)$ is given by $(-2, 1/4, 1, 0, 0)$ and $(-1, 1/4, 0, 1, 0)$. A basis for $\mathcal{N}(A^t)$ is given by $(-1, -1, 0, 1)$. A basis for $\mathcal{R}(A)$ is given by $(1, 2, 0, 3)$, $(0, 4, 0, 4)$, and $(3, 3, 1, 6)$. A basis for $\mathcal{R}(A^t)$ is given by $(1, 0, 2, 1, 3)$, $(0, 4, -1, -1, -3)$, and $(0, 0, 0, 0, 1)$. ■

■ ***Example 2.5.7.*** We now give an example involving other vector spaces. Consider the linear transformation $D : \mathcal{P}^4(\mathbb{R}, \mathbb{R}) \to \mathcal{P}^4(\mathbb{R}, \mathbb{R})$, where $D(p) = p'$; that is, D just differentiates the polynomial. Since the derivative of a function being zero implies that the function is a constant, we get that the null space of D consists of the constant polynomials. They are all scalar multiples of the polynomial 1, so the dimension of $\mathcal{N}(L)$ is 1 and a basis for $\mathcal{N}(L) = \{c\}$ is given by the polynomial 1. By Theorem 2.5.1, the dimension of the range of D is four. But any polynomial which is the derivative of a polynomial of degree less than or equal to 4 has degree less than or equal to 3. Thus the range of D must be contained within the subspace of polynomials of degree less than or equal to 3, which is $\mathcal{P}^3(\mathbb{R}, \mathbb{R})$. Since the range of D and $\mathcal{P}^3(\mathbb{R}, \mathbb{R})$ have the same dimension, they must be equal (see Exercise 2.4.6). Thus the range of D is $\mathcal{P}^3(\mathbb{R}, \mathbb{R})$. It therefore has a basis $1, x, x^2, x^3$. Each $x^i = D(x^{i+1}/(i+1))$, so is in the range of D as required. ■

In Exercises 2.5.1–2.5.4 find bases for each of the four fundamental subspaces associated with the given matrix.

Exercise 2.5.1.

$$A = \begin{pmatrix} 1 & 2 & 0 \\ 2 & 1 & 1 \\ 0 & 3 & -1 \end{pmatrix}$$

Exercise 2.5.2.

$$A = \begin{pmatrix} 1 & 0 & -1 & 0 \\ 0 & 1 & 1 & 1 \\ 3 & 2 & -1 & 2 \end{pmatrix}$$

Exercise 2.5.3.

$$A = \begin{pmatrix} 1 & 2 & 0 & 2 & 1 \\ 0 & 1 & 1 & 1 & 0 \\ 2 & 1 & 1 & 0 & 1 \\ 0 & 4 & 0 & 5 & 1 \end{pmatrix}$$

Exercise 2.5.4.

$$A = \begin{pmatrix} 1 & -1 & 1 & 0 & 2 & 1 \\ 2 & 1 & 2 & 1 & 2 & 1 \\ 3 & 0 & 3 & 1 & 4 & 2 \\ 1 & 2 & 1 & 1 & 0 & 0 \end{pmatrix}$$

Exercise 2.5.5.

Consider the linear transformation $L : \mathbb{R}^3 \to \mathcal{D}(\mathbb{R}, \mathbb{R})$ given by $L(a_1, a_2, a_3) = a_1 \sin t + a_2 \cos t + a_3 \sin(t + \pi/4)$. Find bases for $\mathcal{N}(L)$ and $\mathcal{R}(L)$.

Exercise 2.5.6.

Suppose that $L : \mathbb{R}^5 \to \mathcal{P}^3(\mathbb{R}, \mathbb{R})$ is a linear transformation which maps onto $\mathcal{P}^3(\mathbb{R}, \mathbb{R})$. What is the dimension of the null space $\mathcal{N}(L)$?

Exercise 2.5.7.

Show that the linear transformation $L : \mathbb{R}^4 \to \mathcal{M}(2, 2)$ given by

$$L(\mathbf{e}_1) = \begin{pmatrix} 1 & 0 \\ 0 & 0 \end{pmatrix}, L(\mathbf{e}_2) = \begin{pmatrix} 1 & 1 \\ 0 & 0 \end{pmatrix}, L(\mathbf{e}_3) = \begin{pmatrix} 1 & 1 \\ 1 & 0 \end{pmatrix}, L(\mathbf{e}_4) = \begin{pmatrix} 1 & 1 \\ 1 & 1 \end{pmatrix}$$

is an isomorphism.

Exercise 2.5.8.

Verify that $\mathcal{R}(L)$ and $\mathcal{N}(L)$ are subspaces for any linear transformation L.

Exercise 2.5.9.

(a) Consider the plane in $\mathbb{R}^3$ with basis $(1, 1, 3), (2, 1, 4)$. Find a vector lying on the perpendicular line, and use it to characterize the plane as the null space of a matrix A; that is, the solutions of $A\mathbf{x} = \mathbf{0}$.

(b) Consider the line in $\mathbb{R}^3$ with basis $(1, 0, -2)$. Find a matrix B so that this line is the null space of B.

Exercise 2.5.10.

(a) Find a basis for the subspace S of $\mathbb{R}^4$ spanned by

$$(1, 0, -1, 1), (2, 1, 1, 0), (0, -1, -3, 2)$$

(b) Find a basis for the subspace of vectors perpendicular to this subspace.

(c) Use the basis in (b) to give a matrix A so that S is given by the null space of A.

2.6. SUBSPACE CONSTRUCTIONS

Suppose that S, T are subspaces of a vector space V.

DEFINITION 2.6.1. The **intersection** of S and T is defined as

$$S \cap T = \{\mathbf{v} : \mathbf{v} \in S \text{ and } \mathbf{v} \in T\}$$

The intersection will also be a subspace of V; it will be contained within S and T. For if $\mathbf{v}, \mathbf{w} \in S \cap T$, then this implies that $\mathbf{v}, \mathbf{w} \in S$ and $\mathbf{v}, \mathbf{w} \in T$. Then any linear combination $a\mathbf{v} + b\mathbf{w}$ will be in both S and T since S and T are subspaces. But this means that $a\mathbf{v} + b\mathbf{w} \in S \cap T$, and so $S \cap T$ is a subspace. Since $S \cap T$ is contained in both S and T, its dimension will be less than or equal to each of the dimensions of S and T. Similarly, the intersection of any number (finite or infinite) of subspaces will be a subspace. When we have an equation $A\mathbf{x} = \mathbf{0}$, we are just looking at the intersection of the m subspaces given by the solutions of $\mathbf{A}_i\mathbf{x} = 0, i = 1, \ldots, m$, corresponding to the m rows of A.

The union of subspaces doesn't turn out to be a subspace in general. However, there is a new subspace called the **sum,** which is the smallest subspace that contains the union.

DEFINITION 2.6.2. $S + T = \{\mathbf{v} \in V : \mathbf{v} = \mathbf{s} + \mathbf{t}, \mathbf{s} \in S, \mathbf{t} \in T\}$. $S + T$ is called the sum of S and T.

We now verify that $S+T$ is a subspace of V. If $\mathbf{v}, \mathbf{w} \in S+T$, then $\mathbf{v} = \mathbf{s}_1 + \mathbf{t}_1, \mathbf{w} = \mathbf{s}_2 + \mathbf{t}_2$. Then $a\mathbf{v} + b\mathbf{w} = a(\mathbf{s}_1 + \mathbf{t}_1) + b(\mathbf{s}_2 + \mathbf{t}_2) = (a\mathbf{s}_1 + b\mathbf{s}_2) + (a\mathbf{t}_1 + b\mathbf{t}_2)$. Since S and T are subspaces, we have $a\mathbf{s}_1 + b\mathbf{s}_2 \in S, a\mathbf{t}_1 + b\mathbf{t}_2 \in T$, so $a\mathbf{v} + b\mathbf{w} \in S + T$. Since $S + T$ contains both S and T as subsets, its dimension will be greater than or equal to $\dim S, \dim T$. Note that any subspace which contains the union $S \cup T$ must contain the sum $S + T$ since the subspace is closed under addition.

We now examine how the dimensions of $S \cap T$ and $S + T$ are related to $\dim S$ and $\dim T$.

PROPOSITION 2.6.1. $\dim S + \dim T = \dim(S \cap T) + \dim(S + T)$.

Before proving this proposition we first discuss an idea we will use in the proof. Suppose V is a vector space of dimension n and $\mathbf{v}_1, \ldots, \mathbf{v}_k$ is an independent set of k vectors in V. Then we can extend $\mathbf{v}_1, \ldots, \mathbf{v}_k$ to a basis of V by adding $n - k$ more vectors. If $k = n$, then the $\mathbf{v}_i$ will already be a basis and there is nothing to do. If not, we show how to add one new vector, so that the old set of k vectors and the new one are still independent. By repeating this procedure $n - k$ times we arrive at an independent set of n vectors, which is then a basis. So suppose that $k < n$. Then these k vectors can't span V (which would give $\dim V = k < n = \dim V$), so there is a vector $\mathbf{v} \in V$

which is not a linear combination of $\mathbf{v}_1, \ldots, \mathbf{v}_k$. Then $\mathbf{v}_1, \ldots, \mathbf{v}_k, \mathbf{v}$ will be independent. If there were a linear combination $a_1\mathbf{v}_1 + \cdots + a_k\mathbf{v}_k + a\mathbf{v} = \mathbf{0}$, then we must have $a = 0$ or we could write $\mathbf{v}$ as a linear combination of the $\mathbf{v}_i$. But then the $a_i = 0$ since the $\mathbf{v}_i$ are independent.

We write this construction down as a lemma.

LEMMA 2.6.2. Let $\mathbf{v}_1, \ldots, \mathbf{v}_k$ be k independent vectors in an n dimensional vector space V. Then we can extend these vectors to a basis $\mathbf{v}_1, \ldots, \mathbf{v}_k, \mathbf{v}_{k+1}, \ldots, \mathbf{v}_n$ for V.

We now use the lemma to prove the theorem.

Proof. Let $\mathbf{v}_1, \ldots, \mathbf{v}_k$ be a basis for $S \cap T$. Extend these k vectors to bases for S, T; that is, find $\mathbf{w}_1, \ldots, \mathbf{w}_m$ so $\mathbf{v}_1, \ldots, \mathbf{v}_k, \mathbf{w}_1, \ldots, \mathbf{w}_m$ is basis for S and find $\mathbf{x}_1, \ldots, \mathbf{x}_p$ so that $\mathbf{v}_1, \ldots, \mathbf{v}_k, \mathbf{x}_1, \ldots, \mathbf{x}_p$ is a basis for T. We now claim that

$$\mathbf{v}_1, \ldots, \mathbf{v}_k, \mathbf{w}_1, \ldots, \mathbf{w}_m, \mathbf{x}_1, \ldots, \mathbf{x}_p$$

is a basis for $S + T$. The result follows easily from this since $\dim S = k + m, \dim T = k + p, \dim S \cap T = k, \dim S + T = k + m + p$. To verify the claim we must show that these vectors are independent and span $S + T$. Suppose

$$a_1\mathbf{v}_1 + \cdots + a_k\mathbf{v}_k + b_1\mathbf{w}_1 + \cdots + b_m\mathbf{w}_m + c_1\mathbf{x}_1 + \cdots + c_p\mathbf{x}_p = \mathbf{0}$$

Rewrite this as

$$a_1\mathbf{v}_1 + \cdots + a_k\mathbf{v}_k + b_1\mathbf{w}_1 + \cdots + b_m\mathbf{w}_m = -(c_1\mathbf{x}_1 + \cdots + c_p\mathbf{x}_p)$$

Since the left-hand side is in S and the right-hand side is in T, both sides must be in $S \cap T$. But a vector in $S \cap T$ can be expressed uniquely as a linear combination of the $\mathbf{v}_i$. Regarding it expressed this way as a vector in S, the coefficients of the $\mathbf{w}_i$ will have to be zero. Thus we must have the $b_i = 0$. Then moving the right-hand side back to the left-hand side and using the independence of the $\mathbf{v}_i$ and the $\mathbf{x}_i$ (which together form a basis for T), we get that the a_i and the c_i also must equal 0. Thus the vectors are independent.

We now want to show that they span $S + T$. Given $\mathbf{v} \in S + T$, then write $\mathbf{v} = \mathbf{s} + \mathbf{t}, \mathbf{s} \in S, \mathbf{t} \in T$. Then $\mathbf{s} = a_1\mathbf{v}_1 + \cdots + a_k\mathbf{v}_k + b_1\mathbf{w}_1 + \cdots + b_m\mathbf{w}_m$ and $\mathbf{t} = d_1\mathbf{v}_1 + \cdots + d_k\mathbf{v}_k + c_1\mathbf{x}_1 + \cdots + c_p\mathbf{x}_p$ implies that

$$\begin{aligned}\mathbf{v} &= \mathbf{s} + \mathbf{t}\\ &= (a_1 + d_1)\mathbf{v}_1 + \cdots + (a_k + d_k)\mathbf{v}_k + b_1\mathbf{w}_1 + \cdots + b_m\mathbf{w}_m + c_1\mathbf{x}_1 + \cdots + c_p\mathbf{x}_p\end{aligned}$$

Thus $\mathbf{v}$ is a linear combination of the $\mathbf{v}_i, \mathbf{w}_i, \mathbf{x}_i$ and so these span $S + T$. ∎

Each vector in $S + T$ is expressible as $\mathbf{s} + \mathbf{t}$ with $\mathbf{s} \in S$ and $\mathbf{t} \in T$. This expression need not be unique in general. In fact, $S \cap T$ just measures the lack of uniqueness. For if $\mathbf{w} \in S \cap T$, then $\mathbf{s}+\mathbf{t} = (\mathbf{s}+\mathbf{w})+(\mathbf{t}-\mathbf{w})$ will express the sum in two different ways as a sum of a vector in S and a vector in T. When $S \cap T = \{\mathbf{0}\}$, then the expression will be unique. For if $\mathbf{s}+\mathbf{t} = \mathbf{s}'+\mathbf{t}'$, then $\mathbf{s}-\mathbf{s}' = \mathbf{t}'-\mathbf{t}$ and so both sides will be in $S \cap T$, and thus equal to $\mathbf{0}$. Thus $\mathbf{s} = \mathbf{s}'$ and $\mathbf{t} = \mathbf{t}'$.

DEFINITION 2.6.3. If $S \cap T = \{\mathbf{0}\}$, we say that the sum $S + T$ is a **direct sum** and write it as $S \oplus T$.

We now look at another way of expressing the above results in terms of linear transformations. Let $\mathbf{s}_1, \ldots, \mathbf{s}_m$ be a basis for S and $\mathbf{t}_1, \ldots, \mathbf{t}_n$ be a basis for T. Then $\mathbf{s}_1, \ldots, \mathbf{s}_m, \mathbf{t}_1, \ldots, \mathbf{t}_n$ will span $S+T$ but won't necessarily be a basis. In fact, the proposition implies that it is a basis exactly when $S \cap T = \{\mathbf{0}\}$ (since then we have $\dim S+T = \dim S + \dim T = m+n$). Define $L : \mathbb{R}^{m+n} \to S+T$ by

$$L(a_1, \ldots, a_m, b_1, \ldots, b_n) = a_1\mathbf{s}_1 + \cdots + a_m\mathbf{s}_m + b_1\mathbf{t}_1 + \cdots + b_n\mathbf{t}_n$$

Then note that since L maps $\mathbb{R}^{m+n}$ onto $S+T$, $\dim \mathcal{N}(L) = m + n - \dim S + T = \dim S \cap T$. We will construct an isomorphism between $\mathcal{N}(L)$ and $S \cap T$. We define a linear transformation $M : \mathcal{N}(L) \to S \cap T$ by

$$M(a_1, \ldots, a_m, b_1, \ldots, b_n) = M(\mathbf{a}, \mathbf{b}) = a_1\mathbf{s}_1 + \cdots + a_m\mathbf{s}_m$$

First, note that $(\mathbf{a}, \mathbf{b}) \in \mathcal{N}(L)$ means that $a_1\mathbf{s}_1 + \cdots + a_m\mathbf{s}_m = -(b_1\mathbf{t}_n + \cdots + b_n\mathbf{t}_n)$ and so is in $S \cap T$. Since $\mathcal{N}(L)$ and $S \cap T$ are of the same dimension, we have to show only that $\mathcal{N}(M) = \mathbf{0}$. If $M(\mathbf{a}, \mathbf{b}) = \mathbf{0}$, then $a_1\mathbf{s}_1 + \cdots + a_m\mathbf{s}_m = \mathbf{0}$, and the independence of the $\mathbf{s}_i$ implies that $\mathbf{a} = \mathbf{0}$. Then $(\mathbf{a}, \mathbf{b}) \in \mathcal{N}(L)$ implies $b_1\mathbf{t}_1 + \cdots + b_n\mathbf{t}_n = \mathbf{0}$, and so the independence of the $\mathbf{t}_i$ implies $\mathbf{b} = \mathbf{0}$. Thus the null space of M is $\{\mathbf{0}\}$, and so M is an isomorphism.

This last result gives us a means to find a basis for $S \cap T$ when we have bases for S and T. We can find a basis for $\mathcal{N}(L)$ and then take the image of this basis under the isomorphism M, using the fact that an isomorphism sends a basis to a basis. Let us apply this to subspaces of $\mathbb{R}^m$. Given bases for S and T, we can write them as the columns of a matrix A; the columns will give a spanning set for $S+T$. To find a basis for $S+T$ then just depends on using our algorithm for finding a basis for the column space of A. To find a basis for $S \cap T$ we first use our algorithm to find a basis of $\mathcal{N}(L)$, which is just $\mathcal{N}(A)$ the way that we have set A up, and then take the images of these basis vectors using the linear transformation M. We now give some examples.

■ ***Example 2.6.1.*** Let S be the plane in $\mathbb{R}^4$ with basis $(1, 0, 0, -1)$ and $(1, 2, -1, 0)$, and let T be the plane with basis $(2, 1, 0, 1), (1, -2, 1, -2)$. Then $S+T$ will be spanned by these four vectors and will be the column space of the matrix

$$A = \begin{pmatrix} 1 & 1 & 2 & 1 \\ 0 & 2 & 1 & -2 \\ 0 & -1 & 0 & 1 \\ -1 & 0 & 1 & -2 \end{pmatrix}$$

Since A reduces to

$$R = \begin{pmatrix} 1 & 0 & 0 & 2 \\ 0 & 1 & 0 & -1 \\ 0 & 0 & 1 & 0 \\ 0 & 0 & 0 & 0 \end{pmatrix}$$

we get a basis for $\mathcal{N}(A)$ given by $(-2, 1, 0, 1)$, and M sends this vector to

$$-2(1, 0, 0, -1) + (1, 2, -1, 0) = (-1, 2, -1, 2)$$

which is a basis for $S \cap T$. A basis for $S + T$ is given by

$$\{(1, 0, 0, -1), (1, 2, -1, 0), (2, 1, 0, 1)\}$$

■

■ ***Example 2.6.2.*** Let S, T be planes in $\mathbb{R}^3$ given by the equations $x - y + z = 0$ and $x + 2y - z = 0$. These planes must be identical or intersect in a line. A basis for S is given by $(1, 1, 0), (-1, 0, 1)$ and a basis of T is given by $(-2, 1, 0), (1, 0, 1)$. Since the matrix $A = \begin{pmatrix} 1 & -1 & -2 & 1 \\ 1 & 0 & 1 & 0 \\ 0 & 1 & 0 & 1 \end{pmatrix}$ has rank 3, then $S + T = \mathbb{R}^3$. The null space of this matrix has basis $(-2/3, -1, 2/3, 1)$, and so a basis for the intersection $S \cap T$ is given by $(2/3, 2/3, 0) + (-1, 0, 1) = (-1/3, 2/3, 1)$. This vector could have been found geometrically by taking the cross product of the two normal vectors. It being nonzero implies that $\dim S + T = 3$, which means that it is all of $\mathbb{R}^3$. ■

■ ***Example 2.6.3.*** Let S be the subspace of polynomials $\mathcal{P}^3(\mathbb{R}, \mathbb{R})$ spanned by $1 + x, 2 - x^2 + x^3$, and let T be the subspace spanned by $3 + x^3, x - x^2$. To find bases for $S + T$ and $S \cap T$, we can use the standard isomorphism between $\mathcal{P}^3(\mathbb{R}, \mathbb{R})$ and $\mathbb{R}^4$ to convert this problem to one in $\mathbb{R}^4$. The corresponding matrix whose columns span the image of $S + T$ is $A = \begin{pmatrix} 1 & 2 & 3 & 0 \\ 1 & 0 & 0 & 1 \\ 0 & -1 & 0 & -1 \\ 0 & 1 & 1 & 0 \end{pmatrix}$. The first three columns are a basis for the sum; hence this gives a basis $1 + x, 2 - x^2 + x^3, 3 + x^3$ for $S + T$. The null space of this matrix has a basis given by $(-1, -1, 1, 1)$. Thus a basis for $S \cap T$ is given by $3 + x - x^2 + x^3$. ■

Exercise 2.6.1.

Consider the vectors $(1, 2, 0)$ and $(2, 1, 3)$. Show they are independent. Find a third vector $\mathbf{v}$ so that the three vectors form a basis for $\mathbb{R}^3$. (Hint: Try one of the standard basis vectors.)

Exercise 2.6.2.

Let S be the subspace of $\mathcal{M}(3,3)$ of symmetric matrices and T the subspace of upper triangular matrices. Find bases for $S, T, S \cap T, S + T$.

Exercise 2.6.3.

Let S be the plane in $\mathbb{R}^3$ given as $-x + y + 2z = 0$. Let T be the perpendicular line $c(-1, 1, 2)$. Show that there is a direct sum splitting $S \oplus T = \mathbb{R}^3$ so that each point in $\mathbb{R}^3$ can be written uniquely as a sum of a vector in S and a vector in T. (Hint: Find a basis for S and then use this with $(-1, 1, 2)$ to get a basis for $\mathbb{R}^3$.)

Exercise 2.6.4.

Let S be the subspace of $\mathbb{R}^3$ with basis $(1, 0, -1)$, $(2, 1, 0)$ and T the subspace with basis $(2, 0, 1)$, $(1, 1, -2)$. Find a basis for $S \cap T$. Show that $S + T = \mathbb{R}^3$.

Exercise 2.6.5.

Let S be the subspace of polynomials $\mathcal{P}^3(\mathbb{R}, \mathbb{R})$ spanned by $1 - x + x^2$, $x - x^2 + x^3$ and let T be the subspace spanned by $1 + x$, $x + x^2$, $x^2 + x^3$. Find bases for $S + T$ and $S \cap T$.

2.7. LINEAR TRANSFORMATIONS AND MATRICES

We saw earlier that there was a close connection between linear transformations $L : \mathbb{R}^n \to \mathbb{R}^m$ and m by n matrices. Each such matrix A determines a linear transformation by the formula $L_A(\mathbf{x}) = A\mathbf{x}$. Conversely, each linear transformation L comes from a matrix A as was the case earlier; the jth column of A is just $L(\mathbf{e}_j)$. We now want to generalize this relationship for arbitrary linear transformations. We will also reinterpret the relationship even when the vector spaces involved are $\mathbb{R}^n, \mathbb{R}^m$ in terms of using other bases besides the standard basis. Suppose that V is a vector space of dimension n, W is a vector space of dimension m, and $L : V \to W$ is a linear transformation. Let $\mathbf{v}_1, \ldots, \mathbf{v}_n$ be a basis for V and $\mathbf{w}_1, \ldots, \mathbf{w}_m$ a basis for W. Since L is a linear transformation, the value $L(\mathbf{v})$ on an arbitrary vector $\mathbf{v}$ is determined by the values $L(\mathbf{v}_j)$ on the basis vectors. For $\mathbf{v} = c_1\mathbf{v}_1 + \cdots + c_n\mathbf{v}_n$ with uniquely defined c_j and so

$$L(\mathbf{v}) = L(c_1\mathbf{v}_1 + \cdots + c_n\mathbf{v}_n) = c_1L(\mathbf{v}_1) + \cdots + c_nL(\mathbf{v}_n)$$

Since $\mathbf{w}_1, \ldots, \mathbf{w}_m$ is a basis for W, we can write $L(\mathbf{v}_j) = \sum_{i=1}^{m} a_{ij}\mathbf{w}_i$ for unique a_{ij}.

DEFINITION 2.7.1. The matrix A whose entries a_{ij} satisfy

$$L(\mathbf{v}_j) = \sum_{i=1}^{m} a_{ij}\mathbf{w}_i$$

is called the **matrix that represents L with respect to the bases $\mathcal{V}$ and $\mathcal{W}$**. Its jth column is just the coefficients of $L(\mathbf{v}_j)$ with respect to the basis $\mathbf{w}_1, \ldots, \mathbf{w}_m$. We will denote it by $A = [L]^{\mathcal{V}}_{\mathcal{W}}$, where $\mathcal{V}$ denotes the chosen basis for V and $\mathcal{W}$ denotes the chosen basis for W. When the bases are understood and fixed, we will sometime suppress them in the notation and write this matrix as $[L]$.

We will label the columns of this matrix by the image vectors $L(\mathbf{v}_1), \ldots, L(\mathbf{v}_n)$ and the rows by the basis $\mathbf{w}_1, \ldots, \mathbf{w}_m$. We then represent this information schematically by

$$\begin{array}{c} \\ \mathbf{w}_1 \\ \cdots \\ \mathbf{w}_i \\ \cdots \\ \mathbf{w}_m \end{array} \begin{array}{c} \begin{array}{ccccc} L(\mathbf{v}_1) & \ldots & L(\mathbf{v}_j) & \ldots & L(\mathbf{v}_n) \end{array} \\ \begin{pmatrix} a_{11} & \ldots & a_{1j} & \ldots & a_{1n} \\ \ldots & \ldots & \ldots & \ldots & \ldots \\ a_{i1} & \ldots & a_{ij} & \ldots & a_{in} \\ \ldots & \ldots & \ldots & \ldots & \ldots \\ a_{m1} & \ldots & a_{mj} & \ldots & a_{mn} \end{pmatrix} \end{array}$$

For example, if $L(\mathbf{v}_1) = 2\mathbf{w}_1 - \mathbf{w}_2 + 3\mathbf{w}_3, L(\mathbf{v}_2) = -3\mathbf{w}_1 + \mathbf{w}_2 + \mathbf{w}_3$, then we can represent this information as

$$\begin{array}{c} \\ \mathbf{w}_1 \\ \mathbf{w}_2 \\ \mathbf{w}_3 \end{array} \begin{array}{c} \begin{array}{cc} L(\mathbf{v}_1) & L(\mathbf{v}_2) \end{array} \\ \begin{pmatrix} 2 & -3 \\ -1 & 1 \\ 3 & 1 \end{pmatrix} \end{array}$$

Thus the matrix

$$[L]^{\mathcal{V}}_{\mathcal{W}} = \begin{pmatrix} 2 & -3 \\ -1 & 1 \\ 3 & 1 \end{pmatrix}$$

Let us see that this concept is in fact a generalization from what happens in $\mathbb{R}^n$ and $\mathbb{R}^m$ when we multiply by a matrix A. Let $L_A : \mathbb{R}^n \to \mathbb{R}^m$ be given by $L_A(\mathbf{x}) = A\mathbf{x}$ for some m by n matrix A. Let $\mathcal{E}_n = \{\mathbf{e}_1, \ldots, \mathbf{e}_n\}$ be the standard basis for $\mathbb{R}^n$ and $\mathcal{E}_m = \{\mathbf{e}_1, \ldots, \mathbf{e}_m\}$ be the standard basis for $\mathbb{R}^m$. Note that we are using the usual abuse of notation by denoting the standard basis elements in $\mathbb{R}^n$ and $\mathbb{R}^m$ by the same symbols. Then

$$L_A(\mathbf{e}_j) = A\mathbf{e}_j = \mathbf{A}^j = \sum_{i=1}^{m} a_{ij}\mathbf{e}_i$$

Hence $[L_A]^{\mathcal{E}_n}_{\mathcal{E}_m} = A$, and we depict this by

$$\begin{array}{c} \\ \mathbf{e}_1 \\ \cdots \\ \mathbf{e}_i \\ \cdots \\ \mathbf{e}_m \end{array} \begin{array}{c} \begin{array}{ccccc} L(\mathbf{e}_1) & \dots & L(\mathbf{e}_j) & \dots & L(\mathbf{e}_n) \end{array} \\ \begin{pmatrix} a_{11} & \dots & a_{1j} & \dots & a_{1n} \\ \dots & \dots & \dots & \dots & \dots \\ a_{i1} & \dots & a_{ij} & \dots & a_{in} \\ \dots & \dots & \dots & \dots & \dots \\ a_{m1} & \dots & a_{mj} & \dots & a_{mn} \end{pmatrix} \end{array}$$

We now introduce some alternate notation which is quite useful computationally. Given a collection of n vectors $\{\mathbf{u}_1, \dots, \mathbf{u}_n\}$ in a vector space Z, we can form a row vector $\mathbf{U} = (\mathbf{u}_1 \quad \dots \quad \mathbf{u}_n)$ whose entries are the given vectors. Now if L is a linear transformation defined on Z, we can apply L to each vector $\mathbf{u}_i$ and form a new row vector $(L(\mathbf{u}_1) \quad \dots \quad L(\mathbf{u}_n))$, which we will denote by $L(\mathbf{U})$. Given a row vector $\mathbf{U}$, we can multiply it by an n by k matrix A using the usual rules of matrix multiplication to get another row vector:

$$\mathbf{U}A = (\mathbf{U}\mathbf{A}^1 \quad \dots \quad \mathbf{U}\mathbf{A}^k)$$

where

$$\mathbf{U}\mathbf{A}^j = a_{1j}\mathbf{u}_1 + \cdots + a_{nj}\mathbf{u}_n = \sum_{i=1}^{n} a_{ij}\mathbf{u}_i$$

Now suppose $\mathcal{V} = \{\mathbf{v}_1, \dots, \mathbf{v}_n\}$ is a basis for V and $\mathcal{W} = \{\mathbf{w}_1, \dots, \mathbf{w}_m\}$ is a basis for W. We will have the corresponding row vectors $\mathbf{V} = (\mathbf{v}_1 \quad \dots \quad \mathbf{v}_n), \mathbf{W} = (\mathbf{w}_1 \quad \dots \quad \mathbf{w}_m)$. Then the equations

$$L(\mathbf{v}_j) = \sum_{i=1}^{m} a_{ij}\mathbf{w}_i, A = [L]_{\mathcal{W}}^{\mathcal{V}}$$

can be expressed by a single equation

$$L(\mathbf{V}) = \mathbf{W}A$$

For the jth entry of the left-hand side is $L(\mathbf{v}_j)$ and the jth entry of the right-hand side is found by multiplying the row vector $\mathbf{W}$ by the jth column of A to be $\sum_{i=1}^{m} a_{ij}\mathbf{w}_i$.

We will leave it as an exercise to show that the multiplication of these row vectors of vectors by matrices is associative:

$$(\mathbf{W}A)B = ((\mathbf{w}_1 \quad \dots \quad \mathbf{w}_m)A)B = (\mathbf{w}_1 \quad \dots \quad \mathbf{w}_m)(AB) = \mathbf{W}(AB)$$

L being a linear transformation will imply that if

$$\mathbf{V} = (\mathbf{v}_1 \quad \dots \quad \mathbf{v}_n) = (\mathbf{u}_1 \quad \dots \quad \mathbf{u}_n)C = \mathbf{U}C$$

then

$$L(\mathbf{V}) = (L(\mathbf{v}_1) \quad \dots \quad L(\mathbf{v}_n)) = (L(\mathbf{u}_1) \quad \dots \quad L(\mathbf{u}_n))C = L(\mathbf{U})C$$

This is also left as an exercise and follows from the fact that L sends a linear combination to a corresponding linear combination.

Note that the resulting matrix is dependent on the bases chosen; we will investigate this dependence shortly. Note also that our procedure for finding $[L]^{\mathcal{V}}_{\mathcal{W}}$ is the same procedure which we were using for linear transformations from $\mathbb{R}^n$ to $\mathbb{R}^m$ with the additional restriction that we chose the standard bases $\mathbf{e}_1, \ldots, \mathbf{e}_n$ for $\mathbb{R}^n$ and $\mathbf{e}_1, \ldots, \mathbf{e}_m$ for $\mathbb{R}^m$. We now look at a few examples.

■ ***Example 2.7.1.*** Let $L : \mathbb{R}^3 \to \mathbb{R}^3$ be the linear transformation which projects onto the xy-plane. Then $L(\mathbf{e}_1) = \mathbf{e}_1 = 1 \cdot \mathbf{e}_1 + 0 \cdot \mathbf{e}_2 + 0 \cdot \mathbf{e}_3$, $L(\mathbf{e}_2) = \mathbf{e}_2 = 0 \cdot \mathbf{e}_1 + 1 \cdot \mathbf{e}_2 + 0 \cdot \mathbf{e}_3$, and $L(\mathbf{e}_3) = \mathbf{0} = 0 \cdot \mathbf{e}_1 + 0 \cdot \mathbf{e}_2 + 0 \cdot \mathbf{e}_3$. We depict this as

$$\begin{array}{c} \\ \mathbf{e}_1 \\ \mathbf{e}_2 \\ \mathbf{e}_3 \end{array} \begin{array}{ccc} L(\mathbf{e}_1) & L(\mathbf{e}_2) & L(\mathbf{e}_3) \\ \end{array}$$
$$\begin{array}{c} \mathbf{e}_1 \\ \mathbf{e}_2 \\ \mathbf{e}_3 \end{array} \begin{pmatrix} 1 & 0 & 0 \\ 0 & 1 & 0 \\ 0 & 0 & 0 \end{pmatrix}$$

Thus the matrix $[L]^{\mathcal{E}}_{\mathcal{E}}$ which represents L with respect to the standard basis $\mathcal{E}$ in both copies of $\mathbb{R}^3$ is

$$A = \begin{pmatrix} 1 & 0 & 0 \\ 0 & 1 & 0 \\ 0 & 0 & 0 \end{pmatrix}$$ ■

■ ***Example 2.7.2.*** Let $L : \mathbb{R}^3 \to \mathbb{R}^3$ be the linear transformation which projects $\mathbb{R}^3$ onto the plane $x + y + z = 0$. Note that $(1, 1, 1)$ is the normal vector to this plane and a basis for the plane is given by $(-1, 1, 0)$ and $(-1, 0, 1)$. If we use the bases $\mathbf{v}_1 = \mathbf{w}_1 = (-1, 1, 0)$, $\mathbf{v}_2 = \mathbf{w}_2 = (-1, 0, 1)$, $\mathbf{v}_3 = \mathbf{w}_3 = (1, 1, 1)$, then $L(\mathbf{v}_1) = \mathbf{v}_1 = 1 \cdot \mathbf{v}_1 + 0 \cdot \mathbf{v}_2 + 0 \cdot \mathbf{v}_3$, $L(\mathbf{v}_2) = \mathbf{v}_2 = 0 \cdot \mathbf{v}_1 + 1 \cdot \mathbf{v}_2 + 0 \cdot \mathbf{v}_3$, $L(\mathbf{v}_3) = \mathbf{0} = 0 \cdot \mathbf{v}_1 + 0 \cdot \mathbf{v}_2 + 0 \cdot \mathbf{v}_3$. These formulas use the fact that projection onto a plane leaves any vector in the plane fixed and sends any vector which is perpendicular to the plane to $\mathbf{0}$. Thus we have

$$\begin{array}{ccc} L(\mathbf{v}_1) & L(\mathbf{v}_2) & L(\mathbf{v}_3) \end{array}$$
$$\begin{array}{c} \mathbf{v}_1 \\ \mathbf{v}_2 \\ \mathbf{v}_3 \end{array} \begin{pmatrix} 1 & 0 & 0 \\ 0 & 1 & 0 \\ 0 & 0 & 0 \end{pmatrix}$$

The matrix $[L]^{\mathcal{V}}_{\mathcal{V}}$ which represents L with respect to the basis $\mathcal{V} = \{\mathbf{v}_1, \mathbf{v}_2, \mathbf{v}_3\}$ in both copies of $\mathbb{R}^3$ is just the matrix A in the previous example. Suppose we wanted to use the standard basis $\mathcal{E}$ instead. Then we would have to find $L(\mathbf{e}_j)$ and express it in terms of $\mathbf{e}_1, \mathbf{e}_2, \mathbf{e}_3$. We can use $(\mathbf{e}_1 \quad \mathbf{e}_2 \quad \mathbf{e}_3) = (\mathbf{v}_1 \quad \mathbf{v}_2 \quad \mathbf{v}_3)C$ where $C = \begin{pmatrix} -1/3 & 2/3 & -1/3 \\ -1/3 & -1/3 & 2/3 \\ 1/3 & 1/3 & 1/3 \end{pmatrix}$. Note that C is just the inverse of the matrix with columns given by the $\mathbf{v}_j$. Also, $(\mathbf{v}_1 \quad \mathbf{v}_2 \quad \mathbf{v}_3) = (\mathbf{e}_1 \quad \mathbf{e}_2 \quad \mathbf{e}_3)C^{-1}$. Then

$$\begin{aligned} (L(\mathbf{e}_1) \quad L(\mathbf{e}_2) \quad L(\mathbf{e}_3)) &= (L(\mathbf{v}_1) \quad L(\mathbf{v}_2) \quad L(\mathbf{v}_3))C \\ &= (\mathbf{v}_1 \quad \mathbf{v}_2 \quad \mathbf{v}_3)AC = (\mathbf{e}_1 \quad \mathbf{e}_2 \quad \mathbf{e}_3)C^{-1}AC \end{aligned}$$

Thus the matrix $[L]_{\mathscr{E}}^{\mathscr{E}}$ which represents L with respect to the standard basis in both the domain and range is $C^{-1}AC$, which is $\begin{pmatrix} 2/3 & -1/3 & -1/3 \\ -1/3 & 2/3 & -1/3 \\ -1/3 & -1/3 & 2/3 \end{pmatrix}$. This is just expressing the fact that

$$L(\mathbf{e}_1) = (2/3, -1/3, -1/3), L(\mathbf{e}_2) = (-1/3, 2/3, -1/3), L(\mathbf{e}_3) = (-1/3, -1/3, 2/3)$$

Note that the computation was much simpler with the first pair of bases than it was with the standard bases. One of our tasks in understanding a general linear transformation will be to find bases which are closely related to it so that the matrix which represents the linear transformation with respect to these bases is particularly simple. ■

■ ***Example 2.7.3.*** Let $L : V \to W$ be an isomorphism. Let $\mathbf{v}_1, \ldots, \mathbf{v}_n$ be a basis of V and $\mathbf{w}_1 = L(\mathbf{v}_1), \ldots, \mathbf{w}_n = L(\mathbf{v}_n)$. The fact that L is an isomorphism means that $\mathbf{w}_1, \ldots, \mathbf{w}_n$ will be a basis of W. The matrix of L with respect to these two bases is just the identity matrix. Suppose that a different basis $\mathbf{u}_1, \ldots, \mathbf{u}_n$ for W is chosen. Then the fact that the $\mathbf{u}_i$ and the $\mathbf{w}_i$ both form bases means that there is an **invertible** matrix C with $(\mathbf{w}_1 \ldots \mathbf{w}_n) = (\mathbf{u}_1 \ldots \mathbf{u}_n)C$. Thus $(L(\mathbf{v}_1) \ldots L(\mathbf{v}_n)) = (\mathbf{w}_1 \ldots \mathbf{w}_n) = (\mathbf{u}_1 \ldots \mathbf{u}_n)C$. This means that C is the matrix $[L]_{\mathscr{U}}^{\mathscr{V}}$ which represents L with respect to the bases $\mathscr{V} = \{\mathbf{v}_1, \ldots, \mathbf{v}_n\}$ of V and $\mathscr{U} = \{\mathbf{u}_1, \ldots, \mathbf{u}_n\}$ of W. We can conclude from this argument that if L is an isomorphism from V to W, then the matrix which represents L with respect to any chosen bases is invertible. The converse will be left as an exercise. Moreover, L being an isomorphism means that bases can be chosen so that the matrix will be the identity. ■

■ ***Example 2.7.4.*** Let $D : \mathscr{P}^3(\mathbb{R}, \mathbb{R}) \to \mathscr{P}^3(\mathbb{R}, \mathbb{R})$ be the differentiation linear transformation, $D(p) = p'$. Using the basis $\{1, x, x^2, x^3\}$ in both copies of $\mathscr{P}^3(\mathbb{R}, \mathbb{R})$, then $D(1) = 0, D(x) = 1, D(x^2) = 2x, D(x^3) = 3x^2$. We can write each of these image polynomials in terms of basis $\{1, x, x^2, x^3\}$ to get

$$\begin{array}{c} \\ 1 \\ x \\ x^2 \\ x^3 \end{array} \begin{array}{c} \begin{array}{cccc} D(1) & D(x) & D(x^2) & D(x^3) \end{array} \\ \begin{pmatrix} 0 & 1 & 0 & 0 \\ 0 & 0 & 2 & 0 \\ 0 & 0 & 0 & 3 \\ 0 & 0 & 0 & 0 \end{pmatrix} \end{array}$$

Thus the matrix $[D]_{\mathscr{S}}^{\mathscr{S}}$ with respect to this basis $\mathscr{S} = \{1, x, x^2, x^3\}$ in both copies of $\mathscr{P}^3(\mathbb{R}, \mathbb{R})$ is $\begin{pmatrix} 0 & 1 & 0 & 0 \\ 0 & 0 & 2 & 0 \\ 0 & 0 & 0 & 3 \\ 0 & 0 & 0 & 0 \end{pmatrix}$. ■

■ ***Example 2.7.5.*** Let A be the 2 by 2 matrix $\begin{pmatrix} 1 & 3 \\ 2 & 1 \end{pmatrix}$. Consider the linear transformation $L : \mathcal{M}(2,2) \to \mathcal{M}(2,2)$, $L(B) = AB$. We will use the bases

$$\mathbf{v}_1 = \mathbf{w}_1 = \begin{pmatrix} 1 & 0 \\ 0 & 0 \end{pmatrix}, \mathbf{v}_2 = \mathbf{w}_2 = \begin{pmatrix} 0 & 1 \\ 0 & 0 \end{pmatrix}, \mathbf{v}_3 = \mathbf{w}_3 = \begin{pmatrix} 0 & 0 \\ 1 & 0 \end{pmatrix}, \mathbf{v}_4 = \mathbf{w}_4 = \begin{pmatrix} 0 & 0 \\ 0 & 1 \end{pmatrix}$$

Then

$$L(\mathbf{v}_1) = \begin{pmatrix} 1 & 0 \\ 2 & 0 \end{pmatrix} = \mathbf{v}_1 + 2\mathbf{v}_3, L(\mathbf{v}_2) = \begin{pmatrix} 0 & 1 \\ 0 & 2 \end{pmatrix} = \mathbf{v}_2 + 2\mathbf{v}_4$$

$$L(\mathbf{v}_3) = \begin{pmatrix} 3 & 0 \\ 1 & 0 \end{pmatrix} = 3\mathbf{v}_1 + \mathbf{v}_3, L(\mathbf{v}_4) = \begin{pmatrix} 0 & 3 \\ 0 & 1 \end{pmatrix} = 3\mathbf{v}_2 + \mathbf{v}_4$$

This information can be recorded as

$$\begin{array}{c} \\ \mathbf{v}_1 \\ \mathbf{v}_2 \\ \mathbf{v}_3 \\ \mathbf{v}_4 \end{array} \begin{array}{c} \begin{matrix} L(\mathbf{v}_1) & L(\mathbf{v}_2) & L(\mathbf{v}_3) & L(\mathbf{v}_4) \end{matrix} \\ \begin{pmatrix} 1 & 0 & 3 & 0 \\ 0 & 1 & 0 & 3 \\ 2 & 0 & 1 & 0 \\ 0 & 2 & 0 & 1 \end{pmatrix} \end{array}$$

Thus the matrix $[L]_{\mathcal{V}}^{\mathcal{V}}$, which represents L with respect to these bases, is

$$\begin{pmatrix} 1 & 0 & 3 & 0 \\ 0 & 1 & 0 & 3 \\ 2 & 0 & 1 & 0 \\ 0 & 2 & 0 & 1 \end{pmatrix}$$

This matrix is invertible, which corresponds to the fact that L is an isomorphism. ■

Exercise 2.7.1.

Suppose $\mathbf{v}_1, \ldots, \mathbf{v}_k$ are k vectors in V and C is a k by p matrix. Let $\mathbf{V} = (\mathbf{v}_1 \ldots \mathbf{v}_k)$ be the row vector with entries in V formed from the k vectors $\mathbf{v}_i$. Define $\mathbf{V}C$ to be the 1 by p row vector $\mathbf{U} = (\mathbf{u}_1 \ldots \mathbf{u}_p)$ formed by multiplying the row vector by the matrix by the usual rules: $\mathbf{u}_j = \sum_{i=1}^{k} c_{ij}\mathbf{v}_i$. Show that this operation is associative: $(\mathbf{V}C)D = \mathbf{V}(CD)$.

Exercise 2.7.2.

If $L : V \to W$ is a linear transformation and $\mathbf{V} = (\mathbf{v}_1 \ldots \mathbf{v}_k)$ is a row vector of vectors in V, let $L(\mathbf{V}) = (L(\mathbf{v}_1) \ldots L(\mathbf{v}_k))$ be the row vector of the image vectors in W. Show that if $\mathbf{U} = \mathbf{V}C$ then $L(\mathbf{U}) = L(\mathbf{V})C$.

Exercise 2.7.3.

Suppose that $L : V \to W$ is a linear transformation and A is the matrix which represents L with respect to the bases $\mathbf{v}_1, \ldots, \mathbf{v}_n$ of V and $\mathbf{w}_1, \ldots, \mathbf{w}_m$ of W. Show that if A is invertible then L is an isomorphism. (Hint: If B is the inverse of A, define $M : W \to V$ by $M(\mathbf{W}) = \mathbf{V}B$. Use the facts that $L(\mathbf{V}) = \mathbf{W}A$ and $A^{-1} = B$ to show that $M = L^{-1}$.)

Exercise 2.7.4.

Let $I_0 : \mathcal{P}^2(\mathbb{R}, \mathbb{R}) \to \mathcal{P}^3(\mathbb{R}, \mathbb{R})$ be given by $I_0(p) = q$, where q is the polynomial so that the derivative $q' = p$ and $q(0) = 0$. Using the basis $1, x, x^2$ for $\mathcal{P}^2(\mathbb{R}, \mathbb{R})$ and $1, x, x^2, x^3$ for $\mathcal{P}^3(\mathbb{R}, \mathbb{R})$, give the matrix which represents I_0 with respect to these bases.

Exercise 2.7.5.

Let $R : \mathbb{R}^2 \to \mathbb{R}^2$ be the linear transformation which is given by reflection through the line $x = 2y$. Using the basis $\mathbf{v}_1 = (2, 1), \mathbf{v}_2 = (-1, 2)$ for both the domain and range together with the fact that the reflection through a line sends a vector on the line to itself and sends a vector perpendicular to the line to its negative, give the matrix which represents R with respect to this basis. Write the standard basis in terms of this basis as $(\mathbf{e}_1 \quad \mathbf{e}_2) = (\mathbf{v}_1 \quad \mathbf{v}_2)C$ to find the matrix which represents L with respect to the standard basis in the domain and range.

2.7.1. CHANGE OF BASIS FORMULA

We now restrict our attention to a linear transformation $L : V \to V$ and investigate how $[L]_{\mathcal{V}}^{\mathcal{V}}$ depends on the basis $\mathcal{V}$. We first see how to use a matrix to compare two bases of V.

DEFINITION 2.7.2. The matrix C is called a **transition matrix from the basis $\mathcal{U}$ to the basis $\mathcal{V}$** when

$$\mathbf{U} = \mathbf{V}C; \text{ that is, } \mathbf{u}_j = \sum_{i=1}^{n} c_{ij}\mathbf{v}_i, \; j = 1, \ldots, n$$

The transition matrix C is denoted by $\mathcal{T}_{\mathcal{V}}^{\mathcal{U}}$.

Multiplying the equation

$$\mathbf{U} = \mathbf{V}C$$

by C^{-1} gives

$$\mathcal{T}_{\mathcal{U}}^{\mathcal{V}} = (\mathcal{T}_{\mathcal{V}}^{\mathcal{U}})^{-1}$$

The details are left as an exercise.

Exercise 2.7.6.

Show that $\mathcal{T}_{\mathcal{U}}^{\mathcal{V}} = (\mathcal{T}_{\mathcal{V}}^{\mathcal{U}})^{-1}$.

To find the transition matrix $\mathcal{T}_{\mathcal{U}}^{\mathcal{V}}$ we are just writing the basis vectors in $\mathcal{V}$ in terms of those in $\mathcal{U}$. We can express this relationship as follows:

$$\begin{array}{c} \\ \mathbf{u}_1 \\ \cdots \\ \mathbf{u}_i \\ \cdots \\ \mathbf{u}_n \end{array} \begin{array}{c} \begin{array}{ccccc} \mathbf{v}_1 & \cdots & \mathbf{v}_j & \cdots & \mathbf{v}_n \end{array} \\ \begin{pmatrix} a_{11} & \cdots & a_{1j} & \cdots & a_{1n} \\ \cdots & \cdots & \cdots & \cdots & \cdots \\ a_{i1} & \cdots & a_{ij} & \cdots & a_{in} \\ \cdots & \cdots & \cdots & \cdots & \cdots \\ a_{n1} & \cdots & a_{nj} & \cdots & a_{nn} \end{pmatrix} \end{array}$$

with the jth column giving the coefficients of $\mathbf{v}_j$ with respect to the basis $\mathcal{U}$.

Note that if I denotes the identity linear transformation, then

$$[I]_{\mathcal{V}}^{\mathcal{U}} = \mathcal{T}_{\mathcal{V}}^{\mathcal{U}}$$

In each case the vectors in the basis $\mathcal{U}$ are being written in terms of those in the basis $\mathcal{V}$.

■ ***Example 2.7.6.*** Let $\mathcal{V} = \{\mathbf{v}_1 = (1, -1, 3), \mathbf{v}_2 = (2, 1, 0), \mathbf{v}_3 = (1, 1, 1)\}$ and $\mathcal{E} = \{\mathbf{e}_1, \mathbf{e}_2, \mathbf{e}_3\}$. Then $\mathcal{T}_{\mathcal{E}}^{\mathcal{V}}$ is easy to find, since we can write

$$\begin{aligned} \mathbf{v}_1 &= 1 \cdot \mathbf{e}_1 - 1 \cdot \mathbf{e}_2 + 3 \cdot \mathbf{e}_3 \\ \mathbf{v}_2 &= 2 \cdot \mathbf{e}_1 + 1 \cdot \mathbf{e}_2 + 0 \cdot \mathbf{e}_3 \\ \mathbf{v}_3 &= 1 \cdot \mathbf{e}_1 + 1 \cdot \mathbf{e}_2 + 1 \cdot \mathbf{e}_3 \end{aligned}$$

We represent this information as

$$\begin{array}{c} \\ \mathbf{e}_1 \\ \mathbf{e}_2 \\ \mathbf{e}_3 \end{array} \begin{array}{c} \begin{array}{ccc} \mathbf{v}_1 & \mathbf{v}_2 & \mathbf{v}_3 \end{array} \\ \begin{pmatrix} 1 & 2 & 1 \\ -1 & 1 & 1 \\ 3 & 0 & 1 \end{pmatrix} \end{array}$$

Thus the matrix

$$\mathcal{T}_{\mathcal{E}}^{\mathcal{V}} = \begin{pmatrix} 1 & 2 & 1 \\ -1 & 1 & 1 \\ 3 & 0 & 1 \end{pmatrix}$$

just has the vectors of the basis $\mathcal{V}$ as its columns. To find the matrix $\mathcal{T}_{\mathcal{V}}^{\mathcal{E}}$ we have to write the vectors $\mathbf{e}_1, \mathbf{e}_2, \mathbf{e}_3$ in terms of the basis $\mathcal{V}$. This is more difficult, but we can use the fact that $\mathcal{T}_{\mathcal{V}}^{\mathcal{E}}$ is the inverse of $\mathcal{T}_{\mathcal{E}}^{\mathcal{V}}$. Thus we take the inverse of the matrix to get

$$\mathcal{T}_{\mathcal{V}}^{\mathcal{E}} = \begin{pmatrix} 1/6 & -1/3 & 1/6 \\ 2/3 & -1/3 & -1/3 \\ -1/2 & 1 & 1/2 \end{pmatrix}$$

Alternatively, we can consider the matrix of row vectors $\mathbf{E}$ as the identity matrix by thinking of the vectors themselves as column vectors. Similarly, we can think of the row vector $\mathbf{V}$ as giving us a matrix. Then we will have the matrix equation

$$\mathbf{V} = \mathbf{E} \begin{pmatrix} 1 & 2 & 1 \\ -1 & 1 & 1 \\ 3 & 0 & 1 \end{pmatrix}$$

Then multiplying by the inverse of the matrix on the right gives

$$\mathbf{V} \begin{pmatrix} 1/6 & -1/3 & 1/6 \\ 2/3 & -1/3 & -1/3 \\ -1/2 & 1 & 1/2 \end{pmatrix} = \mathbf{E}$$

and so $\mathcal{T}_{\mathcal{V}}^{\mathcal{E}} = \begin{pmatrix} 1/6 & -1/3 & 1/6 \\ 2/3 & -1/3 & -1/3 \\ -1/2 & 1 & 1/2 \end{pmatrix}$. ■

■ ***Example 2.7.7.*** Let $\mathcal{V} = \{\mathbf{v}_1 = 1 - x + 3x^2, \mathbf{v}_2 = 2 + x, \mathbf{v}_3 = 1 + x + x^2\}$ and $\mathcal{S} = \{1, x, x^2\}$ be bases for $\mathcal{P}^2(\mathbb{R}, \mathbb{R})$. Then to find $\mathcal{T}_{\mathcal{S}}^{\mathcal{V}}$ we need to express the three polynomials $1 - x + 3x^2, 2 + x, 1 + x + x^2$ in terms of $1, x, x^2$. But this is done automatically for us, and we can read off

$$\begin{matrix} & \mathbf{v}_1 & \mathbf{v}_2 & \mathbf{v}_3 \\ 1 \\ x \\ x^2 \end{matrix} \begin{pmatrix} 1 & 2 & 1 \\ -1 & 1 & 1 \\ 3 & 0 & 1 \end{pmatrix}$$

just as in the last example. We similarly find $\mathcal{T}_{\mathcal{V}}^{\mathcal{S}}$ as the inverse of this matrix,

$$\mathcal{T}_{\mathcal{V}}^{\mathcal{S}} = \begin{pmatrix} 1/6 & -1/3 & 1/6 \\ 2/3 & -1/3 & -1/3 \\ -1/2 & 1 & 1/2 \end{pmatrix}$$ ■

■ ***Example 2.7.8.*** In this example we look at the transition matrix between two bases for a subspace of $\mathbb{R}^4$. The bases are $\mathcal{V} = \{(1, 0, -1, 0), (2, -1, 3, 1), (1, 2, 0, 1)\}$ and $\mathcal{U} = \{(3, 4, -1, 2), (8, 0, 3, 3), (5, 7, 4, 5)\}$. When we write the row vectors containing these vectors (as column vectors), we get two matrices

$$\mathbf{V} = \begin{pmatrix} 1 & 2 & 1 \\ 0 & -1 & 2 \\ -1 & 3 & 0 \\ 0 & 1 & 1 \end{pmatrix}, \mathbf{U} = \begin{pmatrix} 3 & 8 & 5 \\ 4 & 0 & 7 \\ -1 & 3 & 4 \\ 2 & 3 & 5 \end{pmatrix}$$

Finding the transition matrix $\mathcal{T}_{\mathcal{U}}^{\mathcal{V}} = T$ then becomes a matter of solving the matrix equation $\mathbf{V} = \mathbf{U}T$. This is solved by solving the corresponding system of equations via Gaussian elimination to give

$$\mathcal{T}_{\mathcal{U}}^{\mathcal{V}} = T = (1/17)\begin{pmatrix} 7 & -13 & 5 \\ 2 & 6 & -1 \\ -4 & 5 & 2 \end{pmatrix}$$ ■

Exercise 2.7.7.

(a) Find the transition matrix $\mathcal{T}_{\mathcal{V}}^{\mathcal{E}}$ from the standard basis $\mathcal{E} = \{\mathbf{e}_1, \mathbf{e}_2, \mathbf{e}_3\}$ to the basis $\mathcal{V} = \{(2,1,3), (1,0,1), (3,1,1)\}$ of $\mathbb{R}^3$.

(b) Find the transition matrix $\mathcal{T}_{\mathcal{V}}^{\mathcal{S}}$ from the standard basis $\mathcal{S} = \{1, x, x^2\}$ to the basis $\mathcal{V} = \{2 + x + 3x^2, 1 + x^2, 3 + x + x^2\}$ of $\mathcal{P}^2(\mathbb{R}, \mathbb{R})$.

Exercise 2.7.8.

Find the transition matrix $\mathcal{T}_{\mathcal{W}}^{\mathcal{V}}$ from the basis

$$\mathcal{V} = \{(1,3,2,1), (4,2,1,5)\} \text{ to the basis } \mathcal{W} = \{(9,7,4,11), (6,8,5,7)\}$$

of a subspace S of $\mathbb{R}^4$.

Let us return to the general linear transformation $L : V \to V$ with $\dim V = n$. Suppose $\mathcal{V} = \{\mathbf{v}_1, \ldots, \mathbf{v}_n\}$ is a basis for V. We can form the corresponding row vector $\mathbf{V} = (\mathbf{v}_1 \ldots \mathbf{v}_n)$. Using the notation introduced earlier, we have $L(\mathbf{V}) = \mathbf{V}A$, where $A = [L]_{\mathcal{V}}^{\mathcal{V}}$ is the matrix which represents L with respect to these two bases. We now investigate what happens as we change bases. Suppose that $\mathbf{U} = (\mathbf{u}_1 \ldots \mathbf{u}_n)$ is a row vector with entries from another basis $\mathcal{U} = \{\mathbf{u}_1, \ldots, \mathbf{u}_n\}$ for V. Let $C = \mathcal{T}_{\mathcal{V}}^{\mathcal{U}}$ be the transition matrix from the basis $\mathcal{U}$ to the basis $\mathcal{V}$. Thus $\mathbf{U} = \mathbf{V}C$, and multiplying by C^{-1} gives $\mathbf{V} = \mathbf{U}C^{-1}$. We compute $[L]_{\mathcal{U}}^{\mathcal{U}}$ as follows:

$$L(\mathbf{U}) = L(\mathbf{V}C) = L(\mathbf{V})C = (\mathbf{V}A)C = \mathbf{U}(C^{-1}AC)$$

This equation means that

$$[L]_{\mathcal{U}}^{\mathcal{U}} = C^{-1}AC = \mathcal{T}_{\mathcal{U}}^{\mathcal{V}}[L]_{\mathcal{V}}^{\mathcal{V}}\mathcal{T}_{\mathcal{V}}^{\mathcal{U}}$$

This latter notation is most useful in remembering the formula: the super- and subscripts on the right-hand side are read from top to bottom and right to left to get the final super- and subscripts on the left. The superscript of one term matches the subscript on the term to the right.

$$\mathcal{U} \to \mathcal{V} \to \mathcal{V} \to \mathcal{U}$$

We now state this relationship for further use.

PROPOSITION 2.7.1. Let $L : V \to V$ be a linear transformation and suppose $\mathcal{V}, \mathcal{U}$ are bases for V. Let $C = \mathcal{T}_{\mathcal{V}}^{\mathcal{U}}$ be the transition matrix from the basis $\mathcal{U}$ to the basis $\mathcal{V}$. Then the matrices $A = [L]_{\mathcal{V}}^{\mathcal{V}}$ and $B = [L]_{\mathcal{U}}^{\mathcal{U}}$ are related by the formula

$$B = C^{-1}AC, \text{ or, equivalently, } [L]_{\mathcal{U}}^{\mathcal{U}} = \mathcal{T}_{\mathcal{U}}^{\mathcal{V}}[L]_{\mathcal{V}}^{\mathcal{V}}T_{\mathcal{V}}^{\mathcal{U}}$$

DEFINITION 2.7.3. A matrix B is said to be **similar** to A if there is an invertible matrix C with $B = C^{-1}AC$.

The relation of two matrices being similar is an equivalence relation on matrices. Thus the proposition says that the effect of changing the basis is to change the matrix which represents L to a similar matrix. In this situation, we want to find the simplest matrix which represents L. This becomes the problem: given a matrix A (which represents L with respect to some bases), find the "simplest" matrix which is similar to A. It turns out that frequently (although not always) there is a diagonal matrix D to which A is similar. This corresponds to finding a basis of vectors $\mathbf{v}_1, \ldots, \mathbf{v}_n$ so that $L(\mathbf{v}_i) = d_i\mathbf{v}_i$. The vectors $\mathbf{v}_i$ will be called eigenvectors and the numbers d_i will be called eigenvalues. Finding this basis will lead us to the eigenvalue–eigenvector problem, which is the main topic in Chapter 4. Finding the simplest similar matrix when it is not diagonal will lead us to the Jordan canonical form for a matrix, which is discussed in Chapter 6.

Exercise 2.7.9.

Let $R : \mathbb{R}^3 \to \mathbb{R}^3$ be reflection in the plane $x + y - z = 0$.

(a) Find the matrix which represents R with respect to the basis given by $(1, 0, 1)$, $(0, 1, 1), (1, 1, -1)$ in both the domain and range. Note that the first two vectors are a basis for this plane and the third vector is perpendicular to the plane.

(b) Find the matrix which represents R with respect to the standard basis.

Exercise 2.7.10.

Let $r : \mathbb{R}^3 \to \mathbb{R}^3$ be the linear transformation which fixes the axis $c(1, 1, 1)$ and rotates each plane perpendicular to this axis by 90 degrees. If $\mathbf{v}_1 = (1/\sqrt{3}, 1/\sqrt{3}, 1/\sqrt{3})$, $\mathbf{v}_2 = (-1/\sqrt{2}, 1/\sqrt{2}, 0)$, $\mathbf{v}_3 = (1/\sqrt{6}, 1/\sqrt{6}, -2/\sqrt{6})$, find the matrix $[r]_{\mathcal{V}}^{\mathcal{V}}, \mathcal{V} = \{\mathbf{v}_1, \mathbf{v}_2, \mathbf{v}_3\}$. Note that $\mathbf{v}_1$ is on the axis of rotation and $\mathbf{v}_2, \mathbf{v}_3$ are perpendicular vectors which are a basis for the plane which is perpendicular to the axis. Use the proposition to find the matrix $[r]_{\mathcal{E}}^{\mathcal{E}}$, where $\mathcal{E} = \{\mathbf{e}_1, \mathbf{e}_2, \mathbf{e}_3\}$ is the standard basis for $\mathbb{R}^3$.

Exercise 2.7.11.

(a) Consider $L : \mathcal{P}^3 \to \mathcal{P}^3$ given by $L(p) = xp' - p$. Using the basis $\mathcal{B}$ of $\mathcal{P}^3$ given by $1, 1 + x, 1 + x + x^2, 1 + x + x^2 + x^3$, find the matrix $[L]_{\mathcal{B}}^{\mathcal{B}}$.

(b) Find the transition matrix $\mathcal{T}_{\mathcal{S}}^{\mathcal{B}}$ which relates the basis $\mathcal{B}$ to the standard basis $\mathcal{S} = \{1, x, x^2, x^3\}$ for $\mathcal{P}^3$.

(c) Find directly the matrix $[L]_{\mathcal{S}}^{\mathcal{S}}$ and also find it by using the formula $[L]_{\mathcal{S}}^{\mathcal{S}} = \mathcal{T}_{\mathcal{S}}^{\mathcal{B}}[L]_{\mathcal{B}}^{\mathcal{B}}\mathcal{T}_{\mathcal{B}}^{\mathcal{S}}$.

(d) Find the matrix $[L]_{\mathcal{S}}^{\mathcal{B}}$.

Exercise 2.7.12.

Show that the relation "B is similar to A" is an equivalence relation; that is, show that it is reflexive (A is similar to A), symmetric (B is similar to A implies A is similar to B), and transitive (B is similar to A and C is similar to B implies C is similar to A).

2.7.2. USING COORDINATES TO TRANSFER PROBLEMS TO $\mathbb{R}^n$

In this section we assume that V is an n-dimensional vector space and $\mathcal{V} = \{\mathbf{v}_1, \ldots, \mathbf{v}_n\}$ is a basis for V. If $\mathbf{v} \in V$, then we may associate a vector $\mathbf{x} = (x_1, \ldots, x_n) \in \mathbb{R}^n$ to $\mathbf{v}$ by the equation

$$\mathbf{v} = \mathbf{V}\mathbf{x} = x_1\mathbf{v}_1 + \cdots + x_n\mathbf{v}_n$$

DEFINITION 2.7.4. If $\mathbf{v} = \mathbf{V}\mathbf{x} = x_1\mathbf{v}_1 + \cdots + x_n\mathbf{v}_n$, where $\mathbf{V}$ is the row vector with entries in the basis $\mathcal{V}$, we call $\mathbf{x}$ the **coordinates** of $\mathbf{v}$ with respect to the basis $\mathcal{V}$ and denote $\mathbf{x} = [\mathbf{v}]_{\mathcal{V}}$.

If we change the basis to a different basis $\mathcal{W} = \{\mathbf{w}_1, \ldots, \mathbf{w}_n\}$ for V and the transition matrix $\mathcal{T}_{\mathcal{W}}^{\mathcal{V}} = T$, then

$$\mathbf{v} = \mathbf{V}\mathbf{x} = (\mathbf{W}T)\mathbf{x} = \mathbf{W}(T\mathbf{x})$$

means

$$[\mathbf{v}]_{\mathcal{W}} = \mathcal{T}_{\mathcal{W}}^{\mathcal{V}}[\mathbf{v}]_{\mathcal{V}}$$

Note in this notation the superscript of the first entry on the right matches the subscript of the second one, similar to our formula for the change in the representing matrix with respect to a basis. We state this as a proposition.

PROPOSITION 2.7.2. If $[\mathbf{v}]_{\mathcal{V}}, [\mathbf{v}]_{\mathcal{W}}$ denote the coordinates of $\mathbf{v}$ with respect to the bases $\mathcal{V}, \mathcal{W}$, respectively, then

$$[\mathbf{v}]_{\mathcal{W}} = \mathcal{T}_{\mathcal{W}}^{\mathcal{V}}[\mathbf{v}]_{\mathcal{V}}$$

■ ***Example 2.7.9.*** Consider the vector space $\mathcal{P}^2(\mathbb{R}, \mathbb{R})$ of polynomials of degree less than or equal to 2. It has a standard basis $\mathcal{S} = \{1, x, x^2\}$. Then

$$[a_0 + a_1x + a_2x^2]_{\mathcal{S}} = (a_0, a_1, a_2)$$

Now consider another basis $\mathcal{B} = \{1, x-1, (x-1)^2\}$. The transition matrix $\mathcal{T}_{\mathcal{S}}^{\mathcal{B}} = \begin{pmatrix} 1 & -1 & 1 \\ 0 & 1 & -2 \\ 0 & 0 & 1 \end{pmatrix}$ is found by just writing the new basis in terms of the standard basis.

By taking the inverse we can find the transition matrix $\mathcal{T}_{\mathcal{B}}^{\mathcal{S}} = \begin{pmatrix} 1 & 1 & 1 \\ 0 & 1 & 2 \\ 0 & 0 & 1 \end{pmatrix}$. Thus

$$
\begin{aligned}
a_0 + a_1 x + a_2 x^2]_{\mathcal{B}} &= \mathcal{T}_{\mathcal{B}}^{\mathcal{S}} [a_0 + a_1 x + a_2 x^2]_{\mathcal{S}} \\
&= \begin{pmatrix} 1 & 1 & 1 \\ 0 & 1 & 2 \\ 0 & 0 & 1 \end{pmatrix} \begin{pmatrix} a_0 \\ a_1 \\ a_2 \end{pmatrix} = \begin{pmatrix} a_0 + a_1 + a_2 \\ a_1 + 2a_2 \\ a_3 \end{pmatrix}
\end{aligned}
$$

This means that the polynomial

$$a_0 + a_1 x + a_2 x^2 = (a_0 + a_1 + a_2)1 + (a_1 + 2a_2)(x-1) + a_2(x-1)^2$$

As a concrete example,

$$3 + 2x - 4x^2 = 1 - 6(x-1) - 4(x-1)^2 \qquad \blacksquare$$

Now suppose that $L : V \to V$ is a linear transformation. We choose a fixed basis $\mathcal{V} = \{\mathbf{v}_1, \ldots, \mathbf{v}_n\}$. Then there will be the following basic formula:

PROPOSITION 2.7.3. If $\mathcal{V}$ is a basis for V, then

$$[L\mathbf{v}]_{\mathcal{V}} = [L]_{\mathcal{V}}^{\mathcal{V}} [\mathbf{v}]_{\mathcal{V}}$$

Proof.

$$L\mathbf{v} = L(\mathbf{V}[\mathbf{v}]_{\mathcal{V}}) = L(\mathbf{V})[\mathbf{v}]_{\mathcal{V}} = (\mathbf{V}[L]_{\mathcal{V}}^{\mathcal{V}})[\mathbf{v}]_{\mathcal{V}} = \mathbf{V}([L]_{\mathcal{V}}^{\mathcal{V}}[\mathbf{v}]_{\mathcal{V}}) \qquad \blacksquare$$

Whenever we have a single basis, we will simplify notation and just write $[L]$ for $L_{\mathcal{V}}^{\mathcal{V}}$ and $[\mathbf{v}]$ for $[\mathbf{v}]_{\mathcal{V}}$. Then the conclusion of the theorem may be restated as

$$[L\mathbf{v}] = [L][\mathbf{v}]$$

We use the notation, deleting super- and subscripts for the fixed basis $\mathcal{V}$, for the remainder of this section. We next note what happens to compositions and linear combinations of linear transformations.

PROPOSITION 2.7.4. If $L, M : V \to V$ are linear transformations, and $[L]$, $[M]$, denote the matrices for these linear transformations with respect to a fixed basis $\mathcal{V}$, then

(1) $[ML] = [M][L]$.

(2) $[aL + bM] = a[L] + b[M]$.

Proof.

(1) $(ML)(\mathbf{V}) = M(L(\mathbf{V})) = M(\mathbf{V}[L]) = M(\mathbf{V})[L] = (\mathbf{V}[M])[L] = \mathbf{V}([M][L])$

implies the result since the defining condition of $[ML]$ is that it satisfies the equation

$$(ML)(\mathbf{V}) = \mathbf{V}[ML]$$

(2) This follows from

$$(aL + bM)(\mathbf{V}) = aL(\mathbf{V}) + bM(\mathbf{V}) = a\mathbf{V}[L] + b\mathbf{V}[M] = \mathbf{V}(a[L] + b[M])$$

■

■ ***Example 2.7.10.*** We look at the vector space $\mathcal{P}^2(\mathbb{R}, \mathbb{R})$ and use the basis $\mathcal{S} = \{1, x, x^2\}$. We first look at the operation of taking derivatives $D(p) = p'$. Then $D(1) = 0, D(x) = 1, D(x^2) = 2x$ means that $[D] = \begin{pmatrix} 0 & 1 & 0 \\ 0 & 0 & 2 \\ 0 & 0 & 0 \end{pmatrix}$. When we look at the second derivative, it is just the composition $D \circ D$, which we denote by D^2. Now consider the differential equation

$$y'' - 2y + 3y = x^2 + 3$$

We can look to see if there are any polynomial solutions to this equation by considering the left-hand side as Ly, where $L = D^2 - 2D + 3I$. Here I means the identity linear transformation, and we consider L as a linear transformation $L : \mathcal{P}^2(\mathbb{R}, \mathbb{R}) \to \mathcal{P}^2(\mathbb{R}, \mathbb{R})$. Note that

$$[D] = \begin{pmatrix} 0 & 1 & 0 \\ 0 & 0 & 2 \\ 0 & 0 & 0 \end{pmatrix}, \quad [D^2] = [D]^2 = \begin{pmatrix} 0 & 0 & 2 \\ 0 & 0 & 0 \\ 0 & 0 & 0 \end{pmatrix}, \quad [I] = \begin{pmatrix} 1 & 0 & 0 \\ 0 & 1 & 0 \\ 0 & 0 & 1 \end{pmatrix}$$

so $[L] = \begin{pmatrix} 3 & -2 & 2 \\ 0 & 3 & -4 \\ 0 & 0 & 3 \end{pmatrix}$. Since $[x^2 + 3] = \begin{pmatrix} 3 \\ 0 \\ 1 \end{pmatrix}$, finding polynomial solutions of this equation $Ly = x^2 + 3$ can get transferred to solving the equation

$$[Ly] = [L][y] = [x^2 + 1]$$

which is the equation

$$\begin{pmatrix} 3 & -2 & 2 \\ 0 & 3 & -4 \\ 0 & 0 & 3 \end{pmatrix} [y] = \begin{pmatrix} 3 \\ 0 \\ 1 \end{pmatrix}$$

This has solution $[y] = (29/27, 4/9, 1/3)$. Thus the polynomial solution to the differential equation is $y = (29/27) + (4/9)x + (1/3)x^2$. ■

■ ***Example 2.7.11.*** Consider the linear transformation $L : \mathcal{M}(2, 2) \to \mathcal{M}(2, 2)$ given by $L(B) = AB - B$, where $A = \begin{pmatrix} 1 & 1 \\ 0 & 1 \end{pmatrix}$. We use the standard basis for $\mathcal{M}(2, 2)$

consisting of

$$B_1 = \begin{pmatrix} 1 & 0 \\ 0 & 0 \end{pmatrix}, \ B_2 = \begin{pmatrix} 0 & 1 \\ 0 & 0 \end{pmatrix}, \ B_3 = \begin{pmatrix} 0 & 0 \\ 1 & 0 \end{pmatrix}, \ B_4 = \begin{pmatrix} 0 & 0 \\ 0 & 1 \end{pmatrix}$$

Since

$$AB_1 - B_1 = \begin{pmatrix} 0 & 0 \\ 0 & 0 \end{pmatrix}, \ AB_2 - B_2 = \begin{pmatrix} 0 & 0 \\ 0 & 0 \end{pmatrix}$$

$$AB_3 - B_3 = \begin{pmatrix} 1 & 0 \\ 0 & 0 \end{pmatrix} = B_1, \ AB_4 - B_4 = \begin{pmatrix} 0 & 1 \\ 0 & 0 \end{pmatrix} = B_2$$

we have

$$[L] = \begin{pmatrix} 0 & 0 & 1 & 0 \\ 0 & 0 & 0 & 1 \\ 0 & 0 & 0 & 0 \\ 0 & 0 & 0 & 0 \end{pmatrix}$$

To determine $\mathcal{N}(L)$, we look for matrices B with $L(B) = \mathbf{0}$. But this equation implies $[L][B] = [\mathbf{0}]$. This reduces the problem to computing $\mathcal{N}([L])$, which becomes a matrix calculation. The matrix $[L]$ is already in reduced normal form, so $\{(1,0,0,0),(0,1,0,0)\}$ is a basis for $\mathcal{N}([L])$. These two vectors then give the coordinates of basis vectors for $\mathcal{N}(L)$, which would have as a basis B_1, B_2. Similarly, basis vectors for $\mathcal{R}([L])$ are $(1,0,0,0),(0,1,0,0)$, which implies that a basis for $\mathcal{R}(L)$ is B_1, B_2. ∎

This last example illustrates the following proposition.

PROPOSITION 2.7.5. Let $L : V \to V$ be a linear transformation, and suppose $\mathcal{V} = \{\mathbf{v}_1, \ldots, \mathbf{v}_n\}$ is a basis for V. Then the linear transformation

$$C : V \to \mathbb{R}^n, C(\mathbf{v}) = [\mathbf{v}]$$

is an isomorphism which restricts to give an isomorphism between $\mathcal{N}(L)$ and $\mathcal{N}([L])$ and an isomorphism between $\mathcal{R}(L)$ and $\mathcal{R}([L])$.

Proof. Since

$$\mathbf{v} = \mathbf{V}\mathbf{x}, \mathbf{w} = \mathbf{V}\mathbf{y} \text{ implies } a\mathbf{v} + b\mathbf{w} = a\mathbf{V}\mathbf{x} + b\mathbf{V}\mathbf{y} = \mathbf{V}(a\mathbf{x} + b\mathbf{y})$$

we have $C(a\mathbf{v} + b\mathbf{w}) = aC(\mathbf{v}) + bC(\mathbf{w})$. Hence C is a linear transformation. It has an inverse given by the linear transformation $F(\mathbf{x}) = \mathbf{V}\mathbf{x}$, so C is an isomorphism. If $\mathbf{v} \in \mathcal{N}(L)$, then

$$L(\mathbf{v}) = \mathbf{0} \text{ implies } [L][\mathbf{v}] = [\mathbf{0}] = \mathbf{0}$$

and so $C(\mathbf{v}) = [\mathbf{v}] = \in \mathcal{N}([L])$. Similarly, if $\mathbf{x} \in \mathcal{N}([L])$, then

$$[L](\mathbf{x}) = \mathbf{0}, \mathbf{x} = [F(\mathbf{x})] = [\mathbf{V}\mathbf{x}] \text{ implies } [L(\mathbf{V}\mathbf{x})] = [\mathbf{0}] \text{ implies } L(F(\mathbf{x})) = \mathbf{0}$$

so $F(\mathbf{x}) \in \mathcal{N}(L)$. Thus C and F restrict to give isomorphisms between $\mathcal{N}(L)$ and $\mathcal{N}([L])$.

The equation

$$C(L\mathbf{v}) = [L\mathbf{v}] = [L][\mathbf{v}]$$

implies that C sends $\mathcal{R}(L)$ to $\mathcal{R}([L])$. If $\mathbf{x} \in \mathcal{R}([L])$, then $\mathbf{x} = [L]\mathbf{y}$, so

$$[L(\mathbf{V}\mathbf{y})] = [(L(\mathbf{V}))\mathbf{y}] = [(\mathbf{V}[L])\mathbf{y}] = [L]\mathbf{y} = \mathbf{x}$$

implying

$$F(\mathbf{x}) = L(\mathbf{V}\mathbf{y}) \in \mathcal{R}(L)$$

■

Here is a diagram that indicates how L and $[L]$ are related, where $[L]$ is denoting multiplication by the matrix $[L]$ in the bottom line. It is called a commutative diagram since

$$CL = [L]C, \text{ i.e., } [L\mathbf{v}] = [L][\mathbf{v}]$$

The proposition is saying that the isomorphism C provides a correspondence between $\mathcal{N}(L), \mathcal{R}(L)$ at the top level with $\mathcal{N}([L]), \mathcal{R}([L])$ at the bottom level. More generally, C gives a means of transferring questions in V to questions in $\mathbb{R}^n$ which are accessible by matrix calculations.

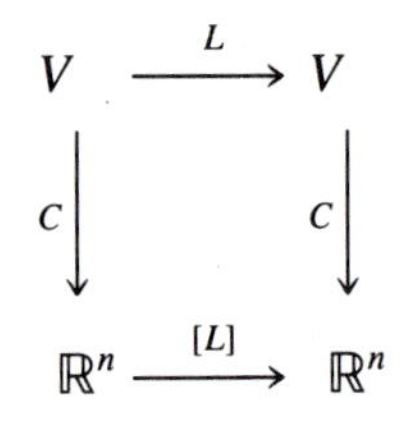

Exercise 2.7.13.

By transferring the problem to one in $\mathbb{R}^n$, find bases for the null space and range of the linear transformation

$$L : \mathcal{P}^3 \to \mathcal{P}^3, \quad L(p) = xp' - p$$

As a first step you should calculate $[L]$. Then find bases for the null space and range of $[L]$, and then use these (via F, the inverse of C) to get the desired bases.

Exercise 2.7.14.

Consider the differential equation

$$y'' - xy' + 3y = x + 1$$

Let $L : \mathcal{P}^1(\mathbb{R}, \mathbb{R}) \to \mathcal{P}^1(\mathbb{R}, \mathbb{R})$ be the linear transformation $D^2 - xD + 3I$, where x denotes multiplication by x and I denotes the identity linear transformation. A polynomial solution of degree less than or equal to 1 would then be a solution of $L(y) = x + 1$. Let $\mathcal{S} = \{1, x\}$ be the standard basis of $\mathcal{P}^1(\mathbb{R}, \mathbb{R})$: all of our matrices and coordinates will be taken with respect to this basis.

(a) Find $[L]$ and $[x + 1]$.

(b) Solve $[L]\mathbf{c} = [x + 1]$ and use the solution to find a polynomial solution of the differential equation.

For simplicity we have restricted our attention here to linear transformations from a vector space V to itself, where we are using the same basis in the domain and range. There are corresponding propositions when we consider $L : V \to W$. We state these next but leave their proofs to the exercises at the end of this chapter.

PROPOSITION 2.7.6. Let $L : V \to W$ be a linear transformation and $\mathcal{V}, \mathcal{W}$ bases for V, W. Denote by $[L]$ the matrix $[L]_{\mathcal{W}}^{\mathcal{V}}$ and, for simplicity, denote $[\mathbf{v}]_{\mathcal{V}} = [\mathbf{v}], [\mathbf{w}]_{\mathcal{W}} = [\mathbf{w}]$. Then

$$[L\mathbf{v}]_{\mathcal{W}} = [L]_{\mathcal{W}}^{\mathcal{V}} \mathbf{v}_{\mathcal{V}} \text{ or, more briefly, } [L\mathbf{v}] = [L][\mathbf{v}]$$

PROPOSITION 2.7.7. Let $L : V \to W, M : W \to X$ be linear transformations and $\mathcal{V}, \mathcal{W}, \mathcal{X}$ be bases for V, W, X. Then

$$[ML]_{\mathcal{X}}^{\mathcal{V}} = [M]_{\mathcal{X}}^{\mathcal{W}} [L]_{\mathcal{W}}^{\mathcal{V}}$$

If all the super- and subscripts denoting the chosen bases are suppressed, this becomes

$$[ML] = [M][L]$$

PROPOSITION 2.7.8. If $L_1, L_2 : V \to W$ are linear transformations, $\mathcal{V}, \mathcal{W}$ are chosen bases of V, W, and $[N] = [N]_{\mathcal{W}}^{\mathcal{V}}$, then

$$[aL_1 + bL_2] = a[L_1] + b[L_2]$$

2.8. APPLICATIONS TO GRAPH THEORY

In this section we give some elementary applications of linear algebra to problems in graph theory. Informally, a graph consists of a collection of points, which we will call vertices, and a collection of line segments joining those points, which we will call edges. We may be interested in giving a direction to each edge, which we will do by putting an arrow on it. When the edges have these arrows, this will give a directed graph. Figure 2.6 depicts four different directed graphs. Graph (a) has four vertices and six edges. Note that graph (b) has an edge going from a vertex to itself, which will be called a simple loop. Note also that graph (d) has a vertex, vertex 9, which has no edges running to or from it.

We now give a formal definition of a graph and a directed graph. All our graphs are finite (i.e., have a finite number of vertices and edges).

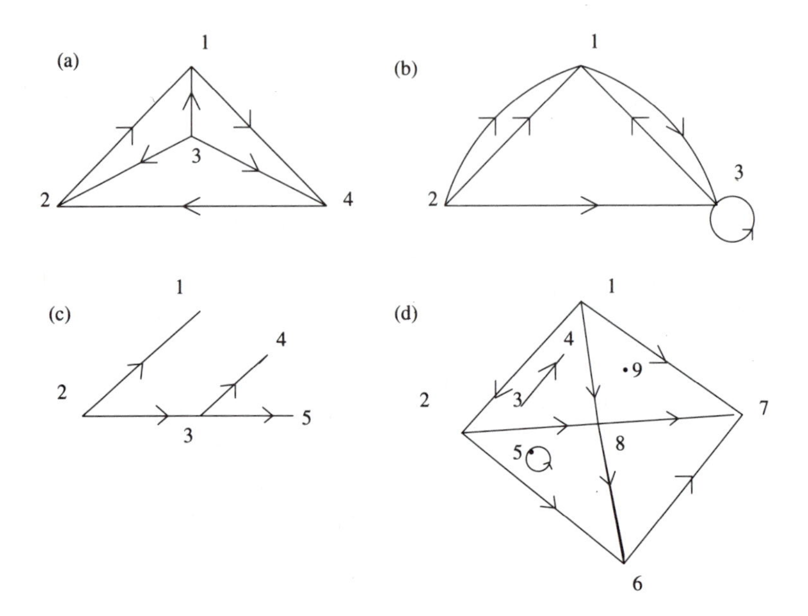

Figure 2.6. Examples of directed graphs

DEFINITION 2.8.1. A (finite) **graph** consists of a set $V = \{v_1, \ldots, v_n\}$ of **vertices** (depicted as points) and a set $E = \{e_1, \ldots, e_m\}$ of **edges** (depicted as curves). Each edge $e \in E$ has related to it a subset of either one or two vertices. We depict this relationship by a curve connecting the vertices, allowing the possibility that the edge runs from a vertex back to itself. For a **directed graph,** also called a **digraph**, one of these vertices is called the initial vertex and the other is the terminal vertex. This relationship is actually two functions $i, t : E \to V$. Here $i(e)$ denotes the initial vertex of e, and $t(e)$ is the terminal vertex of e. We depict this graphically by drawing an arrow on the edge running in the direction from $i(e)$ to $t(e)$. An edge connecting a vertex to itself ($i(e) = t(e)$) is called a **simple loop.**

Associated with each directed graph is a graph that is obtained by ignoring the information concerning the initial and terminal vertex of each edge (i.e., ignoring the arrows). Sometimes we start with a graph, arbitrarily give a direction to each edge to get a directed graph, and then use the directed graph to answer a question about the original graph. Figure 2.6 gives some pictures of directed graphs. We have numbered the vertices for future use. We will consider the directed graph and the associated graph.

In a graph or digraph there are paths which connect up the vertices.

DEFINITION 2.8.2. A **path** in a directed graph is a finite ordered collection of edges $e_1, \ldots, e_k$ so that the terminal vertex of e_i is the initial vertex of e_{i+1}. Here we say the path runs from $i(e_1)$ to $t(e_k)$. It is called a **loop** if the $i(e_1) = t(e_k)$. We use the same definition for a path in a graph with the proviso that we may give a direction to each edge when it occurs (and possibly use it later with the other direction). Alternatively, we can take a directed graph which has the graph associated to it, and then consider as a path in the graph a sequence of signed edges in the directed graph, where the sign is $+1$ if the edge is traversed in the order given in the directed graph and -1 if it is traversed in the other order. A graph is called **connected** if given any two vertices v, w there is a path running from v to w.

Given a graph G and an ordering of its vertices $\mathbf{v}_1, \ldots, \mathbf{v}_n$ there is an associated **connectivity matrix** C. The entry $c_{ij} = 0$ unless there is an edge joining v_i with v_j, in which case it is the number of such edges. For a directed graph the connectivity matrix C has $c_{ij} = 0$ except when there is an edge e with $i(e) = v_i$ and $t(e) = v_j$. Then it counts the number of such edges. Note that the connectivity matrix for a graph is symmetric ($C = C^t$), but it need not be for a directed graph. Here are the connectivity matrices in the four examples in Figure 2.6. We denote the connectivity matrix for the digraph by C_d and the connectivity matrix for the graph by C_g.

(a)

$$C_d = \begin{pmatrix} 0 & 0 & 0 & 1 \\ 1 & 0 & 0 & 0 \\ 1 & 1 & 0 & 1 \\ 0 & 1 & 0 & 0 \end{pmatrix}, C_g = \begin{pmatrix} 0 & 1 & 1 & 1 \\ 1 & 0 & 1 & 1 \\ 1 & 1 & 0 & 1 \\ 1 & 1 & 1 & 0 \end{pmatrix}$$

(b)

$$C_d = \begin{pmatrix} 0 & 0 & 1 \\ 2 & 0 & 1 \\ 1 & 0 & 1 \end{pmatrix}, C_g = \begin{pmatrix} 0 & 2 & 2 \\ 2 & 0 & 1 \\ 2 & 1 & 1 \end{pmatrix}$$

(c)

$$C_d = \begin{pmatrix} 0 & 0 & 0 & 0 & 0 \\ 1 & 0 & 1 & 0 & 0 \\ 0 & 0 & 0 & 1 & 1 \\ 0 & 0 & 0 & 0 & 0 \\ 0 & 0 & 0 & 0 & 0 \end{pmatrix}, C_g = \begin{pmatrix} 0 & 1 & 0 & 0 & 0 \\ 1 & 0 & 1 & 0 & 0 \\ 0 & 1 & 0 & 1 & 1 \\ 0 & 0 & 1 & 0 & 0 \\ 0 & 0 & 1 & 0 & 0 \end{pmatrix}$$

(d)

$$C_d = \begin{pmatrix} 0 & 1 & 0 & 0 & 0 & 0 & 1 & 1 & 0 \\ 0 & 0 & 0 & 0 & 0 & 1 & 0 & 1 & 0 \\ 0 & 0 & 0 & 1 & 0 & 0 & 0 & 0 & 0 \\ 0 & 0 & 0 & 0 & 0 & 0 & 0 & 0 & 0 \\ 0 & 0 & 0 & 0 & 1 & 0 & 0 & 0 & 0 \\ 0 & 0 & 0 & 0 & 0 & 0 & 1 & 0 & 0 \\ 0 & 0 & 0 & 0 & 0 & 0 & 0 & 0 & 0 \\ 0 & 0 & 0 & 0 & 0 & 1 & 1 & 0 & 0 \\ 0 & 0 & 0 & 0 & 0 & 0 & 0 & 0 & 0 \end{pmatrix}, C_g = \begin{pmatrix} 0 & 1 & 0 & 0 & 0 & 0 & 1 & 1 & 0 \\ 1 & 0 & 0 & 0 & 0 & 1 & 0 & 1 & 0 \\ 0 & 0 & 0 & 1 & 0 & 0 & 0 & 0 & 0 \\ 0 & 0 & 1 & 0 & 0 & 0 & 0 & 0 & 0 \\ 0 & 0 & 0 & 0 & 1 & 0 & 0 & 0 & 0 \\ 0 & 1 & 0 & 0 & 0 & 0 & 1 & 1 & 0 \\ 1 & 0 & 0 & 0 & 0 & 1 & 0 & 1 & 0 \\ 1 & 1 & 0 & 0 & 0 & 1 & 1 & 0 & 0 \\ 0 & 0 & 0 & 0 & 0 & 0 & 0 & 0 & 0 \end{pmatrix}$$

Note that the connectivity matrix completely determines the graph or digraph.

■ ***Example 2.8.1.*** Consider the connectivity matrices

$$C_d = \begin{pmatrix} 0 & 1 & 0 & 2 & 0 \\ 1 & 0 & 1 & 0 & 1 \\ 0 & 0 & 0 & 1 & 0 \\ 0 & 0 & 0 & 1 & 0 \\ 0 & 1 & 0 & 1 & 0 \end{pmatrix}, C_g = \begin{pmatrix} 0 & 2 & 0 & 2 & 0 \\ 2 & 0 & 1 & 0 & 2 \\ 0 & 1 & 0 & 1 & 0 \\ 2 & 0 & 1 & 1 & 1 \\ 0 & 2 & 0 & 1 & 0 \end{pmatrix}$$

C_d corresponds to the digraph and C_g corresponds to the associated graph in Figure 2.7. ■

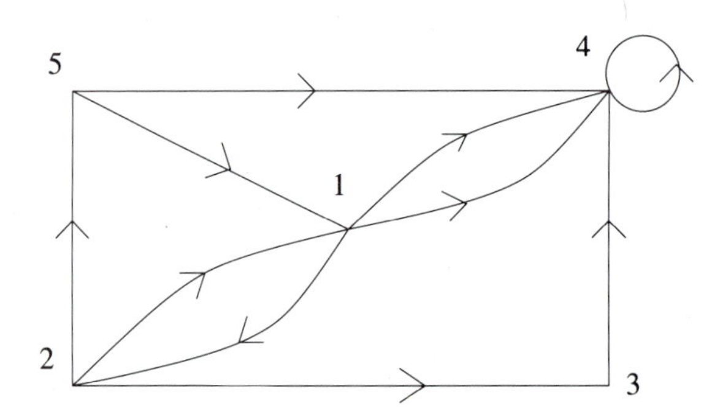

Figure 2.7. Digraph for Example 2.8.1

Exercise 2.8.1.

For each of the directed graphs in Figure 2.8 give its connectivity matrix C_d as well as the connectivity matrix C_g for the associated graph.

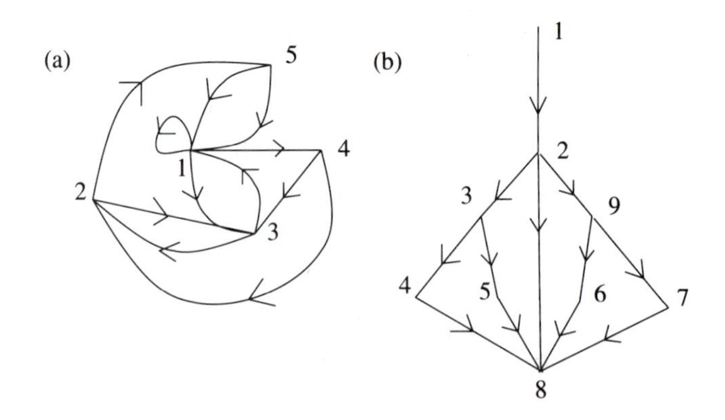

Figure 2.8. Graphs for Exercise 2.8.1

Exercise 2.8.2.

For each of the graphs in Figure 2.9 give its connectivity matrix.

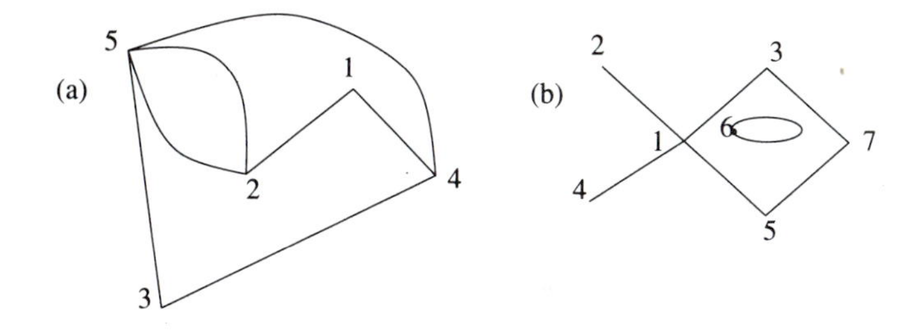

Figure 2.9. Graphs for Exercise 2.8.2

Exercise 2.8.3.

Consider the following connectivity matrix for a directed graph:

$$\begin{pmatrix} 1 & 0 & 1 & 1 & 0 \\ 0 & 0 & 1 & 0 & 1 \\ 1 & 1 & 0 & 0 & 0 \\ 0 & 1 & 1 & 0 & 0 \\ 2 & 0 & 0 & 0 & 0 \end{pmatrix}$$

Draw the corresponding directed graph.

Exercise 2.8.4.

Consider the following connectivity matrix for a graph:

$$\begin{pmatrix} 0 & 1 & 0 & 1 & 0 \\ 1 & 1 & 0 & 0 & 2 \\ 0 & 0 & 0 & 1 & 1 \\ 1 & 0 & 1 & 0 & 1 \\ 0 & 2 & 1 & 1 & 0 \end{pmatrix}$$

Draw the corresponding graph.

One of the natural settings in which graphs arise is a transportation network, such as the streets in a city. Here the vertices would correspond to the intersections and the edges to the streets joining them. If we wanted to distinguish one-way streets from two-way streets, then we could draw a directed graph, where each two-way street joining two vertices would correspond to two directed edges.

A basic question concerning graphs and directed graphs is whether two vertices are connected, and if so, what is the shortest path connecting them. In the context of this question, the graph may have a **weighting,** which is just a nonnegative function defined on the edges. This gives a length $w(e)$ to each edge and the length of a path $e_1, \ldots, e_k$ is $w(e_1) + \cdots + w(e_k)$. We will just use the constant weight function 1 here. Another question is how many paths there are which connect a given pair of vertices. This can be broken down into the subquestion of how many paths there are of a given length.

Since each path of length 1 corresponds to an edge, the answer for paths of length 1 is just given by c_{ij}. There is a path of length 2 from v_i to v_j exactly when there is an edge joining v_i to some other vertex v_k and an edge joining v_k to v_j. There will be such a path involving v_k as the intermediate vertex exactly when both c_{ik} and c_{kj} are not 0, and then the number of them will just be the product $c_{ik}c_{kj}$. Thus the total number of paths of length 2 is $\sum_{k=1}^{n} c_{ik}c_{kj}$. But this is just the ij-entry of C^2. More generally, we have the following proposition.

PROPOSITION 2.8.1. The number of paths of length exactly p in a graph (or digraph) joining v_i to v_j is given by the ij-entry of C^p.

Proof. We prove this by induction. It is true by the definition of the connectivity matrix when $p = 1$. We assume that it is true for $p = q$ and prove it for $p = q + 1$. If there is a path of length $q + 1$ from v_i to v_j then the first q edges in it give a path of length q from v_i to some v_k and the last edge gives a path of length 1 from v_k to v_j. For a fixed k there are b_{ik} paths from v_i to v_j, where by our induction assumption b_{ik} is the ik-entry of C^q. Thus there are $\sum_{k=1}^{n} b_{ik}c_{kj}$ paths of length $q + 1$ joining v_i to v_j. Since this number is the ij-entry of C^{q+1}, this proves the result. ■

COROLLARY 2.8.2. The total number of paths of length less than or equal to p from v_i to v_j is given by the ij-entry of $C + C^2 + \cdots + C^p$.

Exercise 2.8.5.

Prove Corollary 2.8.2.

To check whether a graph is connected is easy for a small graph, but may not be apparent for a more complicated graph. However, if there is a path joining two vertices, there must be a path with at most $n - 1$ edges in it. For if the path contained more than that many edges, it would have to hit some vertex twice, giving a loop which could be omitted from the path to get a path of smaller length which connects the two vertices. Thus we get the following criterion for when a graph is connected.

PROPOSITION 2.8.3. A graph with n vertices is connected iff all the entries of $C + C^2 + \cdots + C^{n-1}$ are nonzero.

Exercise 2.8.6.

For each of the following connectivity matrices of graphs use Proposition 2.8.3 to determine whether it is connected. Also, determine how many paths there are of length less than or equal to 5 which join the first vertex to the second one.

$$
\text{(a) } C = \begin{pmatrix} 0 & 1 & 0 & 0 & 0 \\ 1 & 0 & 0 & 0 & 0 \\ 0 & 0 & 1 & 0 & 0 \\ 0 & 0 & 0 & 1 & 1 \\ 0 & 0 & 0 & 1 & 1 \end{pmatrix}
$$

$$
\text{(b) } C = \begin{pmatrix} 0 & 0 & 0 & 0 & 0 & 0 & 1 & 0 & 0 \\ 0 & 0 & 1 & 1 & 1 & 0 & 0 & 0 & 0 \\ 0 & 1 & 1 & 0 & 0 & 0 & 0 & 1 & 0 \\ 0 & 1 & 0 & 0 & 2 & 0 & 0 & 0 & 1 \\ 0 & 1 & 0 & 2 & 0 & 1 & 0 & 1 & 0 \\ 0 & 0 & 0 & 0 & 1 & 0 & 1 & 0 & 0 \\ 1 & 0 & 0 & 0 & 0 & 1 & 0 & 1 & 0 \\ 0 & 0 & 1 & 0 & 1 & 0 & 1 & 1 & 0 \\ 0 & 0 & 0 & 1 & 0 & 0 & 0 & 0 & 0 \end{pmatrix}
$$

We now look at a second matrix which is assigned to a directed graph which has no isolated vertices and no simple loops, which is the **incidence matrix** N. We assume an ordering $v_1, \ldots, v_n$ of the vertices and an ordering $e_1, \ldots, e_m$ of the edges is given. Here the ij-entry records the relationship between the ith edge and the jth vertex. We have $n_{ij} = 0$ unless either $i(e_i) = v_j$ or $t(e_i) = v_j$. If $i(e_i) = v_j$, then $n_{ij} = -1$; if $t(e_i) = v_j$, then $n_{ij} = 1$. Thus the rows of N record which vertices a given edge runs between and the columns record which edges are running from or toward a given vertex.

■ ***Example 2.8.2.*** Figure 2.10 shows examples of directed graphs and their incidence matrices. To avoid confusion in the labeling of the vertices and edges, we use numbers for vertices and letters for edges. The ordering of the edges is given by alphabetical order.

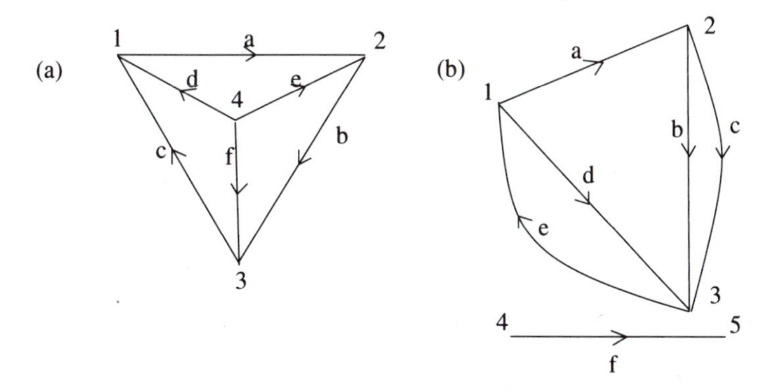

Figure 2.10. Directed graphs for Example 2.8.2

$$\text{(a)} \begin{pmatrix} -1 & 1 & 0 & 0 \\ 0 & -1 & 1 & 0 \\ 1 & 0 & -1 & 0 \\ 1 & 0 & 0 & -1 \\ 0 & 1 & 0 & -1 \\ 0 & 0 & 1 & -1 \end{pmatrix}, \quad \text{(b)} \begin{pmatrix} -1 & 1 & 0 & 0 & 0 \\ 0 & -1 & 1 & 0 & 0 \\ 0 & -1 & 1 & 0 & 0 \\ -1 & 0 & 1 & 0 & 0 \\ 1 & 0 & -1 & 0 & 0 \\ 0 & 0 & 0 & -1 & 1 \end{pmatrix}$$

Note that each row of the incidence matrix has one 1, one -1, and the other entries are 0. Thus the sum of all entries in a row are zero. Since this is true for each row, the sum of all the columns of N will be zero. This means that the vector $(1, 1, \ldots, 1)$ with all entries equal to 1 will be a solution to $N\mathbf{x} = \mathbf{0}$, since multiplying N by this vector just takes the sum of the columns of N. ■

We next consider the question of connectivity of the graph associated to a directed graph in terms of the incidence matrix of the directed graph. We first note that the question of finding a path which connects any two vertices v_i to v_j is equivalent to finding a path which connects the first vertex v_1 to any other vertex v_i. If we can always find such a path, we can get a path from v_i to v_j by first going backward along the path from v_1 to v_i (giving us a path from v_i to v_1) and then traversing the path from v_1 to v_j. We next note that a path from v_1 to v_i corresponds to a sequence of edges $e_1, \ldots, e_k$, together with the sign ± 1 for each edge. The sign is $+1$ if the edge is traversed in its usual order during the path, and is -1 if it is traversed in the opposite direction. The graph is connected if there is such a sequence of signed edges getting from v_1 to v_i for any i. Note that if an edge is traversed twice, then we could find a shorter path which doesn't include one of the occurrences of this edge. Thus we only consider paths which traverse a given edge at most once. Next note that corresponding to such a path is a row vector $\mathbf{p}$ which has $p_j = 0$ if the jth edge is not traversed in the path, $p_j = -1$ if the jth edge is traversed in the wrong direction, and $p_j = 1$ if it is traversed in the correct direction.

■ ***Example 2.8.3*** Figure 2.11 shows a path in a directed graph. The path is given by the signed edges $a, -b, -c$. The corresponding row vector and incidence matrix are

$$\mathbf{p} = (1 \quad -1 \quad -1 \quad 0 \quad 0 \quad 0 \quad 0 \quad 0), N = \begin{pmatrix} -1 & 0 & 1 & 0 & 0 \\ 0 & 0 & 1 & -1 & 0 \\ 0 & 0 & 0 & 1 & -1 \\ 0 & 0 & 1 & 0 & -1 \\ 0 & 1 & 0 & 0 & -1 \\ 0 & -1 & 1 & 0 & 0 \\ -1 & 0 & 0 & 1 & 0 \\ -1 & 1 & 0 & 0 & 0 \end{pmatrix}$$

The useful property that the row vector $\mathbf{p}$ associated to a path connecting e_1 to e_i has is that when we multiply it times N to form $\mathbf{p}N$ we just get the vector $\pm(\mathbf{e}_1 - \mathbf{e}_i)$. This is true because the initial row vector corresponding to the first edge is of this type, and each new edge which is added just cancels off the term corresponding to the vertex where the two edges are joined: $(\mathbf{e}_1 - \mathbf{e}_k) + (\mathbf{e}_k - \mathbf{e}_l) = \mathbf{e}_1 - \mathbf{e}_l$. This uses the fact that when one edge

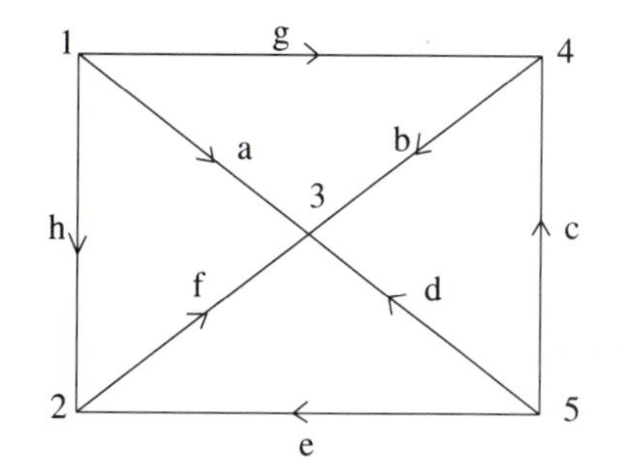

Figure 2.11. Graph for Example 2.8.3

comes into a vertex, the next edge must leave it. Thus the connectivity of the graph is equivalent to the existence of row vectors $\mathbf{p}(i)$ with $\mathbf{p}(i)N = \mathbf{e}_1 - \mathbf{e}_i$ for $i = 2, \ldots, n$. When we take the transpose, we get the condition that $N^t\mathbf{p}(i)^t = \mathbf{e}_1 - \mathbf{e}_i$, where we are now regarding the right-hand side as column vectors.

Thus we see that the graph being connected is equivalent to the vectors $\mathbf{e}_1 - \mathbf{e}_i$ being in $\mathcal{R}(N^t)$. But $\mathcal{N}(N)$ is characterized as being all vectors which are perpendicular to all of $\mathcal{R}(N^t)$. Thus any vector in $\mathcal{N}(N)$ must be perpendicular to each of the vectors $\mathbf{e}_1 - \mathbf{e}_i, i = 2, \ldots, n$. But then $< \mathbf{x}, \mathbf{e}_1 - \mathbf{e}_i >= x_1 - x_i = 0$. Hence $\mathbf{x}$ is multiple of $(1, 1, \ldots, 1) = \mathbf{e}_1 + \cdots + \mathbf{e}_n$. ■

We summarize our results in the following theorem.

THEOREM 2.8.4. Let D be a directed graph without simple loops and G the associated graph. Let N be the incidence matrix of D. Then

1. The vector $\mathbf{s} = \mathbf{e}_1 + \cdots + \mathbf{e}_n = (1, \ldots, 1)$ satisfies $N\mathbf{x} = \mathbf{0}$.
2. The graph G is connected exactly when the solutions of $N\mathbf{x} = \mathbf{0}$ consist of multiples of $\mathbf{s}$.

Exercise 2.8.7.

For the following incidence matrix of a directed graph, determine whether the graph is connected by computing the solutions of $N\mathbf{x} = \mathbf{0}$.

$$\begin{pmatrix} 0 & -1 & 0 & 1 & 0 & 0 & 0 & 0 \\ 0 & 0 & -1 & 0 & 1 & 0 & 0 & 0 \\ 0 & 0 & -1 & 0 & 0 & 0 & 1 & 0 \\ -1 & 0 & 0 & 0 & 1 & 0 & 0 & 0 \\ -1 & 0 & 0 & 0 & 0 & 0 & 0 & 1 \\ 1 & 0 & 0 & 0 & -1 & 0 & 0 & 0 \\ 1 & 0 & 0 & 0 & 0 & -1 & 0 & 0 \end{pmatrix}$$

We next look at how the loops in a graph are related to the null space of N^t. A loop is called **elementary** if each directed edge which occurs in the loop only occurs once (with sign ± 1). If a loop is not elementary, it will contain within it an elementary subloop formed by leaving out certain repeated edges. We will just concentrate on understanding elementary loops. Now each elementary loop may be recorded by the edges which it contains, each with a coefficient of ± 1, where the sign indicates whether the edge is traversed in the preferred direction or in the opposite direction. When the edges are given an order, as we have done by alphabetizing them, this then determines a vector $\mathbf{p}$ with E entries of ± 1, where E denotes the number of edges.

In Example 2.8.3, the vector $(1, -1, 0, 0, 0, 0, 0, -1)$ would indicate the elementary loop which first transverses a, then b in the opposite direction, and then g in the opposite direction. Note that this notation doesn't indicate what the starting vertex is for the loop. This is useful in that whenever there is a loop, we can form other loops from it by just changing the starting vertex but still traversing the edges in the same (cyclical) order. For example, this vector also indicates the loop which first goes over b in the opposite direction, then over g in the opposite direction, then over a. The row vector $\mathbf{p}$ corresponding to each elementary loop will satisfy $\mathbf{p}N = \mathbf{0}$. For multiplying by $\mathbf{p}$ just takes a linear combination of the rows of N. The fact that we have an elementary loop means that nonzero entries of a row corresponding to a vertex will occur with canceling signs from the two terms corresponding to an edge coming into the vertex and then the next edge leaving it.

By taking the transpose, these elementary loops determine elements of $\mathcal{N}(N^t)$. If the graph is connected, Theorem 2.8.4 says that the dimension of $\mathcal{R}(N^t)$ is $V - 1$, where V is the number of vertices of the graph. Thus the rank–nullity theorem implies that the dimension of $\mathcal{N}(N^t)$ is $E - V + 1$. It turns out that we can find a basis for this subspace corresponding to the elementary loops. We will show how to find this basis as a result of the usual Gaussian elimination algorithm. Rather than proving that this works generally, we will just look at an example to illustrate this procedure.

■ ***Example 2.8.4.*** We look at the matrix in the last example. Then $N = O_1 U$, where

$$U = \begin{pmatrix} -1 & 0 & 1 & 0 & 0 \\ 0 & 1 & 0 & 0 & -1 \\ 0 & 0 & 1 & 0 & -1 \\ 0 & 0 & 0 & 1 & -1 \\ 0 & 0 & 0 & 0 & 0 \\ 0 & 0 & 0 & 0 & 0 \\ 0 & 0 & 0 & 0 & 0 \\ 0 & 0 & 0 & 0 & 0 \end{pmatrix}, O_1 = \begin{pmatrix} 1 & 0 & 0 & 0 & 0 & 0 & 0 & 0 \\ 0 & 0 & 0 & 0 & 1 & 0 & 0 & 0 \\ 0 & 0 & 0 & 1 & 0 & 0 & 0 & 0 \\ 0 & 0 & 1 & 0 & 0 & 0 & 0 & 0 \\ 0 & 1 & 1 & -1 & 0 & 0 & 0 & 0 \\ 0 & 0 & 0 & -1 & 1 & 1 & 0 & 0 \\ -1 & 0 & -1 & 1 & 0 & 0 & 1 & 0 \\ -1 & 0 & 0 & 1 & -1 & 0 & 0 & 1 \end{pmatrix}$$

The bottom four rows of O_1 then give a basis of vectors corresponding to elementary loops. In terms of the labeling in the graph, these loops correspond to $bd^{-1}c, d^{-1}ef$, $a^{-1}gc^{-1}d, a^{-1}he^{-1}d$. In this notation, traversing a directed edge in the opposite direction is indicated by the superscript -1. Other elementary loops arise from these four by taking integer linear combinations. For example, the elementary loop coming from the perimeter is $gc^{-1}eh^{-1}$, which corresponds to the vector $(0, 0, -1, 0, 1, 0, 1, -1)$.

This is the linear combination

$$(-1,0,-1,1,0,0,1,0)-(-1,0,0,1,-1,0,0,1) \qquad \blacksquare$$

We close this section by commenting on how this applies to electric circuits, as discussed in Section 1.9. For circuits vertices are called nodes, and edges are called branches. Each branch will have a resistor of positive resistance, and may also have a voltage source. Once we label the currents in each branch of the circuit and the nodes, we can form a directed graph. We are assuming that the circuit leads to a connected graph. We then can take the incidence matrix N of this directed graph. Our analysis is for a circuit with no external current sources at the nodes, but it could be easily modified to allow these if they sum to 0. There is a vector $\mathbf{I}$ whose size corresponds to the number E of edges. There is also a vector $\mathbf{V}_S$ of size E which gives the impressed voltages from sources along each edge. Finally, there is a diagonal E by E matrix R whose positive diagonal entries give the resistances along each edge. Using R and $\mathbf{I}$ we can use Ohm's law to form a vector $\mathbf{V}_R = R\mathbf{I}$ which gives the voltage drops due to resistance along each edge. The difference $\mathbf{V}_S - \mathbf{V}_R$ gives the change in potential along the branches.

When we write down the equations corresponding to the fact that the sum of the currents coming into each node is zero, this gives V equations, which just correspond to the product of a row vector $\mathbf{I}^t$ containing the currents in each edge times each of the columns of the incidence matrix is $\mathbf{0}$. This uses the fact that the columns of N just indicate which edges go from or to a given vertex. These equations are then $\mathbf{I}^t N = \mathbf{0}$. Since the columns of N sum to zero, any one of the equations can be written in terms of the remaining ones. This justifies our procedure of using all but one of the nodes in forming these equations. This gives $V-1$ equations,

$$\overline{N}^t\mathbf{I} = \mathbf{0},$$

where $\overline{N}$ is the matrix formed from N by deleting the last column. Note that the rank of $\overline{N}$ is in fact $V-1$ since the rank of N is $V-1$ and the last column is the negative of the sum of the other columns.

The other equations come from Kirchhoff's law about the voltage drops around a loop, together with Ohm's law which writes the voltage drops due to resistors as $R\mathbf{I}$. To get the other equations we use the fact that we can restrict to elementary loops, and just have to choose a basis of $\mathcal{N}(N^t)$ of such loops. For a small circuit this is fairly easy to do by inspection. For a larger circuit we may use the algorithm given earlier to find a basis of elementary loops. This basis leads to $E-V+1$ more equations. Let us denote by L the $E-V+1$ by E matrix which has as its rows the basis of elementary loops. Note first $\mathcal{N}(L) = \mathcal{R}(N) = \mathcal{R}(\overline{N})$—details are left as an exercise. The remaining equations express the fact that the potential changes around a loop are 0. They are just $L(\mathbf{V}_S) - L(\mathbf{V}_R) = \mathbf{0}$, or $(LR)\mathbf{I} = L(R\mathbf{I}) = L(\mathbf{V}_R) = L(\mathbf{V}_S)$. Thus our equations can be written as

$$\overline{N}^t\mathbf{I} = \mathbf{0}, LR\mathbf{I} = L\mathbf{V}_S \text{ or } \begin{pmatrix} \overline{N}^t \\ LR \end{pmatrix}\mathbf{I} = \begin{pmatrix} \mathbf{0} \\ L\mathbf{V}_S \end{pmatrix}$$

The question arises as to why this system has a unique solution for the current. First note that $L(\mathbf{V}_S - R\mathbf{I}) = \mathbf{0}$. Since $\mathcal{N}(L) = \mathcal{R}(\overline{N})$, there is only a solution if

$\mathbf{V}_S - R\mathbf{I} = \overline{N}\mathbf{x}$ for some $\mathbf{x}$. The vector $\mathbf{x}$ is interpreted as potentials at each node (where the potential is set to 0 at the last node which was omitted in forming $\overline{N}$ from N). The vector $\overline{N}\mathbf{x}$ then measures the change in potential across the branches, and the equation $\overline{N}\mathbf{x} = \mathbf{V}_S - RI$ expresses the change in potential across a branch as the difference of the source voltage on the branch and the voltage drop across the resistor. Then our equations above become

$$R\mathbf{I} + \overline{N}\mathbf{x} = \mathbf{V}_S, \overline{N}^t\mathbf{I} = \mathbf{0}$$

Solutions to this system correspond to solutions to our original system. This may look more complicated since we now have two variables $\mathbf{x}, \mathbf{I}$ to solve for, but it turns out to be relatively easy to solve. Multiply the first equation by $\overline{N}^t R^{-1}$ and subtract from the second to get

$$-\overline{N}^t R^{-1}\overline{N}\mathbf{x} = -\overline{N}^t R^{-1}\mathbf{V}_S.$$

Since the matrix $\overline{N}$ is of rank $V - 1$, a short argument shows that the square matrix $\overline{N}^t R^{-1}\overline{N}$ is invertible. Thus we can solve for a unique $\mathbf{x}$ and back substitute to solve for $\mathbf{I}$. Note that this method using potentials of solving the problem leads to the smaller $V - 1$ by $V - 1$ system $\overline{N}^t R^{-1}\overline{N} = \overline{N}^t R^{-1}\mathbf{V}_S$ than the original program from 1.9 of setting up the equations directly from Kirchhoff's laws and solving the system of E by E system $\begin{pmatrix} N^t \\ LR \end{pmatrix}\mathbf{I} = \begin{pmatrix} \mathbf{0} \\ L\mathbf{V}_S \end{pmatrix}$ which results.

■ ***Example 2.8.5.*** Consider the circuit of Example 1.9.2. Then we have

$$N = \begin{pmatrix} -1 & 1 & 0 & 0 \\ 0 & -1 & 1 & 0 \\ 0 & -1 & 0 & 1 \\ 0 & 0 & -1 & 1 \\ -1 & 0 & 0 & 1 \\ -1 & 0 & 1 & 0 \end{pmatrix}, \overline{N} = \begin{pmatrix} -1 & 1 & 0 \\ 0 & -1 & 1 \\ 0 & -1 & 0 \\ 0 & 0 & -1 \\ -1 & 0 & 0 \\ -1 & 0 & 1 \end{pmatrix},$$

$$L = \begin{pmatrix} 1 & 0 & 1 & 0 & -1 & 0 \\ 0 & 1 & -1 & 1 & 0 & 0 \\ 0 & 0 & 0 & -1 & 1 & -1 \end{pmatrix}, R = \begin{pmatrix} 5 & 0 & 0 & 0 & 0 & 0 \\ 0 & 4 & 0 & 0 & 0 & 0 \\ 0 & 0 & 4 & 0 & 0 & 0 \\ 0 & 0 & 0 & 1 & 0 & 0 \\ 0 & 0 & 0 & 0 & 5 & 0 \\ 0 & 0 & 0 & 0 & 0 & 2 \end{pmatrix}$$

Thus the matrix $\begin{pmatrix} N^t \\ LR \end{pmatrix} = \begin{pmatrix} -1 & 0 & 0 & 0 & -1 & -1 \\ 1 & -1 & -1 & 0 & 0 & 0 \\ 0 & 1 & 0 & -1 & 0 & 1 \\ 5 & 0 & 4 & 0 & -5 & 0 \\ 0 & 4 & -4 & 1 & 0 & 0 \\ 0 & 0 & 0 & -1 & 5 & -2 \end{pmatrix}$ is the matrix we used in our solution in Example 1.9.2. ■

Using the method of potentials above, we could form $\overline{N}^t R^{-1}\overline{N}\mathbf{x} = \overline{N}^t R^{-1}\mathbf{V}_S$,

$$\begin{pmatrix} .9 & -.2 & -.5 \\ -.2 & .7 & -.25 \\ -.5 & -.25 & 1.75 \end{pmatrix} \mathbf{x} = \begin{pmatrix} -2 \\ 2 \\ 0 \end{pmatrix}$$

with solution $\mathbf{x} = (-1.8303, 2.2629, -0.1997)$. Then we solve for

$$I = R^{-1}(V_S - \overline{N}\mathbf{x}) = (1.1814, .6156, .5657, -.1997, -.3661, -.8153).$$

Exercise 2.8.8.

Show that $\mathcal{N}(L) = \mathcal{R}(N) = R(\overline{N})$. Hint: Find an inclusion and use a dimension count.

Exercise 2.8.9.

Verify that the matrix $\overline{N}^t R^{-1}\overline{N}$ is invertible.
Hint: Show that $\{\mathbf{0}\} = \mathcal{N}(\overline{N}) = \mathcal{N}(\overline{N}^t R^{-1}\overline{N})$ by multiplying the equation

$$\overline{N}^t R^{-1}\overline{N}\mathbf{y} = \mathbf{0}$$

on the left by $\mathbf{y}^t$.

2.9. CHAPTER 2 EXERCISES

2.9.1. Determine which of the following subsets of $\mathbb{R}^3$ are subspaces. For each one which is a subspace, give a geometric description of this subspace and find a basis.
(a) $\{\mathbf{x} : x_1x_2 + x_3 = 0\}$
(b) $\{\mathbf{x} : 3x_1 + x_2 - x_3 = 2\}$
(c) $\{\mathbf{x} : \langle \mathbf{x}, (1,3,2)\rangle = 0\}$
(d) $\{\mathbf{x} : \langle \mathbf{x}, \mathbf{x}\rangle = 1\}$

2.9.2. Determine which of the following subsets of $\mathbb{R}^3$ are subspaces. For those that are subspaces, give a geometric description of this subspace and find a basis.
(a) $\{\mathbf{x} : 3x_1 + x_2 - x_3 = 0\}$
(b) $\{\mathbf{x} : x_1^2 + x_2^2 - x_3^2 = 1\}$
(c) $\{\mathbf{x} : x_1^2 + x_2^2 - x_3^2 = 0\}$
(d) $\{\mathbf{x} : \langle \mathbf{x}, (1,2,1)\rangle = 0, \langle \mathbf{x}, (0,2,3)\rangle = 0\}$

2.9.3. Determine which of the following subsets of $\mathbb{R}^4$ are subspaces. For those that are subspaces, find a basis.
(a) $\{\mathbf{x} : x_1 + x_2 = x_3 + x_4\}$
(b) $\{\mathbf{x} : x_1^2 + x_2^2 = x_3^2 + x_4^2\}$
(c) $\{\mathbf{x} : x_1^2 + x_2^2 = 0, x_3^2 + x_4^2 = 0\}$
(d) $\{\mathbf{x} : x_1^2 + x_2^2 = 1, x_3^2 + x_4^2 = 0\}$
(e) $\{\mathbf{x} : (x_1 + x_2)^2 + (x_3 + x_4)^2 = 0\}$

2.9.4. Determine which of the following subsets of $\mathcal{P}^3(\mathbb{R}, \mathbb{R})$ are subspaces. For those that are subspaces, find a basis.
(a) the polynomials of degree greater than 1
(b) the polynomials $a_0 + a_1 x + a_2 x^2 + a_3 x^3$ with $a_0 + a_2 = 0$
(c) the polynomials $a_0 + a_1 x + a_2 x^2 + a_3 x^3$ with $a_0 + a_2 = 1$
(d) the polynomials p with $p(2) - p(3) = 0$

2.9.5. Determine which of the following subsets of $\mathcal{P}^4(\mathbb{R}, \mathbb{R})$ are subspaces. For those that are subspaces, find a basis.
(a) the polynomials of degree 4
(b) the polynomials with $p(0) = 4$
(c) the polynomials with $p(4) = 0$
(d) the polynomials with at least a double root; that is, $p(x) = (x - r)^2 q(x)$ for some r
(e) the polynomials with a double root at 3; that is, $p(x) = (x - 3)^2 q(x)$
(f) the polynomials with $\int_0^1 p(x)\,dx = 0$

2.9.6. Consider the following subsets of 2 by 2 matrices

$$A = \begin{pmatrix} a_{11} & a_{12} \\ a_{21} & a_{22} \end{pmatrix}.$$

Determine which are subspaces. For those that are subspaces, find a basis.
(a) the matrices which are upper triangular: $a_{21} = 0$
(b) the matrices which are invertible
(c) the matrices satisfying $a_{11} + a_{22} = 0$
(d) the matrices satisfying $a_{11} a_{22} = 1$

2.9.7. Let S be the subset of $\mathbb{R}^4$ consisting of solutions of the equations

$$\begin{aligned} 2x - 2y + z - w &= 0 \\ x - y + z &= 0 \end{aligned}$$

(a) Show that S is a subspace of $\mathbb{R}^4$.
(b) Give a matrix A so that $S = \mathcal{N}(A)$.
(c) Find a basis for S.

2.9.8. Let S be the subset of $\mathbb{R}^3$ consisting of solutions of

$$\begin{aligned} x + 2y - 3z &= 0 \\ 3x + 4y + z &= 0 \end{aligned}$$

(a) Show that S is a subspace.
(b) Give a matrix A so that $S = \mathcal{N}(A)$.
(c) Give a matrix B so that $S = \mathcal{N}(B^t)$.
(d) Give a matrix C so that $S = \mathcal{R}(C)$.
(e) Give a matrix D so that $S = \mathcal{R}(D^t)$.

2.9.9. Let S be the subset of $\mathbb{R}^5$ satisfying the equations

$$\begin{aligned} x_1 + x_3 + x_5 &= 0 \\ x_2 + x_4 &= 0 \\ x_1 + x_2 + x_3 + x_4 + x_5 &= 0 \end{aligned}$$

(a) Show that S is a subspace.
(b) Find a basis for S.
(c) Find a matrix A so the S is the column space of A.
(d) Find a matrix A so that S is the row space of A.

2.9.10. Let S be the subspace of $\mathbb{R}^3$ which is spanned by the three vectors $(1, 1, 1)$, $(3, -1, 2)$, $(5, 1, 4)$. Let T be the subset of all vectors which are perpendicular to each of these three vectors.
(a) Show T is a subspace.
(b) Give a basis for S and a basis for T.
(c) Give a matrix A so that $S = \mathcal{R}(A)$.
(d) Give a matrix B so that $T = \mathcal{N}(B)$.

2.9.11. Let S be the subset of $\mathbb{R}^4$ which is spanned by the two vectors $(1, 2, 3, 4)$, $(4, 5, 6, 7)$. Let T be the subset of all vectors which are perpendicular to each of these vectors.
(a) Show T is a subspace.
(b) Give a basis for S and a basis for T.
(c) Give a matrix A so that $S = \mathcal{N}(A)$.
(d) Give a matrix B so that $T = \mathcal{N}(B)$.

2.9.12. Is the union of two subspaces necessarily a subspace? Give a proof or a counter-example.

2.9.13. Show that $(1, 2, 3), (2, 3, 1), (3, 2, 1), (3, 1, 2)$ span $\mathbb{R}^3$. Are they independent? Justify your answer.

2.9.14. Determine whether the following vectors span $\mathbb{R}^4$ and whether they are independent.
(a) $(1, 2, 3, 4), (1, -1, 2, 2), (3, 1, 2, -3), (-4, -3, -2, 1)$
(b) $(1, 2, -2, -1), (3, 2, -2, -3), (4, -2, 2, -4), (5, -2, -2, -1)$

2.9.15. Determine whether the following vectors span the subspace S of vectors in $\mathbb{R}^5$ which are perpendicular to both $(1, 1, 1, 1, 1)$ and $(1, 0, 0, 0, -1)$
(a) $(1, 2, 3, -7, 1), (2, 1, 3, -8, 2), (3, 2, 1, -9, 3)$
(b) $(1, 2, 3, -7, 1), (2, 1, 3, -8, 2), (3, 3, 6, -15, 3)$

2.9.16. Determine whether the following polynomials in $\mathcal{P}^3(\mathbb{R}, \mathbb{R})$ are independent. For those which are not independent, find a basis for the subspace which they span.
(a) $1 - x + x^2 - x^3, 2 - x - x^3, x - x^2, -1 + x^2$
(b) $3 - x + x^2, 3 + 2x + x^3, 1 + x^3, 1 + x + x^2 + x^3$

2.9.17. Determine whether the following polynomials in $\mathcal{P}^4(\mathbb{R}, \mathbb{R})$ are independent. For those which are not independent, find a basis for the subspace which they span.
(a) $1 - x + x^2 - x^3 + x^4, 2 + x^2 + x^4, x - x^3, -3 - 2x^2 + 2x^3 - 2x^4$
(b) $1 - x + x^3, 4 + x^2 + x^4, x + x^2 + x^3, 4 - x^4$

2.9.18. Determine whether the following matrices in $\mathcal{M}(2, 2)$ are independent. If they are not independent, find a basis for the subspace which they span.

(a) $\begin{pmatrix} 1 & -1 \\ 1 & 1 \end{pmatrix}, \begin{pmatrix} 2 & 1 \\ -1 & 3 \end{pmatrix}, \begin{pmatrix} 4 & 2 \\ -2 & 1 \end{pmatrix}, \begin{pmatrix} 3 & -3 \\ 3 & 2 \end{pmatrix}$

(b) $\begin{pmatrix} 1 & 1 \\ 1 & 1 \end{pmatrix}, \begin{pmatrix} 2 & 1 \\ 0 & 2 \end{pmatrix}, \begin{pmatrix} 0 & 1 \\ 2 & 0 \end{pmatrix}$

2.9.19. Let A be an m by n matrix. Show that $A\mathbf{x} = \mathbf{b}$ has a solution iff the rank of A is the rank of the augmented matrix $(A|\mathbf{b})$.

2.9.20. Suppose A is a n by n matrix which is invertible. What is $\mathcal{R}(A)$? What is $\mathcal{N}(A)$?

2.9.21. Suppose A is an n by n matrix which is not invertible. Show that $\mathcal{N}(A)$ contains a nonzero vector $\mathbf{x}$ and there is a vector $\mathbf{b}$ for which one can't solve $A\mathbf{x} = \mathbf{b}$.

2.9.22. Consider the subspace S of $\mathbb{R}^4$ which is spanned by the two vectors $\mathbf{v}_1 = (1, -1, 0, 1)$, $\mathbf{v}_2 = (2, 0, -1, -1)$.

(a) Find necessary and sufficient conditions on a vector $\mathbf{x} = (x_1, x_2, x_3, x_4)$ to be in S.

(b) Find two more vectors $\mathbf{v}_3, \mathbf{v}_4$ so that the four vectors are a basis for $\mathbb{R}^4$.

2.9.23. Consider the subspace S of $\mathbb{R}^4$ which is spanned by the three vectors $\mathbf{v}_1 = (1, 3, -5, 1)$, $\mathbf{v}_2 = (2, 3, -4, -1)$, $\mathbf{v}_3 = (3, 1, -2, -2)$.

(a) Find necessary and sufficient conditions on a vector $\mathbf{x} = (x_1, x_2, x_3, x_4)$ to be in S.

(b) Find another vector $\mathbf{v}_4$ so that the four vectors are a basis of $\mathbb{R}^4$.

2.9.24. Consider the subspace S of $\mathcal{P}^2(\mathbb{R}, \mathbb{R})$ which is spanned by the three polynomials $2 + x + x^2, 3 - 2x + 3x^2, 1 + 4x - x^2$.

(a) Show that these polynomials are dependent.

(b) Find a basis for S.

(c) Find another polynomial p so that your basis from (b) together with p forms a basis for $\mathcal{P}^2(\mathbb{R}, \mathbb{R})$.

2.9.25. Suppose that $f_1, \ldots, f_n \in \mathcal{D}(\mathbb{R}, \mathbb{R})$. Define the Wronskian

$$W(f_1, \ldots, f_n)(0) = \det \begin{pmatrix} f_1(0) & f_2(0) & \cdots & f_n(0) \\ f_1'(0) & f_2'(0) & \cdots & f_n'(0) \\ \cdots & \cdots & \cdots & \cdots \\ f_1^{(n-1)}(0) & f_2^{(n-1)}(0) & \cdots & f_n^{(n-1)}(0) \end{pmatrix}$$

(a) Show that if $W(f_1, \ldots, f_n)(0) \neq 0$, then $f_1, \ldots, f_n$ are independent.

(b) Use this to show that $\cos x, e^x, x$ are independent.

2.9.26. Suppose that $L : V \to R^n$ is a linear transformation.

(a) Suppose that $\mathbf{v}_1, \ldots, \mathbf{v}_n$ are n vectors in V and A is the matrix whose columns are $L(\mathbf{v}_1), \ldots, L(\mathbf{v}_n)$. Show that if $\det A \neq 0$, then $\mathbf{v}_1, \ldots, \mathbf{v}_n$ are independent.

(b) Show that $L : \mathcal{D}(R, R) \to \mathbb{R}^n$ given by $L(f) = (f(0), f'(0), \ldots, f^{(n-1)}(0))$ is a linear transformation.

(c) Use parts (a) and (b) to give a proof of part (a) of the previous exercise.

2.9.27. Determine whether the vector $(4, 0, 8, 2)$ is in the span of the vectors $(2, 1, 3, 2)$, $(1, 2, 1, 2)$, $(0, 1, 1, 0)$. If so, find a, b, c so that

$$(4, 0, 8, 2) = a(2, 1, 3, 2) + b(1, 2, 1, 2) + c(0, 1, 1, 0)$$

2.9.28. Determine whether the polynomial $4 + 8x^2 + 2x^3$ is in the span of the polynomials $2 + x + 3x^2 + 2x^3, 1 + 2x + x^2 + 2x^3, x + x^2$. If so, find a, b, c so that

$$4 + 8x^2 + 2x^3 = a(2 + x + 3x^2 + 2x^3) + b(1 + 2x + x^2 + 2x^3) + c(x + x^2)$$

2.9.29. Determine whether the following vectors are linearly independent. If they are dependent, write one of them as a linear combination of the others.
(a) $(1, 2, 1), (3, 1, 1), (0, 5, 2)$
(b) $(1, 2, 3), (2, 3, 1), (3, 1, 2)$

2.9.30. Determine whether the following vectors are linearly independent. If they are dependent, write one of them as a linear combination of the others.
(a) $(1, 2, 1, 0), (-2, 1, 3, 2), (2, 3, 2, 1)$
(b) $(1, 2, 1, 0), (-2, 1, 3, 2), (1, 7, 6, 2)$

2.9.31. Determine whether the following polynomials are linearly independent. If they are dependent, write one of them as a linear combination of the others.
(a) $2 - x + x^2, -x + x^2, 1 + x + x^2$
(b) $2 - x + x^2, -x + x^2, 1 - x + x^2$

2.9.32. Determine whether the following matrices are linearly independent. If they are dependent, write one of them as a linear combination of the others.

(a) $\begin{pmatrix} 1 & 1 \\ 0 & 1 \end{pmatrix}, \begin{pmatrix} 1 & 0 \\ 1 & 1 \end{pmatrix}, \begin{pmatrix} 1 & 1 \\ 1 & 0 \end{pmatrix}$

(b) $\begin{pmatrix} 1 & 1 \\ 0 & 1 \end{pmatrix}, \begin{pmatrix} 1 & 0 \\ 1 & 1 \end{pmatrix}, \begin{pmatrix} 1 & 1 \\ 1 & 0 \end{pmatrix}, \begin{pmatrix} 0 & 1 \\ 1 & 1 \end{pmatrix}$

2.9.33. Determine whether the following matrices in $\mathcal{M}(2, 2)$ are linearly independent. If they are dependent, write one of them as a linear combination of the others.

(a) $\begin{pmatrix} 1 & 1 \\ 1 & 1 \end{pmatrix}, \begin{pmatrix} 1 & 2 \\ 2 & 1 \end{pmatrix}, \begin{pmatrix} 1 & 2 \\ 3 & 4 \end{pmatrix}$

(b) $\begin{pmatrix} 1 & 0 \\ 0 & 0 \end{pmatrix}, \begin{pmatrix} 1 & 1 \\ 0 & 0 \end{pmatrix}, \begin{pmatrix} 1 & 1 \\ 1 & 0 \end{pmatrix}, \begin{pmatrix} 1 & 1 \\ 1 & 1 \end{pmatrix}, \begin{pmatrix} 1 & 2 \\ 2 & 1 \end{pmatrix}$

2.9.34. (a) Give a basis for the subspace of polynomials spanned by

$$1 - x^2 + x^3, 1 + x - x^2, x - x^3, 1 - x + x^2$$

(b) Give a basis for the subspace of $\mathbb{R}^4$ spanned by the vectors

$$(1, 0, -1, 1), (1, 1, -1, 0), (0, 1, 0, -1), (1, -1, 1, 0)$$

2.9.35. Determine which of the following functions are linear transformations.
(a) $L : \mathbb{R}^3 \to \mathbb{R}^3, L(x, y, z) = (x, x + y, x + y + z)$.
(b) $L : \mathcal{P}^3(\mathbb{R}, \mathbb{R}) \to \mathbb{R}^3, L(p) = p - p(1)$.
(c) $L : \mathcal{M}(2, 2) \to \mathcal{M}(2, 2), L(A) = ABA, B = \begin{pmatrix} 1 & 2 \\ 2 & 1 \end{pmatrix}$.
(d) $L : \mathcal{M}(2, 2) \to \mathbb{R}^4, L(A) = (a_{11} + a_{12}, a_{21} + a_{22}, a_{11} + a_{22}, a_{12} + a_{21})$ where $A = \begin{pmatrix} a_{11} & a_{12} \\ a_{21} & a_{22} \end{pmatrix}$.

2.9.36. Determine which of the following functions are linear transformations.

(a) $L : \mathcal{D}(\mathbb{R}, \mathbb{R}) \to \mathbb{R}$, $L(f) = f(3) - f(2)$.

(b) $L : \mathcal{D}(\mathbb{R}, \mathbb{R}) \to \mathcal{D}(\mathbb{R}, \mathbb{R})$, $L(f)(x) = f(\sin x)$.

(c) $L : \mathcal{D}(\mathbb{R}, \mathbb{R}) \to \mathcal{D}(\mathbb{R}, \mathbb{R})$, $L(f)(x) = \sin(f(x))$.

(d) $L : \mathbb{R} \to \mathbb{R}$, $L(x) = \sin(x)$.

(e) $L : \mathcal{D}(\mathbb{R}, \mathbb{R}) \to \mathcal{D}(\mathbb{R}, \mathbb{R})$, $L(f) = f' - f^2$, where f' denotes differentiation.

2.9.37. Determine which of the following functions are linear transformations. For those which are, determine their null spaces and ranges.

(a) $L : \mathcal{D}(\mathbb{R}, \mathbb{R}) \to \mathcal{D}(\mathbb{R}, \mathbb{R})$, $L(f)(x) = \int_1^x f(t)\,dt$.

(b) $L : \mathcal{P}^2 \to \mathcal{P}^3$, $L(f)(x) = \int_1^x f(t)\,dt$.

2.9.38. Determine which of the following functions are linear transformations. For those which are, determine bases for their null spaces and ranges.

(a) $L : \mathcal{M}(2,2) \to \mathcal{M}(3,3)$, $L\begin{pmatrix} a_{11} & a_{12} \\ a_{21} & a_{22} \end{pmatrix} = \begin{pmatrix} a_{11} & a_{12} & 0 \\ a_{21} & a_{22} & 0 \\ 0 & 0 & 1 \end{pmatrix}$

(b) $L : \mathcal{M}(2,2) \to \mathcal{M}(3,3)$, $L\begin{pmatrix} a_{11} & a_{12} \\ a_{21} & a_{22} \end{pmatrix} = \begin{pmatrix} a_{11} & a_{12} & 0 \\ a_{21} & a_{22} & 0 \\ 0 & 0 & 0 \end{pmatrix}$

2.9.39. Determine which of the following functions are linear transformations. For those which are, determine bases for their null spaces and ranges.

(a) $L : \mathcal{P}^2 \to \mathcal{M}(2,2)$, $L(a_0 + a_1x + a_2x^2) = \begin{pmatrix} a_0 & a_1 \\ a_1 & a_2 \end{pmatrix}$

(b) $L : \mathcal{P}^2(\mathbb{R}, \mathbb{R}) \to \mathcal{D}(\mathbb{R}, \mathbb{R})$,

$$L(a_0 + a_1x + a_2x^2) = a_0\cos(t) + a_1\sin(t) + a_2\cos(1+t)$$

2.9.40. Let $L : V \to W$ be a linear transformation.

(a) Show that if $\dim V > \dim W$, then L is not $1-1$.

(b) Show that if $\dim V < \dim W$, then L is not onto.

2.9.41. Suppose that $L : \mathcal{M}(2,2) \to \mathcal{P}^2(\mathbb{R}, \mathbb{R})$ is a linear transformation. Show that $\mathcal{N}(L) \neq \{\mathbf{0}\}$. Give an example of an L which maps onto $\mathcal{P}^2(\mathbb{R}, \mathbb{R})$ and an example of an L which doesn't map onto. Which has the larger dimensional null space?

2.9.42. Suppose $L : \mathbb{R}^3 \to \mathcal{P}^2$ is a linear transformation.

(a) Show that if L maps onto $\mathcal{P}^2$, then $\mathcal{N}(L) = \{\mathbf{0}\}$.

(b) Show that if $L(\mathbf{e}_1 + \mathbf{e}_2 + \mathbf{e}_3) = \mathbf{0}$, then L does not map onto $\mathcal{P}^2$.

2.9.43. (a) Give an example of vector spaces V, W and a linear transformation $L : V \to W$ so that $\dim V < \dim W$ and L is not $1-1$.

(b) Give an example of vector spaces V, W and a linear transformation $L : V \to W$ so that $\dim V > \dim W$ and L is not onto.

2.9.44. Show that if $\mathbf{v}_1, \ldots, \mathbf{v}_k \in S$, where S is a subspace of V, then $T = \operatorname{span}(\mathbf{v}_1, \ldots, \mathbf{v}_k)$ is a subspace of S. How does $\dim T$ compare to $\dim S$?

2.9.45. Find a maximal independent set among each of the collections of vectors and extend them to a basis of the whole vector space.

(a) $(1, 1, 1), (1, 2, 1), (2, 3, 2)$ in $\mathbb{R}^3$

(b) $(x - 1), (x^2 - 1), (x^2 - x)$ in the space of polynomials of degree less than or equal to 3 which vanish at $x = 1$

2.9.46. Find a maximal independent set among each of the collections of vectors and extend them to a basis of the whole vector space.

(a) $(-1, 0, 1, 0), (2, 3, -3, -2), (1, 2, 3, -6)$ in $\mathbb{R}^4$

(b) $\begin{pmatrix} -1 & 0 \\ 1 & 0 \end{pmatrix}, \begin{pmatrix} 2 & 3 \\ -3 & 2 \end{pmatrix}, \begin{pmatrix} 1 & 2 \\ 3 & -6 \end{pmatrix}$ in $\mathcal{M}(2, 2)$

2.9.47. Show that if $\mathbf{v}_1, \mathbf{v}_2$ are two independent vectors in $\mathbb{R}^3$, then for at least one choice of $i = 1, 2, 3$, the 3 vectors $\mathbf{v}_1, \mathbf{v}_2, \mathbf{e}_i$ will be a basis for $\mathbb{R}^3$.

2.9.48. Use the previous exercise to show that $\mathbf{v}_1 = (a, b, c), \mathbf{v}_2 = (d, e, f)$ are independent vectors in $\mathbb{R}^3$ iff at least one of the determinants

$$\begin{vmatrix} a & d \\ b & e \end{vmatrix}, \begin{vmatrix} a & d \\ c & f \end{vmatrix}, \begin{vmatrix} b & e \\ c & f \end{vmatrix}$$

doesn't equal 0.

2.9.49. Generalize the preceding exercise to show that $n-1$ vectors in $\mathbb{R}^n$ are independent iff when one forms an n by $n - 1$ matrix with the vectors as the columns, then one of the $(n - 1)$ by $(n - 1)$ submatrices formed by deleting a row has a nonzero determinant.

2.9.50. Show that a matrix has rank r if r is size of the largest square submatrix which has nonzero determinant.

2.9.51. Let A be an m by n matrix.

(a) Show that $A\mathbf{v} = \mathbf{0}$ iff $\mathbf{v}$ is perpendicular to each of the rows of A.

(b) Show that $A\mathbf{v} = \mathbf{0}$ iff $\mathbf{v}$ is perpendicular to every vector in the row space of A.

(c) Show that if $\mathbf{v}$ is in the row space, then $A\mathbf{v} = \mathbf{0}$ iff $\mathbf{v} = \mathbf{0}$.

(d) Show that if $\mathbf{v}_1, \ldots, \mathbf{v}_r$ is a basis for the row space, then $A\mathbf{v}_1, \ldots, A\mathbf{v}_r$ is a basis for the column space.

(e) Show that $L : \mathcal{R}(A^t) \to \mathcal{R}(A)$ given by $L(\mathbf{x}) = A\mathbf{x}$ is an isomorphism between the row space and the column space.

2.9.52. Recall that the trace of a matrix is the sum of its diagonal entries. Define $L : \mathcal{M}(3, 3) \to \mathbb{R}$ by $L(A) = \operatorname{tr} A$. Show that L is a linear transformation and give a basis for $\mathcal{N}(L)$.

2.9.53. With reference to the last exercise, show the skew symmetric 3 by 3 matrices form a three-dimensional subspace of the 3 by 3 matrices of trace 0. Find a basis for this subspace and extend it to a basis for the matrices with trace 0.

2.9.54. (a) Consider the n vectors

$$(1, 1, 0, \ldots, 0), (0, 1, 1, 0, \ldots, 0), \ldots, (0, 0, \ldots, 0, 1, 1), (1, 0, \ldots, 0, 1) \in \mathbb{R}^n$$

Show that they are independent iff n is odd.

(b) Show that if $\mathbf{v}_1, \ldots, \mathbf{v}_n$ is a basis for V, then $\mathbf{v}_1 + \mathbf{v}_2, \mathbf{v}_2 + \mathbf{v}_3, \ldots, \mathbf{v}_{n-1} + \mathbf{v}_n, \mathbf{v}_n + \mathbf{v}_1$ is also a basis iff n is odd.

2.9.55. Consider the subspace of $\mathbb{R}^4$ which is spanned by the four vectors $(1, -1, 0, 0)$, $(0, 1, -1, 0), (0, 0, 1, -1), (1, 0, 0, -1)$. Show this is three-dimensional and show that any subcollection of three vectors out of the four will form a basis.

2.9.56. Suppose $\mathbf{v}_1, \ldots, \mathbf{v}_k$ are independent vectors in V, and $L : V \to W$ is a linear transformation. Show that if $\mathcal{N}(L) = \{\mathbf{0}\}$, then $L\mathbf{v}_1, \ldots, L\mathbf{v}_k$ are independent. Give an example to show the converse is not true in general, but holds if $\dim V = k$.

In Exercises 2.9.57–2.9.61 give bases for the four fundamental subspaces associated to the given matrix.

2.9.57. $\begin{pmatrix} 1 & 1 & 1 \\ 0 & 1 & 2 \\ 0 & 0 & 3 \end{pmatrix}$

2.9.58. $\begin{pmatrix} 3 & 2 & 1 & 0 & 3 \\ 2 & 0 & 2 & -1 & -3 \\ 1 & -2 & 3 & -2 & -9 \end{pmatrix}$

2.9.59. $\begin{pmatrix} 1 & 4 & 2 & 3 \\ -2 & -3 & 1 & 1 \\ 0 & 5 & 5 & 7 \\ 3 & 1 & 3 & 4 \end{pmatrix}$

2.9.60. $\begin{pmatrix} 1 & 2 & 3 & 4 \\ 5 & 6 & 7 & 8 \\ 9 & 10 & 11 & 12 \end{pmatrix}$

2.9.61. $\begin{pmatrix} 1 & 2 & -1 & -2 & 1 \\ 0 & 1 & 2 & -1 & -2 \\ 2 & 0 & -10 & 0 & 10 \end{pmatrix}$

2.9.62. Show that if $L : V \to W$ is a linear transformation, then V is finite dimensional iff the nullity $n(L)$ and the rank $r(L)$ are both finite.

2.9.63. Let $V = \mathcal{D}(\mathbb{R}, \mathbb{R})$ be the vector space of infinitely differentiable functions and S the subspace of V spanned by $e^x, e^{-x}, 1, x, x^2$.

(a) Show that $\dim S = 5$. (Hint: If $c_1e^x + c_2e^{-x} + c_3 + c_4x + c_5x^2 = 0$, differentiate three times to get a relation involving e^x and e^{-x}, and use this to show $c_1 = c_2 = 0$.)

(b) Let $L : S \to V$ be $L(f) = f'' - f$. Find a basis for $\mathcal{N}(L)$ and a basis for $\mathcal{R}(L)$.

(c) Find all solutions to $L(f) = 1 + x + x^2$.

2.9.64. Let $V = \mathcal{D}(\mathbb{R}, \mathbb{R})$ be the vector space of infinitely differentiable functions and S the subspace of V spanned by $\sin x, \cos x, \sin 2x, \cos 2x$.

(a) Show that S is four-dimensional by showing that these four functions are independent. (Hint: Start with a general linear combination of these functions being 0 and evaluate at well-chosen points to show that all the coefficients have to be 0.)

(b) Let $L : S \to \mathcal{D}(\mathbb{R}, \mathbb{R})$ be given by $L(f) = f'' + 4f$. Find a basis for $\mathcal{N}(L)$ and a basis for $\mathcal{R}(L)$.

2.9.65. Suppose $L : \mathcal{M}(2, 2) \to \mathcal{P}^2(\mathbb{R}, \mathbb{R})$ is a linear transformation.

(a) Show that $n(L) \neq 0$.

(b) If $n(L) = 2$, show that L does not map onto $\mathcal{P}^2(\mathbb{R}, \mathbb{R})$.

2.9.66. Suppose that $L : P^4(\mathbb{R}, \mathbb{R}) \to \mathcal{M}(2, 2)$ is a linear transformation which has a two-dimensional null space. Show that L does not map onto $\mathcal{M}(2, 2)$.

2.9.67. Suppose that $L : V \to W$ is a linear transformation and V is finite dimensional. Show that $\mathcal{R}(L)$ is also finite dimensional. How does $r(L)$ compare to $\dim V$? Give an example where W is infinite dimensional and $r(L) < \dim V$.

2.9.68. Suppose $L, M : V \to W$ are linear transformations and define $L + M : V \to W$ by $(L + M)(\mathbf{v}) = L(\mathbf{v}) + M(\mathbf{v})$.

(a) Show that $L + M$ is a linear transformation.

(b) Show that $\mathcal{N}(L) \cap \mathcal{N}(M) \subset \mathcal{N}(L + M)$.

(c) Give an example where neither L nor M is surjective, but $L + M$ is surjective.

2.9.69. Let $\mathcal{C}(\mathbb{R}, \mathbb{R})$ denote the continuous functions from the reals to the reals and let $\mathcal{C}^1(\mathbb{R}, \mathbb{R})$ denote those functions from the reals to the reals which have a continuous first derivative. Let $D : \mathcal{C}^1(\mathbb{R}, \mathbb{R}) \to \mathcal{C}(\mathbb{R}, \mathbb{R})$ be $D(f) = f'$.

(a) Show that $\mathcal{R}(D) = \mathcal{C}(\mathbb{R}, \mathbb{R})$.

(b) Give a basis for the null space of D.

(c) Show that $\mathcal{C}(\mathbb{R}, \mathbb{R})$ and $\mathcal{C}^1(\mathbb{R}, \mathbb{R})$ are both infinite dimensional by producing an infinite collection of independent functions in each of them.

2.9.70. Use the concept of rank to show that a square matrix A is invertible iff A' is invertible.

2.9.71. Use the determinant to show that a square matrix A is invertible iff A' is invertible.

2.9.72. Suppose that A is a 3 by 4 matrix. Show that $n(A) \neq n(A')$.

2.9.73. Show that for any nonsquare matrix $n(A) \neq n(A')$.

2.9.74. Construct an isomorphism between the vector space of symmetric 2 by 2 matrices and the vector space of skew symmetric 3 by 3 matrices.

2.9.75. Construct an isomorphism between the vector space of skew symmetric 3 by 3 matrices and the subspace of $\mathbb{R}^5$ which satisfies the equations

$$x_1 + x_2 + x_3 + x_4 + x_5 = 0, x_1 - x_2 + x_3 + x_4 - x_5 = 0.$$

2.9.76. Let $L : \mathbb{R}^4 \to \mathcal{M}(2, 2)$ be given by

$$L(x_1, x_2, x_3, x_4) = \begin{pmatrix} x_1 & (x_2 + x_3) \\ (x_2 - x_3) & -x_4 \end{pmatrix}$$

Show that L is an isomorphism.

2.9.77. Let $L : \mathcal{P}^3(\mathbb{R}, \mathbb{R}) \to \mathcal{M}(2,2)$ be given by $L(a + bx + cx^2 + dx^3) = aI + bA + cA^2 + dA^3$ where $A = \begin{pmatrix} 1 & 1 \\ 0 & 1 \end{pmatrix}$. Find $\mathcal{R}(L)$ and $\mathcal{N}(L)$.

2.9.78. Let $L : \mathcal{M}(3,3) \to \mathcal{M}(3,3)$ be given by $L(A) = 3A - \text{tr}(A)I$, where tr denotes the trace of the matrix, which is the sum of the diagonal entries, and I denotes the identity matrix. Find $\mathcal{N}(L)$ and $\mathcal{R}(L)$.

2.9.79. Let A be an n by n matrix. Show that $B = (A + A^t)/2$ is symmetric and $C = (A - A^t)/2$ is skew symmetric. Show that any matrix is the sum of a symmetric matrix and a skew symmetric matrix. If S denotes the symmetric matrices and T denotes the skew symmetric matrices, show that $\mathcal{M}(n,n)$ is the direct sum of S and T.

Exercises 2.9.80–2.9.83 are connected to seeing how null spaces and ranges behave under composition. In all of them $L : V \to W$ and $M : W \to X$ are linear transformations.

2.9.80. (a) Show that the composition ML is a linear transformation from V to X.
(b) If $V = \mathbb{R}^n, W = \mathbb{R}^m, X = \mathbb{R}^p$, and L, M are linear transformations coming from multiplication by A, B, respectively, show that ML arises from multiplication by BA.

2.9.81. (a) Show that $\mathcal{N}(L) \subset \mathcal{N}(ML)$. Conclude that $n(L) \leq n(ML)$. Use the rank–nullity theorem to show that $r(ML) \leq r(L)$.
(b) Show that $\mathcal{R}(ML) \subset \mathcal{R}(M)$. Conclude that $r(ML) \leq r(M)$.

2.9.82. Suppose L is an isomorphism.
(a) Show that $\mathcal{R}(M) = \mathcal{R}(ML)$ and thus $r(M) = r(ML)$.
(b) Show that L restricts to give an isomorphism between $\mathcal{N}(ML)$ and $\mathcal{N}(M)$. Conclude that $n(ML) = n(M)$.

2.9.83. Suppose that A is an m by n matrix and B is a p by m matrix. Use the preceding exercises to show the following facts:
(a) rank $BA \leq$ rank B, rank A
(b) nullity$A \leq$ nullity BA
(c) If A is invertible, then nullity B = nullity(BA) and rank B = rank BA.
Give an example of 2 by 2 matrices A, B where
(d) rank $A >$ rank BA
(e) rank $B >$ rank BA
(f) nullity $B <$ nullity BA
(g) nullity $A <$ nullity BA
(Hint: For parts (d)–(g), think of the matrices as coming from the linear transformations and define the linear transformations on the standard basis so these inequalities hold. Make these definitions as simple as possible.)

2.9.84. Let S be the subspace of $\mathbb{R}^4$ spanned by $(1,0,1,1), (2,1,0,1), (3,1,1,2)$ and T the subspace spanned by $(1,1,-1,0), (1,1,1,1), (0,0,2,1)$.
(a) Find a basis for S.
(b) Find a basis for T.
(c) Find a basis for $S + T$ and a basis for $S \cap T$.

2.9.85. Let S be the subspace of $\mathbb{R}^4$ spanned by

$$(-2, 1, 3, 1), (2, 2, 1, -5), (1, 3, 2, 1)$$

and T the subspace spanned by

$$(2, 1, 3, 1), (2, 1, 1, 3), (2, 1, 5, -1)$$

(a) Find a basis for S.
(b) Find a basis for T.
(c) Find a basis for $S + T$ and a basis for $S \cap T$.

2.9.86. Let S be the subspace of $\mathcal{P}^4(\mathbb{R}, \mathbb{R})$ spanned by

$$x - x^3, 1 + x + x^2, x + x^3$$

and let T be the subspace spanned by

$$1 - x + x^2 - x^3, x + x^4, 1 - x + x^4, x^2 + x^3 + x^4$$

(a) Find a basis for S and a basis for T.
(b) Find a basis for $S \cap T$ and a basis for $S + T$.

2.9.87. (a) Find a basis for the subspace S of 3 by 3 matrices spanned by the matrices $E(i, j; 1), i \neq j$.
(b) Find a basis for the subspace T spanned by the six permutation matrices.
(c) Find bases for $S + T$ and $S \cap T$.

2.9.88. Show that if S, T are subspaces of V, then $S + T = S$ iff $T \subset S$.

2.9.89. (a) Give a basis for $\mathbb{R}^4$ which contains the two vectors $(3, 1, 1, 2)$ and $(2, 1, 0, 0)$.
(b) If $S = \text{span}((3, 1, 1, 2), (2, 1, 0, 0))$, find a subspace $T = \text{span}(\mathbf{v}, \mathbf{w})$ with $S + T = \mathbb{R}^4$.

2.9.90. Let S be the subspace of $\mathcal{P}^2(\mathbb{R}, \mathbb{R})$ consisting of all polynomials which vanish at 0, and let T be the subspace of polynomials which vanish at 1.
(a) Find a basis for S.
(b) Find a basis for T.
(c) Find a basis for $S + T$ and a basis for $S \cap T$.

2.9.91. Let $L : \mathcal{P}^2(\mathbb{R}, \mathbb{R}) \to \mathcal{P}^3(\mathbb{R}, \mathbb{R})$ be $L(p) = xp - p$. Let $D : \mathcal{P}^3(\mathbb{R}, \mathbb{R}) \to \mathcal{P}^2(\mathbb{R}, \mathbb{R})$ be $D(p) = p'$. Find bases for $\mathcal{N}(L), \mathcal{N}(D), \mathcal{N}(LD), \mathcal{N}(DL)$.

2.9.92. Let $T : \mathcal{P}^4(\mathbb{R}, \mathbb{R}) \to \mathcal{P}^4(\mathbb{R}, \mathbb{R})$ be given by $T(p) = p' - p$.
(a) Show that T is a linear transformation.
(b) Find the matrix which represents T with respect to the basis $1, x, x^2, x^3, x^4$.
(c) Find the null space of T and the range of T.

2.9.93. Let $L : \mathcal{P}^2 \to \mathcal{P}^2$ be given by $L(p) = p + xp' + x^2p''$.
(a) Show that L is a linear transformation.
(b) Find the matrix which represents L with respect to the basis $1, x, x^2$.
(c) Find bases for the null space of L and the range of L.

2.9.94. Let $T : \mathcal{P}^2(\mathbb{R}, \mathbb{R}) \to \mathbb{R}^3$ be given by $T(p) = (p(-1), p(0), p(1))$.
(a) Show that T is a linear transformation.
(b) Find the matrix which represents T with respect to the basis $1, x, x^2$ of $\mathcal{P}^2(\mathbb{R}, \mathbb{R})$ and the standard basis for $\mathbb{R}^3$.
(c) Show that T is an isomorphism.

(d) Find the inverse S of T. Note that $S(a, b, c) = p$ where $p(-1) = a, p(0) = b, p(1) = c$.

2.9.95. Let $L : \mathcal{P}^2(\mathbb{R}, \mathbb{R}) \to \mathcal{P}^2(\mathbb{R}, \mathbb{R})$ be given by $L(p)(x) = p(x-1)$.

(a) Find the matrix which represents L with respect to the basis $1, x, x^2$ in the domain and $1, x-1, (x-1)^2$ in the range.

(b) Find the matrix which represents L with respect to the basis $1, x, x^2$ in both the domain and range.

(c) Show that L is an isomorphism.

2.9.96. Give an example of a linear transformation $L : \mathbb{R}^2 \to \mathbb{R}^2$ so that $\mathcal{N}(L) = \mathcal{R}(L) = \text{span}(\mathbf{e}_1)$.

2.9.97. Suppose $L : V \to V$ and S is a subspace of V which is invariant under L, that is, $L(S) \subset S$. Show that if we choose a basis $\mathbf{v}_1, \ldots, \mathbf{v}_n$ for V which includes a basis for S as the first k vectors, then the matrix which represents L with respect to this basis will have the special form

$$\begin{pmatrix} A & B \\ \mathbf{0} & C \end{pmatrix}$$

What will the matrix be which represents the restriction $L|S : S \to S$ with respect to the basis $\mathbf{v}_1, \ldots, \mathbf{v}_k$?

2.9.98. Find the transition matrix $\mathcal{T}_{\mathcal{V}}^{\mathcal{U}}$ between the the two bases $\mathcal{U} = \{(1, 1, 1), (2, 1, 0), (2, 3, 1)\}$ and $\mathcal{V} = \{(-1, 2, 0), (1, 0, 1), (0, 1, 1)\}$ for $\mathbb{R}^3$.

2.9.99. Find the transition matrix $\mathcal{T}_{\mathcal{W}}^{\mathcal{V}}$ from the basis

$$\mathcal{V} = \{(1, 0, 0, 0), (1, 1, 0, 0), (1, 1, 1, 0), (1, 1, 1, 1)\}$$

to the basis

$$\mathcal{W} = \{(0, 0, 0, 1), (0, 0, 1, 1), (0, 1, 1, 1), (1, 1, 1, 1)\}$$

2.9.100. Find the transition matrix $\mathcal{T}_{\mathcal{W}}^{\mathcal{V}}$ from the basis

$$\mathcal{V} = \{1 - x, 1 + x^2, 1 - x^3\}$$

to the basis

$$\mathcal{W} = \{3 - x - 2x^3, 4 - 3x + x^2, 2 - 2x - x^2 - x^3\}$$

2.9.101. Find the transition matrix $\mathcal{T}_{\mathcal{W}}^{\mathcal{V}}$ from the basis

$$\mathcal{V} = \left\{ \begin{pmatrix} 3 & 2 \\ 2 & 3 \end{pmatrix}, \begin{pmatrix} 1 & 2 \\ 2 & 1 \end{pmatrix}, \begin{pmatrix} -2 & 1 \\ 1 & 5 \end{pmatrix} \right\}$$

to the basis

$$\mathcal{W} = \left\{ \begin{pmatrix} 1 & 0 \\ 0 & 1 \end{pmatrix}, \begin{pmatrix} 1 & 6 \\ 6 & 13 \end{pmatrix}, \begin{pmatrix} 8 & 9 \\ 9 & 15 \end{pmatrix} \right\}$$

2.9.102. Find the transition matrix $\mathcal{T}_{\mathcal{V}}^{\mathcal{U}}$ between the two bases $\mathcal{U} = \{1, 1+x, 1+x+x^2\}$ and $\mathcal{V} = \{x, x-1, x(x-1)\}$ for $\mathcal{P}^2(\mathbb{R}, \mathbb{R})$.

2.9.103. Consider the basis $\mathcal{V} = \{(1,3,2,4),(2,-1,3,2),(0,-1,3,2)\}$ for a subspace of $\mathbb{R}^4$. If the transition matrix $\mathcal{T}_{\mathcal{W}}^{\mathcal{V}} = \begin{pmatrix} 2 & 2 & 1 \\ 1 & 3 & 0 \\ -1 & 4 & 1 \end{pmatrix}$, find the basis $\mathcal{W}$.

2.9.104. Consider the basis $\mathcal{V} = \{1 - x + x^3, x^2 + x^3, 3 - x + x^2 - x^3\}$ of a subspace of $\mathcal{P}^3(\mathbb{R},\mathbb{R})$. If the transition matrix $\mathcal{T}_{\mathcal{W}}^{\mathcal{V}} = \begin{pmatrix} 1 & 1 & 1 \\ 2 & 1 & 0 \\ 3 & 0 & 0 \end{pmatrix}$, find the basis $\mathcal{W}$.

2.9.105. Show that similar matrices have the same rank. Describe how $\mathcal{R}(A)$ and $\mathcal{R}(B)$ are related to each other in terms of the matrix C so that $A = C^{-1}BC$.

2.9.106. Find the matrix which represents the linear transformation $T : \mathcal{M}(2,2) \to \mathcal{M}(2,2)$, $T(A) = A^t$, with respect to the basis

$$\begin{pmatrix} 1 & 0 \\ 0 & 0 \end{pmatrix}, \begin{pmatrix} 0 & 1 \\ 0 & 0 \end{pmatrix}, \begin{pmatrix} 0 & 0 \\ 1 & 0 \end{pmatrix}, \begin{pmatrix} 0 & 0 \\ 0 & 1 \end{pmatrix}$$

2.9.107. Let $L : \mathcal{P}^3(\mathbb{R},\mathbb{R}) \to \mathcal{P}^2(\mathbb{R},\mathbb{R})$ be a linear transformation with $L(1) = x^2$, $L(1 + x) = x$, $L(1 + x + x^2) = 1$, $L(1 + x + x^2 + x^3) = 0$.
(a) Find a formula for $L(p)$ for $p \in \mathcal{P}^3(\mathbb{R},\mathbb{R})$.
(b) Give the matrix of L with respect to the basis $1, x, x^2, x^3$ of $\mathcal{P}^3(\mathbb{R},\mathbb{R})$ and the basis $1, x, x^2$ of $\mathcal{P}^2(\mathbb{R},\mathbb{R})$.

2.9.108. If $L : \mathbb{R}^3 \to \mathbb{R}^2$ is given by multiplication by the matrix $A = \begin{pmatrix} 1 & 1 & 1 \\ 2 & 1 & -1 \end{pmatrix}$, what is the matrix which represents L with respect to the basis $(1,1,1)$, $(1,0,-1)$, $(2,1,1)$ in $\mathbb{R}^3$ and $(1,1)$, $(0,1)$ in $\mathbb{R}^2$.

2.9.109. Suppose that $L : \mathcal{P}^2(\mathbb{R},\mathbb{R}) \to \mathbb{R}^4$ satisfies

$$L(1) = (1,0,2,-1), L(1 + x) = (2,1,0,-1), L(1 + x + x^2) = (0,-1,4,-1)$$

(a) Find a basis for $\mathcal{N}(L)$.
(b) Find a basis for $\mathcal{R}(L)$.
(c) Write a formula for $L(a_0 + a_1 x + a_2 x^2)$.
(d) What is the matrix which represents L with respect to the standard bases $1, x, x^2$ and $\mathbf{e}_1, \mathbf{e}_2, \mathbf{e}_3, \mathbf{e}_4$?

2.9.110. Let $P : \mathbb{R}^3 \to \mathbb{R}^3$ be the linear transformation which projects onto the plane $x = y$. This means that a vector in this plane is sent to itself and a vector which is perpendicular to the plane is sent to $\mathbf{0}$.
(a) Find a basis $\mathbf{v}_1, \mathbf{v}_2$ for this plane.
(b) Find a basis $\mathbf{v}_3$ for the line perpendicular to the plane $x = y$.
(c) Give the matrix which represents P with respect to the bases $\mathbf{v}_1, \mathbf{v}_2, \mathbf{v}_3$ in both domain and range.
(d) Give the matrix C which expresses the standard basis $\mathbf{e}_1, \mathbf{e}_2, \mathbf{e}_3$ in terms of the basis $\mathbf{v}_1, \mathbf{v}_2, \mathbf{v}_3 : \mathbf{E} = \mathbf{V}C$.
(e) Give the matrix which represents P with respect to the standard basis.

2.9.111. Let $T : \mathcal{P}^2(\mathbb{R}, \mathbb{R}) \to \mathcal{P}^2(\mathbb{R}, \mathbb{R})$ be $T(p) = x^2 p'' - p$.

(a) Find the matrix which represents T with respect to the basis $1, x, x^2$.

(b) Show that T is an isomorphism.

2.9.112. Let $\mathcal{L}(\mathbb{R}^3, \mathbb{R}^3)$ denote the vector space of linear transformations from $\mathbb{R}^3 \to \mathbb{R}^3$.

(a) Let L_{ij} denote the linear transformation with $L_{ij}(\mathbf{e}_j) = \mathbf{e}_i$ and $L_{ij}(\mathbf{e}_k) = \mathbf{0}$ if $k \neq j$. Show that the nine linear transformations $L_{ij}, i, j = 1, 2, 3$ form a basis of $\mathcal{L}(\mathbb{R}^3, \mathbb{R}^3)$.

(b) Let $T : \mathcal{L}(\mathbb{R}^3, \mathbb{R}^3) \to M(3, 3)$ be the linear transformation which sends a linear transformation to the matrix which represents the linear transformation with respect to the standard basis. Show that T is an isomorphism.

(c) Let $C : \mathbb{R}^3 \to \mathcal{L}(\mathbb{R}^3, \mathbb{R}^3)$ be defined by $C(\mathbf{x})(\mathbf{y}) = \mathbf{x} \times \mathbf{y}$, where $\mathbf{x} \times \mathbf{y}$ is the cross product of $\mathbf{x}$ and $\mathbf{y}$. Show that $\mathcal{N}(C) = \{\mathbf{0}\}$, so C is $1 - 1$. Find the images of the vectors $\mathbf{e}_1, \mathbf{e}_2, \mathbf{e}_3$ under the composition TC. Show that $\mathcal{R}(TC)$ is the skew symmetric 3 by 3 matrices.

2.9.113. Show that if V is an n-dimensional vector space then the set of linear transformations $\mathcal{L}(V, V)$ from V to itself is isomorphic to $\mathcal{M}(n, n)$.

2.9.114. Prove Proposition 2.7.6.

2.9.115. Prove Proposition 2.7.7.

2.9.116. By using Proposition 2.7.7 together with the facts that $[I]^{\mathcal{U}}_{\mathcal{V}} = \mathcal{T}^{\mathcal{U}}_{\mathcal{V}}$, $[I]^{\mathcal{V}}_{\mathcal{U}} = \mathcal{T}^{\mathcal{V}}_{\mathcal{U}}$ and $L = ILI$, prove the formula

$$[L]^{\mathcal{U}}_{\mathcal{U}} = \mathcal{T}^{\mathcal{V}}_{\mathcal{U}} [L]^{\mathcal{V}}_{\mathcal{V}} \mathcal{T}^{\mathcal{U}}_{\mathcal{V}}$$

2.9.117. Prove Proposition 2.7.8.

Exercises 2.9.118–2.9.127 use the digraphs and their associated graphs from Figure 2.12.

2.9.118. For each of these digraphs give the connectivity matrix, the connectivity matrix of the associated graph, and the incidence matrix of the digraph formed by deleting any simple loops.

2.9.119. Graphs G_1 and G_5 are related by changing the direction of some of the arrows, specifically those for edges b, c, and e.

(a) Describe how this changes the connectivity matrix for the associated graph.

(b) Describe how this changes the incidence matrix.

(c) Write the incidence matrix N_5 for G_5 in terms of the incidence matrix N_1 for G_1 as $N_5 = DN_1$ where D is a certain diagonal matrix.

2.9.120. Generalize the results of the previous exercise for any two digraphs which are related in that the directions of some of the arrows are reversed. Make a conjecture as to how the connectivity and incidence matrices are related and prove your conjecture.

2.9.121. The relationship between graphs G_1 and G_4 is that both the vertices and edges are renumbered to form G_4 from G_1. This renumbering is actually two functions $\sigma : \{1, 2, 3, 4\} \to \{1, 2, 3, 4\}, \sigma(1) = 3, \sigma(2) = 1, \sigma(3) = 4, \sigma(4) = 2,$

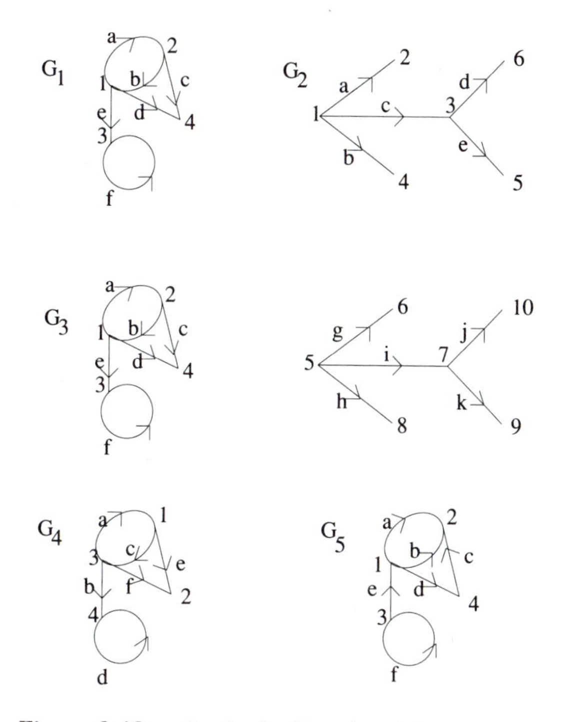

Figure 2.12. Graphs for Exercises 2.9.118–2.9.127

and $\tau : \{a, b, c, d, e, f\} \to \{a, b, c, d, e, f\}$. When these letters are identified with the first six numbers, we can view τ as given by $\tau(1) = 1, \tau(2) = 3, \tau(3) = 5, \tau(4) = 6, \tau(5) = 2, \tau(6) = 4$. Give a matrix S so that $S\mathbf{e}_j = \mathbf{e}_{\sigma(j)}$ and a matrix T so that $T(\mathbf{e}_j) = \mathbf{e}_{\tau(j)}$.

2.9.122. Show that if $A = (a_{ij})$ is the connectivity matrix for G_1 and $B = (b_{ij})$ is the connectivity matrix for G_4, then $b_{\sigma(i)\sigma(j)} = a_{ij}$.

2.9.123. With A and B as in the previous exercise show that $A = S^t BS$. (Hint: Start from the equations

$$a_{ij} = \langle \mathbf{e}_i, A\mathbf{e}_j \rangle \text{ and } b_{\sigma(i)\sigma(j)} = \langle \mathbf{e}_{\sigma(i)}, B\mathbf{e}_{\sigma(j)} \rangle$$

together with the results of the previous two exercises.

2.9.124. Generalize the previous exercises to say how the connectivity matrix changes when the vertices are relabeled via a permutation σ.

2.9.125. Suppose we have two graphs where the second one arises from the first by relabeling vertices and directed edges via permutations σ, τ. Suppose also that (unlike G_1 and G_4) the graphs contain no simple loops. Let A be the incidence matrix for the first graph and B the incidence matrix for the second. If S is the matrix so that $S(\mathbf{e}_j) = \mathbf{e}_{\sigma(j)}$ and T is the matrix so that $T(\mathbf{e}_j) = e_{\tau(j)}$, find an equation relating the four matrices A, B, S, T. Prove your assertion.

2.9.126. Use the theorems on connectivity matrices and incidence matrices to give two different proofs that the graph G_3 is not connected.

2.9.127. The graph G_3 is formed from G_1 and G_2 in a certain way. Describe this relationship and what this implies about the corresponding connectivity and incidence matrices (after deleting simple loops). Formulate a general definition of the sum of two graphs, and determine how the connectivity matrices and incidence matrices of the sum are related to those matrices for the original graphs.

2.9.128. Show that if D is a digraph with associated graph G, and D has no simple loops, then $C_g = C_d + C_d^t$.

2.9.129. Generalize the result in Exercise 2.9.128 to describe how C_g and C_d are related when there may be simple loops.

2.9.130. Let N be an incidence matrix for a connected graph. Show that each step of Gaussian elimination leads to a new matrix whose nonzero rows form an incidence matrix for a connected graph. Show that the nonzero rows of U form the incidence matrix for a connected graph with no loops.

2.9.131. Let A be an m by n matrix of rank n and B be a $(m - n)$ by m matrix of rank $m - n$ so that $BA = 0$. Show that the matrix $\begin{pmatrix} A^t \\ B \end{pmatrix}$ is invertible.

CHAPTER 3

ORTHOGONALITY AND PROJECTIONS

3.1. ORTHOGONAL BASES AND THE *QR* DECOMPOSITION

In $\mathbb{R}^n$, the dot product $\langle \mathbf{x}, \mathbf{y} \rangle$ provides an important additional structure to its structure as a vector space. The dot product is intimately connected to the concept of angle and length of vectors and forms an essential tool in discussing geometric concepts. It is defined by

$$\langle \mathbf{x}, \mathbf{y} \rangle = \sum_{i=1}^{n} x_i y_i$$

The square of the length of a vector $\mathbf{x}$ is given in terms of the dot product by

$$\|\mathbf{x}\|^2 = \langle \mathbf{x}, \mathbf{x} \rangle = x_1^2 + \cdots + x_n^2$$

Recall that the law of cosines gives $\|\mathbf{x} - \mathbf{y}\|^2 = \|\mathbf{x}\|^2 + \|\mathbf{y}\|^2 - 2\|\mathbf{x}\|\,\|\mathbf{y}\| \cos \theta$, where we are using the triangle with two sides coming from the vectors $\mathbf{x}$ and $\mathbf{y}$ emanating from the origin and with angle θ between them. Figure 3.1 illustrates how this follows from the Pythagorean theorem using the shaded triangle.

Using the relationship of length to the dot product, this gives

$$\langle \mathbf{x} - \mathbf{y}, \mathbf{x} - \mathbf{y} \rangle = \langle \mathbf{x}, \mathbf{x} \rangle + \langle \mathbf{y}, \mathbf{y} \rangle - 2\|\mathbf{x}\|\,\|\mathbf{y}\| \cos \theta$$

But the dot product is linear in each term (called bilinear), and so the left-hand side can be expanded out as

$$\langle \mathbf{x}, \mathbf{x} \rangle + \langle \mathbf{y}, \mathbf{y} \rangle - 2\langle \mathbf{x}, \mathbf{y} \rangle$$

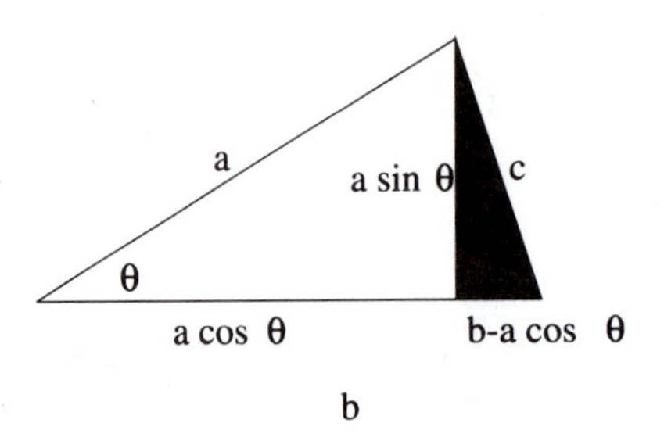

Figure 3.1. Law of cosines

Thus we get the equality

$$\langle \mathbf{x}, \mathbf{y} \rangle = \|\mathbf{x}\| \, \|\mathbf{y}\| \cos \theta$$

If either $\mathbf{x}$ or $\mathbf{y}$ is $\mathbf{0}$, both sides are 0. If both $\mathbf{x}$ and $\mathbf{y}$ are not $\mathbf{0}$, then we can solve for $\cos \theta = \dfrac{\langle \mathbf{x}, \mathbf{y} \rangle}{\|\mathbf{x}\| \, \|\mathbf{y}\|}$. Note that since $|\cos \theta| \leq 1$, we have an inequality

$$|\langle \mathbf{x}, \mathbf{y} \rangle| \leq \|\mathbf{x}\| \, \|\mathbf{y}\|$$

This inequality is called the **Cauchy–Schwarz inequality.**

The dot product has three important properties:

$$(1) \qquad \langle a\mathbf{x} + b\mathbf{y}, \mathbf{z} \rangle = a\langle \mathbf{x}, \mathbf{z} \rangle + b\langle \mathbf{y}, \mathbf{z} \rangle \qquad \text{(bilinearity)}$$
$$\langle \mathbf{x}, a\mathbf{y} + b\mathbf{z} \rangle = a\langle \mathbf{x}, \mathbf{y} \rangle + b\langle \mathbf{x}, \mathbf{z} \rangle$$
$$(2) \qquad \langle \mathbf{x}, \mathbf{y} \rangle = \langle \mathbf{y}, \mathbf{x} \rangle \qquad \text{(symmetry)}$$
$$(3) \qquad \langle \mathbf{x}, \mathbf{x} \rangle \geq 0 \text{ and } = 0 \text{ iff } \mathbf{x} = \mathbf{0} \qquad \text{(positivity(or positive definite))}$$

Each of these properties is straightforward to check. Note that we used them earlier when we verified properties of matrix multiplication, since matrix multiplication is based on the dot product. The second property in (1) follows from the first property (linearity in the first variable) and symmetry. We now use these properties to define the concept of an inner product on a vector space V.

DEFINITION 3.1.1. An **inner product** on a vector space V is a function from $V \times V$ to the real numbers (which we will denote by $\mathbf{x}, \mathbf{y} \rightarrow \langle \mathbf{x}, \mathbf{y} \rangle$) which satisfies the three properties. A vector space with an inner product is called an **inner product space.**

The definition of an inner product is modeled on the dot product in $\mathbb{R}^n$. There are in fact an infinite number of inner products on any vector space. We usually will just be using the existence of an inner product rather than special properties of a particular one in our arguments. However, sometimes a particular inner product will play a special role.

Here is a way to define many different inner products on $\mathbb{R}^n$. Note that for the usual dot product, we can write

$$\mathbf{x} = \sum_{i=1}^{n} x_i \mathbf{e}_i, \mathbf{y} = \sum_{i=1}^{n} y_i \mathbf{e}_i, \mathbf{x} \cdot \mathbf{y} = \sum x_i y_i$$

To define a different inner product, first choose a different basis of $\mathbb{R}^n$, $\mathcal{V} = \{\mathbf{v}_1, \ldots, \mathbf{v}_n\}$. Then write the vectors $\mathbf{x}, \mathbf{y}$ in terms of this basis:

$$\mathbf{x} = \sum_{i=1}^{n} a_i \mathbf{v}_i, \mathbf{y} = \sum_{i=1}^{n} b_i \mathbf{v}_i$$

Then define the inner product

$$\langle \mathbf{x}, \mathbf{y} \rangle_{\mathcal{V}} = \sum_{i=1}^{n} a_i b_i$$

Note that the dot product is just the special case where we are using the standard basis $\mathcal{V} = \{\mathbf{e}_1, \ldots, \mathbf{e}_n\}$. Let us verify that this is indeed an inner product. Let

$$\mathbf{x} = \sum_{i=1}^{n} a_i \mathbf{v}_i, \mathbf{y} = \sum_{i=1}^{n} b_i \mathbf{v}_i, \mathbf{z} = \sum_{i=1}^{n} c_i \mathbf{v}_i$$

(1) $$a\mathbf{x} + b\mathbf{y} = \sum_{i=1}^{n} (aa_i + bb_i)\mathbf{v}_i \qquad \text{(bilinearity)}$$

Hence

$$\langle a\mathbf{x} + b\mathbf{y}, \mathbf{z} \rangle_{\mathcal{V}} = \sum_{i=1}^{n} (aa_i + bb_i)c_i = a\sum_{i=1}^{n} a_i c_i + b\sum_{i=1}^{n} b_i c_i = a\langle \mathbf{x}, \mathbf{z} \rangle_{\mathcal{V}} + b\langle \mathbf{y}, \mathbf{z} \rangle_{\mathcal{V}}$$

Linearity in the second variable is verified similarly.

(2) $$\langle \mathbf{x}, \mathbf{y} \rangle_{\mathcal{V}} = \sum_{i=1}^{n} a_i b_i = \sum_{i=1}^{n} b_i a_i = \langle \mathbf{y}, \mathbf{x} \rangle_{\mathcal{V}} \qquad \text{(symmetry)}$$

(3) $$\langle \mathbf{x}, \mathbf{x} \rangle_{\mathcal{V}} = \sum_{i=1}^{n} a_i^2 \geq 0 \quad \text{and} \quad = 0 \text{ iff } a_i = 0, i = 1, \ldots, n \text{ iff } \mathbf{x} = \mathbf{0} \qquad \text{(positivity)}$$

Here is another way that we can express this inner product. The equation $\mathbf{x} = \sum_{i=1} a_i \mathbf{v}_i$ can be written as a matrix equation $V\mathbf{a} = \mathbf{x}$, where we are writing the coefficients $(a_1, \ldots, a_n)$ as a column vector, writing $\mathbf{x}$ as a column vector, and writing the basis vectors of $\mathcal{V}$ as the columns of the matrix V. Similarly, $V\mathbf{b} = \mathbf{y}$. This gives $\mathbf{a} = W\mathbf{x}, \mathbf{b} = W\mathbf{y}$, with $W = V^{-1}$. Then

$$\langle \mathbf{x}, \mathbf{y} \rangle_{\mathcal{V}} = \sum_{i=1}^{n} a_i b_i = \mathbf{a}^t \mathbf{b} = (W\mathbf{x})^t (W\mathbf{y}) = \mathbf{x}^t (W^t W)\mathbf{x}$$

The matrix $A = W^t W$ which occurs here is a special type of matrix, which is called symmetric positive definite. In terms of the Gaussian elimination, positive definite matrices are characterized by having positive pivots. In general, any inner product on $\mathbb{R}^n$ arises as $\langle \mathbf{x}, \mathbf{y} \rangle_{\mathcal{V}}$ for some basis or, alternatively, as $\mathbf{x}^t A\mathbf{y}$ for a symmetric positive definite matrix A. These results will be proved in Chapter 6.

We now assume that the vector space V is an inner product space and see which geometric concepts in $\mathbb{R}^n$ carry over to V due to the inner product. Because of the third property, we can define the **length** $\|\mathbf{v}\|$ of a vector $\mathbf{v}$ by $\|\mathbf{v}\|^2 = \langle \mathbf{v}, \mathbf{v} \rangle$. Note that (3) means that the only vector with length 0 is the zero vector. Length will have the property that $\|c\mathbf{v}\| = |c|\,\|\mathbf{v}\|$; this is seen by using (1).

Properties (1)–(3) can be used to prove the Cauchy–Schwarz inequality. The inequality holds with both sides being 0 when one of $\mathbf{x}$ or $\mathbf{y}$ is $\mathbf{0}$ so let us assume that neither is $\mathbf{0}$. Then for any real number t we have

$$0 \leq \langle \mathbf{x} - t\mathbf{y}, \mathbf{x} - t\mathbf{y} \rangle = \langle \mathbf{x}, \mathbf{x} \rangle - 2t\langle \mathbf{x}, \mathbf{y} \rangle + t^2\langle \mathbf{y}, \mathbf{y} \rangle$$

Now set $t = \langle \mathbf{x}, \mathbf{y} \rangle / \langle \mathbf{y}, \mathbf{y} \rangle$ and multiply by $\langle \mathbf{y}, \mathbf{y} \rangle$ to get

$$0 \leq \langle \mathbf{x}, \mathbf{x} \rangle \langle \mathbf{y}, \mathbf{y} \rangle - \langle \mathbf{x}, \mathbf{y} \rangle^2$$

Moving $\langle \mathbf{x}, \mathbf{y} \rangle^2$ to the left-hand side and taking square roots gives the inequality. This inequality then implies that $-1 \leq \langle \mathbf{x}, \mathbf{y} \rangle / \|\mathbf{x}\|\,\|\mathbf{y}\| \leq 1$, and so we can define the **angle** $\boldsymbol{\theta}$ (at least up to sign and a multiple of 2π) between the two vectors by

$$\cos\theta = \langle \mathbf{x}, \mathbf{y} \rangle / \|\mathbf{x}\|\,\|\mathbf{y}\|$$

This agrees with what we derived from the law of cosines in $\mathbb{R}^n$. Note that $\theta = \pi/2$ radians corresponds to $\langle \mathbf{x}, \mathbf{y} \rangle = 0$.

DEFINITION 3.1.2. When $\langle \mathbf{x}, \mathbf{y} \rangle = 0$ we say that the vectors $\mathbf{x}$ and $\mathbf{y}$ are **orthogonal**.

By definition all vectors are orthogonal to the zero vector. A vector is orthogonal to itself iff it is the zero vector—this is the positivity property (3) in the definition of an inner product. This can also be stated by saying that the only vector of length 0 is the zero vector.

DEFINITION 3.1.3. We say that a collection $\mathbf{v}_1, \ldots, \mathbf{v}_k$ of vectors are **mutually orthogonal** if any two distinct vectors $\mathbf{v}_i, \mathbf{v}_j, i \neq j$, are orthogonal.

The following fact is quite useful.

PROPOSITION 3.1.1. If $\mathbf{v}_1, \ldots, \mathbf{v}_k$ are mutually orthogonal nonzero vectors, then they are linearly independent.

Proof. Suppose that $c_1\mathbf{v}_1 + \cdots + c_k\mathbf{v}_k = \mathbf{0}$. Then take the inner product of both sides with $\mathbf{v}_j$ to get

$$0 = \langle \mathbf{0}, \mathbf{v}_j \rangle = \sum_{i=1}^{k} c_i \langle \mathbf{v}_i, \mathbf{v}_j \rangle = c_j \langle \mathbf{v}_j, \mathbf{v}_j \rangle$$

Since $\mathbf{v}_j \neq \mathbf{0}$, $\langle \mathbf{v}_j, \mathbf{v}_j \rangle > 0$, and this implies that $c_j = 0$. ∎

Note that if we have n mutually orthogonal nonzero vectors in an n-dimensional vector space, they will have to be a basis.

DEFINITION 3.1.4. A basis consisting of mutually orthogonal vectors is called an **orthogonal basis.** If the vectors are also unit vectors (i.e., their lengths are each 1) then the basis is called an **orthonormal basis.**

Note that the standard basis in $\mathbb{R}^n$ is an orthonormal basis. A confusing aspect of the notation comes when we restrict to $\mathbb{R}^n$ and put the elements of a basis as the columns of a matrix. Then the basis is an orthonormal basis exactly when the matrix which is formed is an orthogonal matrix. If the basis is just an orthogonal basis, then we will have the condition on the matrix that it satisfies $A^tA = D$, where D is a diagonal matrix. We leave the details as an exercise.

Exercise 3.1.1.

Let $\mathcal{V} = \{\mathbf{v}_1, \ldots, \mathbf{v}_n\}$ be a basis for $\mathbb{R}^n$ and let A be the matrix which has these basis vectors as column vectors. Show that $\mathcal{V}$ is an orthogonal basis iff $A^tA = D$ is a diagonal matrix. Show that it is an orthonormal basis iff A is an orthogonal matrix. Show that there are similar statements concerning the rows of A. (Hint: The condition on A^tA should follow directly from the definitions. Use properties of inverses to get the condition on AA^t.)

3.1.1. GRAM–SCHMIDT ORTHOGONALIZATION ALGORITHM AND *QR* DECOMPOSITION

Many of the problems we will be discussing in this chapter are made much easier when we have an orthonormal basis for the relevant subspace. We can get from an orthogonal basis to an orthonormal basis by dividing each basis vector by its length. The harder step is to produce an orthogonal basis from a given basis.

We now give an algorithm, called the Gram–Schmidt orthogonalization algorithm, to get an orthonormal basis for any finite-dimensional inner product space V. Assume that $\mathbf{v}_1, \ldots, \mathbf{v}_n$ is a given basis. We will first form an orthogonal basis $\mathbf{w}_1, \ldots, \mathbf{w}_n$ for V, and the corresponding orthonormal basis will be given by $\mathbf{q}_1, \ldots, \mathbf{q}_n$ with $\mathbf{q}_i = \mathbf{w}_i/\|\mathbf{w}_i\|$. This will be done inductively so that at the end of the ith step, $\mathbf{w}_1, \ldots, \mathbf{w}_i$ will be an orthogonal basis for the subspace spanned by $\mathbf{v}_1, \ldots, \mathbf{v}_i$. Figure 3.2 illustrates how $\mathbf{w}_1, \mathbf{w}_2$ are formed from a basis $\mathbf{v}_1, \mathbf{v}_2$ in $\mathbb{R}^2$.

The initial step of the induction is to define $\mathbf{w}_1 = \mathbf{v}_1$ and $\mathbf{q}_1 = \mathbf{w}_1/\|\mathbf{w}_1\|$. Assume that $\mathbf{w}_1, \ldots, \mathbf{w}_i$ have been defined so that they form an orthogonal basis for the subspace spanned by $\mathbf{v}_1, \ldots, \mathbf{v}_i$ and that $\mathbf{q}_j = \mathbf{w}_j/\|\mathbf{w}_j\|$ for $j = 1, \ldots, i$. Then we will construct $\mathbf{w}_{i+1}$ by starting with $\mathbf{v}_{i+1}$ and adding multiples of the $\mathbf{w}_j$ to it to make it orthogonal to each of the previous $\mathbf{w}_j$. To make a vector $\mathbf{v}$ orthogonal to a vector $\mathbf{w}$, we subtract off an appropriate multiple of $\mathbf{w}$ from $\mathbf{v}$. To find the multiple, note that $\langle \mathbf{v} - t\mathbf{w}, \mathbf{w}\rangle = \langle \mathbf{v}, \mathbf{w}\rangle - t\langle \mathbf{w}, \mathbf{w}\rangle$. Thus if we want the result to be 0, we choose $t = \langle \mathbf{v}, \mathbf{w}\rangle/\langle \mathbf{w}, \mathbf{w}\rangle$. Thus

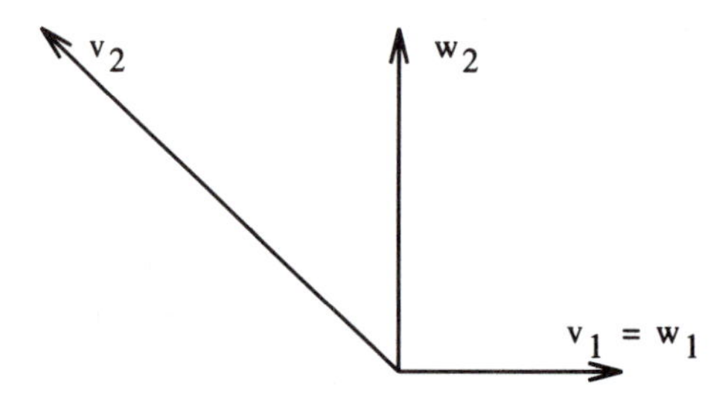

Figure 3.2. Gram–Schmidt algorithm

we will make $\mathbf{v}_{i+1}$ orthogonal to $\mathbf{w}_1$ by subtracting $\dfrac{\langle \mathbf{v}_{i+1}, \mathbf{w}_1 \rangle}{\langle \mathbf{w}_1, \mathbf{w}_1 \rangle}\mathbf{w}_1$ from it. To make it orthogonal to $\mathbf{w}_2$ we subtract a similar multiple of $\mathbf{w}_2$. The important consequence of the $\mathbf{w}_i$ themselves being mutually orthogonal is that when we subtract both vectors, we get a result which is orthogonal to both $\mathbf{w}_1$ and $\mathbf{w}_2$. We thus define

$$\mathbf{w}_{i+1} = \mathbf{v}_{i+1} - (t_1\mathbf{w}_1 + \cdots + t_i\mathbf{w}_i)$$

where

$$t_j = \frac{\langle \mathbf{v}_{i+1}, \mathbf{w}_j \rangle}{\langle \mathbf{w}_j, \mathbf{w}_j \rangle}$$

Then we check that

$$\langle \mathbf{w}_{i+1}, \mathbf{w}_j \rangle = \langle \mathbf{v}_{i+1}, \mathbf{w}_j \rangle - t_j \langle \mathbf{w}_j, \mathbf{w}_j \rangle = 0$$

Moreover, $\mathbf{w}_1, \ldots, \mathbf{w}_{i+1}$ will be an orthogonal basis for the subspace spanned by $\mathbf{v}_1, \ldots, \mathbf{v}_{i+1}$. This completes the induction step of the construction.

Let us look at how the resulting basis $\mathbf{w}_1, \ldots, \mathbf{w}_n$ is related to the original basis. We will have the equations

$$\begin{aligned}
\mathbf{v}_1 &= \mathbf{w}_1 \\
\mathbf{v}_2 &= t_{12}\mathbf{w}_1 + \mathbf{w}_2 \\
\mathbf{v}_3 &= t_{13}\mathbf{w}_1 + t_{23}\mathbf{w}_2 + \mathbf{w}_3 \\
&\vdots \\
\mathbf{v}_n &= t_{1n}\mathbf{w}_1 + t_{2n}\mathbf{w}_2 + \cdots + t_{(n-1)n}\mathbf{w}_{n-1} + \mathbf{w}_n
\end{aligned}$$

Here $t_{ij} = \langle \mathbf{v}_j, \mathbf{w}_i \rangle / \langle \mathbf{w}_i, \mathbf{w}_i \rangle$. In terms of the notation from the last chapter, the row vectors $\mathbf{V}$ and $\mathbf{W}$ determined by the bases $\mathcal{V}$ and $\mathcal{W}$ are related by the equation $\mathbf{V} = \mathbf{W}T$ where T is upper triangular with ones on the diagonal and ij-entry t_{ij} for $i < j$.

Using $\mathbf{w}_j = \|\mathbf{w}_j\|\mathbf{q}_j$, we can express the $\mathbf{v}_i$ in terms of the $\mathbf{q}_i$ by the equations

$$\begin{aligned}
\mathbf{v}_1 &= r_{11}\mathbf{q}_1 \\
\mathbf{v}_2 &= r_{12}\mathbf{q}_1 + r_{22}\mathbf{q}_2 \\
&\vdots \\
\mathbf{v}_n &= r_{1n}\mathbf{q}_1 + r_{2n}\mathbf{q}_2 + \cdots + r_{nn}\mathbf{q}_n
\end{aligned}$$

Here $r_{ii} = \|\mathbf{w}_i\|$ and

$$r_{ij} = t_{ij}\|\mathbf{w}_i\| = \frac{\langle \mathbf{v}_j, \mathbf{w}_i\rangle\|\mathbf{w}_i\|}{\langle \mathbf{w}_i, \mathbf{w}_i\rangle} = \frac{\langle \mathbf{v}_j, \mathbf{w}_i\rangle}{\|\mathbf{w}_i\|} = \langle \mathbf{v}_j, \mathbf{q}_i\rangle \text{ for } i < j$$

In the case that the vectors $\mathbf{v}_i$ are in $\mathbb{R}^m$, we get a matrix equation $A = QR$, where A is an m by n matrix with the n independent vectors $\mathbf{v}_i$ as its columns, Q is an m by n matrix with orthonormal columns $\mathbf{q}_i$, and R is an upper triangular matrix with positive entries on the diagonal. This equation is called the ***QR* decomposition** of the matrix A. The equation expresses the basis $\mathcal{V} = \{\mathbf{v}_1, \ldots, \mathbf{v}_n\}$ in terms of the basis $\mathcal{Q} = \{\mathbf{q}_1, \ldots, \mathbf{q}_n\}$, so $R = \mathcal{T}_{\mathcal{Q}}^{\mathcal{V}}$.

Note that we are assuming that the vectors $\mathbf{v}_i$ are independent. The same procedure can be applied even if the $\mathbf{v}_i$ are not independent. However, if there are dependent vectors, we will get an orthogonal (and orthonormal) basis for the subspace spanned by the $\mathbf{v}_i$ which has fewer vectors that the original collection. What will happen at the first time that $\mathbf{v}_{i+1}$ is a linear combination of the previous $\mathbf{v}_j$ (and thus $\mathbf{w}_j$) is that we will not have to add a new vector $\mathbf{w}_{i+1}$ to give a basis for the subspace spanned by $\mathbf{v}_1, \ldots, \mathbf{v}_{i+1}$. We will still get an equation $\mathbf{V} = \mathbf{Q}R$ relating the original spanning set $\mathbf{V} = \{\mathbf{v}_1, \ldots, \mathbf{v}_n\}$ to the orthonormal basis $\mathbf{Q} = \{\mathbf{q}_1, \ldots, \mathbf{q}_r\}$. If we start off with n vectors that span an r-dimensional subspace, $\mathbf{V}$ will have the original n vectors, $\mathbf{Q}$ will have an orthonormal basis with r vectors, and R will be an r by n matrix that is upper triangular, but now may have some zeros on the diagonal. In the case of $\mathbb{R}^m$ we still have a QR decomposition, but now R is not necessarily a square invertible matrix.

■ ***Example 3.1.1.*** Consider the plane in $\mathbb{R}^3$, $x + y - 3z = 0$. A basis for this plane is given by $\mathbf{v}_1 = (-1, 1, 0)$ and $\mathbf{v}_2 = (3, 0, 1)$. To get an orthonormal basis, we first set $\mathbf{w}_1 = \mathbf{v}_1$ and $\mathbf{q}_1 = \mathbf{w}_1/\|\mathbf{w}_1\| = (-1/\sqrt{2}, 1/\sqrt{2}, 0)$. Note that $r_{11} = \|\mathbf{w}_1\| = \sqrt{2}$. Then $\mathbf{w}_2 = \mathbf{v}_2 - t_{12}\mathbf{w}_1 = \mathbf{v}_2 - \dfrac{\langle \mathbf{v}_2, \mathbf{w}_1\rangle}{\langle \mathbf{w}_1, \mathbf{w}_1\rangle}\mathbf{w}_1 = (3, 0, 1) - (-3/2)(-1, 1, 0) = (3/2, 3/2, 1)$, $r_{22} = \|\mathbf{w}_2\| = \sqrt{22}/2$, and $\mathbf{q}_2 = \mathbf{w}_2/\|\mathbf{w}_2\| = (3/\sqrt{22}, 3/\sqrt{22}, 2/\sqrt{22})$. Here $r_{12} = \langle \mathbf{v}_2, \mathbf{q}_2\rangle = t_{12}\|\mathbf{w}_1\| = -3/\sqrt{2}$. We get $A = QR$ with

$$A = \begin{pmatrix} -1 & 3 \\ 1 & 0 \\ 0 & 1 \end{pmatrix}, \quad Q = \begin{pmatrix} -1/\sqrt{2} & 3/\sqrt{22} \\ 1/\sqrt{2} & 3/\sqrt{22} \\ 0 & 2/\sqrt{22} \end{pmatrix}, \quad R = \begin{pmatrix} \sqrt{2} & -3/\sqrt{2} \\ 0 & \sqrt{22}/2 \end{pmatrix} \quad ■$$

■ ***Example 3.1.2.*** Consider the three-dimensional subspace of $\mathbb{R}^4$ with basis $\mathbf{v}_1 = (1, 1, 0, 1)$, $\mathbf{v}_2 = (2, 0, 0, 1)$, $\mathbf{v}_3 = (0, 0, 1, 1)$. Then $\mathbf{w}_1 = \mathbf{v}_1$, $r_{11} = \|\mathbf{w}_1\| = \sqrt{3}$, and $\mathbf{q}_1 = (1/\sqrt{3})(1, 1, 0, 1)$. Next, $\mathbf{w}_2 = \mathbf{v}_2 - t_{12}\mathbf{w}_1 = (2, 0, 0, 1) - (1)(1, 1, 0, 1) = (1, -1, 0, 0)$, $r_{12} = t_{12}\|\mathbf{w}_1\| = \sqrt{3}$, $r_{22} = \|\mathbf{w}_2\| = \sqrt{2}$, and $\mathbf{q}_2 = (1/\sqrt{2})(1, -1, 0, 0)$. For the next step we compute $t_{13} = 1/3$, $t_{23} = 0$, so $\mathbf{w}_3 = \mathbf{v}_3 - (1/3)\mathbf{w}_1 - (0)\mathbf{w}_2 = (0, 0, 1, 1) - (1/3)(1, 1, 0, 1) = (-1/3, -1/3, 1, 2/3)$, $r_{33} = \|\mathbf{w}_3\| = \sqrt{15}/3$, $r_{13} =$

$\sqrt{3}/3, r_{23} = 0, \mathbf{q}_3 = (1/\sqrt{15})(-1, -1, 3, 2)$. We get $A = QR$ with

$$A = \begin{pmatrix} 1 & 2 & 0 \\ 1 & 0 & 0 \\ 0 & 0 & 1 \\ 1 & 1 & 1 \end{pmatrix}, \quad Q = \begin{pmatrix} 1/\sqrt{3} & 1/\sqrt{2} & -1/\sqrt{15} \\ 1/\sqrt{3} & -1/\sqrt{2} & -1/\sqrt{15} \\ 0 & 0 & 3/\sqrt{15} \\ 1/\sqrt{3} & 0 & 2/\sqrt{15} \end{pmatrix}, \quad R = \begin{pmatrix} \sqrt{3} & \sqrt{3} & \sqrt{3}/3 \\ 0 & \sqrt{2} & 0 \\ 0 & 0 & \sqrt{15}/3 \end{pmatrix}$$

■

■ ***Example 3.1.3.*** In this example, we look at a matrix A whose columns are dependent. Let $\mathbf{v}_1 = (1, 0, 1, 0)$, $\mathbf{v}_2 = (1, 1, 0, 0), \mathbf{v}_3 = (2, 1, 1, 0)$, and let A be the matrix with these vectors as its columns. Then $\mathbf{w}_1 = \mathbf{v}_1 = (1, 0, 1, 0)$, $r_{11} = \|\mathbf{w}_1\| = \sqrt{2}$, $\mathbf{q}_1 = (1/\sqrt{2})(1, 0, 1, 0)$. Next $\mathbf{w}_2 = \mathbf{v}_2 - t_{12}\mathbf{w}_1$, $t_{12} = 1/2$, so $\mathbf{w}_2 = (1, 1, 0, 0) - (1/2)(1, 0, 1, 0) = (1/2, 1, -1/2, 0), r_{12} = 1/\sqrt{2}, r_{22} = \|\mathbf{w}_2\| = \sqrt{6}/2$, $\mathbf{q}_2 = (1/\sqrt{6})(1, 2, -1, 0)$. When we form $\mathbf{w}_3 = \mathbf{v}_3 - t_{13}\mathbf{w}_1 - t_{23}\mathbf{w}_2, t_{13} = 3/2, t_{23} = 1$, so $\mathbf{w}_3 = (2, 1, 1, 0) - (3/2)(1, 0, 1, 0) - (1)(1/2, 1, -1/2, 0) = (0, 0, 0, 0)$. This reflects the linear dependence of the three vectors. We can still form $r_{13} = 3/\sqrt{2}$, $r_{23} = \sqrt{6}/2$ and get $A = QR$, where

$$A = \begin{pmatrix} 1 & 1 & 2 \\ 0 & 1 & 1 \\ 1 & 0 & 1 \\ 0 & 0 & 0 \end{pmatrix}, \quad Q = \begin{pmatrix} 1/\sqrt{2} & 1/\sqrt{6} \\ 0 & 2/\sqrt{6} \\ 1/\sqrt{2} & -1/\sqrt{6} \\ 0 & 0 \end{pmatrix}, \quad R = \begin{pmatrix} \sqrt{2} & 1/\sqrt{2} & 3/\sqrt{2} \\ 0 & \sqrt{6}/2 & \sqrt{6}/2 \end{pmatrix}$$

■

The foregoing examples concern vectors in $\mathbb{R}^n$. The Gram–Schmidt algorithm may also be used without change to find an orthogonal or orthonormal basis for a subspace spanned by a finite collection of vectors in any vector space. We illustrate this with some examples.

■ ***Example 3.1.4.*** We now consider the vector space $\mathcal{P}^2(\mathbb{R}, \mathbb{R})$ with inner product coming from mimicking the inner product in $\mathbb{R}^3$. We define

$$\langle a_0 + a_1x + a_2x^2, b_0 + b_1x + b_2x^2 \rangle = \sum_{i=0}^{2} a_i b_i$$

Then an orthonormal basis for $\mathcal{P}^2(\mathbb{R}, \mathbb{R})$ is given by $1, x, x^2$. Since the inner product is coming so directly from the inner product in $\mathbb{R}^3$, getting an orthonormal basis for a subspace will be essentially the same as the analogous problem in $\mathbb{R}^3$. For example, if we look at the subspace spanned by $-1 + x, 3 + x^2$ analogous to the vectors $(-1, 1, 0)$, $(3, 0, 1)$ in Example 3.1.1, then the same calculations as in that example would give an orthonormal basis of $-1/\sqrt{2} + 1/\sqrt{2}x, 3/\sqrt{22} + 3/\sqrt{22}x + 2/\sqrt{22}x^2$. ■

■ ***Example 3.1.5.*** We again look at $\mathcal{P}^2(\mathbb{R}, \mathbb{R})$, but we use a different inner product. This time we use an inner product which is motivated by the problem of getting the best approximation to a function by a polynomial on the interval $[-1, 1]$, where the square of the distance between two functions f, g is being measured by the integral of the square of the difference over the interval $[-1, 1]$. The inner product corresponding to this distance function is

$$\langle f, g\rangle = \int_{-1}^{1} f(t)g(t)\,dt$$

Then the length squared of a function is

$$\|f\|^2 = \langle f, f\rangle = \int_{-1}^{1} f(t)^2\,dt$$

and the distance squared between two functions f, g is given by the integral

$$\|f - g\|^2 = \int_{-1}^{1} (f(t) - g(t))^2\,dt$$

The basis $1, x, x^2$ no longer gives even an orthogonal basis in this inner product, although 1 and x are orthogonal and x and x^2 are orthogonal. To get an orthogonal basis, we can use the Gram–Schmidt algorithm to make x^2 orthogonal to 1. To do this we replace x^2 by $x^2 - \dfrac{\langle x^2, 1\rangle}{\langle 1, 1\rangle}1 = x^2 - 1/3$. Thus $1, x, x^2 - 1/3$ is an orthogonal basis for $\mathcal{P}^2(\mathbb{R}, \mathbb{R})$. The corresponding orthonormal basis is $1/\sqrt{2}$, $\sqrt{3}x/\sqrt{2}$, $\sqrt{45/8}(x^2 - 1/3)$.

In general, the polynomials in the orthogonal basis for $\mathcal{P}^n(\mathbb{R}, \mathbb{R})$ formed from the basis $1, x, \ldots, x^n$ by the Gram–Schmidt algorithm are related to the well-known **Legendre polynomials.** We will call our polynomials the monic orthogonal polynomials p_n, where the word "monic" indicates that the coefficient of the highest-order term is 1. The Legendre polynomials L_n are obtained from the monic orthogonal polynomials by $L_n(x) = p_n(x)/p_n(1)$. The first three Legendre polynomials are thus 1, x, and $(3/2)(x^2 - 1/3)$. ■

When we are given an orthogonal or an orthonormal basis for a vector space, we may use the inner product to solve the problem of expressing a given vector $\mathbf{v}$ in the vector space in terms of the basis. First let us consider the case of an orthogonal basis $\mathbf{v}_1, \ldots, \mathbf{v}_n$. Then we write $\mathbf{v} = c_1\mathbf{v}_1 + \cdots + c_n\mathbf{v}_n$. Taking the inner product of this with $\mathbf{v}_i$ gives $\langle \mathbf{v}, \mathbf{v}_i\rangle = c_i\langle \mathbf{v}_i, \mathbf{v}_i\rangle$, so $c_i = \langle \mathbf{v}, \mathbf{v}_i\rangle/\langle \mathbf{v}_i, \mathbf{v}_i\rangle$. This gives the formula

$$\mathbf{v} = \frac{\langle \mathbf{v}, \mathbf{v}_1\rangle}{\langle \mathbf{v}_1, \mathbf{v}_1\rangle}\mathbf{v}_1 + \cdots + \frac{\langle \mathbf{v}, \mathbf{v}_n\rangle}{\langle \mathbf{v}_n, \mathbf{v}_n\rangle}\mathbf{v}_n$$

In the case where we have an orthonormal basis this simplifies to $c_i = \langle \mathbf{v}, \mathbf{v}_i\rangle$. Thus if $\mathbf{q}_1, \ldots, \mathbf{q}_n$ is an orthonormal basis, we have

$$\mathbf{v} = \langle \mathbf{v}, \mathbf{q}_1\rangle\mathbf{q}_1 + \cdots + \langle \mathbf{v}, \mathbf{q}_n\rangle\mathbf{q}_n$$

■ ***Example 3.1.6.*** We know that $(-1, 1, 0)$ and $(3/2, 3/2, 1)$ give an orthogonal basis for the plane $x + y - 3z = 0$ from Example 3.1.1. If we consider the vector $(1, 2, 1)$ in this plane, then we can write

$$\begin{aligned}(1, 2, 1) &= \frac{\langle (1, 2, 1), (-1, 1, 0)\rangle}{2}(-1, 1, 0) + \frac{\langle (1, 2, 1), (3/2, 3/2, 1)\rangle}{(22/4)}(3/2, 3/2, 1)\\ &= 1/2(-1, 1, 0) + 1(3/2, 3/2, 1)\end{aligned}$$

■

■ ***Example 3.1.7.*** Using the inner product in Example 3.1.5 for $\mathcal{P}^2(\mathbb{R}, \mathbb{R})$, we can write any polynomial $a + bx + cx^2$ in terms of the orthogonal basis as $a + bx + cx^2 = (a + (1/3)c)1 + bx + c(x^2 - 1/3)$. In this case it is easier to determine the coefficients by just solving the equations directly rather than by using the formula based on the inner product. ■

Exercise 3.1.2.

Give an orthonormal basis for the plane $x + y + z = 0$ in $\mathbb{R}^3$ by first finding one basis and then using the Gram–Schmidt algorithm to change this to an orthonormal basis. Give the QR decomposition for the 3 by 2 matrix A which has as its columns the original basis you found for this plane. Extend your orthonormal basis to a orthonormal basis for $\mathbb{R}^3$ by adding to it a third unit vector which is a normal vector to the plane.

Exercise 3.1.3.

Give the QR decomposition for the matrix $A = \begin{pmatrix} 1 & 2 \\ 0 & 1 \\ 1 & 1 \end{pmatrix}$.

Exercise 3.1.4.

Give the QR decomposition for the matrix $A = \begin{pmatrix} 1 & 2 & 1 \\ 0 & 1 & 1 \\ 1 & 0 & 0 \\ -1 & 1 & 1 \end{pmatrix}$.

Exercise 3.1.5.

Express the vector $(3, 2, 1) \in \mathbb{R}^3$ as a linear combination of the orthogonal vectors $(1, 0, 1)$, $(1, 1, -1)$, $(-1, 2, 1)$.

Exercise 3.1.6.

Let $\mathcal{P}^2(\mathbb{R}, \mathbb{R})$ be given the inner product discussed in Example 3.1.5. Then use the Gram–Schmidt algorithm to find an orthogonal basis for the subspace spanned by $1 + x$ and $1 - x + x^2$. Express the vector $2 + x^2$ as a linear combination of the two polynomials in this orthogonal basis.

Exercise 3.1.7.

Consider $\mathcal{P}^3(\mathbb{R}, \mathbb{R})$ with the inner product as given in Example 3.1.5. Extend the orthogonal polynomials $1, x, x^2 - 1/3$ to an orthogonal basis by first adding the polynomial x^3 to form a basis and then using the Gram–Schmidt algorithm to make this into an orthogonal basis. What you will be finding is the next monic orthogonal polynomial.

3.1.2. HOUSEHOLDER MATRICES: ANOTHER APPROACH TO $A = QR$

Commercial routines handle the computation of the QR decomposition somewhat differently than our algorithm using the Gram–Schmidt algorithm. They tend to use a different algorithm based on Householder matrices which gives better numerical results. These Householder matrices correspond to certain reflections. The final form of the QR decomposition will be different, but will contain within it the information we obtained from Gram–Schmidt of how to write the original column vectors in A in terms of an orthonormal basis for the subspace they span. The columns of the Q used in these QR decompositions will actually be a basis for $\mathbb{R}^m$ if we have n vectors in $\mathbb{R}^m$ and the R matrix will be an m by n matrix. Only the first $r = \text{rank } A$ vectors in Q will be used in the expressions for the columns of A.

We will mainly use the QR decomposition in the case where the columns of A are independent and just take the one coming from the Gram–Schmidt algorithm. However, we do want to explain the QR decomposition based on Householder transformations, which we will call the $\overline{QR}$ decomposition to distinguish the two. Here $\overline{Q} = (Q_1 \quad Q_2)$, where Q_1 is m by r and Q_2 is m by $m-r$. The columns of $\overline{Q}$ form an orthonormal basis for $\mathbb{R}^m$, with the first r columns forming an orthonormal basis for $\mathcal{R}(A)$. The bottom $m - r$ rows of $\overline{R}$ will be zero rows. The idea of the algorithm is to start with the matrix A and find appropriate orthogonal matrices to multiply by A (on the left) to get the required matrix $\overline{R}$; that is, we will find Householder matrices $H_1, \ldots, H_k$, which are special m by m orthogonal matrices, so that $H_k \ldots H_1 A = \overline{R}$. Here $\overline{R}$ is an upper triangular matrix with the bottom $m - r$ rows equal to zero. We will restrict our attention to the simple case where the columns of A are independent, so $r = n$. We do this by induction on n.

We first explain what we mean by a **Householder matrix.** It is just a matrix which represents a **Householder transformation,** which is a reflection. H will reflect through the plane orthogonal to the unit vector $\mathbf{u}$; this means that it sends any vector orthogonal to $\mathbf{u}$ to itself and sends $\mathbf{u}$ to its negative. A picture for vectors in the plane is given in Figure 3.3.

The matrix for doing this is $H = I - 2\mathbf{u}\mathbf{u}^t$.

$$(I - 2\mathbf{u}\mathbf{u}^t)\mathbf{u} = \mathbf{u} - 2\mathbf{u} = -\mathbf{u}$$

We use $\mathbf{u}^t\mathbf{u} = 1$ since $\mathbf{u}$ is a unit vector. If $\mathbf{v}$ is orthogonal to $\mathbf{u}$, then $\mathbf{u}^t\mathbf{v} = 0$ so

$$(I - 2\mathbf{u}\mathbf{u}^t)\mathbf{v} = \mathbf{v} - \mathbf{0} = \mathbf{v}$$

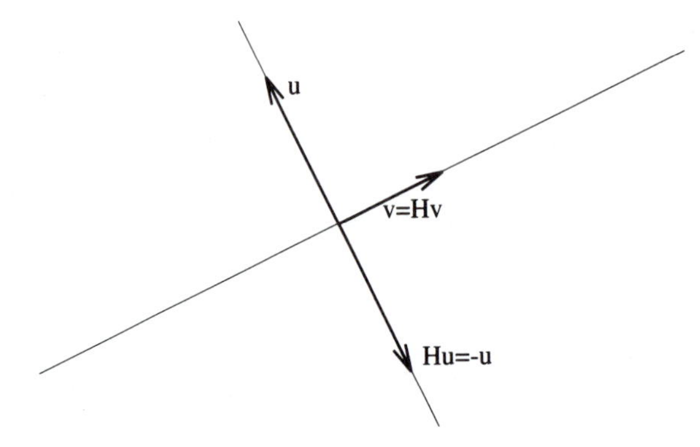

Figure 3.3. Reflection through the line containing **u**

Note that $(I - 2\mathbf{u}\mathbf{u}^t)$ is an orthogonal matrix:

$$(I - 2\mathbf{u}\mathbf{u}^t)^t(I - 2\mathbf{u}\mathbf{u}^t) = (I - 2\mathbf{u}\mathbf{u}^t)(I - 2\mathbf{u}\mathbf{u}^t) = I - 4\mathbf{u}\mathbf{u}^t + 4\mathbf{u}\mathbf{u}^t = I$$

This process of reflection can be used to exchange any two vectors of the same length. For if $\mathbf{v}, \mathbf{w}$ have the same length we can just choose

$$\mathbf{u} = (\mathbf{v} - \mathbf{w})/\|\mathbf{v} - \mathbf{w}\|$$

Since $\|\mathbf{v}\| = \|\mathbf{w}\|$, the vector $\mathbf{x} = (\mathbf{v} + \mathbf{w})$ is orthogonal to $\mathbf{u} = (\mathbf{v} - \mathbf{w})$:

$$\langle \mathbf{v} + \mathbf{w}, \mathbf{v} - \mathbf{w} \rangle = \langle \mathbf{v}, \mathbf{v} \rangle - \langle \mathbf{w}, \mathbf{w} \rangle = \|\mathbf{v}\|^2 - \|\mathbf{w}\|^2 = 0$$

We take the Householder matrix $H_{\mathbf{u}}$ which reflects through the plane perpendicular to $\mathbf{u}$. Thus writing

$$\mathbf{v} = \frac{(\mathbf{v} + \mathbf{w})}{2} + \frac{(\mathbf{v} - \mathbf{w})}{2} = \mathbf{x} + \mathbf{u}, \mathbf{w} = \frac{(\mathbf{v} + \mathbf{w})}{2} - \frac{(\mathbf{v} - \mathbf{w})}{2} = \mathbf{x} - \mathbf{u}$$

the reflection H_u preserves the first term and changes the sign on the second term, so interchanges $\mathbf{v}$ and $\mathbf{w}$. Figure 3.4 illustrates this in the two-dimensional case.

To apply this in our situation, we see that if we take any vector $\mathbf{x}$ in $\mathbb{R}^m$, then there is a Householder transformation H that sends $\mathbf{x}$ to $\|\mathbf{x}\|\mathbf{e}_1$ (or to $-\|\mathbf{x}\|\mathbf{e}_1$). The sign is chosen so that the distance between the two vectors is maximal, since that leads to better results numerically. Figure 3.5 illustrates this in the plane.

We now prove by induction on n the existence of the required Householder transformations $H_1, \ldots, H_k$ so that $H_k \ldots H_1 A = \overline{R}$ as described. If $n = 1$, then the argument shows that there is a Householder matrix H with $H\mathbf{A}^1 = r_{11}\mathbf{e}_1$, where $r_{11} = \pm\|\mathbf{A}^1\|$. We next assume by induction that for a matrix B with fewer than n columns (which are independent) there is an orthogonal matrix H so that $HB = S$, where S is upper triangular. If A is an m by n matrix, we first find a Householder matrix H_1 with $H_1\mathbf{A}^1 = r_{11}\mathbf{e}_1$.

Then $H_1 A = \begin{pmatrix} r_{11} & \mathbf{v} \\ \mathbf{0} & B \end{pmatrix}$.

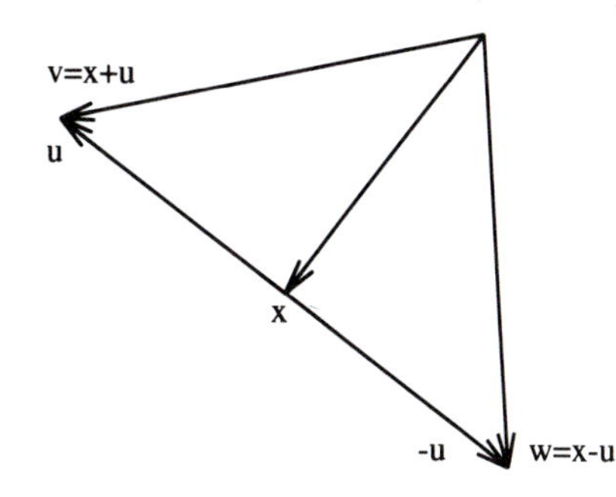

Figure 3.4. Reflection interchanging **v** and **w**

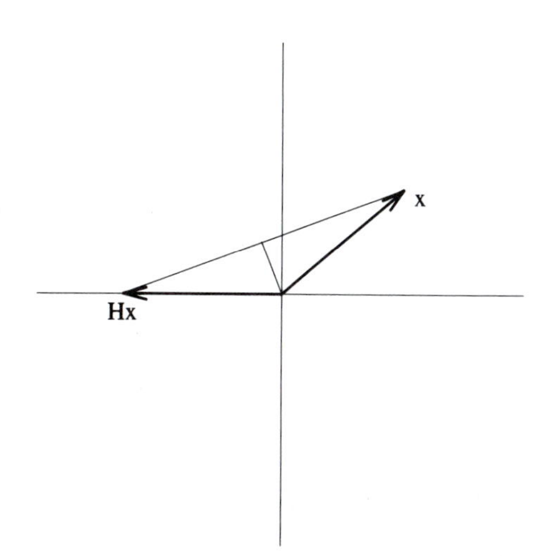

Figure 3.5. Reflection sending **x** to $-\|\mathbf{x}\|\mathbf{e}_1$

Here r_{11} is the $(1, 1)$ entry of the matrix, **v** is an $(n-1)$ row vector so that $(r_{11} \quad \mathbf{v})$ is the first row. The $\mathbf{0}$ is the $(m-1)$ column vector $\mathbf{0}$, and B is an $(m-1)$ by $(n-1)$ matrix. Since multiplying by the invertible matrix H_1 doesn't change the rank of A, the rank of B will be $(n-1)$. Thus we can assume inductively that there is an $(m-1)$ by $(m-1)$ orthogonal matrix C (which is a product of Householder matrices) so that $CB = R_2$. Then if $D = \begin{pmatrix} 1 & \mathbf{0} \\ \mathbf{0} & C \end{pmatrix}$ and $F = DH_1$ we have $FA = \overline{R}$ is upper triangular with nonzero entries on the diagonal. F is a product of Householder matrices. Then if $\overline{Q} = H^t$, we have $A = \overline{Q}\overline{R}$ as required. This is the algorithm that is implemented in MATLAB by typing [Q,R] = qr(A). When MATLAB is run in Example 3.1.2, the output is

$$Q = \begin{pmatrix} -.5774 & .4082 & -.4082 & -.5774 \\ -.5774 & -.8165 & 0 & 0 \\ 0 & 0 & -.8165 & .5774 \\ -.5774 & .4082 & .4082 & .5774 \end{pmatrix},$$

$$R = \begin{pmatrix} -1.7321 & -1.1547 & -1.7321 \\ 0 & .8165 & 1.2247 \\ 0 & 0 & -1.2247 \\ 0 & 0 & 0 \end{pmatrix}$$

The four columns of Q give an orthonormal basis for $\mathbb{R}^4$. The first three columns give an orthonormal basis for the column space of A. The zeros in the last row of R occur since the last column of Q is not involved in writing the columns of A in terms of the orthonormal basis of $\mathbb{R}^4$ given by the columns of Q. This algorithm has two advantages over the one based on the Gram–Schmidt decomposition. First it behaves much better computationally. From a theoretical perspective, it also outputs more information since in $\overline{Q} = (Q_r \quad Q_n)$ the Q_n piece gives an orthonormal basis for the orthogonal complement of the range of A (i.e., those vectors orthogonal to every vector in the range of A), which is $\mathcal{N}(A^t)$.

3.2. ORTHOGONAL SUBSPACES

We now consider the concept of when two subspaces S, T in an inner product space V are orthogonal to each other.

DEFINITION 3.2.1. A subspace S is **orthogonal** to a subspace T if for each $\mathbf{s} \in S$ and $\mathbf{t} \in T$, then $\langle \mathbf{s}, \mathbf{t} \rangle = 0$.

Note that this is a symmetric relation on subspaces; that is, S orthogonal to T implies T is orthogonal to S. We can define the orthogonal complement of S as the subspace of those vectors which are orthogonal to every vector in S.

DEFINITION 3.2.2. For a subspace S, the **orthogonal complement** $S^\perp$ is the subspace

$$S^\perp = \{\mathbf{v} \in V : \langle \mathbf{v}, \mathbf{s} \rangle = 0 \text{ for all } \mathbf{s} \in S\}$$

Let us verify that this is indeed a subspace. Suppose that $\mathbf{v}, \mathbf{w} \in S^\perp$. Then if $a\mathbf{v}+b\mathbf{w}$ is a linear combination of these vectors and $\mathbf{s}$ is any vector in S,

$$\langle a\mathbf{v} + b\mathbf{w}, \mathbf{s} \rangle = a\langle \mathbf{v}, \mathbf{s} \rangle + b\langle \mathbf{w}, \mathbf{s} \rangle = a \cdot 0 + b \cdot 0 = 0$$

Thus $a\mathbf{v} + b\mathbf{w} \in S^\perp$. Note that $S^\perp$ is orthogonal to S and is the largest such subspace; that is, any subspace T which is orthogonal to S is contained in $S^\perp$.

We next give a proposition which tells us how to find the orthogonal complement of a finite-dimensional subspace.

PROPOSITION 3.2.1. Let S be a finite-dimensional subspace of an inner product space V and let $\mathbf{s}_1, \ldots, \mathbf{s}_k$ span S. Then $\mathbf{v} \in S^\perp$ iff $\langle \mathbf{v}, \mathbf{s}_i \rangle = 0, i = 1, \ldots, k$.

Proof. If a vector $\mathbf{v}$ is in $S^{\perp}$, then $\mathbf{v}$ is orthogonal to every vector in S, so $\langle \mathbf{v}, \mathbf{s}_i \rangle = 0$.

If $\langle \mathbf{v}, \mathbf{s}_i \rangle = 0, i = 1, \ldots, k$, then we show $\mathbf{v} \in S^{\perp}$. For any vector $\mathbf{s}$ in S is a linear combination $c_1\mathbf{s}_1 + \cdots + c_k\mathbf{s}_k$. Then $\langle \mathbf{v}, \mathbf{s}_i \rangle = 0$ for all i implies that

$$\langle \mathbf{v}, c_1\mathbf{s}_1 + \cdots + c_k\mathbf{s}_k \rangle = c_1\langle \mathbf{v}, \mathbf{s}_1 \rangle + \cdots + c_k\langle \mathbf{v}, \mathbf{s}_k \rangle = 0$$

■

Thus to find $S^{\perp}$ we just have to solve for those vectors which are orthogonal to each of the basis vectors.

■ ***Example 3.2.1.*** Consider a line ℓ through the origin in $\mathbb{R}^3$. Then the orthogonal complement $\ell^{\perp}$ is the plane through $\mathbf{0}$ which is perpendicular to this line (i.e., has a vector in the line as normal vector). Any line through the origin in this plane would be an orthogonal subspace to ℓ. Note that the sum of the dimensions of the line and its orthogonal complement is the total dimension of $\mathbb{R}^3$. Note also that the orthogonal complement of the plane is the original line. ■

We now prove a couple of general propositions which show the phenomena in this example occur in general.

PROPOSITION 3.2.2. Let S be a finite-dimensional subspace of an inner product space V. Then there is a direct sum decomposition $V = S \oplus S^{\perp}$.

Proof. We first show that $S \cap S^{\perp} = \{\mathbf{0}\}$. If $\mathbf{s} \in S$ and $\mathbf{s} \in S^{\perp}$, then $\langle \mathbf{s}, \mathbf{s} \rangle = 0$. But the only vector which is orthogonal to itself is the zero vector.

We next show that $V = S + S^{\perp}$. Let $\mathbf{v} \in V$ be given. Let $\mathbf{s}_1, \ldots, \mathbf{s}_k$ be an orthonormal basis for S. Consider the vector $\mathbf{s} = \langle \mathbf{v}, \mathbf{s}_1 \rangle \mathbf{s}_1 + \cdots + \langle \mathbf{v}, \mathbf{s}_k \rangle \mathbf{s}_k$. Then $\mathbf{s}$, being a linear combination of the basis vectors for S, is a vector in S. Let $\mathbf{t} = \mathbf{v} - \mathbf{s}$. Then

$$\langle \mathbf{t}, \mathbf{s}_j \rangle = \langle \mathbf{v}, \mathbf{s}_j \rangle - \langle \mathbf{v}, \mathbf{s}_1 \rangle \langle \mathbf{s}_1, \mathbf{s}_j \rangle - \cdots - \langle \mathbf{v}, \mathbf{s}_k \rangle \langle \mathbf{s}_k, \mathbf{s}_j \rangle$$

Since the $\mathbf{s}_i$ form an orthonormal basis, $\langle \mathbf{s}_i, \mathbf{s}_j \rangle = 0$ if $i \neq j$ and equals 1 if $i = j$. So we get $\langle \mathbf{t}, \mathbf{s}_j \rangle = \langle \mathbf{v}, \mathbf{s}_j \rangle - \langle \mathbf{v}, \mathbf{s}_j \rangle = 0$. Thus $\mathbf{t}$ is orthogonal to each of the basis vectors of S and so $\mathbf{t} \in S^{\perp}$. Thus $\mathbf{v} = \mathbf{s} + \mathbf{t}$ is in $S + S^{\perp}$. ■

COROLLARY 3.2.3. Let V be a finite-dimensional inner product space and S a subspace. Then $\dim S + \dim S^{\perp} = \dim V$.

Proof. This follows from the fact that whenever there is a direct sum decomposition $V = S \oplus T$, then $\dim S + \dim T = \dim V$. ■

COROLLARY 3.2.4. Let V be a finite-dimensional inner product space. Then if S is a subspace, $(S^{\perp})^{\perp} = S$.

Proof. First note that any vector in S is orthogonal to any vector in $S^\perp$ by the definition of $S^\perp$. So S is contained in $(S^\perp)^\perp$. But Corollary 3.2.3 implies that $\dim S + \dim S^\perp = \dim V = \dim S^\perp + \dim(S^\perp)^\perp$. Thus S is a subspace of $(S^\perp)^\perp$ of the same dimension, and so they must be equal. ■

■ ***Example 3.2.2.*** Consider the row space of the matrix $A = \begin{pmatrix} 1 & 3 & -1 & 2 \\ 2 & -2 & -1 & 4 \end{pmatrix}$.

This is a two-dimensional subspace of $\mathbb{R}^4$, with basis given by the rows of A. To be in the orthogonal complement of the row space, a vector just has to be orthogonal to each of the rows. The condition for this is to be a solution of $A\mathbf{x} = \mathbf{0}$. Thus the orthogonal complement of the row space of A is just the null space of A. We can find a basis for this orthogonal complement via Gaussian elimination to be $(5/8, 1/8, 1, 0), (-2, 0, 0, 1)$. When the Gram–Schmidt algorithm is applied to this basis, it gives an orthogonal basis for the null space as $(.5270, .1054, .8433, 0), (-.7325, .0563, .4507, .5071)$. Note that when we form a new matrix B with rows given by a basis for the null space of A and then solve $B\mathbf{x} = \mathbf{0}$, we are just solving for vectors in the orthogonal complement of the null space of A, which is the row space of A. The matrix B could be taken as $B = \begin{pmatrix} 5/8 & 1/8 & 1 & 0 \\ -2 & 0 & 0 & 1 \end{pmatrix}$. Here Gaussian elimination would give the basis $(0, -8, 1, 0), (.5, -2.5, 0, 1)$ for the row space. This is, of course, a different basis from the one we started out with for the row space, but it is the one which is obtained using a modified form of Gaussian elimination on A where we make the third and fourth variables the basic variables in finding R. Also, note that the four vectors $(1, 3, -1, 2), (2, -2, -1, 4), (5/8, 1/8, 1, 0), (-2, 0, 0, 1)$ constitute a basis for $\mathbb{R}^4$. The first two vectors are coming from a basis for the row space of A, and the second two come from a basis for the orthogonal complement, which is the null space of A. That these four vectors are a basis for $\mathbb{R}^4$ just illustrates that $\mathbb{R}^4$ is the direct sum of the row space and its orthogonal complement. ■

We now look at examples coming from the four fundamental subspaces associated to a matrix. Suppose A is an m by n matrix. Then we have seen that

(1) $A\mathbf{x} = \mathbf{b}$ has a solution iff $\langle \mathbf{b}, \mathbf{c} \rangle = 0$ for all $\mathbf{c} \in \mathcal{N}(A^t)$.

(2) $A\mathbf{x} = \mathbf{0}$ iff $\langle \mathbf{x}, A^t\mathbf{c} \rangle = 0$ for all $\mathbf{c} \in \mathbb{R}^m$.

Statement (1) is just the statement that the range of A (i.e., all those $\mathbf{b}$ for which there is a solution of $A\mathbf{x} = \mathbf{b}$) coincides with the orthogonal complement of $\mathcal{N}(A^t)$. By Corollary 3.2.4 the orthogonal complement of the range of A is $\mathcal{N}(A^t)$. Statement (2) just says that the null space of A is the orthogonal complement of the range of A^t, (which is the row space of A). Thus Corollary 3.2.4 implies that the orthogonal complement of the null space of A is the row space of A. Let us write out these relationships as a proposition.

PROPOSITION 3.2.5.

(1) $\mathcal{N}(A^t)^\perp = \mathcal{R}(A)$; $\mathcal{R}(A)^\perp = \mathcal{N}(A^t)$.

(2) $\mathcal{N}(A)^\perp = \mathcal{R}(A^t)$; $\mathcal{R}(A^t)^\perp = \mathcal{N}(A)$.

Note that we deduced the second part of each statement from the first part. Note also that by replacing A by A^t, (1) becomes equivalent to (2).

Our proof of the proposition is somewhat indirect, relying on earlier results. To emphasize the use of orthogonality, we now reprove the statement $\mathcal{R}(A) = \mathcal{N}(A^t)^\perp$ more directly. We first show $\mathcal{R}(A) \subset \mathcal{N}(A^t)^\perp$. If $\mathbf{y} \in \mathcal{R}(A), \mathbf{z} \in \mathcal{N}(A^t)$, then we must show $\langle \mathbf{y}, \mathbf{z} \rangle = 0$. But there is an $\mathbf{x} \in \mathbb{R}^n$ with $\mathbf{y} = A\mathbf{x}$. Thus

$$\langle \mathbf{y}, \mathbf{z} \rangle = \langle A\mathbf{x}, \mathbf{z} \rangle = \langle \mathbf{x}, A^t\mathbf{z} \rangle = \langle \mathbf{x}, \mathbf{0} \rangle = 0$$

But the dimension of $\mathcal{R}(A)$ is the rank r of A. Since the dimension of $\mathcal{N}(A^t)$ is $m - r$, the dimension of its orthogonal complement is also r. But when two subspaces have the same dimension and one is contained in the other, they must be equal, showing $\mathcal{R}(A) = \mathcal{N}(A^t)^\perp$.

■ ***Example 3.2.3.*** Our previous example illustrated the relationship between the row space of a matrix and the null space as orthogonal complements. We now look at an example concerning the orthogonal complement of the column space of A. Let $A = \begin{pmatrix} 1 & 0 & 2 \\ 2 & 1 & 5 \\ -1 & 1 & -1 \\ 0 & 1 & 1 \end{pmatrix}$. Gaussian elimination gives the basis $(1, 2, -1, 0), (0, 1, 1, 1)$ for the column space of A. To find the orthogonal complement of the column space, we look at those vectors which are perpendicular to each column. But this is just the condition that the vector is in the null space of A^t, since taking the dot product with a column vector is the same as using the transpose to change the column vector to a row vector and seeing that the product of the row vector with the given column vector is 0. Thus we solve $A^t\mathbf{x} = 0$. Gaussian elimination leads to the basis $(3, -1, 1, 0), (2, -1, 0, 1)$ for the null space of A^t. The four vectors $(1, 2, -1, 0), (0, 1, 1, 1), (3, -1, 1, 0), (2, -1, 0, 1)$ form a basis for $\mathbb{R}^4$, illustrating that $\mathbb{R}^4$ is the direct sum of the column space of A and the null space of A^t. ■

Now consider the linear transformation $L_A : \mathbb{R}^n \to \mathbb{R}^m$ given by multiplying by A, $L_A(\mathbf{x}) = A\mathbf{x}$. We have a direct sum decomposition $\mathbb{R}^n = \mathcal{R}(A^t) \oplus \mathcal{N}(A)$. Note that L_A maps $\mathbb{R}^n$ onto $\mathcal{R}(A)$. Given $\mathbf{b} \in \mathcal{R}(A)$, there is some $\mathbf{v} \in \mathbb{R}^n$ with $L(\mathbf{v}) = \mathbf{b}$. But $\mathbf{v} = \mathbf{r} + \mathbf{n}$, where $\mathbf{r} \in \mathcal{R}(A^t)$ and $\mathbf{n} \in \mathcal{N}(A)$. Then $L_A(\mathbf{v}) = A\mathbf{v} = A(\mathbf{r} + \mathbf{n}) = A\mathbf{r} + A\mathbf{n} = A\mathbf{r} + \mathbf{0} = A\mathbf{r} = L_A(\mathbf{r})$. This implies that when we restrict L_A to $\mathcal{R}(A^t)$ to form a linear transformation which we will denote by L', then $L' : \mathcal{R}(A^t) \to \mathcal{R}(A)$, $L'(\mathbf{r}) = L_A(\mathbf{r})$ will map $\mathcal{R}(A^t)$ onto $\mathcal{R}(A)$. Since these vector spaces each have the same dimension, which is the rank of A, L' is an isomorphism. Thus multiplication by A sends the row space isomorphically onto the column space and sends the orthogonal complement of the row space to zero. Similarly, we can show that multiplication by A^t sends the column space of A isomorphically to the row space of A and sends the orthogonal complement of the column space to zero.

PROPOSITION 3.2.6. Let A be an m by n matrix and let $L_A : \mathbb{R}^n \to \mathbb{R}^m$ be the linear transformation given by $L_A\mathbf{x} = A\mathbf{x}$. In terms of the decomposition $\mathbb{R}^n = \mathcal{R}(A^t) \oplus \mathcal{N}(A)$, let $L' : \mathcal{R}(A^t) \to \mathcal{R}(A)$ denote the restriction of L. Then L' is an isomorphism.

■ ***Example 3.2.4.*** We now illustrate in Figure 3.6 the four fundamental subspaces for A in the case of a 2 by 3 matrix of rank 1. Then $\mathbb{R}^3 = \mathcal{R}(A^t) \oplus \mathcal{N}(A)$, where $\mathcal{R}(A^t)$ is the row space and is given by a line and $\mathcal{N}(A)$ is the plane which is perpendicular to this line. Similarly, $\mathbb{R}^2 = \mathcal{R}(A) \oplus \mathcal{N}(A^t)$, where $\mathcal{R}(A)$ is the column space and is a line and $\mathcal{N}(A^t)$ is the perpendicular line. Multiplication by A sends the row space isomorphically onto the column space and sends the perpendicular plane to $\mathbf{0}$. Analogously, multiplication by A^t sends the column space isomorphically onto the row space and sends the perpendicular line to $\mathbf{0}$.

We now return to Example 3.2.3 to illustrate the proposition. For the matrix A considered there, the row space has a basis $(1, 0, 2), (0, 1, 1)$. When we multiply these two vectors by A, they are sent to the two vectors $(5, 12, -3, 2), (2, 6, 0, 2)$ in the column space. Since we know the column space is two dimensional and these vectors are independent, they are a basis for the column space. These two new basis vectors can be expressed in terms of our earlier basis vectors as

$$(5, 12, -3, 2) = 5(1, 2, -1, 0) + 2(0, 1, 1, 1), (2, 6, 0, 2) = 2(1, 2, -1, 0) + 2(0, 1, 1, 1)$$

■

■ ***Example 3.2.5.*** We now look at an example where we are using a different vector space and inner product. We consider the vector space $\mathcal{P}^2(\mathbb{R}, \mathbb{R})$ and the inner product given by $< p, q >= \int_{-1}^{1} p(t)q(t)\, dt$. We look at the subspace with basis the two polynomials $1, x + x^2$. To find the orthogonal complement for this subspace, we take a polynomial $a_0 + a_1x + a_2x^2$ and compute the inner product with each of these. We get

$$\begin{aligned}\langle a_0 + a_1x + a_2x^2, 1\rangle &= 2a_0 + (2/3)a_2\\ \langle a_0 + a_1x + a_2x^2, x + x^2\rangle &= (2/3)a_0 + (2/3)a_1 + (2/5)a_2\end{aligned}$$

Thus the equations for $a_0 + a_1x + a_2x^2$ to be in the orthogonal complement are $A\mathbf{a} = \mathbf{0}$, where $A = \begin{pmatrix} 2 & 0 & 2/3 \\ 2/3 & 2/3 & 2/5 \end{pmatrix}$, which has solution space with basis $(-5, -4, 15)$. This means that the orthogonal complement of this subspace has basis given by the polynomial $-5 - 4x + 15x^2$. ■

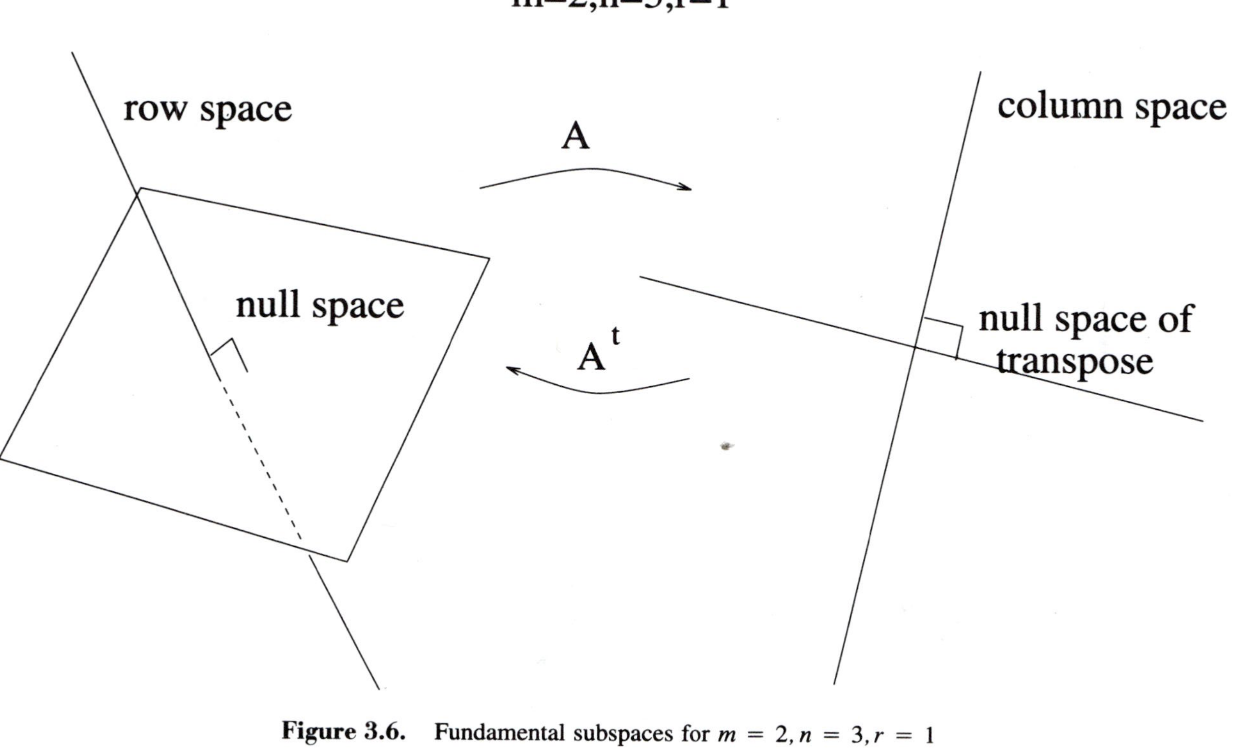

Figure 3.6. Fundamental subspaces for $m = 2, n = 3, r = 1$

Exercise 3.2.1.

For the matrix $A = \begin{pmatrix} 1 & 0 & -1 & 0 \\ 2 & 1 & 0 & 1 \\ 0 & 1 & 2 & 1 \end{pmatrix}$ find bases for the four fundamental subspaces and verify that $\mathcal{N}(A)$ is orthogonal to $\mathcal{R}(A^t)$ and $\mathcal{R}(A)$ is orthogonal to the $\mathcal{N}(A^t)$. Also, show that the image of the basis for the row space under multiplication by A gives a basis for the column space.

Exercise 3.2.2.

Repeat the previous exercise for the matrix

$$A = \begin{pmatrix} 1 & 1 & 1 & 0 & 1 \\ 0 & 1 & 0 & 1 & 0 \\ 2 & 3 & 2 & 1 & 2 \\ 3 & 5 & 3 & 2 & 3 \end{pmatrix}$$

Exercise 3.2.3.

Find a basis for the orthogonal complement of the subspace of $\mathbb{R}^5$ which has a basis given by $(2, 1, 0, 1, 0), (1, 0, -1, 0, 1), (1, 1, 0, 1, 1)$.

Exercise 3.2.4.

Consider the vector space $\mathcal{P}^3(\mathbb{R}, \mathbb{R})$ with an inner product of two polynomials given by $\langle p, q\rangle = \int_{-1}^{1} p(t)q(t)\,dt$. Find a basis for the orthogonal complement of the subspace with basis $1, x$.

Exercise 3.2.5.

Suppose that S is a subspace of a finite-dimensional inner product space V and that $\mathbf{s}_1, \ldots, \mathbf{s}_k$ is an orthonormal basis of S which, when the vectors $\mathbf{t}_1, \ldots, \mathbf{t}_p$ are added, becomes an orthonormal basis for V. Show that $\mathbf{t}_1, \ldots \mathbf{t}_p$ is an orthonormal basis for $S^{\perp}$.

3.3. ORTHOGONAL PROJECTIONS AND LEAST SQUARES SOLUTIONS

Suppose that there is a direct sum decomposition $V = S \oplus T$ of an inner product space V with S and T being orthogonal complements. Then each vector $\mathbf{v} \in V$ can be written uniquely as $\mathbf{s} + \mathbf{t}$, where $\mathbf{s} \in S$ and $\mathbf{t} \in T$.

DEFINITION 3.3.1. With the foregoing hypotheses, define the **orthogonal projection** $P : V \to V$ onto the subspace S by the formula $P(\mathbf{v}) = \mathbf{s}$.

A special case of importance is when $V = \mathbb{R}^n$ with the usual inner product. Here each linear transformation arises by multiplication by a matrix.

DEFINITION 3.3.2. An n by n matrix A is called a **projection matrix** if multiplication by A gives orthogonal projection onto a subspace of $\mathbb{R}^n$.

Note that P is a linear transformation since $\mathbf{v} = \mathbf{s} + \mathbf{t}, \mathbf{w} = \mathbf{s}' + \mathbf{t}'$ implies

$$a\mathbf{v} + b\mathbf{w} = a(\mathbf{s} + \mathbf{t}) + b(\mathbf{s}' + \mathbf{t}') = (a\mathbf{s} + b\mathbf{s}') + (a\mathbf{t} + b\mathbf{t}')$$

and so

$$P(a\mathbf{v} + b\mathbf{w}) = a\mathbf{s} + b\mathbf{s}' = aP(\mathbf{v}) + bP(\mathbf{w})$$

Next note that if $\mathbf{v} = \mathbf{s} + \mathbf{t}$, then

$$\begin{aligned}\|\mathbf{v}\|^2 &= \langle \mathbf{v}, \mathbf{v} \rangle = \langle \mathbf{s} + \mathbf{t}, \mathbf{s} + \mathbf{t} \rangle = \langle \mathbf{s}, \mathbf{s} \rangle + 2\langle \mathbf{s}, \mathbf{t} \rangle + \langle \mathbf{t}, \mathbf{t} \rangle \\ &= \langle \mathbf{s}, \mathbf{s} \rangle + \langle \mathbf{t}, \mathbf{t} \rangle = \|\mathbf{s}\|^2 + \|\mathbf{t}\|^2\end{aligned}$$

In particular, if $\mathbf{v} = \mathbf{s}+\mathbf{t}$ and $\mathbf{s}' \in S$, then $\|\mathbf{v}-\mathbf{s}'\|^2 = \|(\mathbf{s}-\mathbf{s}')+\mathbf{t}\|^2 = \|\mathbf{s}-\mathbf{s}'\|^2+\|\mathbf{t}\|^2 \geq \|\mathbf{t}\|^2$. Moreover, there is an equality only when $\|\mathbf{s} - \mathbf{s}'\|^2 = 0$, which only occurs when $\mathbf{s} - \mathbf{s}' = \mathbf{0}$; that is, $\mathbf{s} = \mathbf{s}'$. This proves the following proposition.

PROPOSITION 3.3.1. If P denotes the orthogonal projection onto a subspace S, then $P(\mathbf{v}) = \mathbf{s}$ is the vector in S which is closest to the vector $\mathbf{v}$.

We discuss now the general problem of projecting onto a subspace. It turns out that this is simplest to describe when the subspace S has an orthonormal basis $\mathbf{q}_1, \ldots, \mathbf{q}_k$. Then our proof of Proposition 3.2.1 that there is a decomposition of V into $S \oplus S^{\perp}$ tells us that to find $P(\mathbf{v}) = \mathbf{s}$ we just take

$$P(\mathbf{v}) = \mathbf{s} = \langle \mathbf{v}, \mathbf{q}_1 \rangle \mathbf{q}_1 + \cdots + \langle \mathbf{v}, \mathbf{q}_k \rangle \mathbf{q}_k$$

There is a similar formula when we just have an orthogonal basis $\mathbf{w}_1, \ldots, \mathbf{w}_k$; then

$$P(\mathbf{v}) = \mathbf{s} = \frac{\langle \mathbf{v}, \mathbf{w}_1 \rangle}{\langle \mathbf{w}_1, \mathbf{w}_1 \rangle}\mathbf{w}_1 + \cdots + \frac{\langle \mathbf{v}, \mathbf{w}_k \rangle}{\langle \mathbf{w}_k, \mathbf{w}_k \rangle}\mathbf{w}_k$$

Of course, this formula is only useful when we have an orthonormal (or orthogonal) basis to begin with. This can be found from any basis by using the Gram–Schmidt algorithm on the basis.

We now work a couple of examples computing the projection of a given vector onto a subspace.

■ ***Example 3.3.1.*** Consider the line through $(1,3,1)$ as our subspace and the vector $\mathbf{b} = (1,1,0)$. Then the projection $P\mathbf{b}$ onto the line is given by

$$P\mathbf{b} = \frac{\langle \mathbf{b}, (1,3,1)\rangle}{\langle (1,3,1),(1,3,1)\rangle}(1,3,1) = (4/11, 12/11, 4/11)$$

This means that $\mathbf{b} = (1,1,0) = (4/11, 12/11, 4/11) + (7/11, -1/11, -4/11)$ and the second vector lies in the orthogonal plane to the line. Then the distance between the point $\mathbf{b}$ and the line, which is just the length of the vector $(7/11, -1/11, -4/11)$, is $\sqrt{66}/11$. Note also that the projection of $\mathbf{b}$ onto the orthogonal plane to the line is just the vector $(7/11, -1/11, -4/11)$ and so the distance between the point and the plane is the length of $(4/11, 12/11, 4/11)$, which is $\sqrt{180}/11$. ■

■ ***Example 3.3.2.*** Let S be the plane $x + y + z = 0$ in $\mathbb{R}^3$ and $\mathbf{b} = (1,2,-1)$. Then a basis for S is given by $(-1,1,0)$ and $(-1,0,1)$. We can use the Gram–Schmidt algorithm to get an orthonormal basis $\mathbf{q}_1 = (-1/\sqrt{2}, 1/\sqrt{2}, 0)$, $\mathbf{q}_2 = (-1/\sqrt{6}, -1/\sqrt{6}, 2/\sqrt{6})$. Then the projection of $\mathbf{b}$ onto S is

$$P\mathbf{b} = \langle \mathbf{b}, \mathbf{q}_1\rangle\mathbf{q}_1 + \langle \mathbf{b}, \mathbf{q}_2\rangle\mathbf{q}_2 = (1/3, 4/3, -5/3)$$

An alternate way to do this calculation is to first project $\mathbf{b}$ onto the orthogonal line through $(1,1,1)$ to get the vector $(2/3, 2/3, 2/3)$ and then subtract this vector from $\mathbf{b}$ to get $(1/3, 4/3, -5/3)$. ■

These last two examples reflect some general facts about projections.

PROPOSITION 3.3.2.
(1) Projection onto a line through a vector $\mathbf{s}$ is just given by the formula

$$P\mathbf{b} = \frac{\langle \mathbf{b}, \mathbf{s}\rangle}{\langle \mathbf{s}, \mathbf{s}\rangle}\mathbf{s}$$

If $\mathbf{s}$ is a unit vector $\mathbf{q}$, then the formula becomes

$$P\mathbf{b} = \langle \mathbf{b}, \mathbf{q}\rangle\mathbf{q}$$

If we are dealing with $\mathbb{R}^n$, this can be rewritten as

$$P\mathbf{b} = \mathbf{q}\langle \mathbf{q}, \mathbf{b}\rangle = \mathbf{q}(\mathbf{q}^t\mathbf{b}) = (\mathbf{q}\mathbf{q}^t)\mathbf{b}$$

Thus the matrix which represents P with respect to the standard basis is $\mathbf{q}\mathbf{q}^t$, where $\mathbf{q}$ is a unit vector in the line.

(2) When we are projecting onto a subspace which has an orthonormal basis $\mathbf{q}_1, \ldots, \mathbf{q}_k$, then

$$P\mathbf{b} = \langle \mathbf{b}, \mathbf{q}_1\rangle\mathbf{q}_1 + \cdots + \langle \mathbf{b}, \mathbf{q}_k\rangle\mathbf{q}_k = P_1\mathbf{b} + \cdots + P_k\mathbf{b}$$

where $P_i\mathbf{b} = \langle \mathbf{b}, \mathbf{q}_i\rangle\mathbf{q}_i$ is the projection of $\mathbf{b}$ onto the line through the vector $\mathbf{q}_i$. In particular, if $V = \mathbb{R}^m$, we get

$$P\mathbf{b} = (\mathbf{q}_1\mathbf{q}_1^t + \cdots + \mathbf{q}_k\mathbf{q}_k^t)\mathbf{b}$$

Thus the matrix which represents P with respect the standard basis is

$$\mathbf{q}_1\mathbf{q}_1^t + \cdots + \mathbf{q}_k\mathbf{q}_k^t = QQ^t$$

where Q is the m by k matrix which has $\mathbf{q}_j$ as the jth column vector.

The last equality follows from the way that matrix multiplication works, since if the columns of A are $\mathbf{A}^1, \ldots, \mathbf{A}^k$ and the rows of B are $\mathbf{B}_1, \ldots, \mathbf{B}_k$, then $AB = \mathbf{A}^1\mathbf{B}_1 + \cdots + \mathbf{A}^k\mathbf{B}_k$. The details are left as an exercise.

The above decomposition $P = P_1 + \cdots + P_k$ is an example of another fact about projections.

PROPOSITION 3.3.3. If S is the orthogonal direct sum of S_1 and S_2 and if P, P_1, P_2 denote the projections onto S, S_1, and S_2, then $P = P_1 + P_2$.

Proof. If $\mathbf{v}$ is any vector in V, we first write $\mathbf{v} = \mathbf{s} + \mathbf{t}$, where $\mathbf{s} \in S$ and $\mathbf{t} \in S^\perp$. Then write $\mathbf{s} = \mathbf{s}_1 + \mathbf{s}_2$, where $\mathbf{s}_i \in S_i$. Then $P\mathbf{v} = \mathbf{s}$, $P_1\mathbf{v} = \mathbf{s}_1$ (since $\mathbf{s}_2$ and $\mathbf{t}$ are both orthogonal to S_1) and $P_2\mathbf{v} = \mathbf{s}_2$ (since $\mathbf{s}_1$ and $\mathbf{t}$ are both orthogonal to S_2). Thus $P\mathbf{v} = \mathbf{s} = \mathbf{s}_1 + \mathbf{s}_2 = P_1\mathbf{v} + P_2\mathbf{v}$. Therefore $P = P_1 + P_2$. ∎

Note that it is critical that the two subspaces S_1 and S_2 are orthogonal as the result does not hold without this assumption. As a special case, take $S = V$ and then P becomes the identity I. Then we get $I = P_1 + P_2$ or $P_2 = I - P_1$. This was used in Example 3.3.2 where we determined the projection onto a plane by using the projection onto the orthogonal line.

We note a couple of properties which projection P onto a subspace S satisfies. First,

$$P^2 = P \tag{1}$$

If $\mathbf{v} = \mathbf{s} + \mathbf{t}$, then $P\mathbf{v} = \mathbf{s}$ and so $P^2\mathbf{v} = P\mathbf{s} = \mathbf{s} = P\mathbf{v}$. Next, there is the following equality:

$$\langle P\mathbf{v}, \mathbf{w}\rangle = \langle \mathbf{v}, P\mathbf{w}\rangle \tag{2}$$

Writing $\mathbf{v} = \mathbf{s} + \mathbf{t}$, $\mathbf{w} = \mathbf{s}' + \mathbf{t}'$, then

$$\langle P\mathbf{v}, \mathbf{w}\rangle = \langle \mathbf{s}, \mathbf{s}' + \mathbf{t}'\rangle = \langle \mathbf{s}, \mathbf{s}'\rangle = \langle \mathbf{s} + \mathbf{t}, \mathbf{s}'\rangle = \langle \mathbf{v}, P\mathbf{w}\rangle$$

It turns out that these two properties actually characterize the linear transformations $P : V \to V$ which are projection onto a subspace, at least when we know that V splits as an orthogonal direct sum of a subspace and its orthogonal complement. Suppose that P satisfies the two properties and that S is the subspace $\mathcal{R}(P)$ and T is its orthogonal complement. Then write $\mathbf{v} = \mathbf{s} + \mathbf{t}$. We must show that $P\mathbf{v} = \mathbf{s}$. Note first that

$\mathbf{s} \in S = \mathcal{R}(P)$ means that $\mathbf{s} = P\mathbf{w}$ for some $\mathbf{w}$. Then $P\mathbf{s} = P(P\mathbf{w}) = P^2(\mathbf{w}) = P\mathbf{w} = \mathbf{s}$ (since $P^2 = P$). Also, $\langle P\mathbf{t}, P\mathbf{t}\rangle = \langle \mathbf{t}, P^2\mathbf{t}\rangle = 0$ since $\mathbf{t}$ is orthogonal to any vector in the range of P (which $P^2\mathbf{t}$ is). But the only vector which is orthogonal to itself is zero, so $P\mathbf{t} = \mathbf{0}$. Thus $P(\mathbf{v}) = P(\mathbf{s} + \mathbf{t}) = P\mathbf{s} + P\mathbf{t} = \mathbf{s}$.

Now suppose that $P\mathbf{v} = A\mathbf{v}$; that is, A is the matrix which represents P with respect to the standard basis of $\mathbb{R}^n$. Then A satisfies the two properties $A^2 = A$ and $\langle A\mathbf{x}, \mathbf{y}\rangle = \langle \mathbf{x}, A\mathbf{y}\rangle$. Writing this second property out in terms of transposes gives $\mathbf{x}^t A^t \mathbf{y} = \mathbf{x}^t A\mathbf{y}$. This equation must hold for all $\mathbf{x}, \mathbf{y}$. If $\mathbf{x} = \mathbf{e}_i$ and $\mathbf{y} = \mathbf{e}_j$, then this gives $a_{ji} = a_{ij}$. Thus A must be symmetric; that is, $A = A^t$. Of course we could have verified both these properties directly, since we know that $A = QQ^t$. Thus $A^t = (QQ^t)^t = QQ^t = A$, and the columns of Q being orthogonal imply $A^2 = (QQ^t)(QQ^t) = Q(Q^tQ)Q^t = QQ^t = A$.

What our argument verifies is that these two properties

$$A = A^t, \qquad A^2 = A$$

actually characterize those matrices which represent projections with respect to the standard basis. Such matrices are called projection matrices.

We record the results of this discussion as a proposition.

PROPOSITION 3.3.4. (1) Let $V = S \oplus T$ be an inner product space which decomposes as the direct sum of orthogonal complements S, T, and let $P : V \to V$ denote orthogonal projection onto S. Then

$$P^2 = P \quad \text{and} \quad \langle P\mathbf{v}, \mathbf{w}\rangle = \langle \mathbf{v}, P\mathbf{w}\rangle$$

(2) If V is an inner product space so that each subspace S gives a splitting $V = S \oplus T$, where $T = S^\perp$, then a linear transformation P satisfying the two properties just given is orthogonal projection onto the subspace $\mathcal{R}(P)$.

(3) Projection matrices are characterized by

$$A^2 = A, A^t = A$$

■ ***Example 3.3.3.*** The matrix A corresponding to projection onto the line through the vector $(1, 1, 1)$ is $\mathbf{q}\mathbf{q}^t$, where $\mathbf{q} = (1/\sqrt{3}, 1/\sqrt{3}, 1/\sqrt{3})$. This matrix is

$$A = \begin{pmatrix} 1/3 & 1/3 & 1/3 \\ 1/3 & 1/3 & 1/3 \\ 1/3 & 1/3 & 1/3 \end{pmatrix}$$

Note that the projection onto the plane $x + y + z = 0$ orthogonal to this line corresponds to the matrix $I - A$, which is

$$B = \begin{pmatrix} 2/3 & -1/3 & -1/3 \\ -1/3 & 2/3 & -1/3 \\ -1/3 & -1/3 & 2/3 \end{pmatrix}$$

Note that both A and B are symmetric and are equal to their own square. B could also have been found by first finding the basis

$$\mathbf{q}_1 = (-1/\sqrt{2}, 1/\sqrt{2}, 0), \mathbf{q}_2 = (-1/\sqrt{6}, -1/\sqrt{6}, 2/\sqrt{6})$$

for the plane, forming $Q = \begin{pmatrix} -1/\sqrt{2} & -1/\sqrt{6} \\ 1/\sqrt{2} & -1/\sqrt{6} \\ 0 & 2/\sqrt{6} \end{pmatrix}$ and then $B = QQ^t$. ■

Now let us return to the equation $A\mathbf{x} = \mathbf{b}$. Recall that there is a direct sum decomposition of orthogonal subspaces $\mathbb{R}^n = \mathcal{R}(A^t) \oplus \mathcal{N}(A)$ and a direct sum decomposition of orthogonal subspaces $\mathbb{R}^m = \mathcal{R}(A) \oplus \mathcal{N}(A^t)$. If $\mathcal{N}(A^t)$ is nonempty, then not all vectors are in the range of A. If $\mathbf{b}$ is not in the range of A, then there is no solution to this equation. However, in some applications we are interested in the closest approximation to a solution which can be found.

DEFINITION 3.3.3. A solution $\mathbf{x}$ to the problem of minimizing $\|A\mathbf{x} - \mathbf{b}\|^2$ is called **a least squares solution** to the problem of solving $A\mathbf{x} = \mathbf{b}$.

Thus what we seek is a vector $\mathbf{x} \in \mathbb{R}^n$ so that $\|A\mathbf{x} - \mathbf{b}\|^2$ is minimal. But the possible vectors $A\mathbf{x}$ vary over the range of A, so we want $A\mathbf{x}$ to be that vector in the range of A which is closest to the vector $\mathbf{b}$. However, we know from Proposition 3.3.1 that the closest vector is the projection $P\mathbf{b}$ of $\mathbf{b}$ onto the range of A. Thus to solve the problem, we need to solve $A\mathbf{x} = P\mathbf{b}$ for $\mathbf{x}$. Of course, there will not be a unique solution unless $\mathcal{N}(A) = \{\mathbf{0}\}$. There will be a unique solution $\mathbf{r}$ in the row space, however. This follows from Proposition 3.2.6, which says that multiplication by A sends the row space isomorphically onto the range of A. If $A\mathbf{r} = P\mathbf{b}$, then all other solutions are of the form $\mathbf{r} + \mathbf{n}$, where $\mathbf{n}$ is any vector in the null space of A. Of all these solutions, the fact that the row space and the null space of A are orthogonal complements means that the solution of minimal length will be the one $\mathbf{r}$ which is in the row space : $\|\mathbf{r}+\mathbf{n}\|^2 = \|\mathbf{r}\|^2 + \|\mathbf{n}\|^2 \geq \|\mathbf{r}\|^2$ with equality iff $\mathbf{n} = \mathbf{0}$.

DEFINITION 3.3.4. The solution $\mathbf{r}$ of $A\mathbf{x} = P\mathbf{b}$ which lies in the row space is called **the least squares solution** of $A\mathbf{x} = \mathbf{b}$, where P is the orthogonal projection onto the range of A. It is the vector $\mathbf{x}$ of minimal length which minimizes $\|A\mathbf{x} - \mathbf{b}\|^2$.

Although the foregoing discussion tells us how to find theoretically the least squares solution of $A\mathbf{x} = \mathbf{b}$, there is a lot of work to be done to find it computationally. We can think of the problem as being broken into three problems:

1. Find the projection $P\mathbf{b}$ of $\mathbf{b}$ onto the range of A.
2. Find one solution $\mathbf{x}$ of $A\mathbf{x} = P\mathbf{b}$.
3. Find the projection $\mathbf{r}$ of $\mathbf{x} = \mathbf{r} + \mathbf{n}$ onto the row space.

We look at a couple of examples.

■ ***Example 3.3.4.*** Let $A = \begin{pmatrix} 1 & 2 \\ 0 & 1 \\ 2 & 4 \end{pmatrix}, \mathbf{b} = \begin{pmatrix} 1 \\ 0 \\ 0 \end{pmatrix}$. We want to find the least squares solution of $A\mathbf{x} = \mathbf{b}$. We first project $\mathbf{b}$ onto the range of A. A basis for the range of A is given by the columns of A since they are independent. We then apply Gram–Schmidt to get an orthonormal basis $\mathbf{q}_1 = (1/\sqrt{5}, 0, 2/\sqrt{5}), \mathbf{q}_2 = (0, 1, 0)$ of $\mathcal{R}(A)$. The row space of A is two-dimensional and is a subspace of $\mathbb{R}^2$, so it is just $\mathbb{R}^2$. Since $\mathcal{N}(A) = \mathbf{0}$, there will be a unique solution of $A\mathbf{x} = P\mathbf{b}$. We first compute $P\mathbf{b} = \langle \mathbf{b}, \mathbf{q}_1 \rangle \mathbf{q}_1 + \langle \mathbf{b}, \mathbf{q}_2 \rangle \mathbf{q}_2 = (1/5, 0, 2/5)$. We then solve $A\mathbf{x} = P\mathbf{b}$ to get $x = (1/5, 0)$. ■

■ ***Example 3.3.5.*** Let $A = \begin{pmatrix} 1 & 0 & 1 \\ 2 & 1 & 0 \\ 3 & 1 & 1 \end{pmatrix}, \mathbf{b} = \begin{pmatrix} 1 \\ 1 \\ 1 \end{pmatrix}$. We will find the least squares solution of $A\mathbf{x} = \mathbf{b}$. We first find bases $(1, 0, 1), (0, 1, -2)$ of the row space and $(1, 2, 3), (0, 1, 1)$ of the column space. The condition to be in the range of A is $b_3 = b_1 + b_2$; thus our $\mathbf{b}$ is not in the range of A. We then use the Gram–Schmidt algorithm (or QR decomposition) to get orthonormal bases

$$\mathbf{q}_1 = (.2673, .5345, .8018), \quad \mathbf{q}_2 = (.7715, -.6172, .1543)$$

for the column space and

$$\mathbf{r}_1 = (.7071, 0, .7071), \quad \mathbf{r}_2 = (.5774, .5774, -.5774)$$

for the row space. Alternatively, we can get $\mathbf{q}_1, \mathbf{q}_2$ from the QR decomposition of A and $\mathbf{r}_1, \mathbf{r}_2$ from the QR decomposition of A^t. Then

$$P\mathbf{b} = \langle \mathbf{b}, \mathbf{q}_1 \rangle \mathbf{q}_1 + \langle \mathbf{b}, \mathbf{q}_2 \rangle \mathbf{q}_2 = (2/3, 2/3, 4/3)$$

We next solve $A\mathbf{x} = P\mathbf{b}$ to get $\mathbf{x} = (2/3, -2/3, 0)$. We then project $\mathbf{x}$ onto the row space to get

$$\mathbf{r} = \langle \mathbf{x}, \mathbf{r}_1 \rangle \mathbf{r}_1 + \langle \mathbf{x}, \mathbf{r}_2 \rangle \mathbf{r}_2 = (1/3, 0, 1/3)$$

Thus $(1/3, 0, 1/3)$ is that vector in the row space which A sends to $P\mathbf{b}$; that is,

$$A\mathbf{r} = (1/3)\mathbf{A}^1 + (1/3)\mathbf{A}^3 = (2/3, 2/3, 4/3)$$ ■

Exercise 3.3.1.

Show that if A is an m by n matrix with columns $\mathbf{A}^j$ and B is an n by p matrix with rows $\mathbf{B}_i$, then $AB = \mathbf{A}^1\mathbf{B}_1 + \cdots + \mathbf{A}^n\mathbf{B}_n$. (Hint: Just look at the ij-entry of both sides.)

Exercise 3.3.2.

Give the matrix which projects onto the line through the vector $(2, 1, 0)$.

Exercise 3.3.3.

Give the matrix which projects onto the plane $2x + y = 0$. (Hint: You can use the answer to the previous exercise.)

Exercise 3.3.4.

Suppose the matrix A has QR decomposition with $A = \begin{pmatrix} 1 & 2 \\ 0 & 1 \\ 1 & 2 \end{pmatrix}$ and $Q = \begin{pmatrix} 1/\sqrt{2} & 0 \\ 0 & 1 \\ 1/\sqrt{2} & 0 \end{pmatrix}$.

Find the matrix which projects onto the range of A. Solve the least squares problem $A\mathbf{x} = (1 \quad 0 \quad 0)^t$. (Hint: Note that the columns of Q give an orthonormal basis for the range of A.)

3.4. THE NORMAL EQUATION AND THE LEAST SQUARES PROBLEM

We now describe another method to solve the least squares problem $A\mathbf{x} = \mathbf{b}$. We denote the matrix which projects onto the range of A by the suggestive notation P. We start by noting that the projection $P\mathbf{b} = A\mathbf{x}$ and so $A\mathbf{x} - \mathbf{b} = (P - I)\mathbf{b}$ lies in the subspace orthogonal to the range of A. Thus $(A\mathbf{x} - \mathbf{b})$ is orthogonal to every vector in the range of A; that is, $\langle A\mathbf{y}, A\mathbf{x} - \mathbf{b}\rangle = 0$ for all $\mathbf{y}$. This implies that

$$\mathbf{y}^t(A^tA\mathbf{x} - A^t\mathbf{b}) = \langle \mathbf{y}, A^tA\mathbf{x} - A^t\mathbf{b}\rangle = 0$$

for all $\mathbf{y}$. But the only vector which is perpendicular to all vectors is the zero vector; hence $A^tA\mathbf{x} - A^t\mathbf{b} = \mathbf{0}$; that is,

$$A^tA\mathbf{x} = A^t\mathbf{b}$$

DEFINITION 3.4.1. The equation

$$A^tA\mathbf{x} = A^t\mathbf{b}$$

is called the **normal equation** for the least squares problem.

Thus any solution of $A\mathbf{x} = P\mathbf{b}$ is a solution of the normal equation $A^tA\mathbf{x} = A^t\mathbf{b}$. Conversely, any solution $\mathbf{x}$ of the normal equation will have $A\mathbf{x} - \mathbf{b}$ orthogonal to the range of A. Thus $\mathbf{b} = A\mathbf{x} + (\mathbf{b} - A\mathbf{x})$ is a decomposition of $\mathbf{b}$ into a vector in the range of A and a vector in the orthogonal complement, and $P\mathbf{b} = A\mathbf{x}$. We have just shown

PROPOSITION 3.4.1. The solutions of $A\mathbf{x} = P\mathbf{b}$ coincide with the solutions of the normal equation $A^tA\mathbf{x} = A^t\mathbf{b}$.

The linear transformation which is multiplication by A^tA is the composition of multiplying by A and multiplying by A^t. Recall that multiplication by A sends the row space isomorphically onto the column space and multiplication by A^t sends the column space isomorphically onto the row space. Thus multiplication by A^tA sends the row space isomorphically onto itself. Since the vector $A^t\mathbf{b}$ is in the row space, which is the range of A^t, there is a unique vector $\mathbf{r}$ in the row space with $A^tA\mathbf{r} = A^t\mathbf{b}$. Note that the null space of A^tA is an $(n - \text{rank } A)$-dimensional subspace which contains the null space of A. Thus the null space of A^tA must be equal to the null space of A. Therefore all other solutions of the normal equation will be of the form $\mathbf{r} + \mathbf{n}$, where $\mathbf{n}$ is in the null space of A. The solution $\mathbf{r}$ is what we called the least squares solution to the equation $A\mathbf{x} = \mathbf{b}$. Note that the projection $P\mathbf{b}$ is just $A\mathbf{x}$, where $\mathbf{x}$ is any solution of the normal equation.

Let us now consider the special case where the null space of A is $\{\mathbf{0}\}$, and so the row space is all of $\mathbb{R}^n$. In this case there will be a unique solution of the normal equation $A^tA\mathbf{x} = A^t\mathbf{b}$. In fact the matrix A^tA will be of rank n and so will be invertible. Thus we can write the solution as

$$\mathbf{x} = (A^tA)^{-1}A^t\mathbf{b}$$

We can find the projection $P\mathbf{b}$ in this case by the equation

$$P\mathbf{b} = A\mathbf{x} = A(A^tA)^{-1}A^t\mathbf{b}$$

This means that the matrix which projects onto the range of A is given in this case by

$$P = A(A^tA)^{-1}A^t$$

The columns of A will be a basis for the column space, and so we will have a QR decomposition for A, where Q is of the same size as A. Then we already know that $P = QQ^t$. We check the consistency with the more complicated formula just given: $A^tA = R^tQ^tQR = R^tR$ since $Q^tQ = I$. Thus $A(A^tA)^{-1}A^t = QR(R^{-1}(R^t)^{-1})(R^tQ^t) = QQ^t$. The normal equations simplify when there is a QR decomposition of A to give $R^tRx = R^tQ^t\mathbf{b}$. Since R is invertible (and thus R^t is invertible), this is equivalent to the equation $Rx = Q^t\mathbf{b}$. This equation is easy to solve since R is upper triangular. We summarize our discussion with the following proposition.

PROPOSITION 3.4.2. Suppose the m by n matrix A is of rank n. Then there will be a unique solution to the least squares problem $A\mathbf{x} = \mathbf{b}$, and it is given by the formula $\mathbf{x} = (A^tA)^{-1}A^t\mathbf{b}$. The projection matrix which projects onto the range of A is $A(A^tA)^{-1}A^t$. If $A = QR$, then the solution will be the solution of $Rx = Q^t\mathbf{b}$ and the matrix which projects onto the range of A will be QQ^t.

Although we have given formulas for the projection matrix in terms of A and the least squares solution as well, it is computationally better to just work with the normal equations directly in the general case and to use the QR decomposition in the case where the rank of A is n.

■ ***Example 3.4.1.*** We find the least squares solution to the equation $A\mathbf{x} = \mathbf{b}$, where

$$A = \begin{pmatrix} 1 & 0 \\ -2 & 1 \\ 1 & 1 \end{pmatrix} \text{ and } \mathbf{b} = \begin{pmatrix} 1 \\ 0 \\ 0 \end{pmatrix}$$

Then the normal equation is

$$\begin{pmatrix} 6 & -1 \\ -1 & 2 \end{pmatrix}\mathbf{x} = \begin{pmatrix} 1 \\ 0 \end{pmatrix}$$

which has solution $\mathbf{x} = (2/11, 1/11)$. The projection

$$P\mathbf{b} = A\mathbf{x} = (2/11, -3/11, 3/11)$$

The matrix which projects onto the range of A can be found using the formula $A(A^tA)^{-1}A^t$, which gives $P = \frac{1}{11}\begin{pmatrix} 2 & -3 & 3 \\ -3 & 10 & 1 \\ 3 & 1 & 10 \end{pmatrix}$. Note that the first column, which is $P\mathbf{e}_1 = P\mathbf{b}$, agrees with the answer just obtained. We could have also solved the equation by first finding the QR decomposition for A, which is $Q = \begin{pmatrix} .4082 & .1231 \\ -.8165 & .4924 \\ .4082 & .8616 \end{pmatrix}$ and $R = \begin{pmatrix} 2.4495 & -.4082 \\ 0 & 1.3540 \end{pmatrix}$, and then solved the equation $R\mathbf{x} = Q^t\mathbf{b}$ for $\mathbf{x}$. We also could have found P via the formula $P = QQ^t$. ■

■ ***Example 3.4.2.*** Let $A = \begin{pmatrix} 1 & 0 & 2 \\ 1 & 1 & 0 \\ 3 & 1 & 4 \end{pmatrix}$ and $\mathbf{b} = \begin{pmatrix} 1 \\ 1 \\ 0 \end{pmatrix}$. Then the normal equation becomes $\begin{pmatrix} 11 & 4 & 14 \\ 4 & 2 & 4 \\ 14 & 4 & 20 \end{pmatrix}\mathbf{x} = \begin{pmatrix} 2 \\ 1 \\ 2 \end{pmatrix}$. One solution found via inspection is $\mathbf{x} = (0, 1/2, 0)$. To find the solution in the row space we need to project this onto the row space. The easiest way to do this is to subtract off the projection onto the null space, which has basis $(-2, 2, 1)$. Thus the projection onto the null space of $\mathbf{x}$ is $(-2/9, 2/9, 1/9)$ and the projection of $\mathbf{x}$ onto the row space is $(2/9, 5/18, -1/9)$, which is the least squares solution to $A\mathbf{x} = \mathbf{b}$. ■

Exercise 3.4.1.

Find the least squares solution to $A\mathbf{x} = \mathbf{b}$, where

$$A = \begin{pmatrix} -1 & 1 \\ 2 & 1 \\ 0 & 1 \end{pmatrix} \quad \text{and} \quad \mathbf{b} = \begin{pmatrix} 1 \\ 0 \\ 0 \end{pmatrix}$$

Give the normal equation as well as the matrix which projects onto the range of A.

Exercise 3.4.2.

Find the least squares solution of $A\mathbf{x} = \mathbf{b}$, where

$$A = \begin{pmatrix} 1 & 1 & 1 \\ 0 & 1 & 2 \\ 1 & 0 & -1 \end{pmatrix} \quad \text{and} \quad \mathbf{b} = \begin{pmatrix} 1 \\ 1 \\ 1 \end{pmatrix}$$

Give the normal equation as well as the matrix which projects onto the range of A. Also, give the vector in the range of A which is closest to $\mathbf{b}$.

3.5. DATA FITTING AND FUNCTION APPROXIMATION

In this section we will apply the ideas involved in the least squares solution to $A\mathbf{x} = \mathbf{b}$ to the problem of finding the best approximation to some given data. Suppose we are given n points in the plane $(x_1, y_1), \ldots, (x_n, y_n)$ and we want to find a function $y = f(x)$ which "best" approximates the given data in the sense that it minimizes the sum of the squares of the distances between (x_i, y_i) and $(x_i, f(x_i))$, $\sum_{i=1}^{n}(y_i - f(x_i))^2$. We will always assume that the x_i are distinct although our discussion may be modified to remove that assumption. In most applications of this type we restrict the types of functions used in the approximation. We will first look at this type of problem in the case where we are approximating by straight lines $f(x) = a_0 + a_1 x$. Then what we wish to minimize over all possible choices of straight lines (i.e., over all possible choices of a_0, a_1) is the square of the distance between the vector $(y_1, \ldots, y_n)$ and the vector $(a_0 + a_1 x_1, \ldots, a_0 + a_1 x_n)$. This latter vector can be thought of as the image $A\mathbf{a}$, where

$$A = \begin{pmatrix} 1 & x_1 \\ 1 & x_2 \\ \cdots & \cdots \\ 1 & x_n \end{pmatrix}, \quad \mathbf{a} = \begin{pmatrix} a_0 \\ a_1 \end{pmatrix}$$

Thus the problem of data fitting becomes a least squares problem $A\mathbf{a} = \mathbf{y}$. Since the values $x_1, \ldots, x_n$ are distinct, the two columns of A are independent, and so there will be a unique solution $\mathbf{a} = (a_0, a_1)$ to this least squares problem. The line $f(x) = a_0 + a_1 x$ which corresponds to this solution is called the **least squares line** or the **regression line** for the data. To find this line we just have to solve the normal equation $A'A\mathbf{a} = A'\mathbf{y}$, which is

$$\begin{pmatrix} n & \sum_{i=1}^{n} x_i \\ \sum_{i=1}^{n} x_i & \sum_{i=1}^{n} x_i^2 \end{pmatrix} \begin{pmatrix} a_0 \\ a_1 \end{pmatrix} = \begin{pmatrix} \sum_{i=1}^{n} y_i \\ \sum_{i=1}^{n} x_i y_i \end{pmatrix}$$

We now work an example to illustrate this technique.

■ ***Example 3.5.1.*** We find the regression line for the data $(1,2), (2,1), (3,2), (4,1)$. We first form the matrix $A = \begin{pmatrix} 1 & 1 \\ 1 & 2 \\ 1 & 3 \\ 1 & 4 \end{pmatrix}$, $\mathbf{y} = \begin{pmatrix} 2 \\ 1 \\ 2 \\ 1 \end{pmatrix}$ and solve the least squares

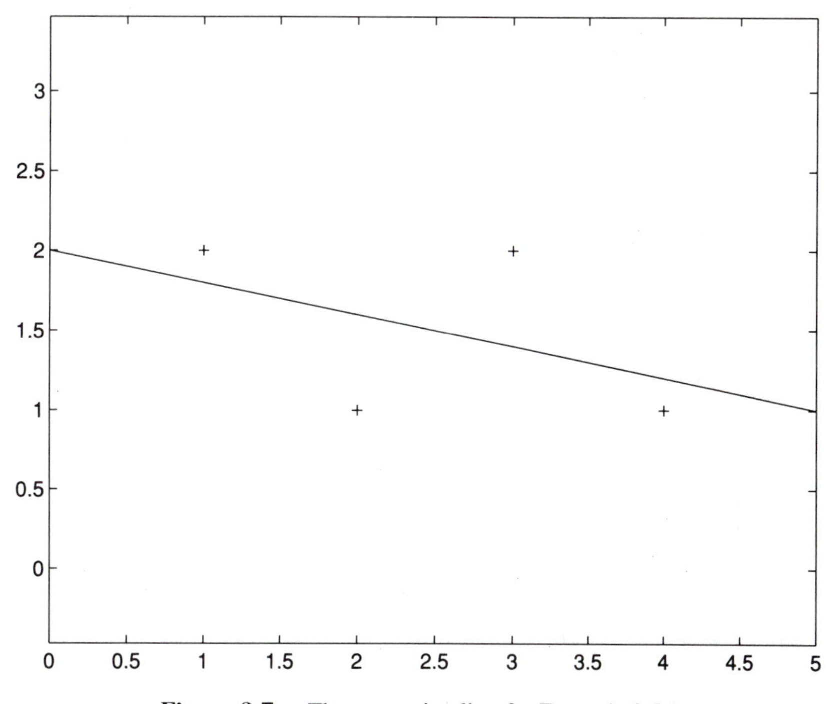

Figure 3.7. The regression line for Example 3.5.1

problem $A\mathbf{a} = \mathbf{y}$. We form the normal equation

$$\begin{pmatrix} 4 & 10 \\ 10 & 30 \end{pmatrix} \begin{pmatrix} a_0 \\ a_1 \end{pmatrix} = \begin{pmatrix} 6 \\ 14 \end{pmatrix}$$

which has solution $a_0 = 2, a_1 = -1/5$. Thus the regression line is $f(x) = 2 - 1/5x$. The regression line and data points are shown in Figure 3.7. ∎

Since the matrix A^tA which occurs on the left-hand side of the normal equation is a 2 by 2 invertible matrix, it is not difficult to derive a general formula for this regression line. It turns out to be somewhat neater first to find the QR decomposition for A and then to rewrite the normal equations in terms of it. Let us denote by $xm = \sum_{i=1}^{n} x_i/n$ and $ym = \sum_{i=1}^{n} y_i/n$. This average value is called the **mean** in statistics. Then when we compute the orthogonal basis we have to compute $t_{12} = \langle(1, \ldots, 1), (x_1, \ldots, x_n)\rangle / \langle(1, \ldots, 1), (1, \ldots, 1)\rangle = xm$. Thus the orthogonal decomposition $A = WT$ analogous to the QR decomposition has

$$W = \begin{pmatrix} 1 & (x_1 - xm) \\ 1 & (x_2 - xm) \\ \vdots & \vdots \\ 1 & (x_n - xm) \end{pmatrix}, \qquad T = \begin{pmatrix} 1 & xm \\ 0 & 1 \end{pmatrix}$$

and so the normal equations can be expressed as $T^tW^tWT\,(a \quad b)^t = T^tW^t\mathbf{y}$. Since T is invertible the T^t term can be canceled from both sides; also W^tW is diagonal, and so the equation takes the form

$$\begin{pmatrix} n & 0 \\ 0 & \sum_{i=1}^{n}(x_i - xm)^2 \end{pmatrix}\begin{pmatrix} a_0 + a_1 xm \\ a_1 \end{pmatrix} = \begin{pmatrix} \sum_{i=1}^{n} y_i \\ \sum_{i=1}^{n}(x_i - xm)y_i \end{pmatrix}$$

Thus

$$a_1 = \frac{\sum_{i=1}^{n}(x_i - xm)y_i}{\sum_{i=1}^{n}(x_i - xm)^2} \quad \text{and} \quad a_0 + a_1 xm = ym; \text{ that is, } a_0 = ym - a_1 xm$$

If we use this formula in the last example, we first compute that $xm = 2.5$, $ym = 1.5$, $a_1 = -1/5$, $a_0 = (1.5) - (-1/5)(2.5) = 2$ as before.

Exercise 3.5.1.

Find the regression line for the data $(1, 3), (2, 1), (3, 5), (4, 2)$.

Exercise 3.5.2.

Find the horizontal line $y = a$ which best fits the data in the previous exercise. (Hint: We are just trying to find the point $a(1, 1, 1, 1)$ on the line through $(1, 1, 1, 1)$ which comes closest to the point $(3, 1, 5, 2)$.)

Instead of asking for a straight line which best fits the data, we could ask for a polynomial of some higher degree. This also will lead to a least squares problem. If the polynomial for which we are looking is of degree k or less, then it will be of the form $f(x) = a_0 + a_1 x + \cdots + a_k x^k$. We take data (x, y) given by n points $(x_1, y_1), \ldots, (x_n, y_n)$ and form the matrix

$$A = \begin{pmatrix} 1 & x_1 & x_1^2 & \ldots & x_1^k \\ 1 & x_2 & x_2^2 & \ldots & x_2^k \\ & & & \vdots & \\ 1 & x_n & x_n^2 & \ldots & x_n^k \end{pmatrix}$$

and the vector

$$\mathbf{y} = \begin{pmatrix} y_1 \\ y_2 \\ \vdots \\ y_n \end{pmatrix}$$

Finding this polynomial which minimizes $\sum_{i=1}^{n}(y_i - f(x_i))^2$ just becomes solving the least squares problem $A\mathbf{a} = \mathbf{y}$. An important fact about the matrix A is that the columns of A are independent as long as $k \leq n - 1$ (using our assumption that the data points x_i are all distinct). For if a linear combination of these columns gives zero, this would determine a polynomial $a_0 + a_1x + \cdots + a_kx^k$ which vanishes at n points, where n is larger than k. But a nonzero polynomial of degree k has at most k distinct roots and so this would imply that the polynomial in question is zero; hence the a_i must all be zero. In particular the matrix A will be invertible when $k = n - 1$. Thus we will be able to solve the equation $A\mathbf{a} = \mathbf{y}$ directly via Gaussian elimination. This gives the following proposition.

PROPOSITION 3.5.1. Given n data points x_i, y_i with distinct x_i, there is a unique polynomial of degree $n - 1$ whose graph passes through these points.

■ ***Example 3.5.2.*** Given the four data points in Example 3.5.1, we find the quadratic $a_0 + a_1x + a_2x^2$ which best approximates the data and the cubic which actually passes through the four points. For the quadratic we set $A = \begin{pmatrix} 1 & 1 & 1 \\ 1 & 2 & 4 \\ 1 & 3 & 9 \\ 1 & 4 & 16 \end{pmatrix}$ and $\mathbf{y} = \begin{pmatrix} 2 \\ 1 \\ 2 \\ 1 \end{pmatrix}$. The normal equation becomes $\begin{pmatrix} 4 & 10 & 30 \\ 10 & 30 & 100 \\ 30 & 100 & 354 \end{pmatrix} \begin{pmatrix} a_0 \\ a_1 \\ a_2 \end{pmatrix} = \begin{pmatrix} 6 \\ 14 \\ 40 \end{pmatrix}$, which has solution $(2, -1/5, 0)$; that is, the line $f(t) = 2 - 1/5x$ still gives the best approximation of all polynomials of degree less than or equal to 2. When we look for the best cubic we use the matrix $A = \begin{pmatrix} 1 & 1 & 1 & 1 \\ 1 & 2 & 4 & 8 \\ 1 & 3 & 9 & 27 \\ 1 & 4 & 16 & 64 \end{pmatrix}$ and $\mathbf{y}$ as before. A is invertible and so we can solve directly $\mathbf{a} = (9, -11.33, 5, -.6667)$ and so the cubic which passes through these points is $9 - 11.33x + 5x^2 - .6667x^3$. ■

Exercise 3.5.3.

For the data in Exercise 3.5.1, find the quadratic which best approximates the data and the cubic which passes through the data.

We now consider an analogous problem to the one just given of finding the best polynomial of degree less than or equal to k to fit n data points, which is to find the polynomial of degree less than or equal to k which best approximates a given function on some interval. We will use the interval $[-1, 1]$ for definiteness and will initially assume the function which we are trying to approximate on this interval is continuous.

Now if we were to select n data points in the interval and try to approximate the values of the function $g(x)$ at those data points by the polynomial values $f(x)$, we would be trying to minimize the difference $S = \sum_{i=1}^{n}(g(x_i) - f(x_i))^2$ by selecting the coefficients of f appropriately. This is the same as minimizing any constant multiple of this sum; in particular we would be trying to minimize $(2/n)S$. If the interval $[-1, 1]$ is divided into n subintervals each of length $2/n$, then $(2/n)S$ is a Riemann sum for the integral $\int_{-1}^{1}(g(x) - f(x))^2\,dx$ and tends toward that integral as the number n of data points gets larger. This is the motivation for defining an inner product $\langle\,,\rangle$ on the vector space $\mathscr{C}([-1, 1], \mathbb{R})$ of continuous functions from the interval $[-1, 1]$ to $\mathbb{R}$ by $\langle f, g\rangle = \int_{-1}^{1} f(t)g(t)\,dt$. With this definition of inner product the vector space $\mathscr{C}([-1, 1], \mathbb{R})$ becomes an inner product space. The positivity property follows from the fact that a nonnegative continuous function on a closed interval has positive integral unless it is identically zero. The assumption of continuity is crucial here; for example, the discontinuous function $f(t)$ which is identically zero except that $f(0) = 1$ is not zero, but its integral from -1 to 1 is still 0. With this definition the square of the distance between two functions f, g is given by $\int_{-1}^{1}(g(t) - f(t))^2\,dt$. Then the problem of approximating a given continuous function $g(x)$ on the interval $[-1, 1]$ by a polynomial $f(x)$ of degree less than or equal to k in the least squares sense means finding a polynomial $f(x)$ in the subspace $\mathscr{P}^k([-1, 1], R)$ of polynomial functions so that $\|g(x) - f(x)\|^2 = \int_{-1}^{1}(g(t) - f(t))^2\,dt$ is minimal. But this problem is solved by projecting $g(x)$ onto the subspace $\mathscr{P}^k([-1, 1], \mathbb{R})$. To do this projection it is easiest to use an orthogonal basis, which is found from the usual basis $1, x, \ldots, x^k$ by the Gram–Schmidt algorithm. These polynomials were discussed earlier; recall that they are related to the Legendre polynomials. The first four monic orthogonal polynomials are $1, x, x^2 - 1/3, x^3 - (3/5)x$. We now apply this to an example.

■ ***Example 3.5.3.*** We find the best approximation by a quadratic in the least squares sense to the continuous function $g(x) = 4/(x^2 - 4)$ on the interval $[-1, 1]$. The answer will be the projection of $g(x)$ onto the subspace $\mathscr{P}^2([-1, 1], \mathbb{R})$. Using the orthogonal basis provided by the first three orthogonal polynomials $1, x, x^2 - 1/3$ gives

$$f(x) = \frac{\langle g(x), 1\rangle}{2}1 + \frac{\langle g(x), x\rangle}{(2/3)}x + \frac{\langle g(x), x^2 - 1/3\rangle}{(8/45)}(x^2 - 1/3)$$

We calculate the integrals $\int_{-1}^{1} 4/(t^2 - 4)\,dt = \int_{-1}^{1} 1/(t - 2)\,dt - \int_{-1}^{1} 1/(t + 2)\,dt = -2\ln(3) = -2.1972$. $\int_{-1}^{1} 4t/(t^2 - 4)\,dt = 0$ since we are integrating an odd function. $\int_{-1}^{1} 4(t^2 - 1/3)/(t^2 - 4)\,dt = 4\int_{-1}^{1} dt + (11/3)\int_{-1}^{1} 4/(t^2 - 4)\,dt = -.0565$. Thus the polynomial $f(x) = -1.0986 - .3178(x^2 - 1/3) = -.9927 - .3178x^2$ is the quadratic which best approximates the function.

Figure 3.8 shows the graphs of $g(x)$, solid line, and the quadratic approximation $f(x)$, dashed line. It was produced using MATLAB. MATLAB was also used to find the quadratic polynomial which best approximates the values of $g(x)$ at the 41 data points given by evaluating $g(x)$ at equally spaced points (of subinterval length .05) between -1 and 1; the desired polynomial is $-.9919 - .3219x^2$, which is very close to the result we obtained earlier. ■

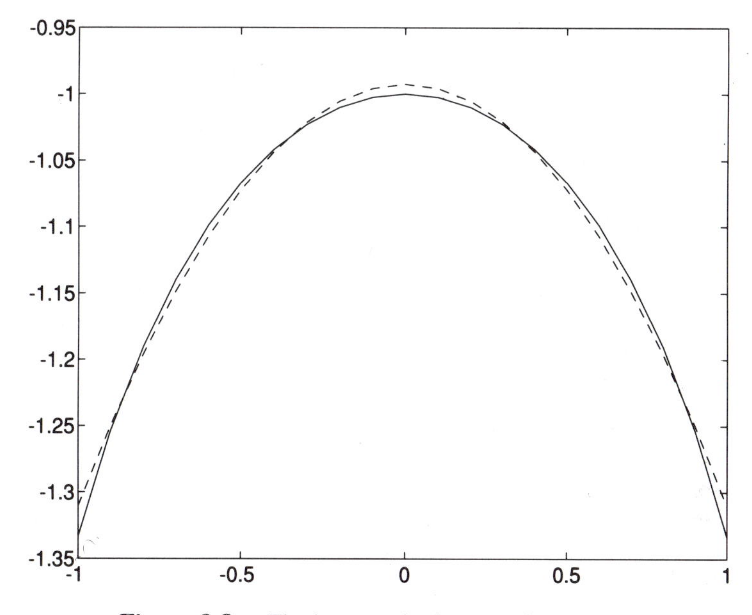

Figure 3.8. The best quadratic approximation to f

Another way in which functions are approximated is by the trigonometric polynomials $\sin mx$ and $\cos mx$. This especially occurs with periodic functions, that is, ones which satisfy $g(x + P) = g(x)$ for some period P. To simplify our discussion we will consider $\mathscr{C}([-\pi, \pi], \mathbb{R})$ so that the period involved is 2π. The inner product which we put on these functions is the same as that earlier, but with the interval of integration changed to $[-\pi, \pi]$; that is, $\langle f, g\rangle = \int_{-\pi}^{\pi} f(t)g(t)\,dt$. With this inner product, the functions $1, \cos x, \sin x, \cos 2x, \sin 2x, \ldots, \cos mx, \sin mx, \ldots$ give an infinite family of functions which are pairwise orthogonal. They are not a basis in the sense which we have defined a basis, for any continuous function is not necessarily a linear combination of a finite number of these functions. However, it is a theorem in analysis that there are constants $a_0, a_1, a_2, \ldots, b_1, b_2, \ldots$ so that $g(x)$ is the limit of a convergent series, called the **Fourier series,**

$$a_0 1 + a_1 \cos x + b_1 \sin x + a_2 \cos 2x + b_2 \sin 2x + \cdots$$

The coefficients a_i and b_i are called the **Fourier coefficients**. Moreover, the fact that this is an orthogonal system of functions allows us to evaluate the Fourier coefficients by the familiar formula

$$a_0 = \frac{1/2}{\pi}\int_{-\pi}^{\pi} g(t)\,dt, a_m = \frac{1}{\pi}\int_{-\pi}^{\pi} g(t)\cos mt\,dt, b_m = \frac{1}{\pi}\int_{-\pi}^{\pi} g(t)\sin mt\,dt$$

The factors 2π and π in the denominators are just the squares of the lengths of the functions 1 and $\cos mx$ (and $\sin mx$). We are usually interested in only taking a finite number of terms in the Fourier series (which is thus a partial sum for the series). If we denote by T_m the subspace which is spanned by the functions $1, \cos x, \sin x, \ldots, \cos mx, \sin mx$, then finding the first $2m+1$ terms of the Fourier series for a function $g(x)$ is equivalent to projecting $g(x)$ onto this subspace.

So far our discussion has centered on continuous functions and approximating them by polynomials or Fourier series. Limits of Fourier series can turn out to be discontinuous. Such functions frequently occur in applications. We will only look at one example. Let $g(x)$ be a "square wave" which has value 0 on the interval $-\pi < t < 0$, and equal to 1 on the interval $0 < t < \pi$. We will arbitrarily define $g(t) = 0$ at $\pm\pi$ and 0, but this makes no difference in the argument below. The function g thus models an off–on phenomenon. We want to calculate the projection of g onto the subspace T_m for some small values of m to see how the Fourier series converges to the function g. We compute the Fourier coefficients

$$\begin{aligned} a_0 &= \frac{1}{2\pi}\int_{-\pi}^{\pi} g(t)1\,dt = \frac{1}{2\pi}\int_0^{\pi} 1\,dt = 1/2 \\ a_m &= \frac{1}{\pi}\int_{-\pi}^{\pi} g(t)\cos mt\,dt = \frac{1}{\pi}\int_0^{\pi} \cos mt\,dt = 0 \\ b_m &= \frac{1}{\pi}\int_{-\pi}^{\pi} g(t)\sin mt\,dt = \frac{1}{\pi}\int_0^{\pi} \sin mt\,dt = \frac{1-\cos m\pi}{m\pi} \end{aligned}$$

Note that $1 - \cos m\pi$ is alternately 0 and 2, so the Fourier series is

$$1/2 + \frac{2}{\pi}(\sin x + (1/3)\sin 3x + (1/5)\sin 5x + \cdots)$$

Denote by f_m the projection onto T_m, which just takes the terms from the series up through the $\sin mx$ term. Then the graph on the left in Figure 3.9 shows f_5, f_{10}, f_{20}, and

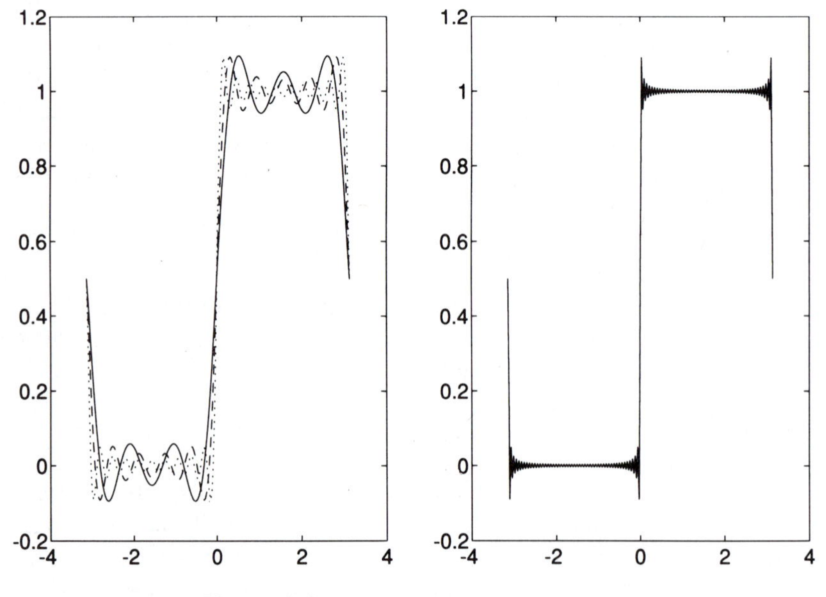

Figure 3.9. Fourier approximations to g

the graph on the right shows f_{99}. The overshooting of the values as we get near the discontinuities is known as the **Gibbs phenomenon.**

Giving a complete description of the vector space which includes general functions such as the square wave so that our integral definition satisfies the properties of an inner product is beyond the level of this book. Informally, the space involved has as its elements not functions, but equivalence classes of functions. For positivity it is necessary to make two functions f, g equivalent if $\int_{-\pi}^{\pi}(f(t) - g(t))^2\, dt = 0$. In particular, functions which differ at only a finite number of points are equivalent. Also, the functions involved must be "square integrable"; that is, $\int_{-\pi}^{\pi} f(t)^2\, dt$ must exist. The vector space involved is usually denoted by $L^2([-\pi, \pi])$ and is an example of what is called a **Hilbert space**. For more information on L^2 and other Hilbert spaces, consult books on real analysis and functional analysis (cf. [6]). Besides ideas from linear algebra, understanding these vector spaces requires concepts from analysis and topology. In particular, convergence plays a critical role.

Exercise 3.5.4.

Find the best linear polynomial $f(x) = a + bx$ approximation to e^x on the interval $[-1, 1]$. Compare your answer to what you would get using the least squares approximation using the 41 data points coming from dividing up the interval $[-1, 1]$ into 40 subintervals.

Exercise 3.5.5.

Find the quadratic polynomial which best approximates x^4 on the interval $[-1, 1]$. Compare your answer to what you would get using the least squares approximation using the 41 data points coming from dividing up the interval $[-1, 1]$ into 40 subintervals.

Exercise 3.5.6.

Give the element of T_2 which best approximates the function $g(x) = |x - \pi|$ on the interval $[-\pi, \pi]$. Find the general Fourier coefficients for the Fourier series for $g(x)$.

3.6. SINGULAR VALUE DECOMPOSITION AND PSEUDOINVERSE

We return to the problem of finding the least squares solution to $A\mathbf{x} = \mathbf{b}$ and try to find the matrix A^+ so that the solution to this problem is given by $\mathbf{x} = A^+\mathbf{b}$. A^+ will be a generalization of the inverse of the matrix A. It will equal the inverse when A is invertible, but can be defined in general.

DEFINITION 3.6.1. The matrix A^+ so that $A^+\mathbf{b}$ is the least squares solution to $A\mathbf{x} = \mathbf{b}$ is called the **pseudoinverse** of A.

In this section we will not actually compute A^+ except in very special circumstances but will reduce the problem of finding it to an eigenvalue–eigenvector problem that we will solve in the next chapter. The idea of how we will find A^+ is to use a special decomposition for a matrix A which is called the singular value decomposition. Our derivation of this decomposition will be through the eigenvalue–eigenvector problem.

DEFINITION 3.6.2. The **singular value decomposition** of A is a decomposition $A = Q_1 \Sigma Q_2'$, where Q_1, Q_2 are orthogonal matrices of sizes m by m and n by n, respectively, and Σ is a diagonal matrix of size m by n. The matrix Σ has the first $r = \text{rank } A$ entries positive and the remaining diagonal entries 0. The nonzero diagonal entries of Σ are called the **singular values** of A.

The first r columns of $Q_2 = (Q_{2r} Q_{2n})$ will form an orthonormal basis of the row space of A and the last $n - r$ columns will form an orthonormal basis of the null space of A. For the matrix $Q_1 = (Q_{1r} Q_{1n})$ the first r columns give an orthonormal basis for the column space of A and the last $m - r$ columns form an orthonormal basis for the null space of A'. The matrix Σ is an m by n matrix with zeros except in the diagonal entries in the first r rows. We will write the r by r submatrix in the upper left-hand corner of Σ as Σ_r and denote its diagonal entries as $\mu_1, \ldots, \mu_r$. For example, if we have a 3 by 4 matrix of rank 2 and $\mu_1 = 2, \mu_2 = 3$, then $\Sigma_r = \begin{pmatrix} 2 & 0 \\ 0 & 3 \end{pmatrix}$ and $\Sigma = \begin{pmatrix} 2 & 0 & 0 & 0 \\ 0 & 3 & 0 & 0 \\ 0 & 0 & 0 & 0 \end{pmatrix}$.

A good way to look at the formula $A = Q_1 \Sigma Q_2'$ is to multiply by Q_2 on the right and get the equivalent equation $AQ_2 = Q_1 \Sigma$. If we denote the columns of Q_2 by $\mathbf{q}_i$, $i = 1, \ldots, n$, and the columns of Q_1 by $\mathbf{q}_i'$, $i = 1, \ldots, m$, then this matrix equation becomes the equations $A\mathbf{q}_i = \mu_i \mathbf{q}_i', i = 1, \ldots, r$, and $A\mathbf{q}_i = \mathbf{0}, i = r + 1, \ldots, n$. What these equations are saying is that the orthonormal bases $\mathbf{q}_1, \ldots, \mathbf{q}_r$ and $\mathbf{q}_1', \ldots, \mathbf{q}_r'$ have been chosen so that multiplication by A sends one basis to certain multiples of the other. Another way of saying this is that Σ_r is the matrix representing the linear transformation given by multiplying by A when regarded as going from the row space to the column space, when we are using the basis $\mathbf{q}_1, \ldots, \mathbf{q}_r$ of the row space and the basis $\mathbf{q}_1', \ldots, \mathbf{q}_r'$ of the column space. When we add in the bases of the two null spaces (of A and A'), the matrix which represents multiplication by A with respect to the bases given by the columns of Q_2 and the columns of Q_1 is Σ. The numbers μ_i will turn out to be positive and can be chosen so that they satisfy $\mu_1 \geq \mu_2 \geq \cdots \geq \mu_r > 0$.

Let us assume that we can find these matrices Q_1, Σ, Q_2 and see what this implies about the pseudoinverse A^+. The linear transformation which comes from multiplying by A^+ is determined by what it does on a basis of $\mathbb{R}^m$. We will choose the basis given by the columns of Q_1. When $i \leq r$, we have $A\mathbf{q}_i = \mu_i \mathbf{q}_i'$. To find the value of $A^+\mathbf{q}_i'$, we solve the least squares problem $A\mathbf{x} = \mathbf{q}_i'$. But the solution to this problem is just $(1/\mu_i)\mathbf{q}_i$. When $i > r$ and we solve the least squares problem $A\mathbf{x} = \mathbf{q}_i'$, we first project $\mathbf{q}_i'$ onto the range of A and get $\mathbf{0}$ and then solve $A\mathbf{x} = \mathbf{0}$ to get $\mathbf{x} = \mathbf{0}$. Thus $A^+\mathbf{q}_i' = \mathbf{0}$ for $i > r$. If we denote by Σ_r^+ the r by r diagonal matrix with diagonal entries $1/\mu_i$ and by Σ^+ the n by m matrix with zeros except in the upper left-hand corner where we have Σ_r^+, then these equations can be written as the matrix equation $A^+Q_1 = Q_2\Sigma^+$, which is equivalent to $A^+ = Q_2\Sigma^+Q_1^t$.

■ ***Example 3.6.1.*** Suppose the singular value decomposition of A is $Q_1\Sigma Q_2^t$, with

$$Q_2 = \begin{pmatrix} 1/\sqrt{2} & 1/\sqrt{3} & 1/\sqrt{6} \\ 1/\sqrt{2} & -1/\sqrt{3} & -1/\sqrt{6} \\ 0 & 1/\sqrt{3} & -2/\sqrt{6} \end{pmatrix}, Q_1 = \begin{pmatrix} 2/\sqrt{5} & -1/\sqrt{5} \\ 1/\sqrt{5} & 2/\sqrt{5} \end{pmatrix}, \Sigma = \begin{pmatrix} 3 & 0 & 0 \\ 0 & 0 & 0 \end{pmatrix}$$

In this case $A = \begin{pmatrix} 6/\sqrt{10} & 6/\sqrt{10} & 0 \\ 3/\sqrt{10} & 3/\sqrt{10} & 0 \end{pmatrix}$. Then the pseudoinverse of A will be the product

$$A^+ = Q_2\Sigma^+Q_1^t = \begin{pmatrix} 1/\sqrt{2} & 1/\sqrt{3} & 1/\sqrt{6} \\ 1/\sqrt{2} & -1/\sqrt{3} & 1/\sqrt{6} \\ 0 & 1/\sqrt{3} & -2/\sqrt{6} \end{pmatrix} \begin{pmatrix} 1/3 & 0 \\ 0 & 0 \\ 0 & 0 \end{pmatrix} \begin{pmatrix} 2/\sqrt{5} & 1/\sqrt{5} \\ -1/\sqrt{5} & 2/\sqrt{5} \end{pmatrix}$$

When computing this product, all that is important is the part Q_{1r} and Q_{2r} of Q_1 and Q_2; that is, $A = Q_{1r}\Sigma_r Q_{2r}^t$ and $A^+ = Q_{2r}\Sigma_r^+ Q_{1r}^t$. Here this gives

$$A^+ = \begin{pmatrix} 1/\sqrt{2} \\ 1/\sqrt{2} \\ 0 \end{pmatrix} (1/3) \begin{pmatrix} 2/\sqrt{5} & 1/\sqrt{5} \end{pmatrix} = \begin{pmatrix} 2/3\sqrt{10} & 1/3\sqrt{10} \\ 2/3\sqrt{10} & 1/3\sqrt{10} \\ 0 & 0 \end{pmatrix}$$ ■

Exercise 3.6.1.

For Example 3.6.1 use the pseudoinverse to find the least squares solution of $A\mathbf{x} = (1 \quad 2)^t$.

In general, finding the singular value decomposition, and from it the pseudoinverse, is not easy. One case where it is particularly simple, however, is when the rank is 1. In this case the possible orthonormal bases for the row space and column space are determined up to sign. Thus we can first find $\mathbf{q}_1$ and then find $\mathbf{q}_1'$ from it by $\mathbf{q}_1' = \dfrac{A\mathbf{q}_1}{\|A\mathbf{q}_1\|}$. With $\mu = \|A\mathbf{q}_1\|$, we have $A\mathbf{q}_1 = \mu\mathbf{q}_1'$. We can let $\mathbf{q}_2, \ldots, \mathbf{q}_{n-1}$ be any orthonormal basis for the null space of A and $\mathbf{q}_2', \ldots, \mathbf{q}_m'$ be any orthonormal basis of the null space of A^t, and we will have the singular value decomposition for A as $Q_1\Sigma Q_2^t$ with Σ being the zero matrix except for the $(1, 1)$-entry, which is μ. We can then find the pseudoinverse as $A^+ = Q_{2r}\Sigma_r{}^+Q^t_{1r}$.

■ ***Example 3.6.2.*** Let $A = \begin{pmatrix} 1 & 2 & 1 \\ 2 & 4 & 2 \\ 5 & 10 & 5 \\ 1 & 2 & 1 \end{pmatrix}$. This is of rank 1. An orthonormal basis for the row space is given by $\mathbf{q}_1 = \begin{pmatrix} 1/\sqrt{6} \\ 2/\sqrt{6} \\ 1/\sqrt{6} \end{pmatrix}$. We compute $A\mathbf{q}_1 = \sqrt{6}\begin{pmatrix} 1 \\ 2 \\ 5 \\ 1 \end{pmatrix}$. Its length $\mu = \|A\mathbf{q}_1\| = \sqrt{186}$, and $\mathbf{q}'_1 = \begin{pmatrix} 1/\sqrt{31} \\ 2/\sqrt{31} \\ 5/\sqrt{31} \\ 1/\sqrt{31} \end{pmatrix}$. To get the remainder of the singular value decomposition, we find the orthonormal bases for $\mathcal{N}(A)$ and $\mathcal{N}(A^t)$. This gives

$$Q_1 = \begin{pmatrix} 1/\sqrt{31} & -2/\sqrt{5} & -1/\sqrt{6} & -1/\sqrt{930} \\ 2/\sqrt{31} & 1/\sqrt{5} & -2/\sqrt{6} & -2/\sqrt{930} \\ 5/\sqrt{31} & 0 & 1/\sqrt{6} & -5/\sqrt{930} \\ 1/\sqrt{31} & 0 & 0 & 30/\sqrt{930} \end{pmatrix}, \quad \Sigma = \begin{pmatrix} \sqrt{186} & 0 & 0 \\ 0 & 0 & 0 \\ 0 & 0 & 0 \\ 0 & 0 & 0 \end{pmatrix},$$

$$Q_2 = \begin{pmatrix} 1/\sqrt{6} & -2/\sqrt{5} & -1/\sqrt{30} \\ 2/\sqrt{6} & 1/\sqrt{5} & -2/\sqrt{30} \\ 1/\sqrt{6} & 0 & 5/\sqrt{30} \end{pmatrix}$$

To find the pseudoinverse just uses

$$A^+ = \mathbf{q}_1(1/\mu)\mathbf{q}_1'^{\,t} = \frac{1}{186}\begin{pmatrix} 1 & 2 & 5 & 1 \\ 2 & 4 & 10 & 2 \\ 1 & 2 & 5 & 1 \end{pmatrix}$$ ■

Exercise 3.6.2.

Find the singular value decomposition of the matrix

$$A = \begin{pmatrix} 1 & 2 & 3 \\ 2 & 4 & 6 \end{pmatrix}$$

(Hint: First find orthonormal bases for the two null spaces and the row space. Then define the basis $\mathbf{q}'_1$ for the column space $\mathbf{q}'_1 = A\mathbf{q}_1/\|A\mathbf{q}_1\|$.)

Exercise 3.6.3.

Find the pseudoinverse for the matrix $\begin{pmatrix} 1 & 1 \\ 2 & 2 \\ 3 & 3 \end{pmatrix}$ and solve the least squares problem $Ax = (1 \quad 1 \quad 1)^t$.

We now discuss one method of finding the singular value decomposition for a matrix. This reduces the problem to one of solving an eigenvalue–eigenvector problem, which is what we will learn to solve in the next chapter. Note that the key problem which we have to solve is the one of finding an orthonormal basis of the row space $\mathbf{q}_1, \ldots, \mathbf{q}_r$ so that $A\mathbf{q}_i = \mu_i \mathbf{q}_i'$, where the $\mathbf{q}_i'$ are an orthonormal basis of the column space. What the solution of the eigenvalue–eigenvector problem will tell us is that for the matrix A^tA, there is an orthonormal basis $\mathbf{q}_1, \ldots, \mathbf{q}_r, \ldots, \mathbf{q}_n$ so that $A^tA\mathbf{q}_i = \sigma_i \mathbf{q}_i$. Since the null space of A^tA is the same as the null space of A, $n - r$ of the values σ_i will be zero. We will order the $\mathbf{q}_i$ so that these are the last ones. Then the first r vectors, $\mathbf{q}_1, \ldots, \mathbf{q}_r$, will be an orthonormal basis for the orthogonal complement of the null space, which is the row space. We next claim that the σ_i are positive for $i \leq r$. To see this, we use

$$0 < \langle A\mathbf{q}_i, A\mathbf{q}_i \rangle = \langle \mathbf{q}_i, A^tA\mathbf{q}_i \rangle = \langle \mathbf{q}_i, \sigma_i \mathbf{q}_i \rangle = \sigma_i$$

since the $\mathbf{q}_i$ are unit vectors. Thus we can define μ_i by $\mu_i = \sqrt{\sigma_i}$. We will reorder the $\mathbf{q}_i$ so that $\mu_1 \geq \mu_2 \geq \cdots \geq \mu_r$. Note that $\langle A\mathbf{q}_i, A\mathbf{q}_i \rangle = \sigma_i$ implies that $\|A\mathbf{q}_i\| = \mu_i$. Thus $\mathbf{q}_i' = (1/\mu_i)A\mathbf{q}_i$ is a unit vector. We next show that these vectors are mutually orthogonal. If $i \neq j$, then

$$\langle \mathbf{q}_i', \mathbf{q}_j' \rangle = (1/\mu_i\mu_j)\langle A\mathbf{q}_i, A\mathbf{q}_j \rangle = (1/\mu_i\mu_j)\langle \mathbf{q}_i, \sigma_j \mathbf{q}_j \rangle = 0$$

since $\langle \mathbf{q}_i, \mathbf{q}_j \rangle = 0$. To find the rest of Q_1, add on a basis $\mathbf{q}_{r+1}', \ldots, \mathbf{q}_m'$ of the null space of A^t.

The numbers σ_i with $A^tA\mathbf{q}_i = \sigma_i \mathbf{q}_i$ are called the eigenvalues of the matrix A^tA and the vectors $\mathbf{q}_i$ satisfying this equation are called the eigenvectors. The key theoretical fact that we will prove in the next chapter which will allow us to solve this problem is that a symmetric matrix (as A^tA is) has real eigenvalues and that it has an orthonormal basis of eigenvectors. We will return to compute some singular value decompositions and pseudoinverses after we have treated the eigenvalue–eigenvector problem in the next chapter.

3.7. CHAPTER 3 EXERCISES

3.7.1. (a) Give the QR decomposition for the matrix $A = \begin{pmatrix} 1 & 0 \\ 1 & 1 \\ 0 & 1 \\ 2 & 1 \end{pmatrix}$.

(b) Use it to find the matrix P which projects onto the range of A.
(c) By solving the equation $A\mathbf{x} = P\mathbf{b}$, solve the least squares problem $A\mathbf{x} = \mathbf{b}$ for $\mathbf{b} = (1 \quad 2 \quad 1 \quad 0)^t$.

3.7.2. (a) Give the QR decomposition for the matrix $B = \begin{pmatrix} 0 & 1 \\ 1 & 1 \\ 1 & 0 \\ 1 & 2 \end{pmatrix}$.

(b) Give an orthonormal basis for the range of A.
(c) By comparing the matrix B to the matrix A in the previous problem, explain how the two orthonormal bases found in the two problems are related.

3.7.3. (a) Give the QR decomposition for the matrix $A = \begin{pmatrix} 1 & 1 & 1 \\ 2 & 1 & 0 \\ 3 & 2 & 1 \end{pmatrix}$.

(b) Give an orthonormal basis for the range of A.

3.7.4. (a) Give the QR decomposition for the matrix

$$A = \begin{pmatrix} 1 & 1 & 2 \\ 0 & 1 & 1 \\ -1 & 1 & 0 \\ 2 & 0 & 2 \end{pmatrix}$$

(b) Explain how the form of this decomposition relates to the independence of the columns of A.

3.7.5. Find an orthonormal basis for the subspace of $\mathbb{R}^3$ which is spanned by $(1, 2, 1)$ and $(0, 1, 1)$.

3.7.6. Find an orthonormal basis for the subspace

$$\text{span}((1, 2, -1, 1, 2), (2, 1, 3, 2, 1), (0, 1, 2, 1, 1))$$

of $\mathbb{R}^5$.

3.7.7. Give the QR decomposition for the matrix $A = \begin{pmatrix} 1 & 3 & 2 \\ -1 & 2 & 0 \\ 0 & 2 & -1 \\ 2 & 2 & 2 \end{pmatrix}$. Use it to find the matrix which projects onto the subspace spanned by the columns of A.

3.7.8. Consider the orthogonal vectors $(1, 0, -1), (1, 1, 1)$ in $\mathbb{R}^3$. Find a third vector so that the three vectors form an orthogonal basis for $\mathbb{R}^3$.

3.7.9. Consider the two orthogonal vectors $(1, -1, 1, -1), (1, 1, 1, 1)$. Extend this basis to an orthogonal basis of $\mathbb{R}^4$.

3.7.10. Consider the plane in $\mathbb{R}^3$ given by the equation $x - 2y + z = 0$.
(a) Find an orthonormal basis for this plane.
(b) Find the matrix which projects $\mathbb{R}^3$ orthogonally onto this plane.
(c) Find a 3 by 3 orthogonal matrix Q so that multiplication by Q sends the xz-plane to this plane.

3.7.11. Consider the plane in $\mathbb{R}^3$ given by the equation $x + 2y + 3z = 0$.
(a) Find an orthonormal basis $\mathbf{q}_1, \mathbf{q}_2$ for this plane.
(b) Find a third vector $\mathbf{q}_3$ so that $\mathbf{q}_1, \mathbf{q}_2, \mathbf{q}_3$ form an orthonormal basis for $\mathbb{R}^3$.

3.7.12. Consider the subspace S of $\mathbb{R}^4$ spanned by the three orthogonal vectors $(1, 0, 0, 1)$, $(1, 1, 0, -1)$, $(0, 0, 1, 0)$. Find the projection of (x, y, z, u) onto S.

3.7.13. Express the vector $(2, 1, 1)$ as a linear combination of the orthogonal vectors $(1, 1, 0), (2, -2, 1), (1, -1, -4)$.

3.7.14. Express the vector $(2, 0, 4, 2)$ as a linear combination of the orthogonal vectors $(1, 1, 1, 1), (1, -1, 1, -1), (-1, -1, 1, 1)$.

Exercises 3.7.15–3.7.19 use the inner product on spaces of polynomials given by

$$\langle f, g\rangle = \int_{-1}^{1} f(t)g(t)\,dt$$

3.7.15. Consider the subspace of $\mathcal{P}^3(\mathbb{R}, \mathbb{R})$ spanned by the two polynomials $1 + x$ and $1 - x^3$. Use the Gram–Schmidt algorithm to find an orthonormal basis for this subspace.

3.7.16. Consider the subspace of $\mathcal{P}^3(\mathbb{R}, \mathbb{R})$ spanned by $1 + x, 1 + x^2, 1 + x^3$. Find an orthogonal basis for this subspace.

3.7.17. Consider the subspace of $\mathcal{P}^3(\mathbb{R}, \mathbb{R})$ spanned by $1 + x, 1 + x^2, x - x^2$. Find an orthogonal basis for this subspace.

3.7.18. Find the orthogonal projection of the polynomial $1 + x + x^2 + x^3$ onto the subspace spanned by $1 - x, x^2 - 2x + 1$. (Hint: First find an orthogonal basis for this subspace.)

3.7.19. Find the orthogonal projection of the general polynomial $a_0 + a_1x + a_2x^2 + a_3x^3$ onto the subspace spanned by $1 - x, x + x^3$. (Hint: First find an orthogonal basis for this subspace.)

3.7.20. Consider the vector space of symmetric 2 by 2 matrices with the inner product defined by $\langle A, B\rangle = \mathrm{tr}(AB)$, where $\mathrm{tr}(C)$ denotes the **trace** of a matrix, which is the sum of the diagonal entries; that is, $\mathrm{tr}(C) = c_{11} + c_{22}$. Thus $\langle A, B\rangle = a_{11}b_{11} + a_{12}b_{21} + a_{21}b_{12} + a_{22}b_{22} = a_{11}b_{11} + 2a_{12}b_{12} + a_{22}b_{22}$, using the fact that A and B are symmetric. Show that this definition satisfies the three properties required of an inner product (bilinearity, symmetry, and positivity) and find an orthonormal basis.

3.7.21. Suppose A is a symmetric n by n matrix which satisfies the additional condition that $\mathbf{x}^t A\mathbf{x} \geq 0$ for all $\mathbf{x}$ and equals 0 iff $\mathbf{x} = \mathbf{0}$. Show that the definition $\langle \mathbf{x}, \mathbf{y}\rangle = \mathbf{x}^t A\mathbf{y}$ satisfies the three properties required of an inner product for $\mathbb{R}^n$.

Exercises 3.7.22–3.7.25 use an inner product on $\mathbb{R}^n$ given by a symmetric matrix A as in the previous exercise.

3.7.22. When $\langle \mathbf{x}, \mathbf{y}\rangle = \mathbf{x}^t A\mathbf{y}$ and $A = \begin{pmatrix} 2 & 1 & 0 \\ 1 & 2 & 1 \\ 0 & 1 & 2 \end{pmatrix}$, find an orthonormal basis for $\mathbb{R}^3$ with this inner product.

3.7.23. When $\langle \mathbf{x}, \mathbf{y} \rangle = \mathbf{x}^t A \mathbf{y}$ and $A = \begin{pmatrix} 1 & 1 & 0 \\ 1 & 2 & 1 \\ 0 & 1 & 3 \end{pmatrix}$, find an orthonormal basis for the subspace span((1, 2, 1), (1, 0, 1)).

3.7.24. When $\langle \mathbf{x}, \mathbf{y} \rangle = \mathbf{x}^t A \mathbf{y}$ and $A = \begin{pmatrix} 1 & 0 & 1 & 2 \\ 0 & 2 & 0 & 1 \\ 1 & 0 & 3 & 0 \\ 2 & 1 & 0 & 8 \end{pmatrix}$, find an orthogonal basis for the subspace

$$\text{span}((1, 2, 0, 0), (1, 0, 2, 1), (2, 1, 1, 0))$$

3.7.25. With the same inner product as in the last exercise, find the orthogonal projection of the vector (1, 1, 1, 1) onto the subspace

$$\text{span}((1, 2, 0, 0), (1, 0, 2, 1), (2, 1, 1, 0))$$

3.7.26. Let $\langle \mathbf{x}, \mathbf{y} \rangle$ be an inner product on $\mathbb{R}^n$. Show that there is an n by n symmetric matrix A so that $\langle \mathbf{x}, \mathbf{y} \rangle = \mathbf{x}^t A \mathbf{y}$.
(Hint: Write

$$\mathbf{x} = x_1 \mathbf{e}_1 + \cdots + x_n \mathbf{e}_n, \mathbf{y} = y_1 \mathbf{e}_1 + \cdots + y_n \mathbf{e}_n$$

and use properties of the inner product to show that $\langle \mathbf{x}, \mathbf{y} \rangle = \mathbf{x}^t A \mathbf{y}$ where $A = (a_{ij})$ satisfies $a_{ij} = \langle \mathbf{e}_i, \mathbf{e}_j \rangle$.)

3.7.27. Let V be an n-dimensional vector space with basis $\mathbf{v}_1, \ldots, \mathbf{v}_n$. Let $a_{ij} = \langle \mathbf{v}_i, \mathbf{v}_j \rangle$. Generalize the last exercise to show that if $\mathbf{v} = c_1 \mathbf{v}_1 + \cdots + c_n \mathbf{v}_n$ and $\mathbf{w} = d_1 \mathbf{v}_1 + \cdots + d_n \mathbf{v}_n$, then $\langle \mathbf{v}, \mathbf{w} \rangle = \mathbf{c}^t A \mathbf{w}$.

Exercises 3.7.28–3.7.33 refer to the result of the previous exercise.

3.7.28. Consider the vector space $\mathcal{P}^2(\mathbb{R}, \mathbb{R})$ and the inner product given by

$$\langle f, g \rangle = \int_{-1}^{1} f(t) g(t)\, dt$$

(a) Using the standard basis $\mathbf{v}_1 = 1, \mathbf{v}_2 = x, \mathbf{v}_3 = x^2$ for $\mathcal{P}^2(\mathbb{R}, \mathbb{R})$ give the matrix A with $a_{ij} = \langle \mathbf{v}_i, \mathbf{v}_j \rangle$.

(b) Use Exercise 3.7.27 and part (a) to compute $\langle 1 - 2x + x^2, 2 + x + x^2 \rangle$ as $\mathbf{c}^t A \mathbf{d}$ for $\mathbf{c} = \begin{pmatrix} 1 \\ -2 \\ 1 \end{pmatrix}, \mathbf{d} = \begin{pmatrix} 2 \\ 1 \\ 1 \end{pmatrix}$. Compare your result with the direct calculation found by computing the integral.

3.7.29. Consider the vector space $\mathcal{P}^2(\mathbb{R}, \mathbb{R})$ and the inner product given by

$$\langle f, g \rangle = \int_{-1}^{1} f(t) g(t)\, dt$$

(a) Using the matrix A in the preceding exercise to compute inner products, find an orthogonal basis for the subspace spanned by $1 + x - x^2, 2 - x^2$.

(b) Use A to find the orthogonal projection of the polynomial x^2 onto this subspace.

3.7.30. Consider the vector space $\mathcal{P}^3(\mathbb{R}, \mathbb{R})$ and the inner product given by

$$\langle f, g \rangle = \int_{-1}^{1} f(t)g(t)\, dt$$

(a) Using the standard basis $\mathbf{v}_1 = 1, \mathbf{v}_2 = x, \mathbf{v}_3 = x^2, \mathbf{v}_4 = x^3$ for $\mathcal{P}^3(\mathbb{R}, \mathbb{R})$ give the matrix A with $a_{ij} = \langle \mathbf{v}_i, \mathbf{v}_j \rangle$.

(b) Use Exercise 3.7.27 and part (a) to compute $\langle 1 - 2x + x^2 + x^3, 2 + x + x^2 - x^3 \rangle$ as $\mathbf{c}^t A\mathbf{d}$ for $\mathbf{c} = \begin{pmatrix} 1 \\ -2 \\ 1 \\ 1 \end{pmatrix}, \mathbf{d} = \begin{pmatrix} 2 \\ 1 \\ 1 \\ -1 \end{pmatrix}$. Compare your result with the direct calculation found by computing the integral.

3.7.31. Consider the vector space $\mathcal{P}^3(\mathbb{R}, \mathbb{R})$ and the inner product given by

$$\langle f, g \rangle = \int_{-1}^{1} f(t)g(t)\, dt$$

(a) Using the matrix A in the preceding exercise to compute inner products, find an orthogonal basis for the subspace spanned by $1 - x + x^2, x - x^3$.

(b) Use A to find the projection of x^2 onto this subspace.

3.7.32. Consider the vector space of continuous functions from the interval $[-\pi, \pi]$ to the reals, $\mathcal{C}([-\pi, \pi], \mathbb{R})$. Using the inner product

$$\langle f, g \rangle = \int_{-\pi}^{\pi} f(t)g(t)\, dt$$

(a) show that the functions $1, \cos mt, \cos nt, \sin mt, \sin nt$ are pairwise orthogonal.

(b) show that the length of 1 is $\sqrt{2\pi}$ and the length of the others is $\sqrt{\pi}$.

3.7.33. For the subspace

$$S = \text{span}(\mathbf{v}_1 = 1, \mathbf{v}_2 = \cos t, \mathbf{v}_3 = \sin t, \mathbf{v}_4 = \cos 2t, \mathbf{v}_5 = \sin 2t)$$

of $C([\pi, \pi], \mathbb{R})$ with the inner product as in the previous exercise,

(a) find the matrix A with $a_{ij} = \langle \mathbf{v}_i, \mathbf{v}_j \rangle$.

(b) use A to find an orthogonal basis for the subspace T of S which is spanned by $2 + \cos 2t, -\sin t + \cos 2t$.

3.7.34. (a) Find an orthogonal basis for the orthogonal complement of the row space of $A = \begin{pmatrix} 0 & 1 & 0 & 1 & 0 \\ 1 & 0 & 1 & 0 & 1 \\ 1 & 1 & 1 & 1 & 1 \end{pmatrix}$.

(b) Find an orthogonal basis for the orthogonal complement of the column space of A.

3.7.35. (a) Find an orthogonal basis for the orthogonal complement of the null space of $A = \begin{pmatrix} 0 & 1 & 0 & 1 & 0 \\ 1 & 0 & 1 & 0 & 1 \\ 1 & 1 & 1 & 1 & 1 \end{pmatrix}$.

(b) Find an orthogonal basis for the orthogonal complement of the null space of A^t.

3.7.36. (a) Find an orthogonal basis for the orthogonal complement of the row space

$$\text{of } A = \begin{pmatrix} 1 & 0 & -1 & 1 & 2 \\ 0 & 1 & 2 & -1 & 0 \\ 2 & 3 & 4 & -1 & 4 \end{pmatrix}.$$

(b) Find an orthogonal basis for the orthogonal complement of the column space of A.

3.7.37. (a) Find an orthogonal basis for the orthogonal complement of the null space

$$\text{of } A = \begin{pmatrix} 1 & 0 & -1 & 1 & 2 \\ 0 & 1 & 2 & -1 & 0 \\ 2 & 3 & 4 & -1 & 4 \end{pmatrix}.$$

(b) Find an orthogonal basis for the orthogonal complement of the null space of A^t.

3.7.38. Suppose V is finite dimensional. Show that if $\mathbf{v}_1, \ldots, \mathbf{v}_k$ is an orthogonal basis for the subspace S of V and $\mathbf{w}_1, \ldots, \mathbf{w}_l$ is an orthogonal basis for $S^\perp$, then $\mathbf{v}_1, \ldots, \mathbf{v}_k, \mathbf{w}_1, \ldots, \mathbf{w}_k$ is an orthogonal basis for V.

3.7.39. Consider the vector space $\mathcal{P}^2(\mathbb{R}, \mathbb{R})$ with the inner product

$$\langle f, g \rangle = \int_{-1}^{1} f(t)g(t)\,dt$$

(a) Find an orthogonal basis for the subspace S spanned by $1 + x, 1 - x^2$.
(b) Find a basis for the orthogonal complement of S.

3.7.40. Consider the vector space $\mathcal{P}^3(\mathbb{R}, \mathbb{R})$ with the inner product

$$\langle f, g \rangle = \int_{-1}^{1} f(t)g(t)\,dt$$

(a) Find an orthogonal basis for the subspace S spanned by $1 + x^2, x + x^3$.
(b) Find an orthogonal basis for the orthogonal complement of S.

3.7.41. Consider $\mathbb{R}^3$ with an inner product given by $\langle \mathbf{x}, \mathbf{y} \rangle = \mathbf{x}^t A \mathbf{y}$, where

$$A = \begin{pmatrix} 1 & 1 & 0 \\ 1 & 2 & 1 \\ 0 & 1 & 5 \end{pmatrix}$$

(a) Find an orthonormal basis for the plane $x + y + z = 0$ using this inner product.
(b) Extend the basis found in part (a) to an orthonormal basis for $\mathbb{R}^3$ using this inner product.

3.7.42. Consider $\mathbb{R}^3$ with an inner product given by $\langle \mathbf{x}, \mathbf{y} \rangle = \mathbf{x}^t A \mathbf{y}$, where

$$A = \begin{pmatrix} 1 & 1 & 0 \\ 1 & 2 & 1 \\ 0 & 1 & 5 \end{pmatrix}$$

(a) Find an orthonormal basis for the plane $x + 2y - z = 0$ using this inner product.
(b) Extend the basis found in part (a) to an orthonormal basis for $\mathbb{R}^3$ using this inner product.

3.7.43. (a) Give the matrix A so that multiplication by A projects orthogonally onto the plane $x - 3y + z = 0$.
(b) Give the matrix B so that multiplication by B projects orthogonally onto the line through $(1, -3, 1)$.
(c) Write the matrix C so that multiplication by C reflects orthogonally through the plane $x - 3y + z = 0$ as a linear combination of A and B.

3.7.44. (a) Give the matrix A so that multiplication by A projects orthogonally onto the plane $3x + y - 4z = 0$.
(b) Give the matrix B so that multiplication by B projects orthogonally onto the line through $(3, 1, -4)$.
(c) Write the matrix C so that multiplication by C reflects orthogonally through the plane $3x + y - 4z = 0$ as a linear combination of A and B.

In Exercises 3.7.45–3.7.48 check the two properties $A^t = A, A^2 = A$ to determine whether the matrix A is a projection matrix. If it is, find a matrix B so that the subspace $\mathcal{R}(A)$ onto which one is projecting is reexpressed as the solutions to $B\mathbf{x} = \mathbf{0}$, that is, as $\mathcal{N}(B)$. (Hint: The rows of B should be orthogonal to the columns of A.)

3.7.45. $A = \begin{pmatrix} 1 & 0 & 1 \\ 0 & 1 & 1 \\ 1 & 1 & 1 \end{pmatrix}.$

3.7.46. $A = \begin{pmatrix} 2/3 & -1/3 & 1/3 \\ -1/3 & 2/3 & 1/3 \\ 1/3 & 1/3 & 2/3 \end{pmatrix}.$

3.7.47. $A = \begin{pmatrix} 1/6 & 0 & 1/3 & 1/6 \\ 0 & 1/6 & -1/6 & 1/3 \\ 1/3 & -1/6 & 5/6 & 0 \\ 1/6 & 1/3 & 0 & 5/6 \end{pmatrix}.$

3.7.48. $A = \begin{pmatrix} .2500 & .1250 & .3750 & .1250 & .1250 \\ .1250 & .3125 & -.0625 & .3125 & .3125 \\ .3750 & -.0625 & .8125 & -.0625 & -.0625 \\ .1250 & .3125 & -.0625 & .3125 & .3125 \\ .1250 & .3125 & -.0625 & .3125 & .3125 \end{pmatrix}.$

3.7.49. Show that if all of the entries of an n by n matrix $A, n > 1$, are greater than or equal to 1, then A is not a projection matrix.

3.7.50. Show that if A is a 2 by 2 matrix which is a projection matrix, then if we form a new matrix B from A by multiplying each off diagonal entry by -1 and leaving the diagonal entries unchanged, then B is a projection matrix. Show that this does not hold in general for n by n matrices when n is bigger than 2.

3.7.51. Solve the least squares problem $A\mathbf{x} = \mathbf{b}$, where

$$A = \begin{pmatrix} 1 & 1 & 0 \\ 2 & 1 & 3 \\ 1 & 0 & 1 \\ 0 & 2 & 2 \end{pmatrix} \quad \text{and} \quad \mathbf{b} = \begin{pmatrix} 1 \\ 1 \\ 1 \\ 1 \end{pmatrix}$$

3.7.52. Find the least squares solution to $A\mathbf{x} = \mathbf{b}$, where

$$A = \begin{pmatrix} 1 & 1 \\ 2 & 1 \\ -1 & 1 \end{pmatrix}, \quad \mathbf{b} = \begin{pmatrix} 1 \\ 0 \\ 0 \end{pmatrix}$$

Give the normal equation for this problem. Give the vector in the range of A which is closest to $\mathbf{b}$.

3.7.53. Find the least squares solution to $A\mathbf{x} = \mathbf{b}$, where

$$A = \begin{pmatrix} 1 & 0 & 1 \\ 0 & 1 & 2 \\ 1 & -1 & -1 \\ 2 & 0 & 0 \end{pmatrix}, \quad \mathbf{b} = \begin{pmatrix} 1 \\ 0 \\ 0 \\ 1 \end{pmatrix}$$

Give the normal equation for this problem. Give the vector in the range of A which is closest to $\mathbf{b}$.

3.7.54. Find the least squares solution to $A\mathbf{x} = \mathbf{b}$, where

$$A = \begin{pmatrix} 2 & 1 & 1 \\ 1 & 3 & -7 \\ 5 & 1 & 7 \\ 1 & 1 & -1 \\ 2 & 2 & -2 \end{pmatrix}, \quad \mathbf{b} = \begin{pmatrix} 1 \\ 0 \\ 0 \\ 0 \\ 0 \end{pmatrix}$$

3.7.55. Find the least squares solution to $A\mathbf{x} = \mathbf{b}$, where

$$A = \begin{pmatrix} 3 & 2 & 0 & 7 \\ 0 & 1 & 1 & -1 \\ -1 & 0 & 2 & -7 \\ 1 & -1 & 1 & -4 \\ 2 & 0 & 1 & -1 \\ 0 & 0 & 1 & -3 \end{pmatrix}, \quad \mathbf{b} = \begin{pmatrix} 1 \\ 0 \\ 0 \\ 0 \\ 0 \\ 0 \end{pmatrix}$$

3.7.56. Suppose that $\mathbf{x}$ is the least squares solution of $A\mathbf{x} = \mathbf{b}$. Form a new matrix B with one additional column which is the sum of the columns of A. Show that if $\mathbf{y}$ is the vector which is formed from $\mathbf{x}$ by adding a zero at the end, then $\mathbf{y}$ is a least squares solution of $B\mathbf{y} = \mathbf{b}$ but is not in the row space of B in general.

3.7.57. Give the matrix which projects onto the line through $(1, 1, 2, 1)$ in $\mathbb{R}^4$. Also, give the matrix which projects onto the subspace which is orthogonal to this line, that is, the subspace given by the equation $x + y + 2z + w = 0$.

3.7.58. Give the matrix which projects onto the subspace S of $\mathbb{R}^4$ spanned by the vectors $(1, 1, 1, 1), (1, 2, -1, -2)$. Give the matrix which projects onto the orthogonal complement of this subspace.

3.7.59. (a) Give the matrix A which projects onto the line through $(1, 2, 1)$.
(b) Give the matrix B which projects onto the line through $(2, 1, 0)$.
(c) Give the matrix C which projects onto the plane with basis $(1, 2, 1), (2, 1, 0)$. Explain why C is not just $A + B$.

3.7.60. Show that $\langle p, q\rangle = p(-1)q(-1)+p(0)q(0)+p(1)q(1)$ defines an inner product for $\mathcal{P}^2(\mathbb{R}, \mathbb{R})$. Apply the Gram–Schmidt algorithm to the basis $1, x, x^2$ to give an orthogonal basis. (Hint: Use the fact that the only polynomial of degree 2 which vanishes at three distinct points is the zero polynomial.)

3.7.61. Suppose that F is a finite-dimensional subspace of a (possibly infinite dimensional) vector space V and $F^\perp$ is the orthogonal complement of F so that we have a direct sum decomposition $V = F \oplus F^\perp$. Let P denote the projection of V to F; that is, $P(f + t) = f$, where $f \in F, t \in F^\perp$. Show that $(F^\perp)^\perp = F$ by showing that if $\mathbf{x}$ is orthogonal to $F^\perp$, then $\mathbf{x}$ must be in F. (Hint: Use the fact that both $\mathbf{x}$ and $P\mathbf{x}$ are orthogonal to $F^\perp$, so $\mathbf{x} - P\mathbf{x}$ is orthogonal to $F^\perp$. On the other hand $\mathbf{x} - P\mathbf{x}$ is in $F^\perp$.)

3.7.62. Suppose that S is a subspace in $\mathbb{R}^n$ and the linear transformation which projects orthogonally onto S is represented by the matrix A with respect to some (unknown) orthonormal basis of $\mathbb{R}^n$. Show that A is symmetric.

3.7.63. Find the regression line for the data

$$(1, 2), (2, 3), (3, 4), (4, 4), (5, 5), (6, 7), (7, 7), (8, 8)$$

3.7.64. (a) Find the regression line for the data $(-2, 1), (-1, 3), (0, 2), (1, 1), (2, 0)$.
(b) Find the quadratic polynomial that best fits this data.

3.7.65. With the same data as in the previous exercise, consider the piecewise linear function f whose graph consists of straight-line segments that join these data points. Using the inner product given by $\langle g, h\rangle = \int_{-2}^{2} g(t)h(t)\,dt$ find the quadratic polynomial $p(t) = a_0 + a_1x + a_2x^2$ which is closest to f. (Hint: Find an orthogonal basis for $\mathcal{P}^2(\mathbb{R}, \mathbb{R})$ using the given inner product.)

3.7.66. (a) Find the regression line for the data

$$(-3, 3), (-2, 4), (-1, 4), (0, 3), (1, 2), (2, 2), (3, 3)$$

(b) Find the cubic polynomial which best fits this data.

3.7.67. For the data in the previous exercise, find the polynomial of degree less than or equal to 6 which passes through these data points.

3.7.68. Consider the function $g(x) = 2/(x + 3)$ on the interval $[-1, 1]$.
(a) Using the inner product $\langle f, g\rangle = \int_{-1}^{1} f(t)g(t)\,dt$ on the continuous functions $\mathcal{C}([-1, 1], \mathbb{R})$ find the quadratic polynomial which best approximates $g(x)$.
(b) Check your answer by dividing the interval up into 40 subintervals, constructing a data set with the coordinates $(x, g(x))$ at 41 end points of these subintervals, and finding the quadratic which best fits this data.

3.7.69. Using the inner product $\int_{-\pi}^{\pi} f(t)g(t)\,dt$, find the trigonometric polynomial of the form $a_0 + a_1 \cos t + b_1 \sin t + a_2 \cos 2t + b_2 \sin 2t$ which best approximates the function $|x|$ on the interval from $-\pi$ to π.

3.7.70. With the same inner product as in the previous exercise, find the trigonometric polynomial of the form $a_0 + a_1 \cos t + b_1 \sin t + a_2 \cos 2t + b_2 \sin 2t$ which best approximates the function e^x on the interval from $-\pi$ to π.

3.7.71. With the same inner product as in the previous exercise, find the trigonometric polynomial of the form $a_0 + a_1 \cos t + b_1 \sin t + a_2 \cos 2t + b_2 \sin 2t$ which best approximates the function with value -1 on $[-\pi, 0]$ and 1 on $[0, \pi]$ on the interval from $-\pi$ to π.

3.7.72. Find the singular value decomposition and pseudoinverse for the matrix $A = \begin{pmatrix} 1 & 3 \\ 3 & 9 \end{pmatrix}$. Use this to find the least squares solution to $A\mathbf{x} = \begin{pmatrix} 1 \\ 1 \end{pmatrix}$.

3.7.73. Find the singular value decomposition and pseudoinverse for the matrix $A = \begin{pmatrix} 1 & 3 & -1 \\ 2 & 6 & -2 \\ 1 & 3 & -1 \\ 4 & 12 & -4 \end{pmatrix}$. Use this to find the least squares solution to $A\mathbf{x} = \begin{pmatrix} 1 \\ 0 \\ 0 \\ 0 \end{pmatrix}$.

3.7.74. Find the singular value decomposition and pseudoinverse for the matrix $A = \begin{pmatrix} 1 & 0 & -1 \\ 2 & 0 & -2 \\ 1 & 0 & -1 \\ 4 & 0 & -4 \end{pmatrix}$. Use this to find the least squares solution to $A\mathbf{x} = \begin{pmatrix} 1 \\ 0 \\ 0 \\ 0 \end{pmatrix}$.

CHAPTER 4

EIGENVALUES AND EIGENVECTORS

4.1. THE EIGENVALUE–EIGENVECTOR PROBLEM

In the last chapter we saw how the pseudoinverse of a matrix A could be found if we could find an orthonormal basis of vectors $\mathbf{v}_i$ so that multiplication by A^tA sends each of these vectors to itself. These vectors were called the eigenvectors of A^tA, and each eigenvector was associated to the particular multiple involved, which was called the eigenvalue. In this chapter we will learn how to solve for the eigenvalues and eigenvectors of a matrix and apply the solution to a variety of other problems. Let us begin by restating the basic definition.

DEFINITION 4.1.1. If A is an n by n matrix, then an **eigenvalue** for A is a number λ so that there is a nonzero vector $\mathbf{v}$ (called an **eigenvector** for A associated to λ) so that $A\mathbf{v} = \lambda\mathbf{v}$. Associated to each eigenvalue λ is its **eigenspace**

$$E(\lambda) = \{\mathbf{v} : A\mathbf{v} = \lambda\mathbf{v}\} = \{\mathbf{v} : (A - \lambda I)\mathbf{v}\} = \mathcal{N}(A - \lambda I)$$

Note that the eigenspace $E(\lambda)$ consists of the eigenvectors for λ together with the zero vector $\mathbf{0}$. The basic equation $A\mathbf{v} = \lambda\mathbf{v}$ can be rewritten as $(A - \lambda I)\mathbf{v} = \mathbf{0}$. For a given λ, its eigenvectors are nontrivial solutions of $(A - \lambda I)\mathbf{v} = \mathbf{0}$. These occur precisely when $\det(A - \lambda I) = 0$. When this equation is regarded as an equation in the variable λ, it becomes a polynomial of degree n in λ. Since a polynomial of degree n has at most n distinct roots, this shows that there are at most n eigenvalues for a given matrix. It also points out a basic difficulty in dealing with eigenvalues. The fundamental theorem of algebra asserts that any polynomial of degree n factors completely as a product of linear factors, *as long as we allow complex numbers;* that is, some of the roots of a polynomial may be complex numbers. This holds whether the coefficients of the polynomial are real or complex numbers. The roots $r_1, \ldots, r_n$ of the polynomial just correspond to the factors of $p(x) = k(x - r_1)\cdots(x - r_n)$. Thus solving the eigenvalue–eigenvector problem of finding the eigenvalues and corresponding eigenvectors of a matrix will require us to use the complex number system. In particular, to correctly state our results when there is a complex eigenvalue, we will have to use the complex vector space $\mathbb{C}^n$, which is just n-tuples $(z_1, \ldots, z_n)$ of complex numbers. Here the addition operation is coordinatewise as in $\mathbb{R}^n$, but we now allow ourselves to multiply by complex numbers. We will discuss this in more detail in Section 4.3. We will see how to get "real" information out of the complex solutions to the problem. Initially, however, we will just concentrate on problems where there are real solutions.

DEFINITION 4.1.2. The polynomial $p(x) = \det(A - xI)$ is called the **characteristic polynomial** of the matrix A. Its roots will be the eigenvalues. The multiplicity of λ as a root of the characteristic polynomial is the **algebraic multiplicity of the eigenvalue λ.** The dimension of $E(\lambda)$ is called the **geometric multiplicity of the eigenvalue λ.**

Some authors use $\det(xI - A)$ for the definition of the characteristic polynomial. The relation with our definition is $\det(xI - A) = (-1)^n \det(A - xI)$. Each form has its own technical advantages; the important point is that they both give the same roots, which are the eigenvalues.

We will be interested in finding a basis for $E(\lambda)$ for each eigenvalue λ and putting together these bases to form a basis for $\mathbb{R}^n$ which consists of eigenvectors. It turns out that many problems associated to A are easier to solve when there is such a basis.

DEFINITION 4.1.3. The n by n matrix A is called **diagonalizable** when there is a basis of eigenvectors $\mathbf{v}_1, \ldots, \mathbf{v}_n$ associated to the eigenvalues $\lambda_1, \ldots, \lambda_n$ for A.

In this definition the eigenvalues are listed with multiplicity; that is, some of the basis vectors may be eigenvectors for the same eigenvalue. This multiplicity will turn out to be both the algebraic and geometric multiplicity when the matrix is diagonalizable. The reason for the terminology is that this means that A is closely related to a diagonal matrix D whose diagonal entries are the eigenvalues of A. The equations $A\mathbf{v}_i = \lambda_i \mathbf{v}_i$ can be written as one matrix equation $AS = SD$, where S is the matrix which has the $\mathbf{v}_i$ as its columns and D is the diagonal matrix with the eigenvalues λ_i on the diagonal. This implies that

$$A = SDS^{-1}$$

We now state for future reference this criterion for diagonalizability.

PROPOSITION 4.1.1. An n by n matrix A is diagonalizable iff there is an invertible matrix S and a diagonal matrix D with

$$A = SDS^{-1}$$

Recall that A, B are similar matrices if there is an invertible matrix S with $A = SBS^{-1}$. We introduced the concept of similar matrices in Section 2.7, where we showed that similarity is an equivalence relation, and any two matrices representing a linear transformation $L : V \to V$ with respect to different choices of bases are similar. Thus A is diagonalizable iff A is similar to a diagonal matrix D, where the matrix S which achieves the similarity just consists of the basis of eigenvectors for A in its columns. Sometimes, diagonalizability is defined in terms of being similar to a diagonal matrix and our definition is given as a criterion. The following lemma is useful in recognizing diagonalizable matrices.

LEMMA 4.1.2. If $\mathbf{v}_1, \ldots, \mathbf{v}_k$ are eigenvectors for A associated to distinct eigenvalues $\lambda_1, \ldots, \lambda_k$, then $\mathbf{v}_1, \ldots, \mathbf{v}_k$ are independent.

Proof. Let $c_1\mathbf{v}_1 + \cdots + c_k\mathbf{v}_k = \mathbf{0}$. We will show that $c_1 = 0$; a similar argument will show that the other c_i are zero. Apply the product

$$(A - \lambda_2 I)(A - \lambda_3 I) \cdots (A - \lambda_k I)$$

to this equation, using the fact that $(A - \lambda_i I)\mathbf{v}_j = (\lambda_j - \lambda_i)\mathbf{v}_j$. Then the first term is sent to $c_1(\lambda_1 - \lambda_2) \cdots (\lambda_1 - \lambda_k)\mathbf{v}_1$, and all the other terms are sent to $\mathbf{0}$. Since the linear combination with which we started was equal to $\mathbf{0}$, we get the equation $c_1(\lambda_1 - \lambda_2) \cdots (\lambda_1 - \lambda_k)\mathbf{v}_1 = \mathbf{0}$. Since the eigenvalues are distinct the only way that this can happen is for $c_1 = 0$. ■

For each eigenvalue there is at least one eigenvector (by the definition of an eigenvalue). This leads to the following useful criterion which guarantees that a matrix is diagonalizable.

PROPOSITION 4.1.3. If the n by n matrix A has n distinct eigenvalues, then A is diagonalizable.

Proof. For each eigenvalue select one eigenvector. Then the lemma guarantees that these give an independent set of n eigenvectors in $\mathbb{R}^n$ (or $\mathbb{C}^n$, if there is a complex eigenvalue involved), and so these constitute a basis of eigenvectors for A. ■

We now look at a few examples.

■ ***Example 4.1.1.*** $A = \begin{pmatrix} 1 & 2 \\ 0 & 3 \end{pmatrix}$. Then the characteristic polynomial of A is

$$\det(A - xI) = \det\begin{pmatrix} 1-x & 2 \\ 0 & 3-x \end{pmatrix} = (1-x)(3-x)$$

Thus the eigenvalues are $1, 3$.

For the eigenvalue 1, the eigenspace consists of solutions of $(A - I)\mathbf{v} = \mathbf{0}$. Since $A - I = \begin{pmatrix} 0 & 2 \\ 0 & 2 \end{pmatrix}$, we see that all solutions are multiples of $\mathbf{v}_1 = (1, 0)$.

For the eigenvalue 3, the eigenspace consists of solutions of $(A - 3I)\mathbf{v} = \mathbf{0}$. Since $A - 3I = \begin{pmatrix} -2 & 2 \\ 0 & 0 \end{pmatrix}$, the solutions are multiples of $\mathbf{v}_2 = (1, 1)$. Thus the matrix A is diagonalizable, and $A = SDS^{-1}$, where

$$S = \begin{pmatrix} 1 & 1 \\ 0 & 1 \end{pmatrix}, D = \begin{pmatrix} 1 & 0 \\ 0 & 3 \end{pmatrix}$$ ■

■ ***Example 4.1.2.*** $A = \begin{pmatrix} 13 & -4 \\ -4 & 7 \end{pmatrix}$. Then $\det(A - xI) = \det\begin{pmatrix} 13-x & -4 \\ -4 & 7-x \end{pmatrix} = (91 - 16) - 20x + x^2 = x^2 - 20x + 75 = (x - 5)(x - 15)$, so the eigenvalues are $5, 15$.

The eigenspace associated to 5 is the solutions of $(A - 5I)\mathbf{v} = \mathbf{0}$. Since $A - 5I = \begin{pmatrix} 8 & -4 \\ -4 & 2 \end{pmatrix}$, the eigenspace has a basis with one vector $(1, 2)$.

The eigenspace associated to 15 is the solutions of $(A - 15I)\mathbf{v} = \mathbf{0}$. Since $A - 15I = \begin{pmatrix} -2 & -4 \\ -4 & -8 \end{pmatrix}$, the eigenspace has a basis with the one vector $(2, -1)$.

Note that these two eigenvectors are orthogonal to each other—this always happens when A is a symmetric matrix. Thus $A = SDS^{-1}$ with

$$S = \begin{pmatrix} 1 & 2 \\ 2 & -1 \end{pmatrix}, D = \begin{pmatrix} 5 & 0 \\ 0 & 15 \end{pmatrix}$$ ■

■ ***Example 4.1.3.*** $A = \begin{pmatrix} 3 & 1 & -2 \\ -2 & 2 & 2 \\ 0 & 1 & 1 \end{pmatrix}$. Then

$$\det(A - xI) = \det\begin{pmatrix} 3-x & 1 & -2 \\ -2 & 2-x & 2 \\ 0 & 1 & 1-x \end{pmatrix}$$

$$= \det\begin{pmatrix} 3-x & 1 & -2-(1-x) \\ -2 & 2-x & 2-(1-x)(2-x) \\ 0 & 1 & 0 \end{pmatrix} \text{ we subtracted } (1-x)\mathbf{A}^2 \text{ from } \mathbf{A}^3$$

$$= -\det\begin{pmatrix} 3-x & -(3-x) \\ -2 & -x^2+3x \end{pmatrix} \text{ we expanded by the bottom row}$$

$$= -(3-x)\det\begin{pmatrix} 1 & -1 \\ -2 & -x^2+3x \end{pmatrix} \text{ we factored out } (3-x) \text{ in the first row}$$

$$= -(3-x)(-x^2+3x-2) = (3-x)(1-x)(2-x)$$

so the eigenvalues are $1, 2, 3$.

This calculation illustrates that in computing the determinant it is frequently easier to use row or column operations first rather than just expanding by cofactors. In particular, our use of column operations led to our finding the factor $(3-x)$ of the characteristic polynomial.

$$x = 1 : A - I = \begin{pmatrix} 2 & 1 & -2 \\ -2 & 1 & 2 \\ 0 & 1 & 0 \end{pmatrix}$$

with null space having basis $(1, 0, 1)$.

$$x = 2 : A - 2I = \begin{pmatrix} 1 & 1 & -2 \\ -2 & 0 & 2 \\ 0 & 1 & -1 \end{pmatrix}$$

with null space having basis $(1, 1, 1)$.

$$x = 3 : A - 3I = \begin{pmatrix} 0 & 1 & -2 \\ -2 & -1 & 2 \\ 0 & 1 & -2 \end{pmatrix}$$

with null space having a basis $(0, 2, 1)$.
Thus $A = SDS^{-1}$ with

$$S = \begin{pmatrix} 1 & 1 & 0 \\ 0 & 1 & 2 \\ 1 & 1 & 1 \end{pmatrix}, D = \begin{pmatrix} 1 & 0 & 0 \\ 0 & 2 & 0 \\ 0 & 0 & 3 \end{pmatrix}$$

■

■ ***Example 4.1.4.*** $A = \begin{pmatrix} 3 & 0 & 0 \\ -2 & 4 & 2 \\ -2 & 1 & 5 \end{pmatrix}$. Then $\det(A - xI) =$

$$\det\begin{pmatrix} 3-x & 0 & 0 \\ -2 & 4-x & 2 \\ -2 & 1 & 5-x \end{pmatrix} = (3-x)(x^2-9x+18) = (3-x)^2(6-x)$$

Here there are just two distinct eigenvalues since 3 is a double root of the characteristic polynomial.

$$x = 3 : A - 3I = \begin{pmatrix} 0 & 0 & 0 \\ -2 & 1 & 2 \\ -2 & 1 & 2 \end{pmatrix}$$

with null space having a basis $(1, 0, 1), (1, 2, 0)$.

Note that this root of algebraic multiplicity 2 led to an eigenspace of dimension 2 (that is, the eigenvalue has geometric multiplicity 2 as well). That will not always be the case as the next example will show.

$$x = 6 : A - 6I = \begin{pmatrix} -3 & 0 & 0 \\ -2 & -2 & 2 \\ -2 & 1 & -1 \end{pmatrix}$$

with null space having a basis $(0, 1, 1)$.

Thus $A = SDS^{-1}$ with

$$S = \begin{pmatrix} 1 & 1 & 0 \\ 0 & 2 & 1 \\ 1 & 0 & 1 \end{pmatrix}, D = \begin{pmatrix} 3 & 0 & 0 \\ 0 & 3 & 0 \\ 0 & 0 & 6 \end{pmatrix}$$

■

■ ***Example 4.1.5.*** $A = \begin{pmatrix} 1 & 1 \\ 0 & 1 \end{pmatrix}$. Then $\det(A - xI) = (1 - x)^2$, so the only eigenvalue is $x = 1$. Then $(A - I) = \begin{pmatrix} 0 & 1 \\ 0 & 0 \end{pmatrix}$, which has null space with basis $(1, 0)$. Thus every eigenvector for the eigenvalue 1 is a multiple of $(1, 0)$. There is *not* a basis of eigenvectors, so this is an example of a matrix which is not diagonalizable. Thus *not all matrices are diagonalizable.* We will see later when we discuss the Jordan canonical form of a matrix that this example is essentially the simplest example of a nondiagonalizable matrix and that any other nondiagonalizable matrix is built up from modifications of this example (in a sense to be made precise later). ■

■ ***Example 4.1.6.*** The last example showed that a fairly simple looking matrix may not be diagonalizable. This example will illustrate the necessity of considering the complex numbers when discussing diagonalizability of real matrices. Let $A = \begin{pmatrix} 0 & -1 \\ 1 & 0 \end{pmatrix}$. Note that multiplication by A just rotates the plane by 90 degrees. Thus no nonzero vector is sent to a multiple of itself, so there are no *real* eigenvalues. However, when we compute $\det(A - xI) = x^2 + 1$, there are two complex roots, which are $\pm i$, where $i = \sqrt{-1}$. We solve for complex eigenvectors associated to these eigenvalues, which will just be vectors in $\mathbb{C}^2$ which are sent to $\pm i$ times themselves when we multiply by A. For i, we look at $(A - iI) = \begin{pmatrix} -i & -1 \\ 1 & -i \end{pmatrix}$ which has null space complex multiples of $(i, 1)$. For $-i$, we look at $A + iI = \begin{pmatrix} i & -1 \\ 1 & i \end{pmatrix}$, which has null space complex multiples of $(-i, 1)$.

Note that these two complex eigenvectors are the conjugates of one another (recall that the **complex conjugate** of the complex number $a + bi$ with $a, b \in \mathbb{R}$ is $a - bi$) and the corresponding eigenvectors are the conjugates of one another as well. This is a consequence of the fact that A is a real matrix (i.e., a matrix with real entries). If $\overline{\lambda}$ denotes the conjugate of λ and $\overline{\mathbf{v}}$ denotes the conjugate of $\mathbf{v}$, then $A\mathbf{v} = \lambda\mathbf{v}$ implies $\overline{A}\overline{\mathbf{v}} = \overline{A\mathbf{v}} = \overline{\lambda\mathbf{v}} = \overline{\lambda}\overline{\mathbf{v}}$. Since A is a real matrix, $\overline{A} = A$ and so $A\overline{\mathbf{v}} = \overline{\lambda}\overline{\mathbf{v}}$. Thus whenever $\mathbf{v}$ is a complex eigenvector with complex eigenvalue λ for a real matrix I, we automatically get $\overline{\mathbf{v}}$ is an eigenvector for the eigenvalue $\overline{\lambda}$. This means that the problem of finding eigenvalues and eigenvectors is actually simplified somewhat when they are complex (for a real matrix) since they come in conjugate pairs.

Returning to our example, we again have $A = SDS^{-1}$, but now S and D are complex matrices:

$$S = \begin{pmatrix} i & -i \\ 1 & 1 \end{pmatrix}, \quad D = \begin{pmatrix} i & 0 \\ 0 & -i \end{pmatrix} \qquad \blacksquare$$

We will return to discuss complex eigenvalues and eigenvectors more thoroughly in Sections 4.3 and 5.1, but for the moment we will just transfer our problem from $\mathbb{R}^n$ to $\mathbb{C}^n$ (i.e., n-tuples of complex numbers) when they occur and solve the problem there. Note that we could see that complex eigenvalues have to occur in conjugate pairs from the point of view of the characteristic polynomial since complex roots of a real polynomial have to occur in complex conjugate pairs.

In Exercises 4.1.1–4.1.5 find the eigenvalues and corresponding eigenvectors and (where possible) give matrices S, D with $A = SDS^{-1}$.

Exercise 4.1.1.

$$A = \begin{pmatrix} 1 & 2 \\ 0 & -1 \end{pmatrix}.$$

Exercise 4.1.2.

$$A = \begin{pmatrix} 1 & 2 \\ 1 & 0 \end{pmatrix}.$$

Exercise 4.1.3.

$$A = \begin{pmatrix} 14 & -4 \\ 1 & 10 \end{pmatrix}.$$

Exercise 4.1.4.

$$A = \begin{pmatrix} 2 & 1 \\ -1 & 2 \end{pmatrix}.$$

Exercise 4.1.5.

$$A = \begin{pmatrix} 2 & 0 & -1 \\ 0 & 2 & 0 \\ -1 & 0 & 2 \end{pmatrix}.$$

4.2. DIAGONALIZABILITY—ALGEBRAIC AND GEOMETRIC MULTIPLICITY

When a matrix A is diagonalizable, then there is an invertible matrix S and a diagonal matrix D so that $A = SDS^{-1}$. The simplest situation is where S and D are real matrices (i.e., matrices with entries in $\mathbb{R}$). Even if A is real, the eigenvalues of A may be complex; then diagonalizability will only mean that there are complex matrices S, D as earlier. The matrix D will contain the eigenvalues on the diagonal, and the matrix S will contain the corresponding eigenvectors. We saw earlier that not all matrices are diagonalizable, for the matrix $\begin{pmatrix} 1 & 1 \\ 0 & 1 \end{pmatrix}$ has only one eigenvalue 1 (with algebraic multiplicity 2) and the eigenspace $E(1)$ of eigenvectors associated to 1 is one dimensional with basis $(1, 0)$. Thus there is not a basis of eigenvectors.

Recall that one situation which guarantees that A is diagonalizable is that it has n distinct eigenvalues. The converse is not true, as any diagonal matrix with repeated entries on the diagonal shows. We defined the geometric multiplicity of an eigenvalue as the dimension of the eigenspace associated with the eigenvalue. We will show that the geometric multiplicity is always less than or equal to the algebraic multiplicity of an eigenvalue, which is the multiplicity of the eigenvalue as a root of the characteristic polynomial. Then a simple counting argument shows that a matrix is diagonalizable iff the geometric multiplicity of each eigenvalue is equal to its algebraic multiplicity. For then there is a basis for each eigenspace; put all these together to form a basis for $\mathbb{R}^n$ (or $\mathbb{C}^n$). There will be n vectors since the sum of the algebraic multiplicities will add up to n. That they are independent will essentially boil down to the fact that the original set of vectors for each eigenspace was independent and the foregoing argument that vectors from different eigenspaces are independent. The details are left as an exercise.

To understand the relationship of the algebraic and geometric multiplicities, we first note some facts about similar matrices. Note that a diagonalizable matrix is just one which is similar to a diagonal matrix. Thus if A is diagonalizable and B is similar to A, then B is also diagonalizable—just use the transitivity of the similarity relation. Similar matrices share a lot of properties. In particular, they have the same characteristic polynomial and thus the same eigenvalues.

$$\det(SAS^{-1} - xI) = \det(S(A - xI)S^{-1}) = \det S \det(A - xI) \det S^{-1} = \det(A - xI)$$

The converse is not true. If two matrices have the same characteristic polynomial, they don't have to be similar. For example, the only matrix which is similar to the identity matrix is the identity matrix itself, but the identity matrix and the matrix $\begin{pmatrix} 1 & 1 \\ 0 & 1 \end{pmatrix}$ both have the same characteristic polynomial. The fact that $p(x) = \det(A - xI)$ implies that the constant term $p(0)$ is just $\det A$. There is another term which can be computed separately as well.

DEFINITION 4.2.1. The **trace** of A, denoted $\operatorname{tr} A$, is the sum of the diagonal entries of A.

Using the permutation definition of the determinant to compute the characteristic polynomial will allow us to identify the trace of A. For when we compute

$$\det\begin{pmatrix} (a_{11} - x) & a_{12} & \dots & a_{1n} \\ a_{21} & (a_{22} - x) & \dots & a_{2n} \\ \vdots & \vdots & \vdots & \vdots \\ a_{n1} & a_{n2} & \dots & (a_{nn} - x) \end{pmatrix}$$

the only way to get a term with x^n in it is to take the product of the diagonal entries. Similarly, this is the only way to get a term with x^{n-1} in it. Thus the characteristic polynomial will be

$$(-1)^n x^n + (-1)^{n-1}(a_{11} + \cdots + a_{nn})x^{n-1} + \cdots + \det A$$

The coefficient of x^n is $(-1)^n$ and the coefficient of x^{n-1} is $(-1)^{n-1}\operatorname{tr} A$.

■ ***Example 4.2.1.*** If $A = \begin{pmatrix} 2 & 3 & 1 \\ 0 & 3 & 1 \\ 1 & 0 & 2 \end{pmatrix}$, then the trace is 7, and so the characteristic polynomial starts off as $-x^3 + 7x^2 + \cdots$. Computing the determinant as 12 tells us the constant term is 12. The full characteristic polynomial here is $-x^3 + 7x^2 - 15x + 12$. ■

Note that for a 2 by 2 matrix, knowing both the trace and the determinant determines the characteristic polynomial completely as $x^2 - \operatorname{tr} Ax + \det A$. For a general matrix, we need more information, but computing the trace and the determinant still provides two checks on the computation of the characteristic polynomial. This is especially important for the trace since it is so easy to compute.

The characteristic polynomial factors as $(c_1 - x)\cdots(c_n - x)$, where the c_i are the eigenvalues. This follows from the fact that the eigenvalues occur as the roots of the characteristic polynomial, together with the fact that the top coefficient is $(-1)^n$. The coefficient of x^{n-1} is $(-1)^{n-1}(c_1 + \cdots + c_n)$, and the constant term is $c_1 \dots c_n$. This leads to the following useful proposition, which provides a useful check on the computations of eigenvalues.

PROPOSITION 4.2.1. The trace of A is the sum of the eigenvalues and the determinant of A is the product of the eigenvalues.

■ ***Example 4.2.2.*** The matrix $A = \begin{pmatrix} 1 & 2 \\ 1 & 2 \end{pmatrix}$ has determinant 0 and trace 3. Since the determinant is 0, 0 will be an eigenvalue. The sum of the eigenvalues is 3, so the other eigenvalue is 3. ■

Since similar matrices have the same characteristic polynomial and the trace is the coefficient (up to sign) of x^{n-1} of the characteristic polynomial, then similar matrices must have the same trace as well. This could also be seen by one of the properties of the trace—$\operatorname{tr}(AB) = \operatorname{tr}(BA)$, which is easy to verify directly. Then

$$\operatorname{tr}(SAS^{-1}) = \operatorname{tr}(AS^{-1}S) = \operatorname{tr}(A)$$

■ ***Example 4.2.3.*** We now look at some of the examples of the previous section in terms of the concepts just introduced. For Example 4.1.2, $A = \begin{pmatrix} 13 & -4 \\ -4 & 7 \end{pmatrix}$. The determinant is 75, and the trace is 20. The sum of the eigenvalues is 20 and the product is 75, leading to the eigenvalues $5, 15$.

For Example 4.1.3, $A = \begin{pmatrix} 3 & 1 & -2 \\ -2 & 2 & 2 \\ 0 & 1 & 1 \end{pmatrix}$. The trace is 6, and the determinant is also 6. These two pieces of information are not sufficient to conclude what the three eigenvalues are. Suppose we knew that 1 is one of the eigenvalues. Then the sum of the other two would be 5, and their product would be 6. From this, we could conclude that the eigenvalues are $1, 2, 3$. ■

We now show that the geometric multiplicity of an eigenvalue has to be less than or equal to the algebraic multiplicity.

PROPOSITION 4.2.2. Let A be an n by n matrix and let μ be an eigenvalue of A. Then the geometric multiplicity of μ is less than or equal to the algebraic multiplicity of μ.

Proof. We will work in $\mathbb{C}^n$, but the argument is identical in $\mathbb{R}^n$ if the eigenvalues all turn out to be real. Suppose that the algebraic multiplicity of the eigenvalue μ is k and we have p independent eigenvectors $\mathbf{v}_1, \ldots, \mathbf{v}_p$. Extend the $\mathbf{v}_i$ to a basis of $\mathbb{C}^n$ and let S be the matrix with this basis as its column vectors. Then there is a matrix equation $AS = S\begin{pmatrix} \mu I_p & B \\ 0 & C \end{pmatrix}$, where B is a p by $(n-p)$ block and C is an $(n-p)$ by $(n-p)$ block. But then A is similar to the matrix on the right and so has the same characteristic polynomial. Expanding by cofactors shows that the characteristic polynomial has the factor $(\mu - x)^p$, and so p must be less than or equal to the algebraic multiplicity of μ. ■

Even when we are just dealing with matrices and their eigenvectors and eigenvalues, it is useful to think of the matrix as giving a linear transformation from $\mathbb{R}^n$ (or $\mathbb{C}^n$) to itself and recast the eigenvalue–eigenvector problem in terms of linear transformations.

DEFINITION 4.2.2. Suppose that V is a vector space of dimension n and $L : V \to V$ is a linear transformation. An **eigenvalue** of L is a number λ so that there is a nonzero vector $\mathbf{v}$ with $L\mathbf{v} = \lambda\mathbf{v}$. As before, such a vector $\mathbf{v}$ would be called an **eigenvector** of L associated to this eigenvalue.

How can the eigenvalues and eigenvectors for L be found? Many times these can be found via geometric descriptions of L, as the following example shows.

■ ***Example 4.2.4.*** Consider the reflection through a plane $x + y + z = 0$ in $\mathbb{R}^3$. For any vector in the plane, the vector will be sent to itself. Thus vectors in the plane will be eigenvectors for the eigenvalue 1. The basis $(1, -1, 0), (1, 0, -1)$ of the plane thus gives two eigenvectors for eigenvalues 1. The normal vector $(1, 1, 1)$ to the plane will be sent to its negative via this reflection, so it will be an eigenvector for the eigenvalue -1. Thus there is a basis of eigenvectors given by $(1, -1, 0), (1, 0, -1)$, $(1, 1, 1)$ corresponding to the eigenvectors $1, 1, -1$. The matrix of L with respect to this basis is $D = \begin{pmatrix} 1 & 0 & 0 \\ 0 & 1 & 0 \\ 0 & 0 & -1 \end{pmatrix}$. ■

Another method to find the eigenvalues and eigenvectors of a linear transformation L is to associate a matrix with L. We now show that the eigenvalues of L are the same as the eigenvalues of the corresponding matrix. To do this, choose a basis $\mathbf{v}_1, \ldots, \mathbf{v}_n$ for V, write $L\mathbf{v}_j = \sum_{i=1}^{n} a_{ij}\mathbf{v}_i$, and associate the matrix $A = (a_{ij})$ to L as before. Of course, the matrix we get will depend on the basis which was chosen. However, if we choose a different basis the new matrix will be of the form $C^{-1}AC$, where C is the transition matrix between the two bases. Since similar matrices have the same characteristic polynomial, the eigenvalues of A will be the same as the eigenvalues of the new matrix. We define the **characteristic polynomial of** L to be the characteristic polynomial of a matrix A which represents L.

If λ is an eigenvalue of L with corresponding eigenvector $\mathbf{v}$ and we choose $\mathbf{v}$ to be the first vector in a basis for V, then the matrix A which represents L with respect to this basis will have first column $\lambda\mathbf{e}_1$ and so λ will be one of the eigenvalues of A. Conversely, if λ is an eigenvalue of A, this means that there is some vector $\mathbf{w} \in \mathbb{R}^n$ with $A\mathbf{w} = \lambda\mathbf{w}$. Then if we take the vector $\mathbf{v} = \mathbf{V}\mathbf{w}$ (i.e., the linear combination of the basis vectors with coefficients coming from $\mathbf{w}$), we get

$$L(\mathbf{v}) = L(\mathbf{V}\mathbf{w}) = L(\mathbf{V})\mathbf{w} = (\mathbf{V}A)\mathbf{w} = \mathbf{V}(A\mathbf{w}) = \mathbf{V}(\lambda\mathbf{w}) = \lambda\mathbf{V}\mathbf{w} = \lambda\mathbf{v}$$

Thus the eigenvalues of L and any associated matrix correspond directly. Moreover, the calculation shows how to translate from an eigenvector of an associated matrix to the corresponding eigenvector of L.

Alternatively, we could use the notation of Section 2.7.2. If $[L]$ denotes the matrix of L with respect to some basis $\mathcal{V}$, and $[\mathbf{v}]$ denotes the coordinates of the vector $\mathbf{v}$ with respect to this basis, then

$$L\mathbf{v} = \lambda\mathbf{v} \Leftrightarrow [L\mathbf{v}] = [\lambda\mathbf{v}] \Leftrightarrow [L][\mathbf{v}] = \lambda[\mathbf{v}]$$

Thus λ is an eigenvalue of L iff it is an eigenvalue of *any* matrix representative $[L]$ of L.

DEFINITION 4.2.3. L is **diagonalizable** if there is a basis for V consisting of eigenvectors of L.

If L is diagonalizable, then the matrix which represents L with respect to the basis from the definition will be a diagonal matrix with the diagonal entries equal to the eigenvalues. Thus the matrix which represents L with respect to any other basis will be similar to a diagonal matrix; that is, it will be diagonalizable.

■ ***Example 4.2.5.*** Continuing with the last example, let us use the foregoing method to determine the eigenvalues and eigenvectors of the reflection map through this plane. Although it is easiest to choose a basis as we did earlier, let us instead just consider using the standard basis $\mathbf{e}_1, \mathbf{e}_2, \mathbf{e}_3$. To see where the reflection sends a given vector, we first decompose it in terms of the normal vector $(1, 1, 1)$ in the plane and a vector in the plane, using our formulas from the last chapter for the projection onto a line. We get

$$\mathbf{e}_1 = \frac{1}{3}(1,1,1) + \frac{1}{3}(2,-1,-1), \mathbf{e}_2 = \frac{1}{3}(1,1,1) + \frac{1}{3}(-1,2,-1),$$

$$\mathbf{e}_3 = \frac{1}{3}(1,1,1) + \frac{1}{3}(-1,-1,2)$$

Then these vectors are sent to

$$L(\mathbf{e}_1) = -\frac{1}{3}(1,1,1) + \frac{1}{3}(2,-1,-1) = \frac{1}{3}(1,-2,-2),$$

$$L(\mathbf{e}_2) = -\frac{1}{3}(1,1,1) + \frac{1}{3}(-1,2,-1) = \frac{1}{3}(-2,1,-2),$$

$$L(\mathbf{e}_3) = -\frac{1}{3}(1,1,1) + \frac{1}{3}(-1,-1,2) = \frac{1}{3}(-2,-2,1)$$

Thus the matrix which represents L with respect to the standard basis is

$$A = \frac{1}{3}\begin{pmatrix} 1 & -2 & -2 \\ -2 & 1 & -2 \\ -2 & -2 & 1 \end{pmatrix}$$ ■

We leave as an exercise for you to show that the eigenvectors of $A = cB$ are the same as for B, but the eigenvalues for A are the product of c with those for B. Thus we will look at

$$B = \begin{pmatrix} 1 & -2 & -2 \\ -2 & 1 & -2 \\ -2 & -2 & 1 \end{pmatrix}$$

By inspection we can see that when we subtract 3 from each diagonal entry, we get a singular matrix. Thus 3 is an eigenvalue for B. A basis of eigenvectors for the eigenvalue 3 is $(1, -1, 0), (1, 0, -1)$. Since the trace is 3, the sum of the eigenvalues must be 3. Thus the other eigenvalue must be -3. An eigenvector for -3 can be found to be $(1, 1, 1)$. These three eigenvectors are still eigenvectors for A, but now the corresponding eigenvalues become $1, 1, -1$. Thus these are the eigenvalues for L. To find the corresponding eigenvectors for L, we have to take linear combinations of the chosen basis using as coefficients the eigenvectors we have found. Since the basis we are using is the standard basis, this process just gives us the vectors back. Thus we recover the result of the last example.

Exercise 4.2.1.

Show that if $\mathbf{v}$ is an eigenvector for B with eigenvalue λ, then $\mathbf{v}$ is also an eigenvector for $A = cB$ with eigenvalue $c\lambda$.

Exercise 4.2.2.

Fill in the details of the argument sketched at the beginning of this section to show that when the geometric multiplicity of each eigenvalue is equal to its algebraic multiplicity, the matrix is diagonalizable.

Exercise 4.2.3.

Verify that $\operatorname{tr} AB = \operatorname{tr} BA$.

Exercise 4.2.4.

For the following matrices, find the eigenvalues, their algebraic multiplicities, and geometric multiplicities. Which of these matrices are diagonalizable?

$$A = \begin{pmatrix} 2.5 & -.5 \\ 1.5 & .5 \end{pmatrix}, B = \begin{pmatrix} 2.5 & -.5 \\ 4.5 & -.5 \end{pmatrix}, C = \begin{pmatrix} 2 & 1 & -1 \\ -4 & -1 & 4 \\ -3 & -1 & 4 \end{pmatrix}$$

Exercise 4.2.5.

1 is an eigenvalue of each of the following matrices. For each matrix compute the trace and determinant. Use the fact that 1 is a root of the characteristic polynomial to find all the eigenvalues.

$$A = \begin{pmatrix} 3 & -2 \\ 1 & 0 \end{pmatrix}, B = \begin{pmatrix} 2.5 & -1 & -1.5 \\ 1.5 & -1 & -1.5 \\ .5 & -1 & .5 \end{pmatrix}$$

Exercise 4.2.6.

Let S be a plane in $\mathbb{R}^3$, and let L be the linear transformation which is projection onto that plane. Verify that 0 is an eigenvalue with geometric multiplicity 1 and that 1 is an eigenvalue with geometric multiplicity 2. What is the characteristic polynomial of L? For the case where the plane is given by the equation $x + y - z = 0$, find a basis of eigenvectors for L so that the matrix which represents L with respect to this basis is $D = \begin{pmatrix} 1 & 0 & 0 \\ 0 & 1 & 0 \\ 0 & 0 & 0 \end{pmatrix}$. L arises from multiplication by a matrix A. Find A. (Hint: Use the relationship of A to D, or equivalently use the equation $AR = RD$, where R is the matrix whose columns are the eigenvectors of L.)

4.3. COMPLEX NUMBERS, VECTORS, AND MATRICES

In this section we want to begin dealing more thoroughly with the fact that eigenvalues of a real matrix or linear transformation may be complex numbers. We will introduce only the more elementary facts here and continue this work in the first section of the next chapter when we discuss the Spectral Theorem. In particular, we will restrict our attention here to the case of matrices.

We begin by reviewing some properties of the complex numbers $\mathbb{C}$. A complex number z is usually written as $a + ib$, where a, b are real and $i = \sqrt{-1}$. We will use i to denote $\sqrt{-1}$, but note that j is also frequently used, particularly in certain areas of engineering. Geometrically, this complex number is identified to the point (a, b) in the plane, and the plane is called the complex plane. Just as the real numbers do, the complex numbers have the operations of addition and multiplication. Addition of complex numbers just corresponds to vector addition in the plane, as illustrated in Figure 4.1:

$$(a_1 + ib_1) + (a_2 + ib_2) = (a_1 + a_2) + i(b_1 + b_2)$$

Multiplication of complex numbers is determined by usual multiplication of real numbers and $i^2 = -1$, together with the rule that multiplication is commutative.

$$(a + ib)(c + id) = ac + iad + ibc + i^2bd = (ac - bd) + i(ad + bc)$$

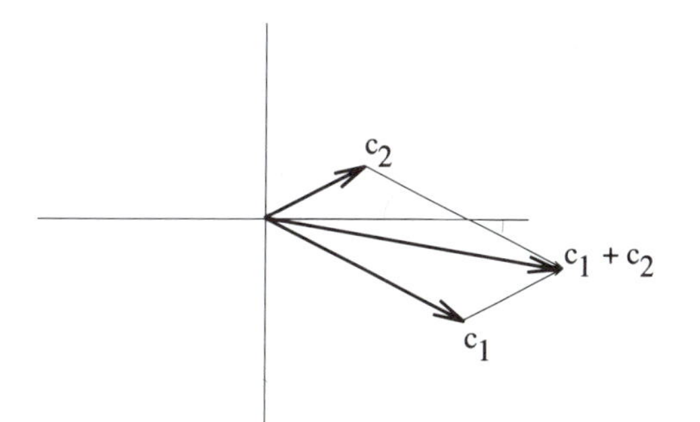

Figure 4.1. Complex addition

From this form, it is not easy to see how this multiplication is represented geometrically. To do this we first change to the polar coordinate representation of the complex number. Recall that for vectors in the plane, the vector (a, b) has polar coordinates (r, θ), where $r^2 = a^2 + b^2$ and $a = r\cos\theta, b = r\sin\theta$. We correspondingly identify the complex number

$$a + ib = r\cos\theta + ir\sin\theta$$

Then when we multiply $a_1 + ib_1$ with $a_2 + ib_2$, we get

$$\begin{aligned}
&(r_1\cos\theta_1 + ir_1\sin\theta_1)(r_2\cos\theta_2 + ir_2\sin\theta_2)\\
&= (r_1r_2)[(\cos\theta_1\cos\theta_2 - \sin\theta_1\sin\theta_2) + i(\sin\theta_1\cos\theta_2 + \cos\theta_1\sin\theta_2)]\\
&= r_1r_2[\cos(\theta_1 + \theta_2) + i\sin(\theta_1 + \theta_2)]
\end{aligned}$$

Thus the distances r_1, r_2 from the origin multiply but the two angles θ_1, θ_2 add. Figure 4.2 illustrates complex multiplication.

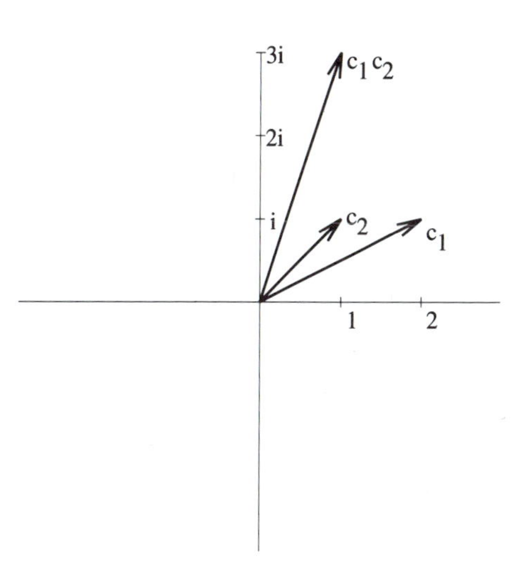

Figure 4.2. Complex multiplication

This last fact is better explained by introducing exponential notation. We denote by $e^{i\theta}$ the unit complex number $\cos\theta + i\sin\theta$. This formula is known as de Moivre's formula. Then our formula for multiplication becomes

$$(r_1 e^{i\theta_1})(r_2 e^{i\theta_2}) = r_1 r_2 e^{i(\theta_1+\theta_2)}$$

This notation is not only convenient but is tied into properties of the exponential and trigonometric functions in terms of power series. Alternatively, we could derive this from solutions of differential equations. From the power series viewpoint, we write the series which converges to e^z as

$$e^z = 1 + z + z^2/2 + \cdots + z^n/n! + \cdots$$

This works for complex numbers z for the same reasons as for real numbers. Recall that the series for $\sin\theta$ and $\cos\theta$ are given by

$$\begin{aligned}\cos\theta &= 1 - \theta^2/2 + \theta^4/4! + \cdots + (-1)^n\theta^{2n}/(2n)! + \cdots \\ \sin\theta &= \theta - \theta^3/3! + \cdots + (-1)^n\theta^{2n+1}/(2n+1)! + \cdots\end{aligned}$$

Then when we substitute $i\theta$ for z into the formula for e^z, we get

$$\begin{aligned}e^{i\theta} &= 1 + i\theta - \theta^2/2! - i\theta^3/3! + \theta^4/4! + \cdots \\ &= (1 - \theta^2/2! + \theta^4/4! - \theta^6/6! + \cdots) + i(\theta - \theta^3/3! + \theta^5/5! - \cdots) \\ &= \cos\theta + i\sin\theta\end{aligned}$$

This calculation has used the facts that $i^{4k+p} = i^p$ and $i^2 = -1, i^3 = -i$. With this notation, we write $e^{a+ib} = e^a e^{ib} = e^a\cos b + ie^a\sin b$. When dealing with functions, we write

$$e^{(a+ib)t} = e^{at}(\cos bt + i\sin bt)$$

which will be a fundamental formula for us when discussing differential equations. We will also use the alternate notation $\exp(z)$ in place of e^z.

Returning to complex arithmetic, we can use the exponential notation to note that the inverse of the nonzero complex number $re^{i\theta}, r \neq 0$, is $(1/r)e^{-i\theta}$. This follows since $r(1/r) = 1$ and $e^{i\theta}e^{-i\theta} = e^{i(\theta-\theta)} = e^0 = 1$. In ordinary notation this becomes

$$(a+ib)^{-1} = \frac{a-ib}{a^2+b^2}$$

The complex numbers $\mathbb{C}$ form a **field.** A set F with two operations, addition and multiplication, is a field if it satisfies the following three properties.

1. F is an abelian group (with identity denoted 0) under the operation of addition.
2. $(F\backslash\{0\})$ forms an abelian group (with identity denoted 1) under the operation of multiplication.
3. There is a distributive law connecting the two operations:

$$z(w_1 + w_2) = zw_1 + zw_2$$

Here we have denoted multiplication by juxtaposition and addition by $+$.

Exercise 4.3.1.

Verify that the complex numbers satisfy the three properties of a field with additive identity $0 = 0 + i0$ and multiplicative identity $1 = 1 + i0$.

Of course, the real numbers also satisfy these properties and so are a field. The advantage which the complex numbers have over the real numbers is that the field of complex numbers is **algebraically closed.** This means that whenever there is a polynomial equation

$$p(z) = z^n + a_{n-1}z^{n-1} + \cdots + a_1 z + a_0 = 0$$

with the a_i being complex numbers, then there will be n complex roots to this equation (counting multiplicities). Alternatively, we may state this by saying that $p(z)$ factors completely over the complex numbers:

$$p(z) = (z - c_1)\cdots(z - c_n)$$

This is not true for the real numbers, with the classic counterexample being $p(x) = x^2 + 1$, which has no real roots. This deficiency is made up for by adding the square root of -1 to the real numbers to get the complex numbers. The amazing fact is that by adjoining the solution to this one equation, we are now able to solve any polynomial equation, even when the coefficients are complex.

The complex numbers have another operation which is quite important, that of taking the complex conjugate, which was introduced briefly in Section 4.4.1. Recall that we defined there the complex conjugate $\overline{z} = a - ib$, when $z = a + ib$. When we use exponential notation this becomes $\overline{re^{i\theta}} = re^{-i\theta}$. We leave it as an exercise to check that this operation is consistent with addition and multiplication:

$$\overline{z + w} = \overline{z} + \overline{w}, \ \overline{zw} = \overline{z}\,\overline{w}$$

Now the formula for the inverse of $z \neq 0$ becomes

$$(1/z) = \frac{\overline{z}}{\overline{z}z}$$

Exercise 4.3.2.

Verify

$$\overline{z + w} = \overline{z} + \overline{w}, \ \overline{zw} = \overline{z}\,\overline{w}$$

The complex numbers $\mathbb{C}$ may also be regarded as a real vector space. The addition is as in $\mathbb{C}$ and the scalar multiplication is just given by complex multiplication, regarding a real number (the scalar) as a special complex number $r = r + 0i$ when multiplying: $r(a+ib) = ra + irb$. As a real vector space, $\mathbb{C}$ is isomorphic to $\mathbb{R}^2$, with the isomorphism $I : \mathbb{R}^2 \to \mathbb{C}$ being $I(a,b) = a + ib$.

It is more useful to think of $\mathbb{C}$ as a complex vector space. The definition of a complex vector space is exactly the same as given for a real vector space in Section 2.1, except that we replace the real numbers by the complex numbers as the scalars.

DEFINITION 4.3.1. A set V which has an operation $+$ so that the set forms an abelian group under addition and an operation of multiplying an element $\mathbf{v} \in V$ by a complex number c to get another element $c\mathbf{v} \in V$ satisfying the associative, distributive, and identity properties (2.1.7)–(2.1.10) is called a **complex vector space** or a vector space over the complex numbers. An element $\mathbf{v} \in V$ is called a vector and the complex number c is called a scalar, and this operation is called scalar multiplication as before.

More generally, we may discuss vector spaces over any field. We will just look at the cases of the field being $\mathbb{R}$ and $\mathbb{C}$ in this book. Just as $\mathbb{R}^n$ forms a real vector space, the n-tuples $(z_1, \ldots, z_n)$ of complex numbers form a complex vector space $\mathbb{C}^n$. The operations are just coordinatewise addition and multiplication:

$$(z_1, \ldots, z_n) + (w_1, \ldots, w_n) = (z_1 + w_1, \ldots, z_n + w_n), c(z_1, \ldots, z_n) = (cz_1, \ldots, cz_n)$$

Exercise 4.3.3.

Check that $\mathbb{C}^n$ forms a complex vector space.

We can form matrices with complex entries. The theory of solving linear equations via Gaussian elimination and finding inverses and determinants parallels exactly the development in Chapter 1 with no significant changes. However, there is a difference in the role of the transpose, and to explain this we need to discuss the appropriate inner product in $\mathbb{C}^n$ which generalizes the usual dot product in $\mathbb{R}^n$. First note that we can extend the notion of conjugation in $\mathbb{C}$ to conjugation in $\mathbb{C}^n$ by just conjugating each entry:

$$\overline{\mathbf{z}} = \overline{(z_1, \ldots, z_n)} = (\overline{z_1}, \ldots, \overline{z_n})$$

We can also extend conjugation to matrices with entries in $\mathbb{C}$ by just conjugating each entry:

$$\overline{A} = (b_{ij}), \qquad \text{where } b_{ij} = \overline{a_{ij}}$$

The operations of conjugation on matrices and vectors are consistent:

$$\overline{A\mathbf{v}} = \overline{A}\overline{\mathbf{v}}$$

Exercise 4.3.4.

Verify that $\overline{A\mathbf{v}} = \overline{A}\overline{\mathbf{v}}$.

Another important relationship concerns that of eigenvalues and eigenvectors. Suppose that A is a complex matrix with eigenvalue λ and eigenvector $\mathbf{v}$; that is, $A\mathbf{v} = \lambda\mathbf{v}$. Then when we take conjugates, we get

$$\overline{A}\overline{\mathbf{v}} = \overline{A\mathbf{v}} = \overline{\lambda\mathbf{v}} = \overline{\lambda}\overline{\mathbf{v}}$$

Thus the eigenvalues of $\overline{A}$ are the conjugates of those for A and the corresponding eigenvectors are also the conjugates. The correspondence of sending $\mathbf{v}$ to $\overline{\mathbf{v}}$ provides an invertible map C between the eigenspace of A associated to the eigenvalue λ and the eigenspace for $\overline{A}$ for the eigenvalue $\overline{\lambda}$. This map is not complex linear but is what is called conjugate linear:

$$C(c\mathbf{v}) = \overline{c}C(\mathbf{v}), C(\mathbf{v}+\mathbf{w}) = C(\mathbf{v}) + C(\mathbf{w})$$

This property is sufficient to imply that the two eigenspaces $E_A(\lambda)$ and $E_{\overline{A}}(\overline{\lambda})$ have the same dimension. Thus the two eigenvalues will have the same geometric dimension. That they have the same algebraic dimension comes from the fact that the characteristic polynomial of one is the complex conjugate of the characteristic polynomial of the other. When A is a real matrix, this becomes the fact that the eigenvalues and eigenvectors of A occur in complex conjugate pairs since $A = \overline{A}$ then, and these eigenvalues have the same algebraic and geometric multiplicities. We state this as a proposition.

PROPOSITION 4.3.1. (1) If $\mathbf{v}$ is an eigenvector of A for the eigenvalue λ, then $\overline{\mathbf{v}}$ is an eigenvector for $\overline{A}$ with eigenvalue $\overline{\lambda}$. The geometric multiplicity (resp., algebraic multiplicity) of λ as an eigenvalue of A is equal to the geometric multiplicity (resp., algebraic multiplicity) of $\overline{\lambda}$ as an eigenvalue for $\overline{A}$. The complex conjugates of eigenvectors of A are eigenvectors of $\overline{A}$. (2) If A is a real matrix, then the nonreal complex eigenvalues and eigenvectors of A occur in complex conjugate pairs. Specifically, if $\mathbf{v}$ is an eigenvector of A with eigenvalue λ, then $\overline{\mathbf{v}}$ is an eigenvector for A with eigenvalue $\overline{\lambda}$.

For a real number r its length squared is given by r^2. But if $z = a+ib = r\exp(i\theta)$ is a complex number, its length squared (determined by identifying it with the point (a, b) in the plane) is not given by $z^2 = r^2\exp(2i\theta)$, but by $\overline{z}z = a^2 + b^2 = r^2$. Note that the real numbers are characterized as a subset of the complex numbers by the condition that $\overline{z} = z$. When we take a point in $\mathbb{C}^n$ and identify it with an element of $\mathbb{R}^{2n}$ and use the usual length in $\mathbb{R}^{2n}$, then the length squared is given by the formula $\overline{z_1}z_1 + \cdots + \overline{z_n}z_n$. For $\mathbb{R}^n$ the length is related to the inner product by $\|\mathbf{v}\|^2 = \langle\mathbf{v},\mathbf{v}\rangle$. We would like to generalize the inner product on $\mathbb{R}^n$ to an inner product on $\mathbb{C}^n$ so that it agrees with the old inner product when all the entries of the vector are real and $\langle \mathbf{v}, \mathbf{v} \rangle$ still gives the length squared. The definition which does this is

$$\langle \mathbf{v}, \mathbf{w} \rangle = \overline{v_1} w_1 + \cdots + \overline{v_n} w_n, \qquad \text{where } \mathbf{v} = (v_1, \ldots, v_n), \mathbf{w} = (w_1, \ldots, w_n)$$

Note that $\langle \mathbf{v}, \mathbf{w} \rangle$ is a complex number. This inner product has the following properties:

(1) $$\langle \mathbf{v}, \mathbf{w} \rangle = \overline{\langle \mathbf{w}, \mathbf{v} \rangle}$$ **Hermitian property**

Note that when $\mathbf{v}$ and $\mathbf{w}$ are real, this is just the symmetry property of the dot product in $\mathbb{R}^n$.

(2) $$\langle \mathbf{v}, a\mathbf{w} + b\mathbf{z} \rangle = a\langle \mathbf{v}, \mathbf{w} \rangle + b\langle \mathbf{v}, \mathbf{z} \rangle$$ **linearity property**

Note that because of (1) we get what is called **conjugate linearity** in the first argument:

$$< a\mathbf{v} + b\mathbf{z}, \mathbf{w} > = \bar{a} < \mathbf{v}, \mathbf{w} > + \bar{b} < \mathbf{z}, \mathbf{w} >$$

(3) $$< \mathbf{v}, \mathbf{v} > \geq 0 \quad \text{and} \quad = 0 \text{ iff } \mathbf{v} = \mathbf{0}$$ **positive definite property**

Note that (3) says more than in the real case since $\langle \mathbf{v}, \mathbf{w} \rangle$ is in general a complex number. It is saying that not only does $\langle \mathbf{v}, \mathbf{v} \rangle$ turn out to be a real number, but it is a positive one unless $\mathbf{v} = \mathbf{0}$.

Exercise 4.3.5.

Verify that the definition of the inner product

$$\langle \mathbf{v}, \mathbf{w} \rangle = \overline{v_1} w_1 + \cdots + \overline{v_n} w_n$$

for $\mathbf{v}, \mathbf{w} \in \mathbb{C}^n$ satisfies properties (1)–(3).

Now define the length $\|\mathbf{v}\|$ of a complex vector by

$$\|\mathbf{v}\|^2 = \langle \mathbf{v}, \mathbf{v} \rangle$$

Note that for $n = 1$ this just gives the length of a complex number z to be $\sqrt{\bar{z}z}$. In these terms, our formula for the inverse of a nonzero complex number becomes

$$(1/z) = \frac{\bar{z}}{\|z\|^2}$$

The complex numbers of length 1 are identified with the unit circle in the complex plane, as shown in Figure 4.3. They can be written as $e^{i\theta}$ and so are given by the angle θ. The complex conjugate of a unit complex number $e^{i\theta}$ is the unit complex number $e^{-i\theta}$ and also lies on the unit circle. Note that it is also the inverse of this complex number.

Recall that for vectors $\mathbf{v}, \mathbf{w} \in \mathbb{R}^n$, the inner product can be rewritten in terms of the transpose by $\langle \mathbf{v}, \mathbf{w} \rangle = \mathbf{v}^t \mathbf{w}$ where we are regarding each vector as a column vector. To get a similar formula for vectors in $\mathbb{C}^n$, we generalize the definition of the transpose.

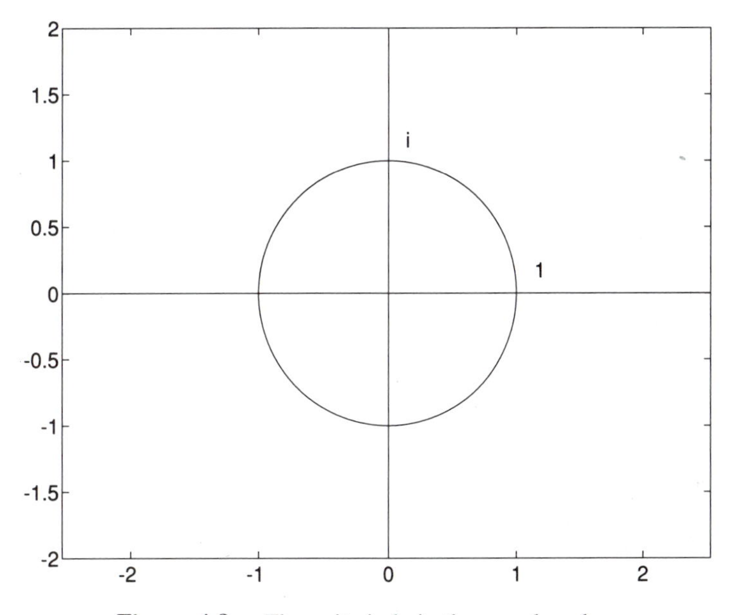

Figure 4.3. The unit circle in the complex plane

DEFINITION 4.3.2. For a complex matrix A we define the **adjoint** A^* by $A^* = (b_{ij})$ where $b_{ij} = \overline{a_{ji}}$. The adjoint is formed by first taking the transpose and then taking the complex conjugate of each entry: $A^* = \overline{A^t}$.

When the matrix has only real entries then the adjoint agrees with the transpose. The formula for the inner product in $\mathbb{C}^n$ can be written in terms of the adjoint by

$$\langle \mathbf{v}, \mathbf{w} \rangle = \mathbf{v}^* \mathbf{w}$$

Here we are regarding $\mathbf{v}, \mathbf{w}$ as column vectors.

The adjoint satisfies analogous properties to that of the transpose. One of these is that $(AB)^* = B^* A^*$. This follows from

$$(AB)^* = \overline{(AB)^t} = \overline{B^t A^t} = \overline{B^t}\,\overline{A^t} = B^* A^*$$

There is also another basic formula (sometimes used as the definition of A^*), using the inner product given for $\mathbb{C}^n$:

$$\langle A\mathbf{v}, \mathbf{w} \rangle = \langle \mathbf{v}, A^* \mathbf{w} \rangle \qquad \text{for all } \mathbf{v}, \mathbf{w} \in \mathbb{C}^n$$

To verify this formula we just write each inner product in terms of the adjoint:

$$\langle A\mathbf{v}, \mathbf{w} \rangle = (A\mathbf{v})^* \mathbf{w} = (\mathbf{v}^* A^*)\mathbf{v} = \mathbf{v}^*(A^* \mathbf{w}) = \langle \mathbf{v}, A^* \mathbf{w} \rangle$$

This generalizes the same formula with transpose instead of adjoint which played a key role in our analysis of the least squares problem in terms of the four fundamental subspaces. There is an analogous theory for complex matrices.

There are special classes of complex matrices which are analogous to symmetric, skew symmetric, and orthogonal real matrices, and reduce to these types when the entries are real. They are the Hermitian matrices: $A^* = A$; the skew Hermitian matrices: $A^* = -A$; and the unitary matrices: $A^* = A^{-1}$. The following table summarizes the relations of these matrices to the inner product in both the real and complex cases.

$\mathbb{R}^n$	*Condition*	$\mathbb{C}^n$
A is symmetric $A = A^t$	$\langle A\mathbf{x}, \mathbf{y}\rangle = \langle \mathbf{x}, A\mathbf{y}\rangle$	A is Hermitian $A = A^*$
A is skew symmetric $A = -A^t$	$\langle A\mathbf{x}, \mathbf{y}\rangle = -\langle \mathbf{x}, A\mathbf{y}\rangle$	A is skew Hermitian $A = -A^*$
A is orthogonal $A^{-1} = A^t$	$\langle A\mathbf{x}, A\mathbf{y}\rangle = \langle \mathbf{x}, \mathbf{y}\rangle$	A is unitary $A^{-1} = A^*$

Exercise 4.3.6.

Classify each of the following matrices as Hermitian, skew Hermitian, or unitary.

(a) $\begin{pmatrix} 2 & 2+i & 3-2i \\ 2-i & 3 & 4 \\ 3+2i & 4 & 1 \end{pmatrix}$

(b) $\begin{pmatrix} i & 5-i & i \\ -5-i & 0 & 2+2i \\ i & -2+2i & 3i \end{pmatrix}$

(c) $\begin{pmatrix} 1/\sqrt{2} & 0 & 1/\sqrt{2} \\ 0 & 1 & 0 \\ 1/\sqrt{2} & 0 & -1/\sqrt{2} \end{pmatrix}$

(d) $\begin{pmatrix} 2 & 0 & 1 \\ 0 & 3 & 2 \\ 1 & 2 & 3 \end{pmatrix}$

(e) $\begin{pmatrix} 1/2+1/2i & -1/2+1/2i & 0 \\ 1/2+1/2i & 1/2-1/2i & 0 \\ 0 & 0 & 1 \end{pmatrix}$

Exercise 4.3.7.

Verify that the sum of two Hermitian matrices is Hermitian. Is the sum of two skew Hermitian matrices skew Hermitian? Is the sum of two unitary matrices unitary?

Exercise 4.3.8.

Show that if A is a Hermitian matrix, then iA is a skew Hermitian matrix.

Exercise 4.3.9.

Show that $\langle A\mathbf{x}, \mathbf{y}\rangle = \langle \mathbf{x}, B\mathbf{y}\rangle$ for all $\mathbf{x}, \mathbf{y} \in \mathbb{C}^n$ iff $B = A^*$. (Hint: For the forward implication, choose $\mathbf{x} = \mathbf{e}_i, \mathbf{y} = \mathbf{e}_j$.)

4.4. COMPUTING POWERS OF MATRICES AND THEIR APPLICATIONS

Assume that A is diagonalizable; that is, $A = SDS^{-1}$. This makes it much easier to compute other quantities associated to A. For example, a power A^k is given by $(SDS^{-1})^k = SD^kS^{-1}$, and to compute D^k, it is just a matter of computing the diagonal entries of D to the kth power. We look at an example.

■ ***Example 4.4.1.*** If $A = SDS^{-1}$ and $S = \begin{pmatrix} 1 & 0 & 2 \\ 1 & 1 & 3 \\ 0 & 0 & 1 \end{pmatrix}$ with

$$S^{-1} = \begin{pmatrix} 1 & 0 & -2 \\ -1 & 1 & -1 \\ 0 & 0 & 1 \end{pmatrix}, D = \begin{pmatrix} 1 & 0 & 0 \\ 0 & 2 & 0 \\ 0 & 0 & .5 \end{pmatrix}$$

that is, $A = \begin{pmatrix} 1 & 0 & -1 \\ -1 & 2 & -2.5 \\ 0 & 0 & .5 \end{pmatrix}$, then to compute A^{20} we don't have to multiply A times itself 20 times. Rather, we compute D^{20} and then multiply by S on the left and S^{-1} on the right (we call this operation **conjugating** by S.) Here

$$D^{20} = \begin{pmatrix} 1 & 0 & 0 \\ 0 & 2^{20} & 0 \\ 0 & 0 & .5^{20} \end{pmatrix} = \begin{pmatrix} 1 & 0 & 0 \\ 0 & 1048576 & 0 \\ 0 & 0 & .00000095 \end{pmatrix}$$

and

$$A^{20} = SD^{20}S^{-1} = \begin{pmatrix} 1 & 0 & -1.99999809 \\ -1048575 & 1048576 & -1048577.99999714 \\ 0 & 0 & .00000095 \end{pmatrix}$$

Note the dominating effect of the second eigenvalue on this computation. When A^{20} is computed on a computer and is automatically scaled, the value is given as

$$10^6 \begin{pmatrix} .000001 & 0 & -.000001999999809 \\ -1.048575 & 1.048575 & -1.04857799999714 \\ 0 & 0 & .00000000000095 \end{pmatrix}$$

When higher and higher powers are taken, the effects of the eigenvalues 1 and .5 become negligible when scaled and so the computer will print out 0 except in the second row and that will become essentially $2^k\,(-1 \quad 1 \quad -1)$ (at least until the numbers become larger than the largest number the computer can handle). ■

■ ***Example 4.4.2.*** We next look at $A = \begin{pmatrix} 1 & -.45 & .9 & -.45 \\ 1 & .045 & .91 & -.945 \\ .5 & -.005 & .51 & -.495 \\ 0 & .045 & -.09 & .055 \end{pmatrix}$.

It is diagonalizable, with $A = SDS^{-1}$, where

$$S = \begin{pmatrix} -.4082 & .0000 & .5774 & .0000 \\ -.8165 & .8944 & .5774 & -.5774 \\ -.4082 & .4472 & .0000 & -.5774 \\ -.0000 & .0000 & .5774 & -.5774 \end{pmatrix}, D = \begin{pmatrix} 1 & 0 & 0 & 0 \\ 0 & .5 & 0 & 0 \\ 0 & 0 & .1 & 0 \\ 0 & 0 & 0 & .01 \end{pmatrix}$$

When we raise A to a power A^k, we can look at this as forming $A^k = (SD^k)S^{-1}$.

Now D^k will tend toward $\begin{pmatrix} 1 & 0 & 0 & 0 \\ 0 & 0 & 0 & 0 \\ 0 & 0 & 0 & 0 \\ 0 & 0 & 0 & 0 \end{pmatrix}$ as k gets larger. Then the limiting value of

$$SD^k \text{ will be} \begin{pmatrix} -.4082 & 0 & 0 & 0 \\ -.8165 & 0 & 0 & 0 \\ -.4082 & 0 & 0 & 0 \\ 0 & 0 & 0 & 0 \end{pmatrix}.$$

In the limit, the first column, which is an eigenvector for the eigenvalue 1, survives, and the other columns become **0**. When we form A^k we have to multiply this by

$$S^{-1} = \begin{pmatrix} -2.4495 & 1.2247 & -2.4495 & 1.2247 \\ -2.2361 & 2.2361 & -2.2361 & .0000 \\ .0000 & .8660 & -1.7321 & .8660 \\ -.0000 & .8660 & -1.7321 & -.8660 \end{pmatrix}$$

Since all the columns of SD^k are essentially zero except the first one for large k, the only thing for A^k which matters in the limit is the first row of S^{-1}, and so the columns of A^k will be tending toward multiples of the first column of SD^k, with the coefficients coming from the various entries in the first row of S^{-1}. The limiting value of A^k will be

$$\begin{pmatrix} 1 & -.5 & 1 & -.5 \\ 2 & -1 & 2 & -1 \\ 1 & -.5 & 1 & -.5 \\ 0 & 0 & 0 & 0 \end{pmatrix}$$

To see how fast A^k approaches this limit (rounded to five digits), we compute

$$A^5 = \begin{pmatrix} 1.0000 & -.5000 & 1.0000 & -.5000 \\ 1.9375 & -.9375 & 1.9375 & -1.0000 \\ .9687 & -.4688 & .9688 & -.5000 \\ -.0000 & .0000 & -.0000 & .0000 \end{pmatrix},$$

$$A^{10} = \begin{pmatrix} 1.0000 & -.5000 & 1.0000 & -.5000 \\ 1.9980 & -.9980 & 1.9980 & -1.0000 \\ .9990 & -.4990 & .9990 & -.5000 \\ .0000 & .0000 & -.0000 & .0000 \end{pmatrix},$$

$$A^{20} = \begin{pmatrix} 1.0000 & -.5000 & 1.0000 & -.5000 \\ 2.0000 & -1.0000 & 2.0000 & -1.0000 \\ 1.0000 & -.5000 & 1.0000 & -.5000 \\ .0000 & -.0000 & .0000 & -.0000 \end{pmatrix}$$

Of course, A^{20} is still not equal to the limiting value. We now print out some more digits; A^{20} =

$$\begin{pmatrix} 1.000000000000 & -.500000000000 & 1.000000000000 & -.500000000000 \\ 1.999998092651 & -.999998092651 & 1.999998092651 & -1.000000000000 \\ .999999046326 & -.499999046326 & .999999046326 & -.500000000000 \\ .000000000000 & -.000000000000 & .000000000000 & -.000000000000 \end{pmatrix} \blacksquare$$

To see this behavior more generally, suppose that the eigenvalues of a matrix A are $\mu_1, \ldots, \mu_n$ with eigenvectors $\mathbf{v}_1, \ldots, \mathbf{v}_n$. Then we have $A = SDS^{-1}$ as earlier and $A^k = SD^kS^{-1}$. Then SD^k will be the matrix with columns $\mu_1^k\mathbf{v}_1, \ldots, \mu_n^k\mathbf{v}_n$ and multiplying by S^{-1} will then just be taking linear combinations of these columns to find the columns of A^k. Now evaluating $A^k\mathbf{b}$ for some $\mathbf{b}$ is a matter of first finding $S^{-1}\mathbf{b}$, that is, finding $\mathbf{c} = (c_1, \ldots, c_n)$ with $S\mathbf{c} = \mathbf{b}$, and then

$$A^k\mathbf{b} = c_1\mu_1^k\mathbf{v}_1 + \cdots + c_n\mu_n^k\mathbf{v}_n$$

From the point of view of a linear transformation what this calculation is saying is that to compute L^k it is easiest to first choose a basis of eigenvectors $\mathbf{v}_i$ where $L\mathbf{v}_i = \mu_i\mathbf{v}_i$, and then $L^k\mathbf{v}_i = \mu_i^k\mathbf{v}_i$. Then express any other vector as a linear combination $\mathbf{v} = c_1\mathbf{v}_1 + \cdots + c_n\mathbf{v}_n$, and

$$L^k\mathbf{v} = c_1\mu_1^k\mathbf{v}_1 + \cdots + c_n\mu_n^k\mathbf{v}_n$$

If the eigenvalues are put in descending order in terms of their lengths, then $c_1\mu_1^k\mathbf{v}_1 + \cdots + c_n\mu_n^k\mathbf{v}_n = \mu_1^k(c_1\mathbf{v}_1 + c_2\alpha_2^k\mathbf{v}_2 + \cdots + c_n\alpha_n^k\mathbf{v}_n)$, where $\alpha_j = \mu_j/\mu_1$ for $j > 1$. If $\|\mu_1\| > \|\mu_j\|$ for $j > 1$, then α_j^k will be approaching 0, and so the first eigenvector and eigenvalue will tend to dominate the image vector for large k. In particular, if all the eigenvalues are less than 1 in absolute value, then A^k will tend to the zero matrix as k grows large. If one of the eigenvalues is greater than one in absolute value, then the value of $A^k\mathbf{b}$ will be unstable and will be approaching infinity (in the sense that its length will be approaching infinity) for certain $\mathbf{b}$. If there is precisely one eigenvalue μ, which is 1, and the others are less than 1 in absolute value, then $A^k\mathbf{b}$ will tend toward $c_1\mathbf{v}_1$ as k goes to infinity, where c_1 is the first entry of $S^{-1}\mathbf{b}$. This situation arises in applications involving Markov chains.

Exercise 4.4.1.

For the matrices A that follow, find the matrices S, D with $A = SDS^{-1}$. Compute A^{10} by the formula $SD^{10}S^{-1}$. Determine what happens to A^k as k goes to infinity. Identify a limiting value of A^k if one exists. If A^k goes to infinity, determine if there is a pattern in its divergence (that is, are the columns tending toward multiples of a single vector).

(a) $A = \begin{pmatrix} 4 & 2 & -2 \\ 3 & 4 & -3 \\ 3 & 2 & -1 \end{pmatrix}$

(b) $A = \begin{pmatrix} 1 & .5 & -.5 \\ .9 & 1 & -.9 \\ .9 & .5 & -.4 \end{pmatrix}$

(c) $A = \begin{pmatrix} .1 & 0 & 0 \\ -.4 & .1 & .4 \\ -.4 & 0 & .5 \end{pmatrix}$

4.4.1. MARKOV CHAINS

A finite **Markov chain** is a probabilistic model of a system which can be in a finite number of states $S_1, \ldots, S_n$. For example, the weather could be said to be in one of the states S_1 = fair, S_2 = cloudy, S_3 = raining. Then we assume that there is some fixed probability, which is given by a number between 0 and 1, for a transition between any one state and another state. These **transition probabilities** are put into a matrix T, where the ij-entry t_{ij} represents the probability of a transition from the jth state to the ith state in one time step. In the weather example, if we assume that the weather is classified as being in one of the three states each day, then the numbers t_{ij} describe the transition probabilities from one day to the next. These probabilities might be determined experimentally from old weather data. Of course, this would be a very simplistic model. The basic assumption in a Markov chain model is that this transition probability depends only on the current state we are in, and not on any past history. Time is measured in discrete increments (usually $0, 1, 2, \ldots$), and we have the same transition probabilities at each time step. A basic rule from probability theory allows us to use this matrix T to determine the probability that we are in a given state after one time step once we know the probabilities that we are in any state now and the transition probability matrix T. If $\mathbf{x} = (x_1, \ldots, x_n)$ represents the probabilities that we are in each of the n states now, then $T\mathbf{x}$ represents the probabilities that we are in each of the states after one transition. For example, suppose the probabilities that the weather is in one of the three states are given by $\mathbf{x} = (1/2, 1/4, 1/4)$, and the transition matrix is given by $T = \begin{pmatrix} 1/2 & 1/3 & 1/6 \\ 1/3 & 1/3 & 1/2 \\ 1/6 & 1/3 & 1/3 \end{pmatrix}$. The $(2, 1)$ entry of T being $1/3$ expresses the fact that the probability is $1/3$ that if it is fair today, then it will be cloudy tomorrow. The vector $T\mathbf{x} = (3/8, 3/8, 1/4)$ expresses the probabilities that we are in each of the three states tomorrow based on the initial probabilities of being in the various states today. For example, the first $3/8$, which gives the probability that it will be fair tomorrow, is found by

$$\begin{aligned} 3/8 = &\text{(prob. fair today } = 1/2\text{)(prob. fair tomorrow if fair today } = 1/2) \\ &+ \text{(prob. cloudy today } = 1/4\text{)(prob. fair tomorrow if cloudy today } = 1/3) \\ &+ \text{(prob. rainy today } = 1/4\text{)(prob. fair tomorrow if rainy today } = 1/6) \end{aligned}$$

The initial vector $\mathbf{x}$ is called a **probability vector.** A probability vector is characterized by having entries between 0 and 1 inclusive so that the sum of the entries is 1. These properties express the fact that the probability that we are in one of the states, no matter which one, is 1 and it is the sum of the probabilities of being in the individual states. The matrix T also has a special property, which is that its entries lie between 0 and 1 and the sum of the entries in any column is 1. Thus the columns of T are probability vectors. The jth column expresses the probabilities that we are in one of the other states if we start in S_j. The sum's being one expresses the fact that no matter what state we are in, the probability is 1 that we will be in one of the states after one transition. This last fact can be reexpressed by saying that the sum of the rows of T is a row vector each of whose entries is one. This has an important consequence for our analysis of a Markov chain. If we look at the matrix $T - I$, then the sum of the rows will be the zero row. This implies that the rows of $T - I$ are dependent, so $\det(T - I) = 0$. Hence 1 is one of the eigenvalues of T.

If $\mathbf{x}$ is a probability vector giving the probabilities of being in the various states, then $T\mathbf{x}$ will also be a probability vector which gives the probabilities of being in the various states after one transition. Then $T^2\mathbf{x} = T(T\mathbf{x})$ will give the probabilities of being in the various states after two transitions, and in general, $T^k\mathbf{x}$ gives the probabilities of being in the various states after k transitions. We will be interested in what happens in the long run, which is found by calculating the limit of $T^k\mathbf{x}$ as k goes to infinity. Our analysis will use a simplifying assumption that the transition matrix is diagonalizable. This is not necessarily the case. An analogous but more complicated analysis can be made in the general case.

When the eigenvalues (in descending order of length) are given by $\mu_1, \ldots, \mu_n$, then we saw that

$$T^k\mathbf{x} = c_1\mu_1^k\mathbf{v}_1 + \cdots + c_n\mu_n^k\mathbf{v}_n$$

where $S\mathbf{c} = \mathbf{x}$ and S is the matrix with the eigenvectors of T when T is diagonalizable. For a probability transition matrix T, we can show that the eigenvalues have to have length less than or equal to 1. Moreover, in many cases (for example, if all entries of some power T^k are positive, called a **regular Markov chain**), it can be shown that the only eigenvalue of length 1 is 1 itself and it has algebraic multiplicity 1. But then $T^k\mathbf{x}$ converges to $c_1\mathbf{v}_1$ as k goes to infinity. This will be a probability vector which is an eigenvector for eigenvalue 1. To find it, just find any eigenvector for the eigenvalue 1, and then normalize it by dividing by the sum of the coordinates. The general theory says that for a regular Markov chain, the eigenspace for the eigenvalue 1 will be one dimensional and will be the multiples of a single probability eigenvector $\mathbf{p}$. This probability vector $\mathbf{p}$ is called a **stable vector** for the Markov chain. Thus $\lim_{k\to\infty} T^k = (\mathbf{p}\cdots\mathbf{p})$. In particular this means that if $\mathbf{x}$ represents any initial probability vector, then $\lim_{k\to\infty} T^k\mathbf{x} = x_1\mathbf{p} + \cdots + x_n\mathbf{p} = (x_1 + \cdots + x_n)\mathbf{p} = \mathbf{p}$. The system tends to the same probability distribution between the states independent of the initial state it is in. This gives an easy way to predict the long-term behavior of the system. We find an eigenvector for the eigenvalue 1, normalize it by dividing by the sum of its entries so that it is a probability eigenvector $\mathbf{p}$, and note that the limiting value of $T^k\mathbf{x}$ is $\mathbf{p}$.

We will give a few examples and then comment some on the general theory.

■ ***Example 4.4.3.*** Our first example is the one given earlier for the weather. We have a transition matrix $T = \begin{pmatrix} 1/2 & 1/3 & 1/6 \\ 1/3 & 1/3 & 1/2 \\ 1/6 & 1/3 & 1/3 \end{pmatrix}$. Note that this is a regular Markov chain since all the entries are positive. This diagonalizes to $T = SDS^{-1}$, where

$$S = \begin{pmatrix} .5923 & .8090 & .3090 \\ .6516 & -.3090 & -.8090 \\ .4739 & -.5000 & .5000 \end{pmatrix}, D = \begin{pmatrix} 1 & 0 & 0 \\ 0 & .2697 & 0 \\ 0 & 0 & -.1030 \end{pmatrix}$$

The probability eigenvector for the eigenvalue 1 is found from the first column by dividing by the sum of its entries : $\mathbf{p} = (.3448, .3793, .2759)$. Thus the model predicts that the long-term probability that the weather will be sunny, no matter what it is now, is .3448. There are similar predictions for the cloudy and rainy states. To see how the system evolves to the stable case, we can compute a few powers.

$$T^5 = \begin{pmatrix} .3457 & .3447 & .3440 \\ .3790 & .3794 & .3796 \\ .2753 & .2760 & .2764 \end{pmatrix}, T^{10} = \begin{pmatrix} .3448 & .3448 & .3448 \\ .3793 & .3793 & .3793 \\ .2759 & .2759 & .2759 \end{pmatrix}$$

Thus this system evolves fairly quickly, at least up to the fourth decimal place. T^5 is fairly close to the limiting value, and T^{10} agrees to four decimal places (actually the first difference is in the sixth decimal place). The reason for this quick convergence is that the other eigenvalues .2697 and $-.1030$ are small in absolute value, and so when they are raised to higher and higher powers, they go quickly to 0. ■

■ ***Example 4.4.4.*** The next example involves a model of the progress of a student through school. Besides the four obvious states, a freshman, sophomore, junior, and senior, there are two more states being considered, that of flunking out of school and of graduating. In a given year we assume that there are four things that might happen: the student might be in the same status in the next academic year, the student might progress to the next grade, the student might flunk out, and the student might graduate. We order the states in the order flunkout, freshman, sophomore, junior, senior, graduate. The states of flunkout and graduate are assumed to be absorbing states—if one is in them, then one is in the same state next year. They will come up in the long-term behavior since the student will (in the long run) either flunk out or graduate. We assign the following transition matrix to this model.

$$T = \begin{pmatrix} 1 & .1 & .08 & .06 & .04 & 0 \\ 0 & .1 & 0 & 0 & 0 & 0 \\ 0 & .8 & .08 & 0 & 0 & 0 \\ 0 & 0 & .84 & .06 & 0 & 0 \\ 0 & 0 & 0 & .88 & .04 & 0 \\ 0 & 0 & 0 & 0 & .92 & 1 \end{pmatrix}$$

This reflects the idea that the further a student progresses, the more likely the student is to progress further and not to flunk out. The eigenspace for the eigenvalue 1 is two dimensional with basis $\mathbf{e}_1, \mathbf{e}_6$, which just corresponds to the two absorbing states. The analysis doesn't apply exactly as before, but what it would say is that in the limit we must be in some linear combination of the eigenvectors for eigenvalue 1. The particular linear combination depends on where we start. This is reflected by the matrices S, D in the diagonalization of $T = SDS^{-1}$.

$$S = \begin{pmatrix} 1.0000 & 0 & -.0301 & -.0314 & -.0329 & -.0344 \\ 0 & 0 & 0 & 0 & 0 & .0001 \\ 0 & 0 & 0 & 0 & .0008 & .0023 \\ 0 & 0 & 0 & .0162 & .0321 & .0476 \\ 0 & 0 & .7217 & .7142 & .7064 & .6981 \\ 0 & 1.0000 & -.6916 & -.6990 & -.7064 & -.7136 \end{pmatrix},$$

$$D = \begin{pmatrix} 1.0000 & 0 & 0 & 0 & 0 & 0 \\ 0 & 1.0000 & 0 & 0 & 0 & 0 \\ 0 & 0 & .0400 & 0 & 0 & 0 \\ 0 & 0 & 0 & .0600 & 0 & 0 \\ 0 & 0 & 0 & 0 & .0800 & 0 \\ 0 & 0 & 0 & 0 & 0 & .1000 \end{pmatrix}$$

We also compute the inverse of S; it will contain information in its first two rows which reflects the particular linear combination of the two states which will occur in each limiting case.

$$S^{-1} = \begin{pmatrix} 1 & .2719 & .1809 & .1028 & .0417 & 0 \\ 0 & .7281 & .8191 & .8972 & .9583 & 1 \\ 0 & -17071.7102 & 1280.3783 & -61.9704 & 1.3857 & 0 \\ 0 & 51749.6511 & -2587.4326 & 61.6055 & 0 & 0 \\ 0 & -52324.6175 & 1308.1154 & 0 & 0 & 0 \\ 0 & 17648.2735 & 0 & 0 & 0 & 0 \end{pmatrix}$$

The limit of SD^k is $\begin{pmatrix} 1 & 0 & 0 & 0 & 0 & 0 \\ 0 & 0 & 0 & 0 & 0 & 0 \\ 0 & 0 & 0 & 0 & 0 & 0 \\ 0 & 0 & 0 & 0 & 0 & 0 \\ 0 & 0 & 0 & 0 & 0 & 0 \\ 0 & 1 & 0 & 0 & 0 & 0 \end{pmatrix}$ and multiplying by S^{-1} gives

$$\begin{pmatrix} 1 & .2719 & .1809 & .1028 & .0417 & 0 \\ 0 & 0 & 0 & 0 & 0 & 0 \\ 0 & 0 & 0 & 0 & 0 & 0 \\ 0 & 0 & 0 & 0 & 0 & 0 \\ 0 & 0 & 0 & 0 & 0 & 0 \\ 0 & .7281 & .8191 & .8972 & .9583 & 1 \end{pmatrix}.$$

Note that in this multiplication only the first two rows of S^{-1} enter and they give the appropriate combinations of the two absorbing states in the limit. For example, the entries in the second column are interpreted as saying that if a student enters as a freshman, the probability is .2719 that the student will flunk out and .7281 that the student will graduate. The fourth column is interpreted as saying that for those students entering their junior year, the probability is .1028 that the student will flunk out and .8972 that the student will graduate. ■

■ ***Example 4.4.5.*** This last example illustrates another facet of Markov chains which did not occur in the previous examples. In each of them there was a limiting value for T^k. This need not be the case in general. An example where there is no limit is when $T = \begin{pmatrix} 0 & 0 & 1 \\ 1 & 0 & 0 \\ 0 & 1 & 0 \end{pmatrix}$. This might model a particularly indecisive person who is trying to decide among three choices and cycles between deciding on each of them, never coming to any ultimate decision. Then $T^2 = \begin{pmatrix} 0 & 1 & 0 \\ 0 & 0 & 1 \\ 1 & 0 & 0 \end{pmatrix}$, $T^3 = I$, and the higher powers just cycle through these three matrices. This Markov chain is a particularly simple example of a *cyclic* Markov chain. ■

Exercise 4.4.2.

Suppose there is a Markov chain where the probability matrix is $T = \begin{pmatrix} 1/2 & 0 & 1/4 \\ 0 & 1/4 & 1/2 \\ 1/2 & 3/4 & 1/4 \end{pmatrix}$. If the initial probability vector is $(1/3, 1/3, 1/3)$, what will the limiting probability vector be? What is the probability that one is in state 1 after 20 steps?

Exercise 4.4.3.

Here we consider a (simplistic) model of population changes among three countries, which we call A, B, C. Suppose that there are migration patterns so that 90% of the population from country A in a given year remains there and the remaining 10% migrate to country B. From country B, 95% remain there and 5% migrate to country A. From country C, 70% remain during the next year, 20% migrate to country A, and 10% migrate to country B. If the initial populations of the three countries are 10 million, 5 million, and 15 million, what will the populations be after 10 years? What will the limiting distribution of populations be?

The examples just presented illustrate some of the features of Markov chains. The full theory of Markov chains belongs to the theory of stochastic processes within probability theory. Our examples are just touching on this theory, and are illustrating how the diagonalizability of the transition matrix can be used to analyze the long-term behavior of the Markov chain. For a more complete treatment of Markov chains from a linear algebra perspective, consult [5, 1].

In the remainder of this section we want to prove a few of the simpler properties of Markov chains, some of which were stated in our previous discussion. T will denote the transition matrix, and $\mathbf{x}$ will be a probability vector. We use $\mathbf{r}$ to denote the row vector, all of whose entries are 1, $\mathbf{r} = (1 \quad 1 \quad \cdots \quad 1)$. We first note the following definitions.

DEFINITION 4.4.1. A probability vector $\mathbf{x}$ is a vector with entries between 0 and 1, so that the sum of the entries is 1. A **transition matrix** T is a square matrix, so that each of its columns is a probability vector.

DEFINITION 4.4.2. A real matrix is called **positive (resp., nonnegative)** iff all its entries are positive (resp., nonnegative). A real vector is called **positive (resp., nonnegative)** iff all its entries are positive (resp., nonnegative).

Here are some basic properties of probability vectors and transition matrices:

1. The product $T\mathbf{x}$ of a transition matrix and a probability vector is a probability vector. This follows easily from the next two statements, which are easy consequences of the rules of matrix multiplication and the definitions.
2. $\mathbf{x}$ is a probability vector iff $\mathbf{x}$ is nonnegative and $\mathbf{rx} = 1$.
3. T is a transition matrix iff T is nonnegative and $\mathbf{r}T = \mathbf{r}$.

Exercise 4.4.4.

Fill in the details to verify these three statements.

The last statement can be restated as $\mathbf{r}(T - I) = \mathbf{0}$. We say that the row vector $\mathbf{r}$ is in the left null space of $T - I$. This is equivalent to $\mathbf{r}^t$ being in the null space of $(T^t - I)$, and hence $\det(T - I) = \det(T - I)^t = 0$. Thus we get

PROPOSITION 4.4.1. A transition matrix for a Markov chain always has 1 as an eigenvalue. A left eigenvector (i.e., satisfying $\mathbf{r}T = \mathbf{r}$) is given by

$$\mathbf{r} = (1 \quad 1 \quad \ldots \quad 1)$$

all of whose entries are 1.

This proposition gives a left eigenvector for the eigenvalue 1. What we are really interested in are the (right) eigenvectors $\mathbf{x}$ with $T\mathbf{x} = \mathbf{x}$. We are particularly interested in when these might be probability vectors. We first note a bound on the eigenvalues of a transition matrix.

DEFINITION 4.4.3. The 1-norm $|\mathbf{x}|$ of a vector $\mathbf{x} = (x_1, \ldots, x_n)$ is the sum

$$\|x_1\| + \cdots + \|x_n\|$$

Note that for a probability vector $\mathbf{x}$, the 1-norm is 1.

PROPOSITION 4.4.2. If T is a transition matrix, and $\mathbf{x}$ is any vector, then

$$|T\mathbf{x}| \leq |\mathbf{x}|$$

If T is a positive matrix and $\mathbf{x}$ is a real vector, then we can have equality only if all the entries of $\mathbf{x}$ are nonnegative or all are nonpositive.

Proof.

$$\begin{aligned} |T\mathbf{x}| &= \|t_{11}x_1 + \cdots + t_{1n}x_n\| + \cdots + \|t_{n1}x_1 + \cdots + t_{nn}x_n\| \\ &\leq t_{11}\|x_1\| + \cdots + t_{1n}\|x_n\| + \cdots + t_{n1}\|x_1\| + \cdots + t_{nn}\|x_n\| \\ &= (t_{11} + \cdots + t_{n1})\|x_1\| + \cdots + (t_{1n} + \cdots + t_{nn})\|x_n\| \\ &= \|x_1\| + \cdots + \|x_n\| = |x| \end{aligned}$$

Moreover, if all the t_{ij} are positive and the entries of $\mathbf{x}$ are real, then the only way we can have equality at the $\leq$ inequality is for all the x_i to have the same sign. ■

PROPOSITION 4.4.3. Suppose T is a positive transition matrix and $\mathbf{x}$ is a real eigenvector for the eigenvalue 1. Then the entries of $\mathbf{x}$ all have the same sign. Thus there is a probability eigenvector $\mathbf{x}$ for the eigenvalue 1.

Proof. The first statement follows from the preceding proposition since we would have $|T\mathbf{x}| = |\mathbf{x}|$. To get the probability eigenvector, just take any eigenvector and normalize by dividing by the sum of its entries. Since they are all the same sign, this sum will not be 0 and the resulting quotient will be a probability eigenvector. ■

We next show that when T is positive there is a unique probability eigenvector for the eigenvalue 1.

PROPOSITION 4.4.4. If T is a positive transition matrix, then the probability eigenvector for the eigenvalue 1 is unique.

Proof. Suppose $\mathbf{x}$ and $\mathbf{y}$ are probability eigenvectors for the eigenvalue 1. If they are not equal, then $\mathbf{u} = \mathbf{x} - \mathbf{y}$ will be an eigenvector for T for the eigenvalue 1 with both positive and negative values. This follows since the sum of the values is $1 - 1 = 0$. ■

PROPOSITION 4.4.5. If T is a positive transition matrix, then the real eigenspace $E(1)$ for the eigenvalue 1 is one-dimensional.

Proof. We know there is a probability eigenvector $\mathbf{x}$ for the eigenvalue 1. Suppose there were another real eigenvector $\mathbf{y}$. Since all of the entries of $\mathbf{y}$ must be of the same sign, we can normalize $\mathbf{y}$ by dividing by the sum of the entries. This will be a probability eigenvector and hence is $\mathbf{x}$ by the uniqueness of probability eigenvectors for the eigenvalue 1. Thus $\mathbf{y}$ is a multiple of $\mathbf{x}$. ■

These results on the eigenspace for the eigenvalue 1 can be generalized to regular matrices. Recall that a transition matrix is called *regular* if there is a number k so that T^k is positive. Note that the eigenvalues of T^k are of the form μ^k, where μ is an eigenvalue of T. Also, any eigenvector for T for the eigenvalue μ will be an eigenvector for T^k for the eigenvalue $\mu^k : T\mathbf{x} = \mu\mathbf{x} \Rightarrow T^k\mathbf{x} = \mu^k\mathbf{x}$. Since any eigenvector $\mathbf{x}$ for T for the eigenvalue 1 will be an eigenvector for T^k for the eigenvalue 1, we see that T^k is positive implies that every eigenvector for T for the eigenvalue 1 must have the same properties that the eigenvectors of a positive transition matrix for the eigenvalue 1 has. Thus every eigenvector for a regular transition matrix for the eigenvalue 1 must be a multiple of a unique probability vector $\mathbf{p}$. In particular, the eigenspace is one-dimensional. More is true, however, for there can be no eigenvalues different from 1 whose length is 1. We won't give a proof here—the interested reader can find the proof in [1].

PROPOSITION 4.4.6. If T is a positive transition matrix and c is an eigenvalue for T, then $\|c\| \leq 1$.

Proof. Let $\mathbf{x}$ be an eigenvector for the eigenvalue c. Then $|T\mathbf{x}| = |c\mathbf{x}| = \|c\||x| \leq |x|$ implies $\|c\| \leq 1$. ■

This last proposition means that the length of the largest eigenvalue is 1. Thus we have seen that the largest length of an eigenvalue is 1, and the only possibility is the eigenvalue 1 itself, which has a one-dimensional eigenspace which consists of multiples of a single probability eigenvector.

THEOREM 4.4.7. Suppose that T is a transition matrix for a regular Markov chain and T is diagonalizable. Then there is a unique positive probability eigenvector $\mathbf{p}$ for the eigenvalue 1 and the $\lim_{k\to\infty} T^k$ is a transition matrix, all of whose columns are $\mathbf{p}$.

Exercise 4.4.5.

Prove Theorem 4.4.7.

Theorem 4.4.7 holds without the diagonalizability hypothesis.

4.4.2. DIFFERENCE EQUATIONS AND RECURRENCE RELATIONS

Finding the limiting value for a Markov chain is just a special case of looking at the limit of A^k as k goes to infinity. In this section we want to discuss some facets of this limiting process. We will assume that there is some initial vector $\mathbf{x}_0$. We then want to study the behavior of the sequence of vectors $\mathbf{x}_0, \mathbf{x}_1, \mathbf{x}_2, \ldots$, where $\mathbf{x}_{i+1} = A\mathbf{x}_i$. This just means that $\mathbf{x}_i = A^k\mathbf{x}_0$. This can be shown by induction on k. When $k = 1$, the statement is just $\mathbf{x}_1 = A\mathbf{x}_0$, which is the defining condition for $\mathbf{x}_1$. If we assume that the statement is true for $k = p$, we may verify it for $k = p + 1$:

$$\mathbf{x}_{p+1} = A\mathbf{x}_p = A(A^p\mathbf{x}_0) = A^{p+1}\mathbf{x}_0$$

A physical interpretation of this sequence is that we are describing the motion of a particle in n-space (usually $n = 3$ or 4, but it may be larger if we are also describing other quantities besides position and time, such as momentum). The interpretation of the index i is that it describes a certain time step. Thus the sequence of vectors gives us a sequence of snapshots at regular time intervals of the particle. An equation $\mathbf{x}_{i+1} = A\mathbf{x}_i$ for the $\mathbf{x}_i$ is called a **difference equation.** The terminology comes from one important use, which is to use certain differences to approximate derivatives. We can then give difference equations as discrete models for differential equations, which can then be solved using a computer.

In this section we will use difference equations to study recurrence relations. We want to elaborate somewhat on our discussion earlier about the limiting value of A^k. We assume that $A = SDS^{-1}$ is diagonalizable and that the eigenvalues $\mu_1, \cdots, \mu_n$ of A are ordered so that $\|\mu_1\| \geq \|\mu_2\| \geq \cdots \geq \|\mu_n\|$. We saw earlier that if $\|\mu_1\| < 1$, then $A^k \to \mathbf{0}$, and hence $\lim_{k\to\infty} \mathbf{x}_k = \mathbf{0}$. The case when $\mu_1 = 1, \|\mu_2\| < 1$ was analyzed in the case of Markov chains earlier. In this case, we let $\mathbf{c} = S^{-1}\mathbf{x}_0$. Then

$$\mathbf{x}_k = SD^kS^{-1}\mathbf{x}_0 = SD^k\mathbf{c} = c_1\mathbf{v}_1 + c_2\mu_2^k\mathbf{v}_2 + \cdots + c_n\mu_n^k\mathbf{v}_n$$

Here $\mathbf{v}_1, \ldots, \mathbf{v}_n$ are the eigenvectors which make up the columns of S. When we take the limit, we get $c_1\mathbf{v}_1$. Thus we only need to know an eigenvector for the eigenvalue 1, and the corresponding vector $\mathbf{c}$ coming from the initial vector $\mathbf{x}_0$. If there are multiple eigenvectors for the eigenvalue 1, then the situation is somewhat more complicated, but there still will be a limit. If there are other eigenvalues of absolute value 1, then the analysis becomes more complicated still and there may no longer be a limit. New phenomena can occur such as periodicity where the vectors $\mathbf{x}_i$ tend to cycle between a number of different vectors. This may happen if there is an eigenvalue which is a pth root of 1. A simple example would be a 2 by 2 matrix where the eigenvalues are ± 1, such as $A = \begin{pmatrix} 0 & 1 \\ 1 & 0 \end{pmatrix}$. Here the $\mathbf{x}_i$ just alternate the two coordinates of the initial vector.

Suppose that $\mathbf{x}_i = (y_{i1}, y_{i2}, \ldots, y_{in})$ and we have $\|\mu_1\| > \|\mu_2\|$. We can form $\alpha_i = \mu_i/\mu_1$, and write $\mathbf{x}_k = \mu_1^k(c_1\mathbf{v}_1 + c_2\alpha_2^k\mathbf{v}_2 + \cdots + c_n\alpha_n^k\mathbf{v}_n)$. If $|\mu_1| > 1$, there can be no limiting value unless $c_1 = 0$. Then the first term will dominate the others and go off to infinity (at least in absolute value). Even in this case, however, we can still say something about the relative values of the coordinates in the limit. These relative values will be the same after we remove the common factor μ_1^k and the constant c_1 (assuming it is not 0). Thus the relative terms will be tending to the same ones as in the eigenvector $\mathbf{v}_1$. Let us look at some examples.

■ ***Example 4.4.6.*** Let $A = \begin{pmatrix} .35 & .15 \\ .15 & .35 \end{pmatrix}$ and $\mathbf{x}_0 = (1, 0)$. Then

$$S = \begin{pmatrix} 1 & -1 \\ 1 & 1 \end{pmatrix}, D = \begin{pmatrix} .5 & 0 \\ 0 & .2 \end{pmatrix}, S^{-1} = \begin{pmatrix} .5 & .5 \\ -.5 & .5 \end{pmatrix}, \mathbf{c} = S^{-1}\mathbf{x}_0 = \begin{pmatrix} .5 \\ -.5 \end{pmatrix}$$

Thus $\lim_{k\to\infty} \mathbf{x}_k = \mathbf{0}$. However, the manner in which this convergence to zero occurs is not random. For

$$\mathbf{x}_k = (.8)^k[(.5, .5) + (.25)^k(.5, -.5)]$$

Thus these vectors are multiples of $(.5, .5) + (.25)^4(.5, -.5)$ and so will be tending to lie closer and closer to the line through $(.5, .5)$. Thus the convergence tends to be along this line.

What has happened for this initial value occurs for most initial values, but not for all of them. Suppose the initial value is $\mathbf{x}_0 = (-1, 1)$; that is, it is an eigenvector for the eigenvalue .2. Then $\mathbf{c} = (0, 1)$ and $\mathbf{x}_k = (.2)^k(-1, 1)$. Thus in this case $\mathbf{x}_k$ converges to $\mathbf{0}$ by moving along the line through $(-1, 1)$. ■

■ ***Example 4.4.7.*** Here we use the same S as earlier and initial vector $\mathbf{x}_0 = (1, 0)$, but the eigenvalues are $1, .1$. The corresponding matrix $A = \begin{pmatrix} .55 & .45 \\ .45 & .55 \end{pmatrix}$. Here

$$\mathbf{x}_k = (.5, .5) + (.1)^k(.5, -.5)$$

Thus the convergence is to $(.5,.5)$ but occurs along the line $(.5,.5) + t(.5,-.5)$. If we were to use initial value $(1,-1)$, then $\mathbf{x}_k = (.1)^k(1,-1)$, and so we would get convergence to $\mathbf{0}$ along this line. ∎

∎ ***Example 4.4.8.*** Here we use the same S and initial vector $(1,0)$ as in the last two examples, but the eigenvalues are now $2,1$. The corresponding matrix is $A = \begin{pmatrix} 1.5 & .5 \\ .5 & 1.5 \end{pmatrix}$. Then

$$\mathbf{x}_k = 2^k[(.5,.5) + 2^{-k}(.5,-.5)]$$

Although this goes off to an infinite limit, it is tending to lie along the line through $(.5,.5)$ and so the ratio of the two coordinates will be tending to 1. ∎

Exercise 4.4.6.

Find the limiting ratio of the first and second coordinates for the difference equation

$$\mathbf{x}_{i+1} = \begin{pmatrix} 1 & 3 \\ 2 & 1 \end{pmatrix}\mathbf{x}_i, \quad \mathbf{x}_0 = \begin{pmatrix} 1 \\ 2 \end{pmatrix}$$

Exercise 4.4.7.

Show that there is no limiting value for $\mathbf{x}_k$ for the difference equation

$$\mathbf{x}_{i+1} = \begin{pmatrix} 0 & 1 \\ 1 & 0 \end{pmatrix}\mathbf{x}_i, \quad \mathbf{x}_0 = (1,0)$$

but that there is a limit if the initial vector $\mathbf{x}_0$ is $(1,1)$.

Exercise 4.4.8.

Show that $\lim_{i\to\infty} \mathbf{x}_i/\|\mathbf{x}_i\|$ exists for the following difference equation and find it:

$$\mathbf{x}_{i+1} = \begin{pmatrix} -1 & -1.5 & 1.5 \\ 1 & 1.5 & -.5 \\ -2 & -2 & 3 \end{pmatrix}\mathbf{x}_i, \quad \mathbf{x}_0 = (1,2,1)$$

Verify that this is an eigenvector for the largest eigenvalue.

We now look at a recursion formula relating elements of a sequence of real numbers and show how information about this sequence can be found by reformulating the problem in terms of a difference equation. We start with a famous example.

■ ***Example 4.4.8.*** The Fibonacci numbers are

$$1, 1, 2, 3, 5, 8, 13, \ldots$$

They are formed from $y_0 = 1, y_1 = 1$ by using the recursion formula

$$y_{i+2} = y_{i+1} + y_i$$

This completely determines all elements of the sequence by generating the next one from the sum of the last two. Their study goes back to the ancient Greek philosophers and they turn up in a wide variety of applications, particularly in the natural sciences. We are interested here in determining the limit of their ratios y_{i+1}/y_i. This limit was known to the Greeks as the **golden ratio** and was said to be the ratio of the sides of a rectangle which is most pleasing to the eye. We reformulate the problem of finding this limit in terms of a difference equation. We first form a vector $\mathbf{x}_i = (y_{i+1}, y_i)$. The initial vector $\mathbf{x}_0 = (1, 1)$ and the first few vectors are

$$(1, 1), (2, 1), (3, 2), (5, 3), (8, 5), (13, 8)$$

These vectors contain the same information as the original sequence, but in a form where we can use our methods to determine the limiting ratio. We next rewrite our recursion formula $y_{i+2} = y_{i+1} + y_i$ in terms of these vectors.

$$\mathbf{x}_{i+1} = \begin{pmatrix} y_{i+2} \\ y_{i+1} \end{pmatrix} = \begin{pmatrix} y_{i+1} + y_i \\ y_{i+1} \end{pmatrix} = \begin{pmatrix} 1 & 1 \\ 1 & 0 \end{pmatrix} \begin{pmatrix} y_{i+1} \\ y_i \end{pmatrix} = A\mathbf{x}_i$$

where $A = \begin{pmatrix} 1 & 1 \\ 1 & 0 \end{pmatrix}$. The matrix $A = SDS^{-1}$ with

$$S = \begin{pmatrix} \frac{1+\sqrt{5}}{2} & -1 \\ 1 & \frac{1+\sqrt{5}}{2} \end{pmatrix}, \; D = \begin{pmatrix} \frac{1+\sqrt{5}}{2} & 0 \\ 0 & \frac{1-\sqrt{5}}{2} \end{pmatrix}$$

The decimal approximations to these eigenvalues are $1.6180, -.6180$. Since the coefficient c_1 in $\mathbf{c} = S^{-1}\mathbf{x}_0$ is not zero, in the limit the ratio of the first entry to the second will be that of the eigenvector for the largest eigenvalue. Thus we get

$$\lim_{k \to \infty} y_{i+1}/y_i = \frac{1 + \sqrt{5}}{2}$$

Note that the limiting ratio in this example is the largest eigenvalue. This will be true whenever a limiting ratio exists. For we can find it by choosing any initial vector with $c_1 \neq 0$. If we choose the eigenvector $\mathbf{v}_1$ as the initial vector, then multiplication by A will just multiply each entry by this eigenvalue, and so the ratio in this case is the eigenvalue. ■

We can generalize the technique of the last example. The main step is going from a recursion formula telling how to generate the next element of a sequence from preceding ones to a difference equation. We will just deal with the case where there is a recursion formula of the form

$$y_{i+n} = a_1 y_{i+n-1} + a_2 y_{i+n-2} + \cdots + a_n y_i$$

To determine all the terms of the sequence y_i, we need to know the initial ones $y_{n-1}, \ldots, y_0$. We can form a vector $\mathbf{x}_0 = (y_{n-1}, \ldots, y_0)$ from these. We also form a sequence of vectors $\mathbf{x}_i = (y_{i+n-1}, \ldots, y_i)$. The recursion formula is then translated into the matrix equation $\mathbf{x}_{k+1} = A\mathbf{x}_k$ with matrix

$$A = \begin{pmatrix} a_1 & a_2 & \ldots & a_{n-1} & a_n \\ 1 & 0 & \ldots & 0 & 0 \\ 0 & 1 & \ldots & 0 & 0 \\ \vdots & \vdots & \vdots & \vdots & \vdots \\ 0 & 0 & \ldots & 1 & 0 \end{pmatrix}$$

By analyzing the matrix equation we can determine what is happening to the original sequence, and in particular can answer questions about limiting ratios.

■ ***Example 4.4.9.*** Consider the analog of the Fibonacci sequence, where we start off with three numbers 1,1,1 and then each number is the sum of the preceding three numbers. Thus this sequence starts off

$$1, 1, 1, 3, 5, 9, 17, 31, \ldots$$

This translates into a matrix equation $x_{k+1} = A\mathbf{x}_k$, where $A = \begin{pmatrix} 1 & 1 & 1 \\ 1 & 0 & 0 \\ 0 & 1 & 0 \end{pmatrix}$. Here the eigenvalues are 1.8393 and two complex conjugate eigenvalues $-.4196 \pm .6063i$ which have length less than 1. Thus the limiting ratio will be 1.8393. ■

In Exercises 4.4.9–4.4.12 rewrite the recursion formula in a matrix difference equation and find the limiting ratio of y_{i+1}/y_i if it exists.

Exercise 4.4.9.

$1, 1, 3, 5, 11, \ldots; \ y_{i+2} = y_{i+1} + 2y_i$

Exercise 4.4.10.

$1, 2, 3, 10, 22, 51, \ldots; \ y_{i+3} = y_{i+2} + 2y_{i+1} + 3y_i$

Exercise 4.4.11.

$0, 1, 2, 3, 3, 4, 6, 9, 12, 16, 22, \ldots; \ y_{i+4} = y_{i+3} + y_i$

Exercise 4.4.12.

$0, 1, 2, 3, 3, 4, 5, 7, 8, 11, \ldots; \ y_{i+4} = y_{i+2} + y_i$

4.4.3. THE POWER METHOD FOR EIGENVALUES

We want to describe a method for computing the largest eigenvalue of a matrix A. It comes from looking at the difference equation $\mathbf{x}_{i+1} = A\mathbf{x}_i$. In the situation where there is a largest eigenvalue μ_1 with $\|\mu_1\| > \|\mu_i\|, i = 2, \ldots, n$, we have seen that as k goes to infinity, $\mathbf{x}_i$ tends toward a multiple of an eigenvector for the largest eigenvalue. To eliminate the effect of the eigenvalue itself in the limit, we can scale by always taking a unit vector. Thus we form a sequence of vectors

$$\mathbf{z}_0 = \mathbf{x}_0/\|\mathbf{x}_0\|, \mathbf{z}_1 = A\mathbf{z}_0/\|A\mathbf{z}_0\|, \ldots, \mathbf{z}_{i+1} = A\mathbf{z}_i/\|A\mathbf{z}_i\|$$

Equivalently, we could rephrase this sequence as

$$\mathbf{z}_i = \mathbf{x}_i/\|\mathbf{x}_i\|$$

Since the $\mathbf{x}_i$ will be tending to a multiple of an eigenvector for the largest eigenvalue μ_1, the $\mathbf{z}_i$ will be tending toward a unit eigenvector for this eigenvalue. Hence we can find an eigenvector for the largest eigenvalue by this means. The eigenvalue itself is then found by just comparing entries of $A\mathbf{z}_i$ with $\mathbf{z}_i$ for large i. Of course, we are only claiming that this method works when there is in fact a largest eigenvalue (of algebraic multiplicity 1) and the matrix is diagonalizable. The diagonalizability is not actually required, but the assumption on the largest eigenvalue is essential. There are also numerical issues such as the effect of round-off error which we will not discuss. Actually, round-off error can be a help in such a calculation since if we hit (initially or later) onto a vector that has coefficient 0 with respect to the eigenvector $\mathbf{v}_1$, a small round-off error may take it later to a vector where this coefficient is no longer 0 and so the method may yet work in practice when it shouldn't from a purely theoretical viewpoint. Another important point computationally is the rate of convergence. This is related to the ratio of the other eigenvalues to the largest one. Roughly, we get faster convergence as these ratios are smaller. We also point out that the method only gives information about one of the eigenvalues and corresponding eigenvector. There are more elaborate techniques to compute the other eigenvalues. We content ourselves here with one example and a few exercises to illustrate the method.

■ ***Example 4.4.10.*** Let $A = \begin{pmatrix} -2 & -3 & 3 \\ 2 & 3 & -1 \\ -4 & -4 & 6 \end{pmatrix}$. We take an initial vector $\mathbf{x}_0 = (1, 0, 0)$. Here is a short MATLAB program which computes normalized powers as described above until the difference between $\mathbf{z}_k$ and $\mathbf{z}_{k-1}$ is less than 10^{-10} together with the output. The program has a condition which stops the program if there is not sufficient convergence in 1000 steps.

```
x=[1 0 0]';
A=[-2 -3 3;2 3 -1;-4 -4 6];
clear X
X(:,1)=x;
X(:,2)=A*X(:,1)/norm(A*X(:,1));
```

```
k=2;
while (norm(X(:,k)-X(:,k-1)) > 10^(-10))&(k<1000)
X(:,k+1)=A*X(:,k)/norm(A*X(:,k));
k=k+1;
end
if k<1000
k,X(:,k)
p=1;
while (abs(X(p,k)) < .01)
p=p+1;
end
v=X(:,k);w=A*X(:,k);
maxeigval=w(p,1)/v(p,1)
else
disp(['the power method fails to converge in 1000 steps'])
break
end

k =

    35


ans =

  -0.44721359552078
   0.00000000005206
  -0.89442719098950


maxeigval =

   4.00000000000000
```

From this we see that $(.4472, 0, .8944)$ is an approximate eigenvector for the eigenvalue 4. The actual eigenvalues of this matrix are 4,2,1. ■

Exercise 4.4.13.

Use the power method to find the largest eigenvalue and a corresponding eigenvector for the matrix $A = \begin{pmatrix} 1 & 2 & 3 \\ 4 & 5 & 6 \\ 7 & 8 & 9 \end{pmatrix}$. Compare your answer with what is obtained by computing the characteristic polynomial and solving for the eigenvalues and an eigenvector for the largest eigenvalue.

Exercise 4.4.14.

Use the power method to find the largest eigenvalue of

$$\begin{pmatrix} 3 & -6 & 5 & -1 \\ -1 & 4 & 1 & 1 \\ -2 & -4 & 10 & 2 \\ -1 & -2 & 1 & 3 \end{pmatrix}$$

Exercise 4.4.15.

Use the power method to find the largest eigenvalue of

$$\begin{pmatrix} 1 & -1 & 2 & 1 & 4 \\ 0 & 2 & 2 & 1 & -1 \\ 2 & 0 & 1 & 1 & 2 \\ -3 & 1 & 2 & 3 & 0 \\ 4 & 1 & 2 & 1 & -1 \end{pmatrix}$$

and a corresponding eigenvector. Compute these eigenvalues using a regular computer program for the eigenvalues and explain your answer.

4.4.4. ROOTS AND EXPONENTIALS OF MATRICES

Another use of the diagonalization of a matrix is that it allows us to find roots and thus fractional powers of a matrix, using the corresponding roots of a number. By $A^{1/p}$ we mean a matrix B such that $B^p = A$.

We can use the description of complex multiplication in terms of polar coordinates to find roots of a complex number. Note that $(r\exp(i\theta))^n = r^n\exp(in\theta)$. Thus an nth root of $r\exp(i\theta)$ is given by $r^{1/n}\exp(i\theta/n)$. There are actually n distinct nth roots. For the number 1 has as its nth roots the n numbers $\exp(2\pi ik/n)$, where $0 \le k \le n-1$. These are n points which are equally spaced on the unit circle in the complex plane. If we multiply any nth root of a number by one of these then we will get a new nth root. Furthermore, all the nth roots occur from taking one nth root and multiplying by the n nth roots of unity. When we want to find the nth roots of a diagonalizable matrix A, we reduce the problem to one of finding the roots of a diagonal matrix D. This is a matter of finding the nth roots of each of the eigenvalues. Let us look at an example which has complex eigenvalues.

■ ***Example 4.4.11.*** $A = \begin{pmatrix} 1 & -1 \\ 1 & 1 \end{pmatrix}$. Then the eigenvalues of A are $1+i$, $1-i$, and the corresponding eigenvectors are $(i, 1)$ and $(-i, 1)$. We write $1+i = \sqrt{2}\exp(i\pi/4)$ and $1-i = \sqrt{2}\exp(-i\pi/4)$. Thus to find a 10th root of A we just have to take a 10th root of each of these eigenvalues, which gives $2^{1/20}\exp(i\pi/40)$ and $2^{1/20}\exp(-i\pi/40)$, then conjugate the diagonal matrix with these entries by $S = \begin{pmatrix} i & -i \\ 1 & 1 \end{pmatrix}$. In this case we get $B = \begin{pmatrix} 1.0321 & -.0812 \\ .0812 & 1.0321 \end{pmatrix}$. ■

Note that there are 99 other 10th roots of this matrix because we have 10 choices of each of the 10th roots of the eigenvalues. We can find 10 of these roots geometrically by noting multiplication by A represents a dilation by $\sqrt{2}$ followed by a rotation by $\pi/4$. The 10th root which we found was just the dilation by $2^{1/20}$ followed by the rotation by $\pi/40$. We could have rotated by $\pi/40 + 2k\pi/10$, $k = 0, \ldots, 9$ and thus we can find 10 of the 10th roots this way.

Once we can find roots of a matrix we can also find rational powers. To find an arbitrary power, we must know how to find exponentials since arbitrary powers are written in terms of them. We also would have to know how to find logarithms—a discussion of this would take us too far afield (of course, computationally we are only finding approximate answers anyway, and so we can express irrational powers as limits of rational powers). We do want to discuss exponentiating a square matrix, however. This is just done through the power series,

$$\exp(A) = I + A + A^2/2! + \cdots + A^n/n! + \cdots$$

although there is a little analysis involved to establish convergence. When the entries are square matrices instead of numbers, we have to be somewhat more careful. Since matrix multiplication does not commute, $\exp(A + B) \neq \exp(A)\exp(B)$ in general. However, this does happen when A and B commute with one another. We will use this for the case when A is a multiple of the identity (and thus commutes with any other matrix). When the matrix is diagonalizable, $\exp(A)$ is easily computed by computing exp of each of the eigenvalues. This uses the fact that for a diagonal matrix D, $\exp(D)$ is a diagonal matrix with the diagonal entries being $\exp(d)$ for the original diagonal entries d. For example, for the A earlier,

$$\exp(A) = S\begin{pmatrix} \exp(1+i) & 0 \\ 0 & \exp(1-i) \end{pmatrix}S^{-1} = \begin{pmatrix} 1.4687 & -2.2874 \\ 2.2874 & 1.4687 \end{pmatrix}$$

Exercise 4.4.16.

Find a cube root of the matrix $A = \begin{pmatrix} -3.5 & 4.5 \\ 4.5 & -3.5 \end{pmatrix}$.

Exercise 4.4.17.

Find *all* square roots of $A = \begin{pmatrix} 9 & -10 \\ 5 & -6 \end{pmatrix}$.

Exercise 4.4.18.

Show that the nondiagonalizable matrix $\begin{pmatrix} 0 & 1 \\ 0 & 0 \end{pmatrix}$ does not have a square root. (Hint: Suppose its square root is given by $\begin{pmatrix} a & b \\ c & d \end{pmatrix}$. Write out the equations which this gives and show that there is no solution.)

Exercise 4.4.19.

Find $\exp(A)$ for $A = \begin{pmatrix} -4 & 6 \\ -3 & 5 \end{pmatrix}$.

4.4.5. LEONTIEF INPUT–OUTPUT MODEL REVISITED

In Chapter 1 we introduced the Leontief input–output model of an economy. We assume an economy with a number of industries, each of which uses input from itself and other industries to produce a product. We form a consumption matrix C whose entries tell us how much input is required to produce a unit of output. Given a demand vector $\mathbf{d}$, we would like to form a production vector $\mathbf{p}$ so that the net output matches the demand. This leads to the equation

$$(I - C)\mathbf{p} = \mathbf{d}$$

We want to solve this equation via

$$\mathbf{p} = (I - C)^{-1}\mathbf{d}$$

As long as 1 is not an eigenvalue of C, then the matrix $I - C$ will be invertible, and so we can solve this equation. The problem is that the solution may have negative entries and so does not provide a practical answer to our problem. Thus we would like to know some sufficient conditions which guarantee that when we start with a nonnegative matrix C and a nonnegative demand vector $\mathbf{d}$, there is a solution which is also nonnegative. It is clearly necessary to have all entries of $(I - C)^{-1}$ to be nonnegative. For if one was negative, say, the ij-entry, then when the demand vector is $\mathbf{e}_j$, the production vector would have a negative ij-entry. Thus we can rephrase the problem of finding conditions on C so that $(I - C)^{-1}$ is nonnegative.

As usual we assume that C is diagonalizable. Let the eigenvalues of C be denoted $\mu_1, \ldots, \mu_n$. Then the eigenvalues of $I - C$ will be $1 - \mu_1, \ldots, 1 - \mu_n$. As long as the eigenvalues satisfy $\|\mu_i\| < 1$, we can form $(I - C)^{-1}$ in terms of the power series

$$I + C + C^2 + \cdots + C^k + \cdots$$

To see that this series in fact converges, we evaluate it on a vector $\mathbf{v}_i$ which is eigenvector for the eigenvalue μ_i. The answer is

$$(1 + \mu_i + \mu_i^2 + \cdots + \mu_i^k + \cdots)\,\mathbf{v}_i = \frac{1}{1 - \mu_i}\mathbf{v}_i$$

When we multiply by $(I - C)$, this vector is sent to back to $\mathbf{v}_i$. Thus $(I - C)^{-1}$ agrees with this series in terms of how it evaluates the basis vectors $\mathbf{v}_1, \ldots, \mathbf{v}_n$, and thus they agree for all vectors. Since this series will have nonnegative terms, we get that $(I - C)^{-1}$ is nonnegative. Thus we conclude

PROPOSITION 4.4.8. Suppose the consumption matrix C is diagonalizable and has entries $0 \leq c_{ij} < 1$ and the eigenvalues $\mu_1, \ldots, \mu_n$ of C satisfy $\|\mu_i\| < 1$. Then the entries of $(I - C)^{-1}$ will be nonnegative, and so we can find a nonnegative production vector $\mathbf{p}$ to meet any nonnegative demand vector $\mathbf{d}$.

The theory of nonnegative matrices ([1],[5]) can be used to show that the conclusion of Proposition 4.4.8 holds precisely when the largest eigenvalue, which is known to be nonnegative, is less than 1. This will hold if $\sum_i c_{ij} < 1, j = 1, \ldots, n$, using the argument in the proof of Proposition 4.4.2.

4.5. LINEAR DIFFERENTIAL EQUATIONS

In this section we discuss how to use linear algebra to solve what is called a **first order linear system** of differential equations. Our treatment will be limited to certain types of equations where the influence of the eigenvalue–eigenvector analysis of a matrix leads to the solutions of these equations. For a more complete treatment, consult a text on differential equations such as [2]. We study a vector-valued function $\mathbf{x}(t)$ defined on some interval of real numbers, which is really just n real-valued functions $x_1(t), \ldots, x_n(t)$; that is, $\mathbf{x}(t) = (x_1(t), \ldots, x_n(t))$. By a first-order linear system of differential equations, we mean an equation of the form

$$\mathbf{x}'(t) = A(t)\mathbf{x}(t) + \mathbf{b}(t)$$

Here $A(t)$ is a matrix-valued function and $\mathbf{b}(t)$ is a vector-valued function. We will restrict our interest here to an important special case when $A(t) = A$ is constant and $b(t) = \mathbf{0}$. This special case is crucial to the general case and illustrates well the use of eigenvalue–eigenvector analysis of A. Thus the equation becomes

$$\mathbf{x}'(t) = A\mathbf{x}(t)$$

When this is written out in terms of the coordinate functions $x_1(t), \ldots, x_n(t)$, then we will get a system of equations

$$x_1'(t) = a_{11}x_1(t) + \cdots + a_{1n}x_n(t)$$

$$\vdots$$

$$x_n'(t) = a_{n1}x_1(t) + \cdots + a_{nn}x_n(t)$$

As with ordinary systems of linear equations, it is more convenient to work with this in the matrix form, although after we have found the solution, we may want to interpret the results in terms of the individual coordinate functions.

In the case when $n = 1$, the system just becomes a single equation

$$x'(t) = ax(t)$$

Here the general solution is known to be

$$x(t) = c\exp(at)$$

We would like to give a similar form for the solution of the vector-valued equation, but we first need to discuss some general principles of dealing with vector-valued and matrix-valued functions.

If $p(x)$ is a polynomial with real coefficients and A is an n by n matrix, then we can form $p(A)$, which will be an n by n real matrix. More generally, if $f(x)$ is a function which is given by a power series (such as $\exp(x)$) then we can form $f(A)$ by evaluating the series by substituting A for the variable and make sense out of this, at least when we have conditions to guarantee convergence. If $g(t)$ is a function that sends a real number t to an n by n matrix $g(t)$ and f is a function that is defined on n by n matrices, then we can form the composition $f(g(t))$. The important example of this that we will use is $\exp(At)$. This is given by the power series

$$\exp(At) = I + tA + \frac{t^2A^2}{2!} + \cdots + \frac{t^kA^k}{k!} + \cdots$$

It can be shown by arguments analogous to those given to the usual exponential function of a real variable that this series converges for any square matrix. Moreover, it can be shown that the series for the derivative of the function $\exp(At)$ (when identifying the matrix image with a point in $\mathbb{R}^{n^2}$) is given by the series obtained by differentiating the given series term by term. Thus the derivative of $\exp(At)$ is the matrix $A\exp(At)$. This is a matrix-valued function of t. If $\mathbf{v}$ is any column vector in $\mathbb{R}^n$, $\mathbf{x}(t) = \exp(At)\mathbf{v}$ is a solution of the differential equation $\mathbf{x}' = A\mathbf{x}$ since

$$\mathbf{x}'(t) = [\exp(At)]'\mathbf{v} = A\exp(At)\mathbf{v} = A\mathbf{x}(t)$$

We will now explain why all solutions can be obtained in this manner. The treatment of the general n by n system $\mathbf{x}' = A\mathbf{x}$ is just modeled on the case $n = 1$ discussed earlier. Given an n by n matrix A, we have seen that the function $\exp(At)\mathbf{v}$ is a solution for any vector $\mathbf{v}$. The problem is to find a method to compute enough linearly independent functions $\exp(At)\mathbf{v}$ to form a basis for all solutions of $\mathbf{x}' = A\mathbf{x}$. We will use the following important theorem from differential equations. This is just a special case of a much more general theorem.

THEOREM 4.5.1. Existence and Uniqueness Theorem. There exists a unique solution of the initial value problem $\mathbf{x}' = A\mathbf{x}, \mathbf{x}(0) = \mathbf{x}_0$.

We can now apply some results from linear algebra to see how to find the general solution from this theorem. We first note that the solutions of $\mathbf{x}' = A\mathbf{x}$ form a vector space, which is a subspace of the vector space of all vector-valued infinitely differentiable functions $V = \mathfrak{D}(\mathbb{R}, \mathbb{R}^n)$. Let us call the subspace of solutions S. That S is a subspace can be verified by checking that it is closed under addition and scalar multiplication. If $\mathbf{x}, \mathbf{y}$ are solutions, then $\mathbf{x}' = A\mathbf{x}$ and $\mathbf{y}' = A\mathbf{y}$ implies that $(\mathbf{x}+\mathbf{y})' = \mathbf{x}'+\mathbf{y}' = A\mathbf{x} + A\mathbf{y} = A(\mathbf{x} + \mathbf{y})$, so that $\mathbf{x} + \mathbf{y}$ is a solution. Also $(c\mathbf{x})' = c\mathbf{x}' = cA\mathbf{x} = A(c\mathbf{x})$ implies that $c\mathbf{x}$ is a solution if $\mathbf{x}$ is. We claim that S is a vector space of dimension n. We prove this by giving a basis for S. We describe how to find the general basis.

THEOREM 4.5.2. Let $\mathbf{v}_1, \ldots, \mathbf{v}_n$ be a basis for $\mathbb{R}^n$. By Theorem 4.5.1 there is a solution $\mathbf{x}_i$ of $\mathbf{x}' = A\mathbf{x}$ satisfying the initial values $\mathbf{x}_i(0) = \mathbf{v}_i$. Then $\mathbf{x}_1, \ldots, \mathbf{x}_n$ is a basis for S.

Proof. We have to show that these functions are linearly independent and they span S. For linear independence suppose $c_1\mathbf{x}_1 + \cdots + c_n\mathbf{x}_n = \mathbf{0}$, where we mean the vector-valued function $\mathbf{0}$ on the right. Then evaluating both sides of this equation at $t = 0$ gives $c_1\mathbf{v}_1 + \cdots + c_n\mathbf{v}_n = \mathbf{0}$, and the linear independence of the $\mathbf{v}_i$ implies all the $c_i = 0$. To see that these span S, let $\mathbf{x}$ be an element of S and let $\mathbf{x}(0) = \mathbf{v}$. Since the $\mathbf{v}_i$ span $\mathbb{R}^n$, we can find constants c_i so that $\mathbf{v} = c_1\mathbf{v}_1+\cdots+c_n\mathbf{v}_n$. Then $c_1\mathbf{x}_1+\cdots+c_n\mathbf{x}_n$ is also a solution of $\mathbf{x}' = A\mathbf{x}$ which has the same value at $t = 0$. By the uniqueness part of the theorem we have $\mathbf{x} = c_1\mathbf{x}_1 + \cdots + c_n\mathbf{x}_n$. ■

Thus the theorem tells us that to find the general solution of $\mathbf{x}' = A\mathbf{x}$ we need to be able to find a basis $\mathbf{v}_1, \ldots, \mathbf{v}_n$ of $\mathbb{R}^n$ for which we can solve the initial value problem $\mathbf{x}' = A\mathbf{x}$, $\mathbf{x}(0) = \mathbf{v}_i$. If the matrix A is diagonalizable, this is easy to do. We will have a basis of eigenvectors $\mathbf{v}_1, \ldots, \mathbf{v}_n$ (for eigenvalues μ_i) of $\mathbb{R}^n$ (or $\mathbb{C}^n$) and $\exp(At)\mathbf{v}_i$ will be a solution satisfying $\mathbf{x}(0) = \mathbf{v}_i$. Also,

$$\exp(At) = \exp(\mu It)\exp[(A - \mu I)t]$$

implies that

$$\exp(At)\mathbf{v}_i = \exp(\mu_i t)\exp[(A - \mu_i I)t]\mathbf{v}_i = \exp(\mu_i t)\mathbf{v}_i$$

The first equality uses the fact that $\mu_i It$ and $(A - \mu_i I)t$ commute, so $\exp(At) =$

$$\exp[\mu_i It + (A - \mu_i I)t] = \exp(\mu_i It)\exp[(A - \mu_i I)t] = \exp(\mu_i t)\exp[(A - \mu_i I)t].$$

The last equality comes from the fact that $(A - \mu_i I)\mathbf{v}_i = 0$, so in the power series for $\exp[(A - \mu_i I)t]\mathbf{v}_i$ all the terms except the first one, which is $I\mathbf{v}_i = \mathbf{v}_i$, are zero. If all the eigenvalues are real numbers, we get real-valued solutions by this method. When there are complex eigenvalues, the method gives us complex-valued solutions. We will see how to get real-valued solutions from these. Thus we get

THEOREM 4.5.3. If A is diagonalizable with eigenvalues $\mu_1, \ldots, \mu_n$ and eigenvectors $\mathbf{v}_1, \ldots, \mathbf{v}_n$, then the general solution to $\mathbf{x}' = A\mathbf{x}$ is

$$\mathbf{x}(t) = c_1\mathbf{v}_1\exp(\mu_1 t) + \cdots + c_n\mathbf{v}_n\exp(\mu_n t)$$

The solution to the initial value problem $\mathbf{x}' = A\mathbf{x}, \mathbf{x}(0) = \mathbf{x}_0$ is found by solving the system $c_1\mathbf{v}_1 + \cdots + c_n\mathbf{v}_n = \mathbf{x}_0$ for $\mathbf{c}$.

Since this is such an important theorem, we give two alternate derivations. The first is to simplify the equations by changing variables. We will make the substitution

$$\mathbf{x} = S\mathbf{y}$$

where S is a matrix yet to be determined. Then the equation

$$\mathbf{x}'(t) = A\mathbf{x}(t)$$

becomes

$$S\mathbf{y}'(t) = ASy(t)$$

This is equivalent to

$$\mathbf{y}'(t) = S^{-1}ASy(t)$$

If A is diagonalizable, we may choose S so $S^{-1}AS = D = \text{diag}(\mu_1, \ldots, \mu_n)$. Here $S = (\mathbf{v}_1 \cdots \mathbf{v}_n)$ is the matrix whose columns are eigenvectors. Then this last equation becomes

$$\begin{aligned} y_1'(t) &= \mu_1 y_1(t) \\ y_2'(t) &= \mu_2 y_2(t) \\ &\vdots \\ y_n'(t) &= \mu_n y_n(t) \end{aligned}$$

These equations can then be solved independently to give

$$y_1(t) = c_1 \exp(\mu_1 t), \ldots, y_n = c_n \exp(\mu_n t)$$

Then

$$\mathbf{x}(t) = S\mathbf{y}(t) = c_1\mathbf{v}_1 \exp(\mu_1 t) + \cdots + c_n\mathbf{v}_n \exp(\mu_n t)$$

For a slightly different derivation, start out with a basis $\mathbf{v}_1, \ldots, \mathbf{v}_n$ of eigenvectors for the eigenvalues $\mu_1, \ldots, \mu_n$. Then the function $\mathbf{x}(t)$ may be written as $\mathbf{x}(t) = c_1(t)\mathbf{v}_1 + \cdots + c_n(t)\mathbf{v}_n$. When we substitute into the differential equation, we get

$$\mathbf{x}'(t) = c_1'(t)\mathbf{v}_1 + \cdots + c_n'(t)\mathbf{v}_n = A(c_1(t)\mathbf{v}_1 + \cdots + c_n(t)\mathbf{v}_n) = c_1(t)\mu_1\mathbf{v}_1 + \cdots + c_n(t)\mu_n\mathbf{v}_n$$

Since $\mathbf{v}_1, \ldots, \mathbf{v}_n$ is a basis, this implies

$$\begin{aligned} c_1'(t) &= \mu_1 c_1(t) \\ c_2'(t) &= \mu_2 c_2(t) \\ &\vdots \\ c_n'(t) &= \mu_n c_n(t) \end{aligned}$$

As in the last derivation, this leads to

$$c_1(t) = c_1 \exp(\mu_1 t), \ldots, c_n(t) = c_n \exp(\mu_n t)$$

and the general solution as before.

■ ***Example 4.5.1.*** $A = \begin{pmatrix} 13 & -4 \\ -4 & 7 \end{pmatrix}$. We have

$$S = \begin{pmatrix} 1 & 2 \\ 2 & -1 \end{pmatrix}, \; D = \begin{pmatrix} 5 & 0 \\ 0 & 15 \end{pmatrix}$$

Thus the general solution to $\mathbf{x}' = A\mathbf{x}$ is

$$\mathbf{x}(t) = c_1 \exp(5t)\begin{pmatrix} 1 \\ 2 \end{pmatrix} + c_2 \exp(15t)\begin{pmatrix} 2 \\ -1 \end{pmatrix}$$

If we are given the initial value problem $\mathbf{x}(0) = \begin{pmatrix} 1 \\ 1 \end{pmatrix}$, then we can solve $S\mathbf{c} = \begin{pmatrix} 1 \\ 1 \end{pmatrix}$ for $\mathbf{c} = \begin{pmatrix} 3/5 \\ 1/5 \end{pmatrix}$ and get the solution

$$\mathbf{x}(t) = \frac{3}{5}\exp(5t)\begin{pmatrix} 1 \\ 2 \end{pmatrix} + \frac{1}{5}\exp(15t)\begin{pmatrix} 2 \\ -1 \end{pmatrix}$$ ■

■ ***Example 4.5.2.*** $A = \begin{pmatrix} 3 & 1 & -2 \\ -2 & 2 & 2 \\ 0 & 1 & 1 \end{pmatrix}$. We have

$$S = \begin{pmatrix} 1 & 1 & 0 \\ 0 & 1 & 2 \\ 1 & 1 & 1 \end{pmatrix}, \; D = \begin{pmatrix} 1 & 0 & 0 \\ 0 & 2 & 0 \\ 0 & 0 & 3 \end{pmatrix}$$

We get the general solution

$$\mathbf{x}(t) = c_1 \exp(t)\begin{pmatrix} 1 \\ 0 \\ 1 \end{pmatrix} + c_2 \exp(2t)\begin{pmatrix} 1 \\ 1 \\ 1 \end{pmatrix} + c_3 \exp(3t)\begin{pmatrix} 0 \\ 2 \\ 1 \end{pmatrix}$$

If we have initial value $\mathbf{x}(0) = (0, 1, 0)$, then solving $S\mathbf{c} = \mathbf{x}(0)$ for $\mathbf{c} = (-1, 1, 0)$ gives the solution

$$\mathbf{x}(t) = \begin{pmatrix} -\exp(t) + \exp(2t) \\ \exp(2t) \\ -\exp(t) + \exp(2t) \end{pmatrix}$$ ■

Now suppose that A is a real n by n matrix. When we have a complex eigenvalue $a + bi$ and eigenvector $\mathbf{v} + \mathbf{w}i$, we will get the solution $(\mathbf{v} + \mathbf{w}i)\{\exp[(a + bi)t]\}$. But

$$\exp[(a + bi)t] = \exp(at)\cos(bt) + i\exp(at)\sin(bt)$$

When we multiply this out we get

$$[\mathbf{v}\exp(at)\cos(bt) - \mathbf{w}\exp(at)\sin(bt)] + i[\mathbf{v}\exp(at)\sin(bt) + \mathbf{w}\exp(at)\cos(bt)]$$

Then both the real and imaginary parts of this function will be solutions and so we get two solutions

$$\mathbf{v}\exp(at)\cos(bt) - \mathbf{w}\exp(at)\sin(bt), \mathbf{v}\exp(at)\sin(bt) + \mathbf{w}\exp(at)\cos(bt)$$

The fact that the two complex eigenvectors corresponding to complex conjugate eigenvalues are independent (since the eigenvalues are distinct) implies that the real and imaginary parts of the eigenvectors from one of them will also have to be independent. For if $\mathbf{v}$ and $\mathbf{w}$ were dependent, it would imply that we could solve for one of them in terms of the other. Then $\mathbf{w} = c\mathbf{v}$ implies $\mathbf{v} + i\mathbf{w} = (1 + ci)\mathbf{v}$ and $\mathbf{v} - i\mathbf{w} = (1 - ci)\mathbf{v}$. Since both are multiples of $\mathbf{v}$, this implies that they are dependent, contradicting the independence of $\mathbf{v} + i\mathbf{w}$ and $\mathbf{v} - i\mathbf{w}$ The independence of $\mathbf{v}, \mathbf{w}$ will then imply that the two real solutions which we obtained are independent. Thus out of one complex solution we have found two linearly independent real solutions. Of course, the other complex solution using the conjugate eigenvalue will give the same two real solutions, so we are not really getting two solutions out of one.

Note that if $a > 0$ this solution goes to infinity as t goes to infinity. If $a = 0$ then these solutions are either constant (if the eigenvalue is *0* itself) or oscillate (if the eigenvalue is pure imaginary) since they involve the trigonometric functions. If $a < 0$, then these solutions converge to 0 as t goes to infinity. Thus the long-term behavior of these solutions just depends on a.

■ ***Example 4.5.3.*** $A = \begin{pmatrix} 0 & -1 \\ 1 & 0 \end{pmatrix}$. Here $S = \begin{pmatrix} 1 & 1 \\ i & -i \end{pmatrix}$ and $D = \begin{pmatrix} i & 0 \\ 0 & -i \end{pmatrix}$. This gives the general solution $\mathbf{x}(t) = c_1 \exp(it)\begin{pmatrix} 1 \\ i \end{pmatrix} + c_2 \exp(-it)\begin{pmatrix} 1 \\ -i \end{pmatrix}$. Here the c_i are complex constants. If we had an initial value problem $\mathbf{x}(0) = (1, 2)$, then we would solve $S\mathbf{c} = \mathbf{x}(0)$ to get $\mathbf{c} = (1/2 - i, 1/2 + i)$. We could then write the solution as $\mathbf{x}(t) = \begin{bmatrix} (1/2 - i)\exp(it) + (1/2 + i)\exp(-it) \\ (1 + i/2)\exp(it) + (1 - i/2)\exp(-it) \end{bmatrix}$. Rewriting $\exp(it) = \cos t + i \sin t$ and $\exp(-it) = \cos t - i \sin t$ and multiplying gives $\mathbf{x}(t) = \begin{pmatrix} \cos t + 2\sin t \\ 2\cos t - \sin t \end{pmatrix}$.

Alternatively, we could have first decomposed our eigenvalue i as $0 + 1i$ and eigenvector $(1, i)$ as $(1, 0) + (0, 1)i$ and then used the preceding discussion to write the general solution as $\mathbf{x}(t) =$

$$c_1\left[\begin{pmatrix} 1 \\ 0 \end{pmatrix}\cos t - \begin{pmatrix} 0 \\ 1 \end{pmatrix}\sin t\right] + c_2\left[\begin{pmatrix} 1 \\ 0 \end{pmatrix}\sin t - \begin{pmatrix} 0 \\ 1 \end{pmatrix}\cos t\right] = c_1\begin{pmatrix} \cos t \\ -\sin t \end{pmatrix} + c_2\begin{pmatrix} \sin t \\ \cos t \end{pmatrix}$$

To solve the above initial value problem we set $t = 0$ and evaluate to get $\begin{pmatrix} c_1 \\ c_2 \end{pmatrix} = \begin{pmatrix} 1 \\ 2 \end{pmatrix}$, giving the solution $\mathbf{x}(t) = \begin{pmatrix} \cos t + 2\sin t \\ -\sin t + 2\cos t \end{pmatrix}$ as before. ■

■ ***Example 4.5.4.*** If A is not diagonalizable, then we can still find the general solution by the method of Theorem 4.5.2, but now we have to use the Jordan form of the matrix A to figure out how to evaluate $\exp(At)\mathbf{v}_i$ on a basis $\mathbf{v}_1, \ldots, \mathbf{v}_n$. Here is a simple example; we will discuss this further when we talk about the Jordan form. Suppose $A = \begin{pmatrix} 0 & 1 \\ 0 & 0 \end{pmatrix}$. Then $\mathbf{e}_1$ is an eigenvector for the eigenvalue 0 and so $\mathbf{e}_1 \exp(0t) = \mathbf{e}_1$ is one solution to $\mathbf{x}' = A\mathbf{x}$. To get a second solution, compute $\exp(At)\mathbf{e}_2$. This is made easy by the fact that $A^2\mathbf{e}_2 = \mathbf{0}$ so $A^k\mathbf{e}_2 = \mathbf{0}$ for $k \geq 2$. Thus

$$\exp(At)\mathbf{e}_2 = (I + At)\mathbf{e}_2 = \mathbf{e}_2 + tA\mathbf{e}_2 = \begin{pmatrix} t \\ 1 \end{pmatrix}$$

Thus the general solution is

$$\mathbf{x} = c_1 \begin{pmatrix} 1 \\ 0 \end{pmatrix} + c_2 \begin{pmatrix} t \\ 1 \end{pmatrix}$$

The point of the Jordan form is that it says that even if the matrix is not diagonalizable, it is still similar to a matrix like the one earlier, so we can find a basis $\mathbf{v}_1, \ldots, \mathbf{v}_n$ on which we can compute $\exp(At)\mathbf{v}_i$ by just adding up the first few terms in the series (after an appropriate shift) since all the later terms will vanish. ■

Exercise 4.5.1.

Find the general solution of $\mathbf{x}' = A\mathbf{x}$ for $A = \begin{pmatrix} 1 & 12 \\ 3 & 1 \end{pmatrix}$.

Exercise 4.5.2.

Solve the initial value problem $\mathbf{x}' = A\mathbf{x}$ for $A = \begin{pmatrix} -3 & 2 \\ -1 & -1 \end{pmatrix}$ with $\mathbf{x}(0) = (2, -1)$.

Exercise 4.5.3.

Solve the initial value problem $\mathbf{x}' = A\mathbf{x}$ for $A = \begin{pmatrix} 3 & 1 & -1 \\ 1 & 3 & -1 \\ 3 & 3 & -1 \end{pmatrix}$ with $\mathbf{x}(0) = (1, -2, -1)$.

4.5.1. HIGHER-ORDER EQUATIONS AND SYSTEMS

One common way that differential equations arise is not as systems but as higher-order equations for a single real-valued function $y(t)$. A linear nth-order differential equation is one of the form

$$a_n(t)y^{(n)}(t) + a_{n-1}y^{(n-1)}(t) + \cdots + a_1(t)y'(t) + a_0y(t) = b(t)$$

There are complications which arise at points where $a_n(t)$ vanishes, so we will assume for our discussion here that it is nonvanishing. This allows us to divide by it and get an equivalent equation which has $a_n(t) = 1$. Henceforth we will just assume that $a_n(t) = 1$. Moving all the terms but $y^{(n)}(t)$ to the right-hand side gives

$$y^{(n)}(t) = -a_{n-1}y^{(n-1)}(t) - \cdots - a_1(t)y'(t) - a_0y(t) + b(t)$$

This equation can be changed to a first-order system by considering the vector-valued function

$$\mathbf{x}(t) = (y(t), y'(t), \ldots, y^{(n-1)}(t))$$

When we differentiate $\mathbf{x}(t)$ we get

$$\begin{aligned}\mathbf{x}'(t) &= (y'(t), y''(t), \ldots, y^{(n-1)}(t), y^{(n)}(t)) \\ &= (y'(t), y''(t), \ldots, y^{(n-1)}(t), -a_{n-1}y^{(n-1)}(t) - \cdots - a_1(t)y'(t) - a_0y(t) + b(t)) \\ &= A(t)\mathbf{x}(t) + \mathbf{b}(t)\end{aligned}$$

where

$$A(t) = \begin{pmatrix} 0 & 1 & 0 & \ldots & 0 & 0 \\ 0 & 0 & 1 & \ldots & 0 & 0 \\ \vdots & \vdots & \vdots & \vdots & \vdots & \vdots \\ 0 & 0 & 0 & \ldots & 0 & 1 \\ -a_0(t) & -a_1(t) & -a_2(t) & \ldots & -a_{n-2}(t) & -a_{n-1}(t) \end{pmatrix}, \mathbf{b}(t) = \begin{pmatrix} 0 \\ 0 \\ \vdots \\ 0 \\ b(t) \end{pmatrix}$$

The initial values necessary for a system specify the vector-valued function $\mathbf{x}(t)$ at a point, which we have chosen to be 0. In terms of our definition of $\mathbf{x}(t)$, we have $\mathbf{x}(0) = (y(0), y'(0), \ldots, y^{(n-1)}(0))$.

We now specialize to the case where the functions $a_i(t)$ are constants and $b(t) = 0$. This is called a **constant coefficient linear nth-order homogeneous equation**. Applying the existence and uniqueness theorem for systems will imply that the corresponding system has a unique solution once we specify an initial vector $\mathbf{x}(0) = (y(0), y'(0), \ldots, y^{(n-1)}(0))$. By looking at the first component $y(t)$ of $\mathbf{x}(t)$ we conclude that this nth-order initial value problem has a unique solution.

We can deduce the form of the solution $y(t)$ from that of $\mathbf{x}(t)$, at least when the matrix A is diagonalizable. First we note what the characteristic polynomial for the matrix A is.

$$\det(A - xI) = \det\begin{pmatrix} -x & 1 & 0 & \ldots & 0 & 0 \\ 0 & -x & 1 & \ldots & 0 & 0 \\ \vdots & \vdots & \vdots & \vdots & \vdots & \vdots \\ 0 & 0 & 0 & \ldots & -x & 1 \\ -a_0 & -a_1 & -a_2 & \ldots & -a_{n-2} & -a_{n-1} - x \end{pmatrix}$$

To compute this determinant, we successively perform column operations to make all but the last diagonal entry 0. We would first add x times the last column to the column $(n-1)$, then add x times column $(n-1)$ to column $(n-2)$, and so on, finally adding x times the second column to the first. Then the first $n-1$ diagonal entries will be 0, and the entry in the $(n, 1)$ position will be $-(a_0 + a_1 x + \ldots + a_{n-1}x^{n-1} + x^n)$. To see this we work inductively on how the columns change. After one step, column $(n-1)$ is $-a_{n-2} + x(-a_{n-1} - x) = -(a_{n-2} + a_{n-1}x + x^2)$. What we prove by (finite) induction is that after k steps, the last entry of column $(n-k)$ is given by $-(a_{n-k-1} + a_{n-k}x + \cdots + a_{n-1}x^k + x^{k+1})$. Then when we do the $k+1$ step, it changes the last entry of column $n-k-1$ to

$$\begin{aligned} &-a_{n-k-2} - x(a_{n-k-1} + a_{n-k}x + \cdots + a_{n-1}x^k + x^{k+1}) \\ &= -a_{n-k-2} - a_{n-k-1}x - a_{n-k}x^2 - \cdots - a_{n-1}x^{k+1} - x^{k+2} \end{aligned}$$

Thus at the last step (which is the $n-1$ step), we have that the $(n, 1)$ entry is $-(a_0 + a_1 x + \cdots + a_{n-1}x^{n-1} + x^n)$. We now compute the determinant. We can either use expansion by cofactors, using the first row, then the second, etc., or use the permutation definition of the determinant. The permutation definition is particularly nice since in each of the first $n-1$ rows, there is only one nonzero entry, which is a 1. With this method, we get that the characteristic polynomial of A is

$$(-1)^n(x^n + a_{n-1}x^{n-1} + \cdots + a_1 x + a_0)$$

Thus up to sign, the characteristic polynomial of A is just given by taking the original equation and substituting x^k for $y^{(k)}$. Equating this to 0 gives what is called the **characteristic equation** of nth-order differential equation. It is easy to find one eigenvector for each eigenvalue λ, for when we form $A - \lambda I$, the first $n-1$ rows give equations—with variables called $(s_1, \ldots, s_n)$—

$$\begin{aligned} -\lambda s_1 + s_2 &= 0 \\ -\lambda s_2 + s_3 &= 0 \\ &\vdots \\ -\lambda s_{n-1} + s_n &= 0 \end{aligned}$$

Thus we get the solution $(1, \lambda, \lambda^2, \ldots, \lambda^{n-1})$. In the case that there are n distinct eigenvalues, the matrix A will be diagonalizable, and so we can write out the form of the general solution to the associated system $\mathbf{x}'(t) = A\mathbf{x}(t)$ as

$$\mathbf{x}(t) = c_1 e^{\lambda_1 t}\mathbf{v}_1 + \cdots + c_n e^{\lambda_n t}\mathbf{v}_n$$

For a given initial value $\mathbf{x}(0)$ the constants $\mathbf{c}$ are found by solving $S\mathbf{c} = \mathbf{x}(0)$, where S is the matrix with the corresponding eigenvectors as columns.

By taking the first coordinate function $y(t)$ we get the following theorem.

THEOREM 4.5.4. Consider the differential equation

$$y^{(n)}(t) + a_{n-1}y^{(n-1)}(t) + \cdots + a_1 y'(t) + a_0 y(t) = 0$$

If the polynomial $x^n + a_{n-1}x^{n-1} + \cdots + a_1 x + a_0$ has n distinct roots $\lambda_1, \ldots, \lambda_n$, then the general solution to this equation is

$$y(t) = c_1 e^{\lambda_1 t} + \cdots + c_n e^{\lambda_n t}$$

If we are given initial values $[y(0), y'(0), \ldots, y^{n-1}(0)] = \mathbf{v}$, then the unique solution to this initial value problem is found by solving $S\mathbf{c} = \mathbf{v}$, where S is the matrix whose jth column vector is $(1, \lambda_j, \lambda_j^2, \ldots, \lambda_j^{n-1})$.

■ ***Example 4.5.5.*** Consider the second-order differential equation

$$y''(t) + 3y'(t) + 2y(t) = 0$$

The characteristic equation is $x^2 + 3x + 2 = 0$, which has solutions $-2, -1$. Thus the general solution is

$$y(t) = c_1 e^{-2t} + c_2 e^{-t}$$

If we had initial values $y(0) = 3, y'(0) = 5$, then we would have to solve

$$\begin{pmatrix} 1 & 1 \\ -2 & -1 \end{pmatrix} \mathbf{c} = \begin{pmatrix} 3 \\ 5 \end{pmatrix}$$

to get $c_1 = -8, c_2 = 11$. Thus the solution to this initial value problem is

$$y(t) = -8e^{-2t} + 11e^{-t}$$ ■

■ ***Example 4.5.6.*** Consider the second-order differential equation

$$y''(t) + 2y'(t) + 2y(t) = 0$$

The characteristic equation is $x^2 + 2x + 2 = 0$, which has solutions $1 \pm i$. From this we can write the general (complex) solution as

$$y(t) = c_1 e^{(1+i)t} + c_2 e^{(1-i)t}$$

where c_1, c_2 are complex numbers. By writing out $e^{(1+i)t} = e^t(\cos t + i \sin t)$, the general real-valued solution is

$$y(t) = d_1 e^t \cos t + d_2 e^t \sin t$$

where d_1, d_2 are real numbers. In general, any time when $a \pm ib$ are complex conjugate roots of the characteristic equation, we can substitute $e^{at} \cos bt, e^{at} \sin bt$ for the terms $e^{(a+ib)t}, e^{(a-ib)t}$ in writing out the general solution as given in the theorem. To solve the initial value problem $y(0) = 2, y'(0) = -1$, we can substitute $t = 0$ in the preceding equation to get one equation $2 = d_1$. To get a second equation, we differentiate once to get

$$y'(t) = d_1(e^t \cos t - e^t \sin t) + d_2(e^t \sin t + e^t \cos t)$$

Substituting $t = 0$ into this equation gives $-1 = d_1 + d_2$, leading to $d_2 = -3$. Thus the solution to this initial value problem is

$$y(t) = 2e^t \cos t - 3e^t \sin t$$

We could also have found this by using the complex form $c_1e^{(1+i)t} + c_2^{(1-i)t}$ of the general solution and then solved the equation $\begin{pmatrix} 1 & 1 \\ 1+i & 1-i \end{pmatrix}\mathbf{c} = \begin{pmatrix} 2 \\ -1 \end{pmatrix}$ to get $c_1 = 1 + 1.5i, c_2 = 1 - 1.5i$. This can then be converted to the form above by expanding out the terms $e^{(1\pm i)t}$ as before. ■

■ ***Example 4.5.7.*** Consider the fourth-order equation

$$y^{(4)} - y = 0$$

The characteristic equation is $x^4 - 1 = 0$, which has solutions $\pm 1, \pm i$. Thus the general complex solution is

$$y(t) = c_1e^t + c_2e^{-t} + c_3e^{it} + c_4e^{-it}$$

The general real solution is

$$y(t) = d_1e^t + d_2e^{-t} + d_3\cos t + d_4\sin t$$

If we have an initial value problem

$$y(0) = 2, y'(0) = -3, y''(0) = 2, y'''(0) = -1$$

then we can solve this from the complex form by solving the matrix equation

$$\begin{pmatrix} 1 & 1 & 1 & 1 \\ 1 & -1 & i & -i \\ 1 & 1 & -1 & -1 \\ 1 & -1 & -i & i \end{pmatrix}\mathbf{c} = \begin{pmatrix} 2 \\ -3 \\ 2 \\ -1 \end{pmatrix}$$

to get $\mathbf{c} = (0, 2, .5i, -.5i)$. Thus the solution is

$$y(t) = 2e^{-t} + .5ie^{it} - .5ie^{-it} = 2e^{-t} - \sin t$$

To arrive at this answer from the real form of the general solution, we differentiate the general solution three times and plug in the value at $t = 0$ to get the equation

$$\begin{pmatrix} 1 & 1 & 1 & 0 \\ 1 & -1 & 0 & 1 \\ 1 & 1 & -1 & 0 \\ 1 & -1 & 0 & -1 \end{pmatrix}\mathbf{d} = \begin{pmatrix} 2 \\ -3 \\ 2 \\ -1 \end{pmatrix}$$

The solution is $\mathbf{d} = (0, 2, 0, -1)$ which then gives the answer just obtained. ■

Exercise 4.5.4.

Solve the initial value problem

$$y''(t) + y'(t) - 2y(t) = 0, \; y(0) = 1, \; y'(0) = 2$$

Exercise 4.5.5.

Solve the initial value problem

$$y'''(t) - 2y''(t) - y'(t) + 2y(t) = 0, \ y(0) = 1, \ y'(0) = 0, y''(0) = -1$$

Exercise 4.5.6.

Solve the initial value problem

$$y'''(t) - 2y''(t) + y'(t) - 2y(t) = 0, \ y(0) = 1, \ y'(0) = 0, y''(0) = -1$$

Exercise 4.5.7.

Solve the initial value problem

$$y^{(4)}(t) + 4y''(t) + 3y(t) = 0, \ y(0) = 3, \ y'(0) = -1, \ y''(0) = 0, y'''(0) = -2$$

The foregoing methods give the general solution when there are n distinct solutions of the characteristic equation. If there is a repeated root, then the matrix A will not turn out to be diagonalizable. Thus we have to find other solutions besides those of the form $e^{\lambda t}$. It can be shown that there will be m independent solutions of the form $e^{\lambda t}, te^{\lambda t}, \ldots, t^{m-1}e^{\lambda t}$ when the algebraic multiplicity of the eigenvalue λ is m and that the general solution can then be found as a linear combination of such solutions where λ ranges over the eigenvalues. For example, if the roots are $1, 1, 1, 2, 2$, for a fifth-order equation, then the general solution is

$$c_1e^t + c_2te^t + c_3t^2e^t + c_4e^{2t} + c_5te^{2t}$$

This could be justified by using the Jordan canonical form for A in the $\mathbf{x}'(t) = A\mathbf{x}(t)$ form of the equation, but there are much simpler justifications. We will look at this from another view in the next section.

4.5.2. LINEAR DIFFERENTIAL OPERATORS

We want to discuss the material in the last section from another point of view, which emphasizes other ideas of linear algebra. First we recall that $\mathfrak{D}(\mathbb{R}, \mathbb{R})$ denotes the vector space of infinitely differentiable functions from the reals to the reals. The operation of differentiation provides a linear transformation $D : \mathfrak{D}(\mathbb{R}, \mathbb{R}) \to \mathfrak{D}(\mathbb{R}, \mathbb{R})$ given by $D(y) = y'$. In this context it is common to call this a **linear operator** rather than a linear transformation. We can form other linear operators from D by taking compositions. Thus the composition $D \circ D$ of D with itself sends a function to its second derivative. We will denote this by D^2. In general we will denote by D^k the linear transformation which differentiates a function k times. That these are linear transformations follows from the fact that the composition of linear transformations is a linear transformation. We can form various linear combinations of these differentiation operators using functions as coefficients and these will also be linear operators. We will restrict ourselves here to the case when the coefficients are constants.

DEFINITION 4.5.1. A constant coefficient nth-order linear differential operator is a linear operator of the form

$$L = a_n D^n + a_{n-1} D^{n-1} + \cdots + a_1 D + a_0, \qquad a_n \neq 0$$

In the definition the term a_0 denotes the operation on functions of multiplying by a_0. Thus

$$L(y)(t) = a_n y^{(n)}(t) + a_{n-1} y^{n-1}(t) + \cdots + a_1 y'(t) + a_0 y(t)$$

From this viewpoint solving the differential equation

$$a_n y^{(n)}(t) + a_{n-1} y^{n-1}(t) + \cdots + a_1 y'(t) + a_0 y(t) = 0$$

becomes equivalent to finding the null space $\mathcal{N}(L)$.

Our existence and uniqueness theorem can then be translated into the following theorem. We again normalize $a_n = 1$ for convenience as we can always reduce to this case. This is the analog of Theorem 4.5.2 and its proof is similar.

THEOREM 4.5.5. The solutions of

$$y^{(n)}(t) + a_{n-1} y^{n-1}(t) + \cdots + a_1 y'(t) + a_0 y(t) = 0$$

form an n-dimensional subspace of $\mathfrak{D}(\mathbb{R}, \mathbb{R})$. If $\mathbf{v}_1, \ldots, \mathbf{v}_n$ are a basis for $\mathbb{R}^n$, and if $y_1(t), \ldots, y_n(t)$ denote solutions so that $(y_i(0), y_i'(0), \ldots, y_i^{(n-1)}(0)) = \mathbf{v}_i$, then the general solution is

$$y(t) = c_1 y_1(t) + \cdots + c_n y_n(t)$$

To apply this theorem we have to find n solutions whose initial value vectors form an independent set in $\mathbb{R}^n$. To see how to get these solutions we use the view of the solutions as $\mathcal{N}(L)$. We look at the polynomial $x^n + a_{n-1}x^{n-1} + \cdots + a_1 x + a_0$. If its roots are $\lambda_1, \ldots, \lambda_n$, then it factors as $(x - \lambda_1)\cdots(x - \lambda_n)$. But this means that the differential operator L factors as a composition $L = (D - \lambda_1)\cdots(D - \lambda_n)$. Here the constant terms again denote the operator which is multiplication by this constant.

That we have this factorization depends on the fact that differentiation commutes with multiplication by a constant, which is just the statement that the derivative of ky is k times the derivative of y, where k is a constant and y is a function. We next note that if $(D - \lambda_i)y = 0$, then $Ly = 0$. This uses the fact that the factors $(D - \lambda_i)$ and $(D - \lambda_j)$ commute; that is, $(D - \lambda_i)(D - \lambda_j) = (D - \lambda_j)(D - \lambda_i)$. We check this by applying each to a function y to get $y'' - (\lambda_i + \lambda_j)y' + \lambda_i\lambda_j y$. Thus we can write this composition in the form $L = P(D - \lambda_i)$, where P is the composition of the other terms except $(D - \lambda_i)$.

We then use the fact that $(D - \lambda_i)y = 0$ and $P(0) = 0$ to show that $(D - \lambda_i)y = 0$ implies $Ly = 0$. This tells us that we can find solutions by first finding solutions of $(D - \lambda_i)y = 0$. This equation is just $y' - \lambda_i y = 0$, whose general solution is known to be a multiple of $e^{\lambda_i t}$. The initial value vector $[y(0), y'(0), \ldots, y^{(n-1)}(0)]$ for this solution is $(1, \lambda_i, \lambda_i^2, \ldots, \lambda_i^{n-1})$. If there are n distinct roots, then we get n solutions $e^{\lambda_1 t}, \ldots, e^{\lambda_n t}$. These will be real solutions if the λ_i are real numbers; otherwise these are complex solutions. These n vectors form a basis for $\mathbb{R}^n$ (or $\mathbb{C}^n$) since we can check that the matrix $\begin{pmatrix} 1 & 1 & \cdots & 1 \\ \lambda_1 & \lambda_2 & \cdots & \lambda_n \\ \vdots & \vdots & \vdots & \vdots \\ \lambda_1^{n-1} & \lambda_2^{n-1} & \cdots & \lambda_n^{n-1} \end{pmatrix}$ is invertible. The argument for this is given in Section 3.5 concerning data fitting where the transpose to this matrix occurs in the problem of finding a polynomial of degree $n - 1$ which passes through n data points (x_i, y_i) with distinct x-coordinates.

We then recover the result which was proved by a different method in the last section that if there are n distinct roots $\lambda_1, \ldots, \lambda_n$ of the characteristic equation, then the general solution is $c_1 e^{\lambda_1 t} + \cdots + c_n e^{\lambda_i t}$. This solution may be complex valued if any λ_i is but we can always exchange a pair of complex valued solutions $e^{(a+ib)t}, e^{(a-ib)t}$ in our formula for the general solution by the corresponding real valued solutions $e^{at} \cos bt, e^{at} \sin bt$.

The question still remains as to how we should handle repeated roots of the characteristic equation. Suppose the distinct roots are $\mu_1, \ldots, \mu_k$ and their multiplicities are $m_1, \ldots, m_k$. For each root μ_i there is one solution $e^{\mu_i t}$. We would like to find m_i independent solutions for these roots. We claim that

$$e^{\mu_i t}, te^{\mu_i t}, t^2 e^{\mu_i t}, \ldots, t^{m_i - 1} e^{\mu_i t}$$

are all solutions. To see this, we factor out all factors corresponding to μ_i and rewrite $L = P_i(D - \mu_i)^{m_i}$. Thus it suffices to see that these m functions are solutions to $(D - \mu_i)^{m_i} y = 0$ and they are independent.

LEMMA 4.5.6. The functions $e^{\mu t}, te^{\mu t}, t^2 e^{\mu t}, \ldots, t^{m-1} e^{\mu t}$ are m independent solutions to $(D - \mu)^m y = 0$.

Proof. We prove this by induction on m. When $m = 1$ this is just the statement that $e^{\mu t}$ is a solution to $y' - \mu y = 0$. Assume it is true for $m = p$ and let $m = p + 1$. Then we know by the inductive assumption that $e^{\mu t}, te^{\mu t}, t^2 e^{\mu t}, \ldots, t^{p-1} e^{\mu t}$ are p independent solutions of $(D - \mu)^p y = 0$. Since $(D - \mu)^{p+1} = (D - \mu)(D - \mu)^p$, any solution of $(D - \mu)^p y = 0$ will be a solution of $(D - \mu)^{p+1} y = 0$. It thus remains to see why when we add the solution $t^p e^{\mu t}$, this will give us one more independent solution of $(D - \mu)^{p+1} y = 0$. To see this we factor $(D - \mu)^{p+1} = (D - \mu)^p (D - \mu)$. Then

$$(D - \mu)(t^p e^{\mu t}) = pt^{p-1} e^{\mu t} + t^p \mu e^{\mu t} - \mu t^p e^{\mu t} = pt^{p-1} e^{\mu t}$$

Thus

$$(D-\mu)^{p+1}(t^p e^{\mu t}) = (D-\mu)^p[(D-\mu)(t^p e^{\mu t})] = (D-\mu)^p(pt^{p-1}e^{\mu t}) = 0$$

To see why these $p+1$ solutions are independent, we can take a linear combination

$$c_1 e^{\mu t} + c_2 t e^{\mu t} + \cdots + c_p t^{p-1} e^{\mu t} + c_{p+1} t^p e^{\mu t} = (c_1 + c_2 t + \cdots + c_{p+1} t^p) e^{\mu t} = 0$$

We now apply $(D-\mu)^p$ to this equation. All terms but the last one are sent to 0 by our inductive hypothesis. The last term is sent to $c_{p+1}(D-\mu)^p(t^p e^{\mu t})$. We saw earlier that $(D-\mu)(t^p e^{\mu t}) = pt^{p-1}e^{\mu t}$. We leave it as an exercise to show inductively that $(D-\mu)^k(t^p e^{\mu t}) = p(p-1)\cdots(p-k+1)t^{p-k}e^{\mu t}$. Thus $c_{p+1}(D-\mu)^p(t^p e^{\mu t}) = c_{p+1}p!e^{\mu t}$ and so this can vanish only if $c_{p+1} = 0$. Thus we are led to

$$c_1 e^{\mu t} + c_2 t e^{\mu t} + \cdots + c_p t^{p-1} e^{\mu t} = 0$$

which implies that $c_1 = \cdots = c_p = 0$ by the inductive assumption. ■

Finally, we have to see that the whole collection of solutions

$$e^{\mu_1 t}, \ldots, t^{m_1-1}e^{\mu_1 t}, \ldots, e^{\mu_k t}, \ldots, t^{m_k-1}e^{\mu_k t}$$

are independent. Since there are n of them and the solution space is n-dimensional, they will be a basis for all solutions. To see this, set

$$(c_{11}e^{\mu_1 t} + \cdots + c_{1m_1}t^{m_1-1}e^{\mu_1 t}) + \cdots + (c_{k1}e^{\mu_k t} + \cdots + c_{km_k}t^{m_k-1}e^{\mu_k t}) = 0$$

We write $L = (D-\mu_i)P_i$ where P_i indicates the product of the other terms. Applying P_i to this linear combination sends all the terms except those involving $t^{m-1}e^{\mu_i t}$ to zero by the lemma. But $P_i = \overline{P_i}(D-\mu_i)^{m_i-1}$ and applying $(D-\mu_i)^{m_i-1}$ to $t^{m_i-1}e^{\mu_i t}$ gives $(m-1)!e^{\mu_i t}$. Using $(D-\mu_j)e^{\mu_i t} = (\mu_i - \mu_j)e^{\mu_i t}$, we then get

$$\overline{P_i}e^{\mu_i t} = \prod_{i \neq j}(\mu_i - \mu_j)^{m_j}e^{\mu_i t}$$

Putting these calculations together, we get

$$P_i[(c_{11}e^{\mu_1 t} + \cdots + c_{1m_1}t^{m_1-1}e^{\mu_1 t}) + \cdots + (c_{k1}e^{\mu_k t} + \cdots + c_{km_k}t^{m_k-1}e^{\mu_k t})] =$$
$$c_{im_i}m_i!\prod_{i \neq j}(\mu_i - \mu_j)^{m_j}e^{\mu_i t} = 0$$

From this, we conclude $c_{im_i} = 0$. We can delete this term from the sum and repeat this argument using lower powers of the coefficient of the term $(D-\mu_i)$ to get that all $c_{is} = 0$. More formally, we can use an inductive argument on the power m_i. The argument works for $m_i = 1$ and serves to reduce the power m_i by one by showing the coefficient of the highest term is zero. By now noting that i is arbitrary in the argument, we see that all coefficients $c_{ij} = 0$, proving independence of these functions.

We have thus proved the following theorem.

THEOREM 4.5.7. The general (complex valued) solution of the differential equation

$$y^{(n)} + a_{n-1}y^{(n-1)} + \cdots + a_1 y' + a_0 y = 0$$

is of the form

$$y(t) = (c_{11}e^{\mu_1 t} + \cdots + c_{1m_1}t^{m_1-1}e^{\mu_1 t}) + \cdots + (c_{k1}e^{\mu_k t} + \cdots + c_{km_k}t^{m_k-1}e^{\mu_k t})$$

where the polynomial

$$x^n + a_{n-1}x^{n-1} + \cdots + a_1 x + a_0 = (x - \mu_1)^{m_1}\cdots(x - \mu_k)^{m_k}$$

and $\mu_1, \ldots, \mu_k$ are the distinct roots of these polynomials. The constants here are complex numbers. If all the μ_i are real numbers, then the general real-valued solution is of the same form with the c_i being real numbers. If some $\mu_i = a_i + ib_i, b_i \neq 0$, then the pair of terms $t^p e^{\mu_i t}, t^p e^{\overline{\mu_i} t}$ in the preceding sum can be replaced by terms $t^p e^{a_i t}\cos b_i t, t^p e^{a_i t}\sin b_i t$ in writing the general real-valued solution and the coefficients will be real.

■ ***Example 4.5.8.*** Consider the equation

$$y'' + 2y + 1 = 0,\ y(0) = 4,\ y'(0) = -3$$

The roots of the characteristic equation $x^2+2x+1 = (x+1)^2 = 0$ are $-1, -1$. Hence the general solution is

$$y(t) = c_1 e^{-t} + c_2 t e^{-t}$$

To solve the initial value problem, we note that the initial values at $t = 0$ of these two basic solutions are $(1, -1), (0, 1)$. Thus we need to solve the matrix equation

$$\begin{pmatrix} 1 & 0 \\ -1 & 1 \end{pmatrix}\mathbf{c} = \begin{pmatrix} 4 \\ -3 \end{pmatrix}$$

This has solution $c = (4, 1)$, so the solution is

$$y(t) = 4e^{-t} + te^{-t}$$ ■

■ ***Example 4.5.9.*** Consider the equation

$$y^{(4)} + 2y^{(2)} + y = 0, (y(0), y'(0), y''(0), y'''(0)) = (2, -1, 0, 1)$$

The roots of the characteristic equation

$$x^4 + 2x^2 + 1 = (x^2 + 1)^2 = 0$$

are $\pm i$, with each root having algebraic multiplicity 2. Thus the general real-valued solution is

$$c_1 \cos t + c_2 \sin t + c_3 t \cos t + c_4 t \sin t$$

To solve the initial value problem, we form the matrix with the initial values of these basic solutions, which is

$$\begin{pmatrix} 1 & 0 & 0 & 0 \\ 0 & 1 & 1 & 0 \\ -1 & 0 & 0 & 2 \\ 0 & -1 & -3 & 0 \end{pmatrix}$$

and solve

$$\begin{pmatrix} 1 & 0 & 0 & 0 \\ 0 & 1 & 1 & 0 \\ -1 & 0 & 0 & 2 \\ 0 & -1 & -3 & 0 \end{pmatrix} \mathbf{c} = \begin{pmatrix} 2 \\ -1 \\ 0 \\ 1 \end{pmatrix}$$

The solution is $\mathbf{c} = (2, -1, 0, 1)$ and so the solution to the initial value problem is

$$y(t) = 2\cos t - \sin t + t \sin t$$

■

■ ***Example 4.5.10.*** Our last example essentially combines the last two examples in looking at the sixth-order equation

$$y^{(6)} + 2y^{(5)} + 3y^{(4)} + 4y^{(3)} + 3y'' + 2y' + y = 0$$

$$(y(0), y'(0), y''(0), y^{(3)}(0), y^{(4)}(0), y^{(5)}(0)) = (1, 0, -1, 2, 1, 0)$$

The characteristic equation is

$$x^6 + 2x^5 + 3x^4 + 4x^3 + 3x^2 + 2x + 1 = (x^2 + 1)^2(x + 1)^2 = 0$$

Thus the roots are $-1, \pm i$, each with algebraic multiplicity 2. The general solution is

$$y(t) = c_1 e^{-t} + c_2 t e^{-t} + c_3 \cos t + c_4 \sin t + c_5 t \cos t + c_6 t \sin t$$

To solve the initial value problem we form the matrix

$$\begin{pmatrix} 1 & 0 & 1 & 0 & 0 & 0 \\ -1 & 1 & 0 & 1 & 1 & 0 \\ 1 & -2 & -1 & 0 & 0 & 2 \\ -1 & 3 & 0 & -1 & -3 & 0 \\ 1 & -4 & 1 & 0 & 0 & -4 \\ -1 & 5 & 0 & 1 & 5 & 0 \end{pmatrix}$$

of initial values of the six basis solutions. We then solve

$$\begin{pmatrix} 1 & 0 & 1 & 0 & 0 & 0 \\ -1 & 1 & 0 & 1 & 1 & 0 \\ 1 & -2 & -1 & 0 & 0 & 2 \\ -1 & 3 & 0 & -1 & -3 & 0 \\ 1 & -4 & 1 & 0 & 0 & -4 \\ -1 & 5 & 0 & 1 & 5 & 0 \end{pmatrix} \mathbf{c} = \begin{pmatrix} 1 \\ 0 \\ -1 \\ 2 \\ 1 \\ 0 \end{pmatrix}$$

to get $\mathbf{c} = (2, 1, -1, 2, -1, -1)$ Thus the solution to the initial value problem is

$$y(t) = 2e^{-t} + te^{-t} - \cos t + 2\sin t - t\cos t - t\sin t$$

■

In Exercises 4.5.8–4.5.11 solve the initial value problems.

Exercise 4.5.8.

$y'' - 4y' + 4y = 0, (y(0), y'(0)) = (1, -1).$

Exercise 4.5.9.

$$y^{(4)} - 2y^{(3)} + 5y'' - 8y' + 4y = 0, (y(0), y'(0), y''(0), y'''(0)) = (1,0,0,0).$$

Exercise 4.5.10.

$$y''' - 3y'' + 3y - y = 0, (y(0), y'(0), y''(0)) = (-2,1,0).$$

Exercise 4.5.11.

$$y^{(5)} - 4y^{(4)} + 2y^{(3)} + 4y'' + 8y' - 16y = 0,$$
$$(y(0), y'(0), y''(0), y^{(3)}(0), y^{(4)}(0)) = (1,2,4,8,16)$$

4.6. CHAPTER 4 EXERCISES

In Exercises 4.6.1–4.6.14 find the eigenvalues. For each eigenvalue λ, give a basis for the eigenspace $E(\lambda)$. When A is diagonalizable, give matrices S, D with $A = SDS^{-1}$. If A is not diagonalizable, identify those eigenvalues where the algebraic and geometric multiplicities differ.

4.6.1. $A = \begin{pmatrix} 6 & -2 \\ -2 & 9 \end{pmatrix}.$

4.6.2. $A = \begin{pmatrix} -1 & 2 & -4 \\ 4 & 1 & 16 \\ 0 & 0 & 9 \end{pmatrix}.$

4.6.3. $A = \begin{pmatrix} 1 & 3 & -3 \\ -3 & 7 & -3 \\ -6 & 6 & -2 \end{pmatrix}.$

4.6.4. $A = \begin{pmatrix} -4 & -2 & 3 \\ 2 & 1 & -2 \\ -4 & -2 & 3 \end{pmatrix}.$

4.6.5. $A = \begin{pmatrix} -1 & -1 & 2 \\ 2 & 2 & -2 \\ -2 & -1 & 3 \end{pmatrix}.$

4.6.6. $A = \begin{pmatrix} 0 & 0 & 1 \\ 0 & 1 & 0 \\ -1 & 0 & 2 \end{pmatrix}.$

4.6.7. $A = \begin{pmatrix} 0 & 1 & 2 \\ -1 & 0 & 1 \\ -2 & -1 & 0 \end{pmatrix}$.

4.6.8. $A = \begin{pmatrix} 0 & 1 & 1 \\ -1 & 0 & 1 \\ -1 & -1 & 0 \end{pmatrix}$.

4.6.9. $A = \begin{pmatrix} 1/\sqrt{3} & -1/\sqrt{2} & -1/\sqrt{6} \\ 1/\sqrt{3} & 1/\sqrt{2} & -1/\sqrt{6} \\ 1/\sqrt{3} & 0 & 2/\sqrt{6} \end{pmatrix}$.

4.6.10. $\begin{pmatrix} (1 - 1/\sqrt{(6)}) & 1/\sqrt{6} & 0 \\ 1/\sqrt{6} & (1 + 1/\sqrt{6}) & 0 \\ 1/\sqrt{6} & 1/\sqrt{6} & 1 \end{pmatrix}$.

4.6.11. $A = \begin{pmatrix} 4 & 0 & -2 & -2 \\ 0 & 10 & -2 & 2 \\ -2 & -2 & 5 & -3 \\ -2 & 2 & -3 & 5 \end{pmatrix}$.

4.6.12. $\begin{pmatrix} 7 & -1 & -3 & 1 \\ -1 & 7 & 1 & -3 \\ -3 & 1 & 7 & -1 \\ 1 & -3 & -1 & 7 \end{pmatrix}$.

4.6.13. $\begin{pmatrix} 7 & -1 & -3 & 1 \\ -1 & 7 & 1 & -3 \\ -3 & 1 & 7 & 1 \\ 1 & -3 & -1 & 7 \end{pmatrix}$.

4.6.14. $\begin{pmatrix} 2 & 4 & 2 & 0 \\ 1 & 1 & -1 & 3 \\ 2 & 0 & 2 & 4 \\ -3 & 1 & 3 & 3 \end{pmatrix}$.

4.6.15. Compute the trace and determinant of each of the following matrices.

(a) $\begin{pmatrix} 1 & 2 \\ -2 & 1 \end{pmatrix}$, (b) $\begin{pmatrix} 1 & i & -1 \\ 0 & -1 & i \\ 1 & 0 & 2+i \end{pmatrix}$

4.6.16. Compute the trace and determinant of each of the following matrices.

(a) $\begin{pmatrix} 2 & 0 & i \\ -1 & 1 & i \\ 0 & i & 1 \end{pmatrix}$, (b) $\begin{pmatrix} -2 & i & -2 & 1 \\ 0 & 2i & -i & 1 \\ 3 & 2 & 2i & 1 \\ i+1 & i-1 & i & i \end{pmatrix}$

4.6.17. For the 3 by 3 matrix A its trace is 4 and its determinant is -4. Show that if its eigenvalues are integers, then A is diagonalizable.

4.6.18. For the 3 by 3 matrix A its distinct eigenvalues are $1, -1$, and eigenvectors for distinct eigenvalues are orthogonal. If a basis for the eigenspace $E(1)$ is given by $(1,0,1), (0,1,0)$, find A.

4.6.19. For the 3 by 3 diagonalizable matrix A its distinct eigenvalues are $1, -1$, and eigenvectors for distinct eigenvalues are orthogonal. The eigenspace for the eigenvalue -1 has basis $(1,1,1)$. Find A.

4.6.20. Show the determinant of A is 0 iff 0 is an eigenvalue of A.

4.6.21. The 3 by 3 matrix A has only two distinct eigenvalues. The trace of A is 4 and the determinant is 2. What are the eigenvalues of A?

4.6.22. Show that a 3 by 3 matrix with determinant 0 and trace 0 is diagonalizable as long as there is some other eigenvalue besides 0.

4.6.23. Suppose that the entries of A are positive.
(a) Show that the trace of A is positive.
(b) Show that if A is an invertible 2 by 2 matrix then its eigenvalues are positive if $\det A$ is positive.

4.6.24. Show that $A + rI$ has the same eigenvectors as A. How are the eigenvalues related?

4.6.25. Consider an n by n matrix with all entries constant b. Show that the eigenvalues of this matrix are 0 with multiplicity $n - 1$ and nb with multiplicity 1. Find a basis of eigenvectors for this matrix.

4.6.26. Consider an n by n matrix with all off-diagonal entries b and diagonal entries a. By using the previous two exercises, find the eigenvalues and eigenvectors for this matrix. Compute the determinant of the matrix.

4.6.27. Show that the eigenvalues of A^{-1} are the reciprocals of the eigenvalues of A. How are the eigenvectors related?

4.6.28. Show that the eigenvalues of A^t are the same as the eigenvalues of A. How are the eigenvectors related?

4.6.29. Show that the eigenvalues of $\overline{A}$ are the conjugates of the eigenvalues of A. How are the eigenvectors related?

4.6.30. Find the inverse of each of the following complex matrices.

$$\text{(a)} \begin{pmatrix} 2+i & 1-i \\ 2-i & 1+i \end{pmatrix}, \quad \text{(b)} \begin{pmatrix} 1 & i & -i \\ 0 & 2i & 1 \\ 1 & i & 0 \end{pmatrix}$$

4.6.31. Find the inverse of each of the following complex matrices.

$$\text{(a)} \begin{pmatrix} -i & i \\ 1 & 2 \end{pmatrix}, \quad \text{(b)} \begin{pmatrix} i & i-1 & i+1 \\ 1 & i & 0 \\ -1 & i & i \end{pmatrix}$$

4.6.32. Determine whether the following vectors in $\mathbb{C}^3$ are independent. If they are dependent, write one as a linear combination of the others.
(a) $(2, i, -1), (2 - i, 0, 1), (i, i, 1)$
(b) $(2 - i, 2 + i, 1), (i, -1 + i, 1 + i), (4 - i, 3 + 2i, i)$

4.6.33. Determine whether the following vectors in $\mathbb{C}^3$ are independent. If they are dependent, write one as a linear combination of the others.
(a) $(i, -i, 2), (1, i, 1), (-1 + i, 2 - 2i, 4 + i)$
(b) $(i, -i, 2), (1, i, 1), (-1 + i, 2 - 2i, 4)$

4.6.34. Use the Gram–Schmidt algorithm with the standard Hermitian inner product in $\mathbb{C}^3$ to give an orthonormal basis for the subspace of $\mathbb{C}^3$ with basis $(1 + i, 1 - i, i), (1, i, -i)$.

4.6.35. Use the Gram–Schmidt algorithm with the standard Hermitian inner product in $\mathbb{C}^3$ to give an orthonormal basis for the subspace of $\mathbb{C}^3$ with basis $(1, i, 1 + i), (-i, i, 0)$.

4.6.36. Use the Gram–Schmidt algorithm with the standard Hermitian inner product in $\mathbb{C}^4$ to give an orthonormal basis for the subspace of $\mathbb{C}^4$ with basis $(1, 0, -i, 0), (1, 0, i, 1), (2 + i, -1, 2 - i, 1)$.

4.6.37. Verify that the product of two unitary matrices is unitary. Is the product of two Hermitian matrices Hermitian? Is the product of two skew Hermitian matrices skew Hermitian?

4.6.38. Show that if A is a skew Hermitian matrix, then $\exp(A)$ is a unitary matrix.

4.6.39. Classify each of the following matrices as either Hermitian, skew Hermitian, unitary, or none of the above.

$$A = \begin{pmatrix} 1 & 3 & 4 \\ 3 & 2 & 3 \\ 4 & 3 & 5 \end{pmatrix}, B = \begin{pmatrix} i & 3 & 0 \\ -3 & 3i & i \\ 0 & i & 0 \end{pmatrix}, C = \begin{pmatrix} 1/\sqrt{2} & 1/\sqrt{3} & -1/\sqrt{6} \\ 0 & 1/\sqrt{3} & 2/\sqrt{6} \\ 1/\sqrt{2} & -1/\sqrt{3} & 1/\sqrt{6} \end{pmatrix},$$

$$D = \begin{pmatrix} .5 + .5i & .5 + .5i \\ .5 + .5i & -.5 - .5i \end{pmatrix}, E = \begin{pmatrix} i & 2 & -i \\ -2 & i & 0 \\ -i & 0 & 1 \end{pmatrix}$$

4.6.40. Classify each of the following matrices as either Hermitian, skew Hermitian, unitary, or none of the above.

$$A = \begin{pmatrix} 1 & 1 & 1 \\ -1 & 1 & 0 \\ -1 & 0 & 1 \end{pmatrix}, B = \begin{pmatrix} 0 & 1 & 0 \\ -1 & 0 & 1 \\ 0 & -1 & 0 \end{pmatrix}, C = \begin{pmatrix} 1/\sqrt{2} & -1/\sqrt{3} & -1/\sqrt{6} \\ 1/\sqrt{2} & 1/\sqrt{3} & 1/\sqrt{6} \\ 0 & 1/\sqrt{3} & -2/\sqrt{6} \end{pmatrix},$$

$$D = \begin{pmatrix} 4 & 2 + i & 3 - i \\ 2 - i & 5 & 1 \\ 3 + i & 1 & 3 \end{pmatrix}, E = \begin{pmatrix} i & i & i \\ i & i & i \\ i & i & i \end{pmatrix}$$

4.6.41. Show that a symmetric matrix all of whose entries are pure imaginary is a skew Hermitian matrix.

4.6.42. Prove by induction that $(SDS^{-1})^k = SD^kS^{-1}$.

4.6.43. Suppose A is an n by n matrix which is diagonalizable with eigenvalues $1, \lambda_2, \ldots, \lambda_n$, where $\|\lambda_i\| < 1$. Show that $\lim_{k\to\infty} A^k$ exists and that its columns are either zero or are eigenvectors for the eigenvalue 1.

4.6.44. Let P be the plane in R^3 spanned by $\mathbf{v}_1 = (1, 1, 1)$ and $\mathbf{v}_2 = (1, 1, -2)$. Give a matrix A so that multiplication by A fixes the axis perpendicular to this plane and rotates the plane by an angle of $\pi/4$. (Hint: First find an orthonormal basis $\mathbf{q}_1, \mathbf{q}_2, \mathbf{q}_3$ so that $\mathbf{q}_1$ is the axis of rotation and $\mathbf{q}_2, \mathbf{q}_3$ span P. Then write $A(\mathbf{q}_1\mathbf{q}_2\mathbf{q}_3) = (\mathbf{q}_1\mathbf{q}_2\mathbf{q}_3)B$ for appropriate B and solve for A. B is just the matrix of the linear transformation with respect to the basis given by the $\mathbf{q}_i$.)

4.6.45. Suppose A is diagonalizable.

(a) Show A^k is diagonalizable.

(b) Give an example of a 2 by 2 matrix A which is not diagonalizable but A^2 is diagonalizable.

4.6.46. Determine whether each of the following statements is true or false. Justify your answers.

(a) If A is a Hermitian matrix with real entries, then A is symmetric.

(b) If A is a 3 by 3 matrix with eigenvalues $1, 2, 3$, then A is diagonalizable.

(c) If A is a 3 by 4 matrix of rank 2, then A^tA has 0 as an eigenvalue of geometric multiplicity 2.

4.6.47. If $\mathbf{u}$ is a unit vector and $Q = I - 2\mathbf{u}\mathbf{u}^t$, show that Q is an orthogonal matrix which transforms $\mathbf{u}$ to $-\mathbf{u}$ and leaves the vectors orthogonal to $\mathbf{u}$ fixed (i.e., is a reflection through the subspace which is orthogonal to $\mathbf{u}$). What do you conclude about the eigenvalues and eigenvectors of Q?

4.6.48. Give 2 by 2 matrices A, B, C, D so the linear transformations coming from multiplication by A, B, C, D achieve the desired geometric effect. (Hint: Find the eigenvalues and eigenvectors of these linear transformations.)

(a) Multiplication by A rotates the plane by $\pi/4$.

(b) Multiplication by B projects orthogonally onto the line $x = y$.

(c) Multiplication by C reflects through the line $x = y$.

(d) Multiplication by D is the composition of multiplication by A followed by multiplication by C. Describe this geometrically as a single reflection.

4.6.49. A charitable organization accepts donations from a group of potential contributors in amounts of \$20 and \$50 . Of those who contribute nothing one year, the next year 1/2 will contribute \$20, 1/4 will contribute \$50, and 1/4 will contribute nothing. Of those contributing \$20, the next year 1/2 will contribute \$20, 1/4 don't contribute, and 1/4 will contribute \$50. Of those contributing \$50, 3/4 will contribute \$50 again, and 1/4 will contribute \$20 the next year.

(a) Give a transition matrix representing this information.

(b) What is the probability that a potential contributor will be a \$50 contributor in the long run (i.e., many years later)?

In Exercises 4.6.50–4.6.54 the given matrix is a transition matrix for a Markov chain. For each problem determine

(a) the probability of moving from state 2 to state 3 in 10 steps

(b) the limiting probability distribution(if it exists) given that one starts out with equal probability of being in any of the states

4.6.50. $\begin{pmatrix} .5 & .5 & .1 \\ 0 & .5 & .1 \\ .5 & 0 & .8 \end{pmatrix}$

4.6.51. $\begin{pmatrix} 1/3 & 1/2 & 0 \\ 1/2 & 0 & 1 \\ 1/6 & 1/2 & 0 \end{pmatrix}$

4.6.52. $\begin{pmatrix} .1 & .2 & .3 & .4 \\ .2 & .3 & .4 & .1 \\ .3 & .4 & .1 & .2 \\ .4 & .1 & .2 & .3 \end{pmatrix}$

4.6.53. $\begin{pmatrix} 1 & 0 & 0 & 0 \\ 0 & 1/3 & 1/3 & 1/6 \\ 0 & 1/2 & 1/3 & 1/3 \\ 0 & 1/6 & 1/3 & 1/2 \end{pmatrix}$

4.6.54. $\begin{pmatrix} 1 & 1/2 & 1/3 & 1/4 & 1/5 \\ 0 & 1/2 & 1/3 & 1/4 & 1/5 \\ 0 & 0 & 1/3 & 1/4 & 1/5 \\ 0 & 0 & 0 & 1/4 & 1/5 \\ 0 & 0 & 0 & 0 & 1/5 \end{pmatrix}$

4.6.55. Suppose the n by n matrix T is the transition matrix for a Markov chain and all its entries are positive. Show that if T is symmetric, then T^k tends toward the matrix with all entries equal to $1/n$. (Hint: The eigenvectors of T and T^t are equal.)

4.6.56. A gambler with \$2000 enters a game where she has to continue gambling until she is wiped out or triples her money. In each game she either wins \$1000 with a probability of .51 or loses \$1000 with a probability of .49. The matrix

$$T = \begin{pmatrix} 1 & .49 & 0 & 0 & 0 & 0 & 0 \\ 0 & 0 & .49 & 0 & 0 & 0 & 0 \\ 0 & .51 & 0 & .49 & 0 & 0 & 0 \\ 0 & 0 & .51 & 0 & .49 & 0 & 0 \\ 0 & 0 & 0 & .51 & 0 & .49 & 0 \\ 0 & 0 & 0 & 0 & .51 & 0 & 0 \\ 0 & 0 & 0 & 0 & 0 & .51 & 1 \end{pmatrix}$$

gives the transition matrix for the Markov chain which models this, where there are seven states which measure how many thousand dollars she has (from 0 to 6) at a given time. What is the probability that

(a) she has \$4000 after 10 games
(b) she eventually loses all of her money

4.6.57. Suppose that μ is an eigenvalue different from 1 for the transition matrix T of a Markov chain. Show that if $\mathbf{x}$ is an eigenvector for μ, then the sum of the entries of $\mathbf{x}$ is zero. (Hint: Use $\mathbf{r}$ and the fact that $\mathbf{rx}$ is the sum of the entries of $\mathbf{x}$ together with $\mathbf{r}T = \mathbf{r}, T\mathbf{x} = \mu\mathbf{x}$.)

4.6.58. Consider the difference equation

$$\mathbf{u}_{k+1} = A\mathbf{u}_k, \mathbf{u}_0 = \begin{pmatrix} 1 \\ 0 \\ 0 \end{pmatrix}, A = \begin{pmatrix} 1/4 & 1/3 & 0 \\ 3/4 & 1/3 & 1/2 \\ 0 & 1/3 & 1/2 \end{pmatrix}$$

What is $\lim_{k\to\infty} \mathbf{u}_k$?

4.6.59. Consider the difference equation

$$\mathbf{u}_{k+1} = A\mathbf{u}_k, \mathbf{u}_0 = \begin{pmatrix} 1 \\ 0 \\ 0 \end{pmatrix}, A = \begin{pmatrix} 5.5 & 7.5 & -10 \\ 1 & 2 & -2 \\ 3 & 4.5 & -5.5 \end{pmatrix}$$

What is $\lim_{k\to\infty} \mathbf{u}_k$?

4.6.60. (a) Consider the difference equation

$$\mathbf{u}_{k+1} = A\mathbf{u}_k,\ \mathbf{u}_0 = \begin{pmatrix} 1 \\ 0 \\ 0 \end{pmatrix}, A = \begin{pmatrix} 7 & 10.5 & -13.5 \\ 4 & 8 & -9 \\ 6 & 10.5 & -12.5 \end{pmatrix}$$

What is $\lim_{k\to\infty} \mathbf{u}_k$?

(b) Repeat this problem with initial vector $\mathbf{u}_0 = \begin{pmatrix} 0 \\ 1 \\ 0 \end{pmatrix}$.

4.6.61. Consider the difference equation

$$\mathbf{u}_{k+1} = A\mathbf{u}_k,\ \mathbf{u}_0 = \begin{pmatrix} 1 \\ 0 \\ 0 \end{pmatrix}, A = \begin{pmatrix} 1 & -1.5 & 0.5 \\ -8 & -16 & 19 \\ -6 & -13.5 & 15.5 \end{pmatrix}$$

(a) Show that $\lim_{k\to\infty} \mathbf{u}_k$ does not exist.
(b) Show that $\lim_{k\to\infty} \mathbf{u}_{2k}$ exists and evaluate it.
(c) Show that $\lim_{k\to\infty} \mathbf{u}_{2k+1}$ exists and evaluate it.

4.6.62. Consider the difference equation

$$\mathbf{u}_{k+1} = A\mathbf{u}_k,\ \mathbf{u}_0 = \begin{pmatrix} 1 \\ 0 \\ 0 \end{pmatrix}, A = \begin{pmatrix} 33 & 53.5 & -70 \\ 64 & 112.5 & -142 \\ 66 & 114 & -145 \end{pmatrix}$$

(a) Show that $\lim_{k\to\infty} \mathbf{u}_k$ does not exist.
(b) Show that $\lim_{k\to\infty} \mathbf{u}_{2k}$ exists and evaluate it.
(c) Show that $\lim_{k\to\infty} \mathbf{u}_{2k+1}$ exists and evaluate it.

For Exercises 4.6.63–4.6.68, rephrase the recursion formula as a difference equation $\mathbf{x}_{k+1} = A\mathbf{x}_k$ with an initial vector $\mathbf{x}_0$ and find the limiting ratio $\lim_{k\to\infty} y_{k+1}/y_k$ if it exists.

4.6.63. $1, 1, 5, 13, 41, 121, \ldots$; $y_{k+2} = 2y_{k+1} + 3y_k$.

4.6.64. $1, 2, -1, -5, -13, -29, -61, \ldots$; $y_{k+2} = -2y_{k+1} + 3y_k$.

4.6.65. $1, 1, 0, -16, 80, -384, \ldots$; $y_{k+2} = -4y_{k+1} + 4y_k$.

4.6.66. $1, 1, 2, 6, 12, 28, \ldots$; $y_{k+3} = 2y_{k+2} + y_{k+1} + y_k$.

4.6.67. $2, 3, 1, -3, -11, -27, \ldots$; $y_{k+3} = 3y_{k+2} - 2y_{k+1}$.

4.6.68. $4, 2, 1, 1, 2, 4, 7, 11, 16, \ldots$; $y_{k+3} = 3y_{k+2} - 3y_{k+1} + y_k$.

4.6.69. Show that for the matrix

$$A = \begin{pmatrix} a_1 & a_2 & \ldots & a_{n-1} & a_n \\ 1 & 0 & \ldots & 0 & 0 \\ 0 & 1 & \ldots & 0 & 0 \\ \vdots & \vdots & \vdots & \vdots & \vdots \\ 0 & 0 & \ldots & 1 & 0 \end{pmatrix}$$

which arises in changing a recursion formula to a difference equation, the characteristic polynomial of A is $(-1)^n(x^n - a_1x^{n-1} - a_2x^{n-2} - \cdots - a_n)$.

In Exercises 4.6.70–4.6.74 use the power method to find the largest eigenvalue for the given matrix. Where possible, compare your answer with what is obtained by computing the characteristic polynomial by hand. Use another computation package to find the actual eigenvalues, comparing the answers obtained by that given by the power method. Use this information to explain any times the power method fails to converge.

4.6.70. $\begin{pmatrix} 1 & 1 \\ -2 & 4 \end{pmatrix}$

4.6.71. $\begin{pmatrix} -17 & -33 & 41 \\ -20 & -37 & 46 \\ -24 & -45 & 56 \end{pmatrix}$

4.6.72. $\begin{pmatrix} 12 & -3 & 3 \\ 6 & 2 & 4 \\ 6 & -1 & 7 \end{pmatrix}$

4.6.73. $\begin{pmatrix} 2 & -1/3 & 1/3 \\ 4 & -4/3 & -2/3 \\ 4 & -7/3 & 1/3 \end{pmatrix}$

4.6.74. $\begin{pmatrix} -13 & -10 & 0 & 8 & 10 \\ 9 & 7 & 0 & -5 & -7 \\ -20 & -14 & 2 & 11 & 14 \\ 14 & 10 & 0 & -7 & -10 \\ -19 & -14 & 0 & 11 & 14 \end{pmatrix}$

4.6.75. Consider an economy with three industries in which output of one unit from each industry requires equal inputs of c units from each industry; that is, the consumption matrix is given by cK, where K is the 3 by 3 matrix with all entries 1. Here we assume that c lies between 0 and 1.

(a) For what values of c will $I - C$ be invertible?

(b) For what values of c will $(I - C)^{-1}$ have only nonnegative entries?

(c) To meet a demand of 1000 units from each industry, how many units must each industry manufacture?

4.6.76. (a) Redo Exercise 4.6.75 for an economy with five industries.

(b) Redo the last problem for an economy with n industries.

4.6.77. Show that the following sequences $\mathbf{u}_k, \mathbf{v}_k$ coming from an initial unit vector $\mathbf{x}_0$ and a matrix A are the same:

$$\mathbf{x}_{k+1} = A\mathbf{x}_k, \mathbf{u}_k = \mathbf{x}_k/\|\mathbf{x}_k\|, \mathbf{v}_0 = \mathbf{x}_0, \mathbf{v}_{k+1} = A\mathbf{v}_k/\|A\mathbf{v}_k\|$$

4.6.78. Consider the matrix $A = \begin{pmatrix} -1 & 1 & -1 \\ -4 & 3 & -4 \\ -2 & 1 & -2 \end{pmatrix}$.

(a) Compute A^8.

(b) Compute $\exp(A)$.

4.6.79. Evaluate $\exp(At)$ for $A = \begin{pmatrix} -5 & 3 & 3 \\ -4 & 8 & -4 \\ 13 & -7 & 5 \end{pmatrix}$.

4.6.80. Compute $\exp(At)$ where $A = \begin{pmatrix} 3 & 0 \\ 0 & 3 \end{pmatrix} + \begin{pmatrix} 0 & 1 \\ 0 & 0 \end{pmatrix}$.

In Exercises 4.6.81–4.6.85, find the general solution of the system $\mathbf{x}' = A\mathbf{x}$ and the solution of the given initial value problem.

4.6.81. $A = \begin{pmatrix} 2 & -1 \\ 1 & 2 \end{pmatrix}, \mathbf{x}(0) = (1, 3)$.

4.6.82. $A = \begin{pmatrix} 1 & 3 \\ 3 & 1 \end{pmatrix}, \mathbf{x}(0) = (1, 2)$.

4.6.83. $A = \begin{pmatrix} -1 & -2 & 2 \\ 2 & 4 & -2 \\ -2 & -1 & 3 \end{pmatrix}, \mathbf{x}(0) = (1, 0, 0)$.

4.6.84. $A = \begin{pmatrix} 1.5 & 1 & .5 \\ -.5 & 1 & .5 \\ .5 & -1 & 1.5 \end{pmatrix}, \mathbf{x}(0) = (0, 1, 0)$.

4.6.85. $A = \begin{pmatrix} 0 & -1 & 1 & 0 \\ 6 & 7 & -3 & -6 \\ 6 & 5 & -1 & -6 \\ 2 & 1 & -1 & -2 \end{pmatrix}, \mathbf{x}(0) = (1, 0, 1, 0)$.

In Exercises 4.6.86–4.6.91 find the general solution of the differential equation and the solution to the initial value problem.

4.6.86. $y'' + y' - 2y = 0,\ y(0) = 1, y'(0) = 0.$

4.6.87. $y''' + y'' + y' + y = 0,\ y(0) = 1, y'(0) = 1, y''(0) = 1.$

4.6.88. $y^{(4)} + 2y''' + 2y'' + 2y' + y = 0,\ (y(0), y'(0), y''(0), y'''(0)) = (1, -1, 2, 1).$

4.6.89. $y''' - 6y'' + 12y' - 8y = 0,\ (y(0), y'(0), y''(0)) = (-2, 1, 0).$

4.6.90. $y^{(4)} + 4y''' + 8y'' + 16y' + 16y = 0, (y(0), y'(0), y''(0), y'''(0)) = (1, 0, 0, 1).$

4.6.91. $y^{(4)} - y'' - 2y' + 2y = 0,\ (y(0), y'(0), y''(0), y'''(0)) = (1, 0, 1, 0).$

4.6.92. Suppose that for the system $\mathbf{x}' = A\mathbf{x}$, we have that A is diagonalizable with all eigenvalues having real part less than 0. Show that for any solution, $\lim_{t\to\infty} \mathbf{x}(t) = \mathbf{0}$.

4.6.93. Suppose that for the system $\mathbf{x}' = A\mathbf{x}$, we have that A is diagonalizable with all eigenvalues except one having real part less than 0.

(a) Suppose the last eigenvalue is 0. Show that either $\lim_{t\to\infty} \mathbf{x}(t) = \mathbf{v}$, where $\mathbf{v}$ is an eigenvector for eigenvalue 0, or $\lim_{t\to\infty} \mathbf{x}(t) = \mathbf{0}$. Describe how one decides which case occurs in terms of the initial values.

(b) Suppose the last eigenvalue is $2i$. Show that $\lim_{t\to\infty} \mathbf{x}(t)$ doesn't exist for most initial values, but when it does exist the limit is $\mathbf{0}$. Show that $\lim_{t\to\infty} \|\mathbf{x}(t)\| = 1$ for most solutions and describe those where this doesn't hold.

4.6.94. Suppose that for the system $\mathbf{x}' = A\mathbf{x}$, we have that A is diagonalizable and the eigenvalues are ordered $\lambda_1 > \cdots > \lambda_n > 1$.

(a) Show that $\lim_{t\to\infty} \mathbf{x}(t)$ does not exist for any solution.

(b) Suppose that $A = SDS^{-1}$, with the ordered eigenvalues on the diagonal of D. If the initial value $\mathbf{x}(0)$ satisfies $S\mathbf{c} = \mathbf{x}(0)$ with $c_1 \neq 0$, show that the solution of the initial value problem satisfies $\lim_{t\to\infty} \mathbf{x}(t)/\|\mathbf{x}(t)\|$ is an eigenvector for the eigenvalue λ_1.

4.6.95 (a) Find the general solution of $\mathbf{x}' = A\mathbf{x}$ and the possible limiting ratios of

$$x_3(t)/x_2(t) \text{ as } t \to \infty \text{ for } A = \begin{pmatrix} 1 & 1 & -2 \\ 1 & 1 & 2 \\ -2 & 2 & 2 \end{pmatrix}.$$

(b) Find the limiting ratio $x_3(t)/x_2(t)$ for the solution satisfying the initial value $x(0) = (1, -1, 1)$.

4.6.96. Reduce the second-order system $\mathbf{x}'' = A\mathbf{x}$ to a first-order system $\mathbf{y}' = B\mathbf{y}$ by using the substitution $y_1 = x_1, y_2 = x_2, y_3 = x_1', y_4 = x_2'$. Solve this first-order system and use its solution to find the general solution of $\mathbf{x}'' = A\mathbf{x}$. Here

$$A = \begin{pmatrix} -10 & -6 \\ -6 & -10 \end{pmatrix}.$$

4.6.97. For the same matrix A as in the previous problem find a square root B of A. Rewrite the equation in the form $(D^2 - B^2)\mathbf{x} = (D - B)(D + B)\mathbf{x} = \mathbf{0}$, and then use solutions of $(D-B)\mathbf{x} = \mathbf{0}$ and $(D+B)\mathbf{x} = 0$ to find the general solution of the equation $\mathbf{x}'' = A\mathbf{x}$.

CHAPTER 5

THE SPECTRAL THEOREM AND APPLICATIONS

5.1. COMPLEX VECTOR SPACES AND HERMITIAN INNER PRODUCTS

In this chapter we will prove the Spectral Theorem, which says that certain types of linear transformations (and matrices) are diagonalizable. Moreover, the basis of eigenvectors may be chosen to form an orthonormal basis. Since some eigenvalues may be complex, we will be forced to deal with complex matrices and complex vector spaces, if only $\mathbb{C}^n$. This section will continue the introduction to those ideas which we gave in Section 4.3. In Section 4.3 we introduced $\mathbb{C}^n$ and verified that it satisfied the properties of a complex vector space. We also introduced the inner product $\langle \mathbf{v}, \mathbf{w} \rangle = \mathbf{v}^*\mathbf{w} = \overline{v}_1 w_1 + \cdots + \overline{v}_n w_n$. We showed that this inner product has the following properties:

(1) $$\langle \mathbf{v}, \mathbf{w} \rangle = \overline{\langle \mathbf{w}, \mathbf{v} \rangle} \qquad \textbf{(Hermitian property)}$$

(2) $$\langle \mathbf{v}, a\mathbf{w} + b\mathbf{z} \rangle = a\langle \mathbf{v}, \mathbf{w} \rangle + b\langle \mathbf{v}, \mathbf{z} \rangle \qquad \textbf{(linearity property)}$$

Note that because of (1) we get what is called **conjugate linearity** in the first argument:

$$\langle a\mathbf{v} + b\mathbf{z}, \mathbf{w} \rangle = \overline{a}\langle \mathbf{v}, \mathbf{w} \rangle + \overline{b}\langle \mathbf{z}, \mathbf{w} \rangle$$

(3) $$\langle \mathbf{v}, \mathbf{v} \rangle \geq 0 \quad \text{and} \quad = 0 \quad \text{iff } \mathbf{v} = \mathbf{0} \qquad \textbf{(positive definite property)}$$

DEFINITION 5.1.1. An inner product on a complex vector space that satisfies properties (1)–(3) is called a **Hermitian inner product.**

In this chapter we will always assume that any complex vector space we are working with has a Hermitian inner product. Given such an inner product, then an orthonormal basis is defined as before in terms of the inner product. Note that the basis $\mathbf{e}_1, \ldots, \mathbf{e}_n$ is an orthonormal basis for $\mathbb{C}^n$.

The main reason we will deal with complex vector spaces is that they are much better to work with in proving theorems since the field $\mathbb{C}$ is algebraically closed, that is, polynomials with complex coefficients factor into linear factors. This is especially important for eigenvalue–eigenvector problems when there is a complex eigenvalue. Whenever all the eigenvalues of a matrix or linear transformation are real, we can continue to work within $\mathbb{R}^n$ or the real vector space V. When there are complex eigenvalues, however, we are forced to replace the vector space and linear transformation by complex ones and then use the information obtained from studying them to get information about our original problem. We will largely handle this informally, but do want to indicate some of what is going on in this replacement.

From the matrix point of view, this is just a matter of regarding $\mathbb{R}^n$ as a subset of $\mathbb{C}^n$ and n by n real matrices as special cases of n by n complex matrices. We then will allow ourselves to use complex matrices when diagonalizing the given real matrix; that is, we will be writing $A = SDS^{-1}$, where S, D are complex matrices when there is a complex eigenvalue. Another fairly straightforward situation occurs when we are studying the vector space $\mathfrak{D}(\mathbb{R}, \mathbb{R}^n)$ in connection with solving the differential equation $\mathbf{x}' = A\mathbf{x}$. By $\mathfrak{D}(\mathbb{R}, \mathbb{C}^n)$ we mean the differentiable functions from the reals to $\mathbb{C}^n$. The operations of addition of functions and multiplication by a scalar just use the operations in $\mathbb{C}^n$ where we are adding and multiplying the values of the function. When we are studying the vector space $\mathfrak{D}(\mathbb{R}, \mathbb{R}^n)$ in connection with solving the differential equation $\mathbf{x}' = A\mathbf{x}$, we sometimes will extend the problem to the complex setting. Our method for solving this equation depends on finding the eigenvalues and corresponding eigenvectors. When one of these is complex, what we are doing is regarding A as a special type of complex matrix with all entries real and then looking in the vector space $\mathfrak{D}(\mathbb{R}, \mathbb{C}^n)$ for complex-valued solutions of the equation. The differentiation operator D is then extended to $D : \mathfrak{D}(\mathbb{R}, \mathbb{C}^n) \to \mathfrak{D}(\mathbb{R}, \mathbb{C}^n)$ in a standard way. We then can use the information on complex-valued solutions to translate a pair of complex-valued solutions back into corresponding real-valued solutions and solve the original problem. Contained in this vector space are the solutions to differential equations such as $x'' + x = 0$, where the solutions are linear combinations of $\exp(ix)$ and $\exp(-ix)$. Out of these two solutions we get the real-valued solutions $\cos x, \sin x$. Our work on differential equations can be construed to say that the vector space of complex solutions to the complex differential equation $\mathbf{x}' = A\mathbf{x}$, where A may now be a complex matrix, has dimension n, and if $\mathbf{v}_1, \ldots, \mathbf{v}_n$ are any vectors in $\mathbb{C}^n$ which are given as initial values, then there is a basis of complex solutions $\mathbf{x}_1, \ldots, \mathbf{x}_n$ which have these vectors as initial values.

When discussing special classes of functions such as ones which are periodic and satisfy integrability conditions, then there are important inner products on $\mathfrak{D}(\mathbb{R}, \mathbb{C})$ such as

$$\langle f, g \rangle = \int_{-\pi}^{\pi} \overline{f(t)} g(t)\, dt$$

Here Fourier series arise in the context of expressing a given element of the complex inner product space $\mathfrak{D}(\mathbb{R}, \mathbb{C})$ in terms of the orthogonal collection

$$1, \exp(ix), \exp(-ix), \ldots, \exp(inx), \exp(-inx), \ldots$$

The situation is somewhat more complicated in the case of a general linear transformation $L : V \to V$ between a real vector space V and itself. Suppose one of the eigenvalues (in the sense of being a root of the characteristic polynomial) is not real. We may then replace V by its complexification $V_{\mathbb{C}}$, which is the analog of going from $\mathbb{R}^n$ to $\mathbb{C}^n$. As a set, $V_{\mathbb{C}}$ can be identified with the set $V \times V$ of ordered pairs of vectors in V. We define the operation of addition as $(\mathbf{v}_1, \mathbf{v}_2) + (\mathbf{w}_1, \mathbf{w}_2) = (\mathbf{v}_1 + \mathbf{w}_1, \mathbf{v}_2 + \mathbf{w}_2)$. We also need to see how to multiply a complex number by an element of $V_{\mathbb{C}}$. Motivated by the identification of $\mathbb{C}$ with $\mathbb{R}^2$, we define

$$(a + ib)(\mathbf{v}_1, \mathbf{v}_2) = (a\mathbf{v}_1 - b\mathbf{v}_2, b\mathbf{v}_1 + a\mathbf{v}_2)$$

In both cases the motivation for the definitions is to think of $(\mathbf{v}_1, \mathbf{v}_2)$ as $\mathbf{v}_1 + i\mathbf{v}_2$ and use usual operations on real vectors and the formula $i^2 = -1$. We leave it as an exercise to check that with these definitions, $V_{\mathbb{C}}$ satisfies all of the properties required to be a complex vector space when V is a real vector space. Note that we can regard $V \subset V_{\mathbb{C}}$ if we identify V with those vectors in $V_{\mathbb{C}}$ whose second coordinate is $\mathbf{0}$. There is a complex conjugation operation in $V_{\mathbb{C}}$ which is given by $\overline{(\mathbf{v}_1, \mathbf{v}_2)} = (\mathbf{v}_1, -\mathbf{v}_2)$. This is consistent with the complex conjugation in $\mathbb{C}$ in the sense that $\overline{z\mathbf{w}} = \overline{z}\,\overline{\mathbf{w}}$. If we start with a basis $\mathbf{v}_1, \ldots, \mathbf{v}_n$ for the real vector space V, then these vectors (when considered as lying in $V_{\mathbb{C}}$) will continue to be a basis for the complex vector space $V_{\mathbb{C}}$.

Next suppose that $L : V \to V$ is a linear transformation between real vector spaces. Then there is a corresponding complex linear transformation $L_{\mathbb{C}} : V_{\mathbb{C}} \to V_{\mathbb{C}}$, which restricts to $V \subset V_{\mathbb{C}}$ to give the original L. It is just defined by $L_{\mathbb{C}}(\mathbf{v}_1, \mathbf{v}_2) = (L\mathbf{v}_1, L\mathbf{v}_2)$. $L_{\mathbb{C}}$ is (complex) linear:

$$L_{\mathbb{C}}(z\mathbf{w}) = zL_{\mathbb{C}}(\mathbf{w}), L_{\mathbb{C}}(\mathbf{w} + \mathbf{y}) = L_{\mathbb{C}}\mathbf{w} + L_{\mathbb{C}}\mathbf{y}$$

when $z \in \mathbb{C}, \mathbf{w}, \mathbf{y} \in V_{\mathbb{C}}$.

Exercise 5.1.1.

Verify that when V is a real vector space, $V_{\mathbb{C}}$ satisfies all of the properties for a complex vector space.

Exercise 5.1.2.

Verify that with the definition of conjugation in $V_{\mathbb{C}}$ just given, conjugation satisfies the properties

$$\overline{z\mathbf{w}} = \overline{z}\,\overline{\mathbf{w}}, \quad \overline{\mathbf{w} + \mathbf{y}} = \overline{\mathbf{w}} + \overline{\mathbf{y}}$$

when $z \in \mathbb{C}, \mathbf{w}, \mathbf{y} \in V_{\mathbb{C}}$.

Exercise 5.1.3.

Verify that when $L : V \to V$ is a linear transformation between a real vector space V and itself, $L_{\mathbb{C}} : V_{\mathbb{C}} \to V_{\mathbb{C}}$ is complex linear.

Exercise 5.1.4.

Show that if $\mathbf{v}_1, \ldots, \mathbf{v}_n$ is a basis for the real vector space V, then these same vectors, when regarded as lying in $V_{\mathbb{C}}$, form a basis for the complex vector space $V_{\mathbb{C}}$. Moreover, show that if A is the real n by n matrix which represents $L : V \to V$ with respect to the basis $\mathcal{V} = \{\mathbf{v}_1, \ldots, \mathbf{v}_n\}$, then this same matrix still represents $L_{\mathbb{C}}$ with respect to $\mathcal{V}$ when it is regarded as a basis for $V_{\mathbb{C}}$.

Suppose that $L : V \to V$ is a linear transformation between a complex inner product space V and itself.

DEFINITION 5.1.2. The **adjoint** $L^* : V \to V$ is defined by the formula

$$\langle L\mathbf{v}, \mathbf{w} \rangle = \langle \mathbf{v}, L^*\mathbf{w} \rangle \qquad \text{for all } \mathbf{v}, \mathbf{w} \in V$$

We need to see that this definition actually characterizes a unique linear transformation. Recall that if we are given a vector $\mathbf{v}$ in V and an orthonormal basis $\mathbf{v}_1, \ldots, \mathbf{v}_n$ for V, then $\mathbf{v} = \langle \mathbf{v}_1, \mathbf{v} \rangle \mathbf{v}_1 + \cdots + \langle \mathbf{v}_n, \mathbf{v} \rangle \mathbf{v}_n$. Thus a vector in V is completely determined when we give its inner products with an orthonormal basis. Let $\mathbf{v}_1, \ldots, \mathbf{v}_n$ be an orthonormal basis for V (such a basis can be obtained from any other basis by the Gram–Schmidt orthogonalization procedure). To see that this formula determines a unique function $L^* : V \to V$, we just have to note that it is a linear transformation that determines every inner product $\langle \mathbf{v}_i, L^*\mathbf{w} \rangle = \langle L\mathbf{v}_i, \mathbf{w} \rangle$. That L^* is in fact a linear transformation follows from

$$\begin{aligned}
\langle \mathbf{v}_i, L^*(\mathbf{w} + \mathbf{u}) \rangle &= \langle L\mathbf{v}_i, \mathbf{w} + \mathbf{u} \rangle \\
&= \langle L\mathbf{v}_i, \mathbf{w} \rangle + \langle L\mathbf{v}_i, \mathbf{u} \rangle \\
&= \langle \mathbf{v}_i, L^*\mathbf{w} \rangle + \langle \mathbf{v}_i, L^*\mathbf{u} \rangle \\
\langle \mathbf{v}_i, L^*(c\mathbf{w}) \rangle &= \langle L\mathbf{v}_i, c\mathbf{w} \rangle \\
&= c\langle L\mathbf{v}_i, \mathbf{w} \rangle \\
&= c\langle \mathbf{v}_i, L^*\mathbf{w} \rangle
\end{aligned}$$

The middle equalities follow from linearity properties of the inner product.

Since a linear transformation is determined by its values on a basis, L^* is determined by the inner products $\langle \mathbf{v}_i, L^*\mathbf{v}_j \rangle$. We will have $L^*(\mathbf{V}) = \mathbf{V}B$; that is, L^* is represented by B with respect to this basis. Now $L^*\mathbf{v}_j = \sum_{k=1}^{n} b_{kj}\mathbf{v}_k$, and so

$$\langle \mathbf{v}_i, L^*\mathbf{v}_j \rangle = \langle \mathbf{v}_i, \sum_{k=1}^{n} b_{kj}\mathbf{v}_k \rangle = \langle \mathbf{v}_i, b_{ij}\mathbf{v}_i \rangle = b_{ij}$$

Now suppose that L is represented by the matrix A with respect to this orthonormal basis. Then $L\mathbf{v}_i = \sum_{k=1}^{n} a_{ki}\mathbf{v}_k$, and so using property (2) of the inner product as well as the fact that we have an orthonormal basis gives

$$b_{ij} = \langle L\mathbf{v}_i, \mathbf{v}_j\rangle = \sum_{k=1}^{n} \overline{a}_{ki}\langle \mathbf{v}_k, \mathbf{v}_j\rangle = \overline{a}_{ji}$$

Thus we have proved the following proposition.

PROPOSITION 5.1.1. Suppose that $L : V \to V$ is a linear transformation between a complex inner product space V and itself. Let $\mathcal{V}$ be an orthonormal basis for V, and let $A = [L]_{\mathcal{V}}^{\mathcal{V}}$ be the matrix which represents L with respect to this basis. Then the matrix B which represents the adjoint L^* with respect to this basis is $B = A^*$; that is,

$$[L^*]_{\mathcal{V}}^{\mathcal{V}} = ([L]_{\mathcal{V}}^{\mathcal{V}})^*$$

Although we have given the definition of L^* in the context of a complex inner product space since that is where we will mainly use it, it applies just as well in the context of a real inner product space. In the real vector space context, the adjoint L^* is represented by the transpose of the matrix which represents L when an orthonormal basis is used.

Note that $(L^*)^* = L$. For $(L^*)^*$ is characterized by the equation

$$\langle L^*\mathbf{v}, \mathbf{w}\rangle = \langle \mathbf{v}, (L^*)^*\mathbf{w}\rangle$$

But

$$\langle L^*\mathbf{v}, \mathbf{w}\rangle = \overline{\langle \mathbf{w}, L^*\mathbf{v}\rangle} = \overline{\langle L\mathbf{w}, \mathbf{v}\rangle} = \langle \mathbf{v}, L\mathbf{w}\rangle$$

using property (2) of a Hermitian inner product and the definition of L^*.

DEFINITION 5.1.3. A linear transformation $L : V \to V$ is called **self-adjoint** if $L = L^*$: L will then be represented by a Hermitian matrix with respect to an orthonormal basis. Call L **skew-adjoint** if $L^* = -L$: L will then be represented by a skew Hermitian matrix with respect to an orthonormal basis. Call L **unitary** if $L^* = L^{-1}$: L will be represented by a unitary matrix with respect to an orthonormal basis.

An equivalent characterization of L being unitary is that

$$\langle L\mathbf{v}, L\mathbf{w}\rangle = \langle \mathbf{v}, \mathbf{w}\rangle$$

since

$$\langle L\mathbf{v}, L\mathbf{w}\rangle = \langle \mathbf{v}, L^*L\mathbf{w}\rangle = \langle \mathbf{v}, \mathbf{w}\rangle$$

for all $\mathbf{v}, \mathbf{w}$ iff $L^*L = I$, where I denotes the identity linear transformation.

Note the relation of these definitions to certain classes of complex numbers. Within the complex numbers the real numbers are characterized as those complex numbers $z = a + ib$ with $\bar{z} = z$; the analog here is the self-adjoint linear transformations satisfying $L^* = L$. A complex number of the form $0+ib$ is called pure imaginary (this includes the case $b = 0$). The pure imaginary numbers $z = a + ib$ are characterized by $\bar{z} = -z$; the analog here is the skew-adjoint linear transformations satisfying $L^* = -L$. The complex numbers of length 1 (also called unit complex numbers) are those numbers $z = a + ib$ satisfying $\bar{z}z = z\bar{z} = a^2 + b^2 = 1$; the analog here is the unitary linear transformations satisfying $L^*L = LL^* = I$. We will show that the properties of these three types of linear transformations are reflected in their eigenvalues.

Note that if we start with a complex matrix A and let L_A be the linear transformation corresponding to multiplication by A, then the matrix of L_A with respect to the standard basis of $\mathbb{C}^n$, which is an orthonormal basis, is just A itself. Then L_A is self-adjoint iff A is Hermitian, L_A is skew-adjoint iff A is skew Hermitian, and L_A is unitary iff A is unitary. The statement that there is an orthonormal basis of eigenvectors for L_A just means that if we form a matrix U with these vectors as the column vectors, we will have the equation $AU = UD$, where U is unitary and D is the diagonal matrix with the corresponding eigenvalues down the diagonal. Thus $A = UDU^{-1}$; that is, A is diagonalizable and the diagonalizing matrix can be chosen to be unitary. We wish to show that this is true for Hermitian, skew Hermitian, and unitary matrices (and thus is true for symmetric, skew symmetric, and orthogonal real matrices). We will prove it by showing that there is always an orthonormal basis of eigenvectors for a linear transformation which is self-adjoint, skew-adjoint, or unitary. The first step is to make an observation about the eigenvalues and eigenvectors of such linear transformations.

PROPOSITION 5.1.2.

(1) (a) If L is self-adjoint then its eigenvalues are real.
 (b) If L is skew-adjoint then its eigenvalues are pure imaginary.
 (c) If L is unitary, then its eigenvalues are complex numbers of length 1, that is, $\exp(i\theta)$ for some θ.

(2) If L is self-adjoint, skew-adjoint, or unitary, then eigenvectors corresponding to distinct eigenvalues are mutually orthogonal.

Proof.

(1) Let $\mathbf{v}$ be an eigenvector for the eigenvalue c.
 (a) If $L\mathbf{v} = c\mathbf{v}$, then

$$\bar{c}\langle \mathbf{v}, \mathbf{v}\rangle = \langle c\mathbf{v}, \mathbf{v}\rangle = \langle L\mathbf{v}, \mathbf{v}\rangle = \langle \mathbf{v}, L\mathbf{v}\rangle = c\langle \mathbf{v}, \mathbf{v}\rangle$$

 thus $c = \bar{c}$ and so c is real. Note that this means that the eigenvalues of a Hermitian (and in particular a real symmetric) matrix are real.
 (b) If $L\mathbf{v} = c\mathbf{v}$, then

$$\bar{c}\langle \mathbf{v}, \mathbf{v}\rangle = \langle c\mathbf{v}, \mathbf{v}\rangle = \langle L\mathbf{v}, \mathbf{v}\rangle = -\langle \mathbf{v}, L\mathbf{v}\rangle = -c\langle \mathbf{v}, \mathbf{v}\rangle$$

thus $\overline{c} = -c$, and so c is pure imaginary. Note that this means that the eigenvalues of a skew Hermitian (and in particular a real skew symmetric) matrix are pure imaginary.

(c) If $L\mathbf{v} = c\mathbf{v}$, then

$$\langle \mathbf{v}, \mathbf{v} \rangle = \langle L\mathbf{v}, L\mathbf{v} \rangle = \langle c\mathbf{v}, c\mathbf{v} \rangle = c\overline{c}\langle \mathbf{v}, \mathbf{v} \rangle$$

and so $\|c\| = 1$. Note that this means that the eigenvalues of a unitary (and in particular a real orthogonal) matrix are unit complex numbers, that is, complex numbers of length 1 of the form $e^{i\theta} = \cos\theta + i\sin\theta$. If an eigenvalue is real, then this means that it must be ± 1.

(2) Here are the arguments for self-adjoint and unitary linear transformations; the argument for a skew-adjoint linear transformation is essentially the same as the one for a self-adjoint one. If L is unitary and $\mathbf{v}, \mathbf{w}$ are eigenvectors for eigenvalues c, d, then $\langle \mathbf{v}, \mathbf{w} \rangle = \langle L\mathbf{v}, L\mathbf{w} \rangle = \langle c\mathbf{v}, d\mathbf{w} \rangle = \overline{c}d\langle \mathbf{v}, \mathbf{w} \rangle$ implies $(\overline{c}d - 1)\langle \mathbf{v}, \mathbf{w} \rangle = 0$. Then c, d distinct implies $(\overline{c}d - 1)$ is not zero, so $\langle \mathbf{v}, \mathbf{w} \rangle = 0$. If L is self-adjoint, and $\mathbf{v}, \mathbf{w}$ are eigenvectors for the eigenvalues c, d, then

$$\begin{aligned}\overline{c}\langle \mathbf{v}, \mathbf{w} \rangle &= \langle c\mathbf{v}, \mathbf{w} \rangle = \langle L\mathbf{v}, \mathbf{w} \rangle \\ &= \langle \mathbf{v}, L\mathbf{w} \rangle = \langle \mathbf{v}, d\mathbf{w} \rangle = d\langle \mathbf{v}, \mathbf{w} \rangle\end{aligned}$$

Thus $(\overline{c} - d)\langle \mathbf{v}, \mathbf{w} \rangle = 0$. But c, d are real and distinct, so $\overline{c} - d$ is not zero. Hence $\langle \mathbf{v}, \mathbf{w} \rangle = 0$. ■

Exercise 5.1.5.

Suppose that $L : V \to V$ is a linear transformation from the complex vector space V to itself, and A is the matrix which represents L with respect to an orthonormal basis.

(a) Show that L is unitary iff A is unitary.

(b) Show that L is self-adjoint iff A is Hermitian.

(c) Show that L is skew-adjoint iff A is skew Hermitian.

(d) Show by example (choose $V = \mathbb{C}^2$ and L given by multiplication by a real matrix) that none of these conclusions hold if the basis is not an orthonormal basis.

5.2. THE SPECTRAL THEOREM

We now want to discuss linear transformations L from an n-dimensional complex vector space V to itself which are self-adjoint, skew-adjoint, or unitary. Proposition 5.1.2 shows that if there are n distinct eigenvalues for L, we can find n mutually orthogonal eigenvectors. By making each of them unit vectors, we can get an orthonormal basis of eigenvectors for L. When L arises from a matrix A, this shows that there is a unitary matrix U so that $A = UDU^{-1}$. We now wish to see that this conclusion still holds even if the eigenvalues of L are not distinct.

THEOREM 5.2.1. Spectral Theorem (Complex vector space version)
Linear transformation version. Suppose V is a finite-dimensional complex vector space with a Hermitian inner product, and $L : V \to V$ is a linear transformation which is self-adjoint, skew adjoint, or unitary. Then there is an orthonormal basis of V consisting of eigenvectors for L.

Complex matrix version. If A is a Hermitian, skew Hermitian, or unitary n by n matrix, then there is a unitary matrix U and a diagonal matrix D so that $A = UDU^{-1}$; that is, A is diagonalizable by a unitary matrix.

Proof. We prove the result for the linear transformation version. The matrix version follows from this by applying it to the linear transformation given by multiplication by a matrix A.

Suppose L is self-adjoint, skew-adjoint, or unitary. We will show there is an orthonormal basis consisting of eigenvectors for L. First note that if $\mathbf{v}$ is an eigenvector for L with eigenvalue c and if $\langle \mathbf{v}, \mathbf{w} \rangle = 0$, then $\langle \mathbf{v}, L\mathbf{w} \rangle = 0$; that is, L sends the orthogonal complement of the line through $\mathbf{v}$ to itself. If L is unitary, this follows from

$$0 = \langle \mathbf{v}, \mathbf{w} \rangle = \langle L\mathbf{v}, L\mathbf{w} \rangle = \langle c\mathbf{v}, L\mathbf{w} \rangle = \overline{c}\langle \mathbf{v}, L\mathbf{w} \rangle$$

Since the eigenvalues for unitary matrices have length 1, then $c \neq 0$, so this implies that $\langle \mathbf{v}, L\mathbf{w} \rangle = 0$.

We now treat the self-adjoint case; the skew adjoint case is handled similarly. If L is self-adjoint, then

$$\langle \mathbf{v}, L\mathbf{w} \rangle = \langle L\mathbf{v}, \mathbf{w} \rangle = \langle c\mathbf{v}, \mathbf{w} \rangle = \overline{c}\langle \mathbf{v}, \mathbf{w} \rangle = 0$$

Thus L splits into a linear transformation of each summand of $V = \text{span}(\mathbf{v}) \oplus \text{span}(\mathbf{v})^{\perp}$. Moreover, the adjoint L^* also preserves each summand, since

$$\langle \mathbf{v}, L^*\mathbf{w} \rangle = \langle L\mathbf{v}, \mathbf{w} \rangle = 0$$

implies that $L^*\mathbf{w}$ is still in $\text{span}(\mathbf{v})^{\perp}$ if $\mathbf{w}$ is. Also,

$$\langle L^*\mathbf{v}, \mathbf{w} \rangle = \langle \mathbf{v}, L\mathbf{w} \rangle = 0$$

for all $\mathbf{w} \in \text{span}(\mathbf{v})^{\perp}$ since $L\mathbf{w}$ is still in the orthogonal complement of $\mathbf{v}$. Thus L^* must send $\text{span}(\mathbf{v}) = (\text{span}(\mathbf{v})^{\perp})^{\perp}$ to itself. This implies that $\mathbf{v}$ is also an eigenvector for L^* for some eigenvalue d. The equation

$$\overline{c}\langle \mathbf{v}, \mathbf{v} \rangle = \langle L\mathbf{v}, \mathbf{v} \rangle = \langle \mathbf{v}, L^*\mathbf{v} \rangle = \langle \mathbf{v}, d\mathbf{v} \rangle = d\langle \mathbf{v}, \mathbf{v} \rangle$$

implies that $\overline{c} = d$; that is, the eigenvalues are conjugate.

The result now follows by induction on the dimension n of V. If the dimension is 1, there is nothing to prove. Assuming that the theorem is true for vector spaces of dimension less than n, suppose that the dimension of V is n. Then let c be an eigenvalue for L and $\mathbf{v}$ an eigenvector, which we can take to be a unit vector. Let $W = \text{span}(\mathbf{v})^{\perp}$ and let M denote the restriction of L to W; we showed earlier that this makes sense since L sends W to W. Now the adjoint M^* of M is just the restriction of L^* to W, and so L self-adjoint (skew-adjoint, unitary) implies that M is also. By the induction assumption, there is an orthonormal basis of eigenvectors for M. When we add $\mathbf{v}$ to this basis, we then get an orthonormal basis of V consisting of eigenvectors for L. ■

We might ask what the most general condition is which guarantees that the foregoing conclusions hold. It is easy to check that when the conclusion holds, $LL^* = L^*L$ in the first case (since each map is just a multiplication by a constant on the eigenvectors of L, which are the eigenvectors of L^*). In the second case $AA^* = A^*A$.

DEFINITION 5.2.1. We call L a **normal linear transformation** if $LL^* = L^*L$. Analogously, we call A a **normal matrix** if $AA^* = A^*A$.

The Spectral Theorem will hold under the hypothesis that L (or A) is normal. The proof follows the same outline just given, where the key step is to show that the normality condition implies that L (and L^*) preserve both span($\mathbf{v}$) and span$(\mathbf{v})^{\perp}$ for an eigenvector $\mathbf{v}$, and thus we may proceed by induction. It can be shown directly that any linear transformation which has an orthonormal basis of eigenvectors will be normal.

There is an interesting general consequence of the foregoing proof. For any normal linear transformation L, the eigenvectors of the adjoint L^* are the same as for L, but the eigenvalues are the conjugates of those for L. Without the normality hypothesis, the eigenvalues of any linear transformation $L : V \to V$ and those of its adjoint will be conjugate sets of complex numbers. However, for a general L, there will not be the same set of eigenvectors for L and L^*.

There is one special case of the Spectral Theorem where a conclusion about diagonalizability of real linear transformations and real matrices comes out of the proof. That is when L is self-adjoint (or A is a real symmetric matrix). Then we know that the eigenvalues are real, and so we can find a real eigenvector and take the orthogonal complement in the real vector space V. The argument then produces an orthonormal basis of eigenvectors for the real vector space V.

THEOREM 5.2.2. Spectral Theorem (Real vector space version)

Self-adjoint transformation version. If $L : V \to V$ is a self-adjoint linear transformation from finite dimensional real inner product space V to itself, then there is an orthonormal basis of eigenvectors for L.

Symmetric matrix version. If A is a real n by n symmetric matrix, then there is an orthogonal matrix S and a diagonal matrix D so that $A = SDS^{-1}$.

There is also a version of the Spectral Theorem in terms of projection matrices (and projection linear transformations). First note that if S_i denotes the space spanned by the eigenvectors for the ith eigenvalue c_i, then the restriction of the linear transformation L to S_i just gives multiplication by c_i. If P_i denotes the orthogonal projection onto the subspace S_i, then L agrees with c_iP_i on this subspace. If we take the sum $c_1P_1 + \cdots + c_kP_k$ corresponding to all the distinct eigenvalues $c_1, \ldots, c_k$, then L and this sum will agree on the orthonormal basis of eigenvectors and thus must be equal.

THEOREM 5.2.3. Spectral Theorem (Projection version)

Complex version. If $L : V \to V$ is a normal linear transformation ($LL^* = L^*L$) from a finite-dimensional complex vector space V with Hermitian inner product to itself, then there are projection matrices P_i which project onto orthogonal subspaces S_i (eigenspaces for the distinct eigenvalues c_i) so that $L = c_1P_1 + \cdots + c_kP_k$. Moreover, these eigenspaces are orthogonal, and their sum is V, so $P_1 + \cdots + P_k$ is the identity linear transformation. If A is a normal matrix, then the same conclusion holds with A replacing L and where P_i are projection matrices.

Self-adjoint real linear transformation version. Suppose $L : V \to V$ is a self-adjoint linear transformation from the finite dimensional real inner product space V to itself. If the distinct eigenvalues are $c_1, \ldots, c_k$ and P_i denotes the orthogonal projection onto the ith eigenspace, then

$$L = c_1P_1 + \cdots + c_kP_k$$

Moreover these eigenspaces are orthogonal and their sum is V, so that $P_1 + \cdots + P_k$ is the identity.

Symmetric matrix version. A real symmetric matrix A may be written as $A = c_1P_1 + \cdots + c_kP_k$, where the c_i are the distinct eigenvalues and the P_i are the matrices which project onto the ith eigenspace. These eigenspaces are orthogonal, and their direct sum is $\mathbb{R}^n$, so that $P_1 + \cdots + P_k$ is the identity matrix.

■ ***Example 5.2.1.*** The matrix $A = \begin{pmatrix} 7 & 4 & -5 \\ 4 & -2 & 4 \\ -5 & 4 & 7 \end{pmatrix}$ is symmetric. Its eigenvalues are $6, 12, -6$, and corresponding eigenvectors are $(1, 1, 1), (-1, 0, 1), (1, -2, 1)$. These are pairwise orthogonal. When they are made into unit vectors by dividing by their lengths and made the columns of a matrix, it gives an orthogonal matrix S so that $A = SDS^{-1}$, with $S = \begin{pmatrix} 1/\sqrt{3} & -1/\sqrt{2} & 1/\sqrt{6} \\ 1/\sqrt{3} & 0 & -2/\sqrt{6} \\ 1/\sqrt{3} & 1/\sqrt{2} & 1/\sqrt{6} \end{pmatrix}$ and $D = \begin{pmatrix} 6 & 0 & 0 \\ 0 & 12 & 0 \\ 0 & 0 & -6 \end{pmatrix}$. We can write A as a linear combination of projection matrices

$$\begin{aligned} A &= 6P_6 + 12P_{12} - 6P_{-6} \\ &= 6\begin{pmatrix} 1/3 & 1/3 & 1/3 \\ 1/3 & 1/3 & 1/3 \\ 1/3 & 1/3 & 1/3 \end{pmatrix} + 12\begin{pmatrix} 1/2 & 0 & -1/2 \\ 0 & 0 & 0 \\ -1/2 & 0 & 1/2 \end{pmatrix} - 6\begin{pmatrix} 1/6 & -1/3 & 1/6 \\ -1/3 & 2/3 & -1/3 \\ 1/6 & -1/3 & 1/6 \end{pmatrix} \end{aligned}$$

■

■ ***Example 5.2.2.*** The matrix $A = \begin{pmatrix} 0 & 1 & 1 \\ -1 & 0 & 2 \\ -1 & -2 & 0 \end{pmatrix}$ is skew symmetric. Note that skew symmetric matrices have 0 along the diagonal. All the eigenvalues are pure imaginary. Since the characteristic polynomial is of odd degree, it must have a real root. This implies that one of the eigenvalues is 0. The other two have to add up to 0 since the sum of all the eigenvalues is equal to the sum of the diagonal entries, which is 0. The eigenvalues turn out to be $0, \sqrt{5}i, -\sqrt{5}i$. The corresponding eigenvectors, which are pairwise orthogonal using the Hermitian inner product on $\mathbb{C}^n$, are

$$(2,-1,1), (-.4-.4899i, .2-.9798i, 1), (-.4+.4899i, .2+.9798i, 1)$$

When we divide by their lengths and make them the columns of a matrix, we get $A =$ SDS^{-1} with $S = \begin{pmatrix} 2/\sqrt{6} & -.2582-.3162i & -.2582+.3162i \\ -1/\sqrt{6} & .1291-.6325i & .1291+.6325i \\ 1/\sqrt{6} & .6455 & .6455 \end{pmatrix}$. ■

■ ***Example 5.2.3.*** The matrix $A = \begin{pmatrix} 1/\sqrt{3} & 2/\sqrt{6} & 0 \\ 1/\sqrt{3} & -1/\sqrt{6} & -1/\sqrt{2} \\ 1/\sqrt{3} & -1/\sqrt{6} & 1/\sqrt{2} \end{pmatrix}$ is orthogonal. Its eigenvalues are complex numbers of length 1, one of which must be real since the characteristic polynomial is of odd degree. The eigenvalues are $-1, .9381+.3464i, .9381-.3464i$. The eigenvectors are $(-.5176, 1, .4142)$, $(.1691+.9463i, -.3267+.4898i, 1)$, $(.1691-.9463i, -.3267-.4898i, 1)$.

Normalizing these vectors and putting them into the columns of a matrix gives a unitary matrix

$$S = \begin{pmatrix} -.4314 & .1122+.6280i & .1122-.6280i \\ .8335 & -.2168+.3251i & -.2168-.3251i \\ .3452 & .6636 & .6636 \end{pmatrix}$$

with $A = SDS^{-1}$. Here D is the diagonal matrix

$$\begin{pmatrix} -1 & 0 & 0 \\ 0 & .9381+.3464i & 0 \\ 0 & 0 & .9381-.3464i \end{pmatrix}$$ ■

■ ***Example 5.2.4.*** The matrix $A = \begin{pmatrix} 1/2-1/2i & 1/2+1/2i & 0 \\ 0 & 0 & 1 \\ 1/2+1/2i & 1/2-1/2i & 0 \end{pmatrix}$ is a unitary matrix. Thus its eigenvalues are complex numbers of length 1. They are $-.9114+.4114i, .4114-.9114i, 1$. Note that the two complex eigenvalues are not conjugates as in the case of a 3 by 3 real matrix. Solving for the eigenvectors, which are orthogonal, and making them unit vectors gives a unitary matrix S and a diagonal matrix D with $A = SDS^{-1}$. Here

$$S = \begin{pmatrix} -.06 \quad -.2788i & .7651 & 1/\sqrt{3} \\ -.6177+.2788i & -.3825+.247i & 1/\sqrt{3} \\ .6777 & -.3825-.247i & 1/\sqrt{3} \end{pmatrix}$$

and

$$D = \begin{pmatrix} -.9114 + .4114i & 0 & 0 \\ 0 & .4114 - .9114i & 0 \\ 0 & 0 & 1 \end{pmatrix}$$ ■

Exercise 5.2.1.

For each matrix A give a unitary matrix S and a diagonal matrix D with $A = SDS^{-1}$.

(a) $\begin{pmatrix} 5 & -2 & -1 \\ -2 & 5 & 1 \\ -1 & 1 & 8 \end{pmatrix}$

(b) $\begin{pmatrix} 0 & 2 & 1 \\ -2 & 0 & -1 \\ -1 & 1 & 0 \end{pmatrix}$

(c) $\begin{pmatrix} -1 & -1 - 12i & 2 + 12i \\ -1 + 12i & 2 & -1 - 12i \\ 2 - 12i & -1 + 12i & -1 \end{pmatrix}$

(d) $\begin{pmatrix} 1/\sqrt{2} & 1/\sqrt{2} & 0 \\ 1/\sqrt{2} & -1/\sqrt{2} & 0 \\ 0 & 0 & 1 \end{pmatrix}$

Exercise 5.2.2.

For the matrices in the previous exercise, write the matrix as a linear combination of projection matrices.

Exercise 5.2.3.

Show that $A + A^*$ is Hermitian and $A - A^*$ is skew Hermitian. Verify that any matrix is the sum of a Hermitian matrix and a skew Hermitian matrix. Use this to show that any matrix can be written as a linear combination of projection matrices.

Exercise 5.2.4.

Give an example of a normal matrix which is not Hermitian, skew Hermitian, or unitary. (Hint: The required matrix should be diagonalizable by a unitary matrix, and its eigenvalues cannot be all real or all pure imaginary or all unit complex numbers.)

Exercise 5.2.5.

Show that the following statements are equivalent.

(a) $L : V \to V$ is normal.

(b) $\langle L^*L\mathbf{v}, \mathbf{w}\rangle = \langle LL^*\mathbf{v}, \mathbf{w}\rangle$ for all $\mathbf{v}, \mathbf{w} \in V$.

(c) $\langle L^*L\mathbf{v}_i, \mathbf{v}_j\rangle = \langle LL^*\mathbf{v}_i, \mathbf{v}_j\rangle$ for *some* basis $\mathbf{v}_1, \ldots, \mathbf{v}_n$ of V.

(d) $\langle L\mathbf{v}_i, L\mathbf{v}_j\rangle = \langle L^*\mathbf{v}_i, L^*\mathbf{v}_j\rangle$ for *some* basis $\mathbf{v}_1, \ldots, \mathbf{v}_n$ of V.

Exercise 5.2.6.

Suppose $L : V \to V$ is diagonalizable with an orthonormal basis $\mathbf{v}_1, \ldots, \mathbf{v}_n$ of eigenvectors corresponding to the eigenvalues c_i.

(a) Show that L^* is also diagonalizable with the same basis of eigenvectors and eigenvalues $\overline{c_i}$.

(b) Use part (d) of the preceding exercise to show that L is normal.

Exercise 5.2.7.

Follow the outline in this exercise to show that for any square matrix A there is a unitary matrix S and an upper triangular matrix T with $A = STS^{-1}$. Outline: Prove the result by induction on the size n of A. The result is trivial if $n = 1$. Assume the result is true if $n < k$ and prove it for $n = k$. First find an eigenvalue λ and eigenvector $\mathbf{v}$ for A, where $\mathbf{v}$ is a unit vector. Then let $\mathbf{v} = \mathbf{v}_1$ and extend it to an orthonormal basis for $\mathbb{C}^n$. Let S_1 be the matrix with this basis as its columns and verify that $\mathbf{v}_1$ being an eigenvector implies that $AS_1 = S_1B$, where $B = \begin{pmatrix} \lambda & D \\ \mathbf{0} & C \end{pmatrix}$ and where C is $(k-1)$ by $(k-1)$. Then apply the inductive hypothesis to give a unitary matrix U with $U^{-1}CU = T_1$, a triangular $(k-1)$ by $(k-1)$ matrix. Let $S_2 = \begin{pmatrix} 1 & \mathbf{0} \\ \mathbf{0} & U \end{pmatrix}$ and verify that $S_2^{-1}BS_2 = T$, a triangular matrix. Check that if $S = S_1S_2$, then $A = STS^{-1}$.

5.3. SINGULAR VALUE DECOMPOSITION AND PSEUDOINVERSE

Now that we have learned about the properties of real symmetric matrices, let us return to the problem of finding the singular value decomposition and the pseudoinverse for a matrix A. Recall that we wish to find orthogonal matrices $Q_1 = (Q_{1r} \quad Q_{1n})$ and $Q_2 = (Q_{2r} \quad Q_{2n})$ and a matrix Σ with the **singular values** down the diagonal and all the off-diagonal entries 0 so that $A = Q_1\Sigma Q_2^t$. The singular values are ordered

$\mu_1 \geq \ldots \geq \mu_r > 0 = \ldots = 0$. Here r is the rank of A. The columns of Q_{2r} form an orthonormal basis of the row space of A, the columns of Q_{2n} form an orthonormal basis of the null space of A, the columns of Q_{1r} form an orthonormal basis for the range of A, and the columns of Q_{1n} form an orthonormal basis for the null space of A^t.

We saw earlier that $A^tA = Q_2\Sigma^t Q_1^t Q_1 \Sigma Q_2^t = Q_2\Sigma^t\Sigma Q_2^t$. Thus the squares of the singular values occur as the eigenvalues of A^tA, and the matrix Q_2 occurs as the matrix which contains the eigenvectors which correspond to those eigenvalues. A^tA is a symmetric matrix, and the Spectral Theorem states that its eigenvalues are real, and there is an orthonormal basis of eigenvectors, allowing us to find Q_2. Since A^tA has rank r, there must be r nonzero eigenvalues, which we can order $\sigma_1 > \ldots > \sigma_r > 0$, and the remaining eigenvalues are 0. The relation between the σ_i and μ_i is $\sigma_i = \mu_i^2$. Recall that we showed in our discussion of the singular value decomposition in Chapter 3 that these nonzero eigenvalues were in fact positive.

Thus we can find an orthonormal basis $\mathbf{q}_1, \ldots, \mathbf{q}_r$ of eigenvectors for the nonzero eigenvalues. Since these vectors are orthogonal to the eigenvectors for the eigenvalue 0, which is just the null space of A, these will be a basis for the row space. We can then select an orthonormal basis for the null space. For example, we could apply the Gram–Schmidt process to any basis. Call this basis $\mathbf{q}_{r+1}, \ldots, \mathbf{q}_n$. These n vectors will comprise the columns of Q_2. To find Q_1, the part Q_{1n} just comes from finding an orthonormal basis for the null space of A^t. It thus remains to find Q_{1r}. Define $\mathbf{q}_i' = A\mathbf{q}_i/\mu_i$, $i = 1, \ldots, r$. Note that the μ_i^2 are just the nonzero eigenvalues of A^tA. Then the $\mathbf{q}_i'$ form an orthonormal basis of vectors in the range of A. That these are orthogonal unit vectors follows from

$$\begin{aligned}\langle \mathbf{q}_i', \mathbf{q}_j' \rangle &= (1/\mu_i\mu_j)\langle A\mathbf{q}_i, A\mathbf{q}_j \rangle = (1/\mu_i\mu_j)\langle \mathbf{q}_i, A^tA\mathbf{q}_j \rangle \\ &= (\mu_j/\mu_i)\langle \mathbf{q}_i, \mathbf{q}_j \rangle = \begin{cases} 0 & i \neq j \\ 1 & i = j \end{cases}\end{aligned}$$

By the way that Q_1, Σ, Q_2 are defined, we will have the required equation $A = Q_1\Sigma Q_2^t$. A simpler form of this which may be used is $A = Q_{1r}\Sigma_r Q_{2r}^t$. This is a lot easier to find by our methods, although standard computational techniques find the full singular value decomposition. Here $\Sigma_r = \text{diag}(\mu_1, \ldots, \mu_r)$, where by this notation we mean the diagonal matrix with entries $\mu_1, \ldots, \mu_r$. Now the pseudoinverse A^+ is found using the simpler form of the singular value decomposition as $A^+ = Q_{2r}\Sigma_r^+ Q_{1r}^t$. Recall that $A^+\mathbf{b}$ is the vector $\mathbf{x}_r$ in the row space which is the least squares solution to $A\mathbf{x} = \mathbf{b}$ of minimal length.

■ ***Example 5.3.1.*** $A = \begin{pmatrix} 1 & 2 & 0 & 1 \\ 1 & 0 & 2 & 1 \\ 2 & 2 & 2 & 2 \end{pmatrix}$. Then $A^tA = \begin{pmatrix} 6 & 6 & 6 & 6 \\ 6 & 8 & 4 & 6 \\ 6 & 4 & 8 & 6 \\ 6 & 6 & 6 & 6 \end{pmatrix}$. This has characteristic polynomial $x^2(x^2-28x+96) = x^2(x-24)(x-4)$. Thus the nonzero eigenvalues of A^tA are 24, 4, and the rank of A is 2. We solve for an orthonormal basis for the null space A^tA and get $\mathbf{q}_3 = 1/\sqrt{12}(-3, 1, 1, 1)$ and $\mathbf{q}_4 = 1/\sqrt{6}(0, -1, -1, 2)$.

We compute the eigenvector for the eigenvalue 24 and get $\mathbf{q}_1 = (1/2)(1,1,1,1)$. The eigenvector for the eigenvalue 4 is computed as $\mathbf{q}_2 = (1/\sqrt{2})(0,-1,1,0)$, and thus

$$Q_2 = \begin{pmatrix} 1/2 & 0 & -3/\sqrt{12} & 0 \\ 1/2 & -1/\sqrt{2} & 1/\sqrt{12} & -1/\sqrt{6} \\ 1/2 & 1/\sqrt{2} & 1/\sqrt{12} & -1/\sqrt{6} \\ 1/2 & 0 & 1/\sqrt{12} & 2/\sqrt{6} \end{pmatrix}$$

To get Q_1 we first compute $\mathbf{q}_1' = A\mathbf{q}_1/\sqrt{24} = (1/\sqrt{6})(1,1,2)$ and $\mathbf{q}_2' = A\mathbf{q}_2/2 = (1/\sqrt{2})(-1,1,0)$. We get $\mathbf{q}_3'$ by finding a basis for the null space of A^t, or just finding a third unit vector which is orthogonal to $\mathbf{q}_1', \mathbf{q}_2'$. We solve for $\mathbf{q}_3' = (1/\sqrt{3})(-1,-1,1)$. Thus

$$Q_1 = \begin{pmatrix} 1/\sqrt{6} & -1/\sqrt{2} & -1/\sqrt{3} \\ 1/\sqrt{6} & 1/\sqrt{2} & -1/\sqrt{3} \\ 2/\sqrt{6} & 0 & 1/\sqrt{3} \end{pmatrix} \quad \text{and} \quad \Sigma = \begin{pmatrix} \sqrt{24} & 0 & 0 & 0 \\ 0 & 2 & 0 & 0 \\ 0 & 0 & 0 & 0 \end{pmatrix}$$

To find the pseudoinverse A^+, we form

$$Q_{2r}\Sigma_r^+ Q_{1r}^t = A^+ = \begin{pmatrix} .0417 & .0417 & .0833 \\ .2917 & -.2083 & .0833 \\ -.2083 & .2917 & .0833 \\ .0417 & .0417 & .0833 \end{pmatrix} \qquad \blacksquare$$

■ ***Example 5.3.2.*** In this example we concentrate on finding the pseudoinverse and using it to solve the least squares problem $A\mathbf{x} = \mathbf{b}$, where

$$A = \begin{pmatrix} 1 & 2 & 1 \\ 0 & 3 & 3 \\ 1 & 3 & 2 \\ 2 & 0 & -2 \end{pmatrix}, \ \mathbf{b} = \begin{pmatrix} 1 \\ 0 \\ 0 \\ 0 \end{pmatrix}$$

We first compute $B = A^tA = \begin{pmatrix} 6 & 5 & -1 \\ 5 & 22 & 17 \\ -1 & 17 & 18 \end{pmatrix}$. We then write $B = SDS^{-1}$, with

$$S = \begin{pmatrix} .0988 & .8105 & .5774 \\ .7513 & .3197 & -.5774 \\ .6525 & -.4908 & .6525 \end{pmatrix}, \ D = \begin{pmatrix} 37.4222 & 0 & 0 \\ 0 & 8.5778 & 0 \\ 0 & 0 & 0 \end{pmatrix}$$

Thus

$$Q_{2r} = \begin{pmatrix} .0988 & .8105 \\ .7513 & .3197 \\ .6525 & -.4908 \end{pmatrix}, \quad \Sigma_r = \begin{pmatrix} \sqrt{37.4222} & 0 \\ 0 & \sqrt{8.5778} \end{pmatrix} = \begin{pmatrix} 6.1174 & 0 \\ 0 & 2.9288 \end{pmatrix}$$

We then compute

$$\mathbf{q}_1' = A\mathbf{q}_1/\mu_1 = (.3684, .6884, .5979, -.1810),$$
$$\mathbf{q}_2' = A\mathbf{q}_2/\mu_2 = (.3275, -.1753, .2691, .8886)$$

Thus

$$Q_{1r} = \begin{pmatrix} .3684 & .3275 \\ .6884 & -.1753 \\ .5979 & .2691 \\ -.1810 & .8886 \end{pmatrix}$$

and

$$A^+ = Q_{2r}\Sigma_r^+ Q_{1r}^t = \begin{pmatrix} .0966 & -.0374 & .0841 & .2430 \\ .0810 & .0654 & .1028 & .0748 \\ -.0156 & .1028 & .0187 & -.1682 \end{pmatrix}$$

The solution to the least squares problem is

$$A^+\mathbf{e}_1 = (.0966, .0810, -.0156)$$

■

Exercise 5.3.1.

Find the singular value decomposition and pseudoinverse of

$$A = \begin{pmatrix} 1 & 2 \\ 2 & 4 \\ 3 & 6 \end{pmatrix}$$

Exercise 5.3.2.

Find the singular value decomposition and pseudoinverse for

$$A = \begin{pmatrix} 1 & 2 & 0 \\ 0 & 1 & 1 \end{pmatrix}$$

Exercise 5.3.3.

Find the pseudoinverse of $A = \begin{pmatrix} 1 & -1 & 0 \\ 2 & 1 & 3 \\ -1 & 2 & 1 \\ 0 & 1 & 1 \end{pmatrix}$ and use it to solve the least squares problem $A\mathbf{x} = \mathbf{b}$ with $\mathbf{b} = (1, 1, 1, 1)$.

5.4. THE ORTHOGONAL GROUP—ROTATIONS AND REFLECTIONS IN $\mathbb{R}^3$

Suppose A is a 3 by 3 real orthogonal matrix. Then its eigenvalues must be of unit length. Also, the eigenvectors for distinct eigenvalues must be orthogonal to each other. Moreover, since a cubic polynomial must have at least one real root, there will be at least one real eigenvalue, which must be ± 1.

If all the eigenvalues are real, then there are four cases, depending on how many are -1. The analysis in each case depends on the fact that a linear transformation is completely determined by its value on a basis, so we show that two linear transformations agree by finding a basis on which they agree.

1. If they are all 1, then A is the identity, since there will be a basis of vectors (the eigenvectors for the eigenvalue 1) which A sends to itself.
2. If one eigenvalue is 1 and the other two are -1, then A is a rotation which fixes the axis in the direction of the eigenvector with eigenvalue 1 and rotates the orthogonal plane by π. Suppose $\mathbf{v}_1, \mathbf{v}_2, \mathbf{v}_3$ is an orthonormal basis of eigenvectors, with $\mathbf{v}_1$ the eigenvector for the eigenvalue 1 and $\mathbf{v}_2, \mathbf{v}_3$ eigenvectors for the eigenvalue -1. The effect of this rotation is to send $\mathbf{v}_2, \mathbf{v}_3$ to their negatives and to send $\mathbf{v}_1$ to itself.
3. When all are -1, A is just the preceding rotation followed by the reflection through the plane. This follows since this map agrees with A on the foregoing basis. Alternatively, A can be thought of as reflection through the origin.
4. Finally, when one is -1 and the other two are 1, then the eigenspace for the eigenvalue 1 is a plane and A is the reflection through this plane. If $\mathbf{v}_1, \mathbf{v}_2, \mathbf{v}_3$ is an orthonormal basis with $\mathbf{v}_1$ the eigenvector for the eigenvalue -1, and $\mathbf{v}_2, \mathbf{v}_3$ eigenvectors for the eigenvalue 1, then A agrees with this reflection on this basis.

Let us now consider the case where there is a complex eigenvalue which is not real. A complex eigenvalue will be of the form $\cos\theta + i\sin\theta$, and we have an eigenvector $\mathbf{v} + i\mathbf{w}$. Note that the conjugate $\cos\theta - i\sin\theta$ is also an eigenvalue with eigenvector $\mathbf{v} - i\mathbf{w}$. But

$$A(\mathbf{v} + i\mathbf{w}) = (\mathbf{v}\cos\theta - \mathbf{w}\sin\theta) + i(\mathbf{v}\sin\theta + \mathbf{w}\cos\theta)$$

so multiplication by A sends the plane spanned by $\mathbf{v}, \mathbf{w}$ to itself. We claim that $\mathbf{v}$ and $\mathbf{w}$ have to be orthogonal to each other and must be of the same length. Since $\mathbf{v} + i\mathbf{w}$ and $\mathbf{v} - i\mathbf{w}$ are eigenvectors for distinct eigenvalues, they are orthogonal. But the inner product $\langle \mathbf{v} + i\mathbf{w}, \mathbf{v} - i\mathbf{w}\rangle = (\langle \mathbf{v}, \mathbf{v}\rangle - \langle \mathbf{w}, \mathbf{w}\rangle) - 2i\langle \mathbf{v}, \mathbf{w}\rangle$, so we conclude that $\langle \mathbf{w}, \mathbf{w}\rangle = \langle \mathbf{v}, \mathbf{v}\rangle$ and $\langle \mathbf{v}, \mathbf{w}\rangle = 0$. Now if $\mathbf{v}, \mathbf{w}$ are chosen to be unit vectors, then the plane with orthonormal basis $\mathbf{w}, \mathbf{v}$ is sent to itself by rotating by an angle θ. For $A\mathbf{w} = \cos\theta\mathbf{w} + \sin\theta\mathbf{v}$, $A\mathbf{v} = -\sin\theta\mathbf{w} + \cos\theta\mathbf{v}$. The matrix representing the rotation of this plane with respect to the basis $\mathbf{w}, \mathbf{v}$ is just the standard rotation matrix $\begin{pmatrix} \cos\theta & -\sin\theta \\ \sin\theta & \cos\theta \end{pmatrix}$. The orthogonal line is either sent to itself identically, in which case L represents a rotation with axis given by the orthogonal line and angle θ, or the other eigenvalue is -1 and so L represents this rotation followed by a reflection through the plane.

■ ***Example 5.4.1.*** The matrix $A = \begin{pmatrix} .8536 & -.1464 & -.5 \\ -.1464 & .8536 & -.5 \\ .5 & .5 & .7071 \end{pmatrix}$ is orthogonal.

Its eigenvalues are $1, 1/\sqrt{2} + 1/\sqrt{2}i, 1/\sqrt{2} - 1/\sqrt{2}i$. The last two are $\cos\pi/4 \pm$

$i\sin\pi/4$. The eigenvectors are $(1/\sqrt{2}, -1/\sqrt{2}, 0)$ and $(0,0,1) \pm (1/\sqrt{2}, 1/\sqrt{2}, 0)i$.

Thus if $S = \begin{pmatrix} 1/\sqrt{2} & 1/\sqrt{2} & 0 \\ -1/\sqrt{2} & 1/\sqrt{2} & 0 \\ 0 & 0 & 1 \end{pmatrix}$ and $C = \begin{pmatrix} 1 & 0 & 0 \\ 0 & 1/\sqrt{2} & -1/\sqrt{2} \\ 0 & 1/\sqrt{2} & 1/\sqrt{2} \end{pmatrix}$, we have $A = SCS^{-1}$. Multiplication by A gives a rotation of 45 degrees of planes perpendicular to the axis through the vector $(1/\sqrt{2}, -1/\sqrt{2}, 0)$. ■

■ ***Example 5.4.2.*** $A = \begin{pmatrix} 2/3 & 1/3 & 2/3 \\ 1/3 & 2/3 & -2/3 \\ 2/3 & -2/3 & -1/3 \end{pmatrix}$ has determinant -1. There is an orthonormal basis of eigenvectors

$$(1/\sqrt{2}, 1/\sqrt{2}, 0), (1/\sqrt{3}, -1/\sqrt{3}, 1/\sqrt{3}), (-1/\sqrt{6}, 1/\sqrt{6}, 2/\sqrt{6})$$

where the first two are eigenvectors for the eigenvalue 1 and the last one is an eigenvector for the eigenvalue -1. Thus multiplication by A is a reflection through the plane with basis the first two vectors; a normal vector to this plane is given by the last vector, so its equation is $-x + y + 2z = 0$. ■

The n by n orthogonal matrices form a group under matrix multiplication, which is called the **orthogonal group** and is denoted $O(n)$. We saw earlier that elements of $O(3)$ correspond geometrically to compositions of rotations and reflections. To see that this is a group, we check that when we multiply two orthogonal matrices we get an orthogonal matrix: if $A^tA = I, B^tB = I$, then $(AB)^t(AB) = (B^tA^t)AB = B^t(A^tA)B = B^tB = I$. The identity matrix I is in $O(n)$ and serves as the identity for this group operation. The inverse of A is A^t, which is also orthogonal. Since matrix multiplication is associative, the group operation obeys the associative law. Note that this group is not abelian. Sitting inside $O(n)$ is a subgroup of all orthogonal matrices which have determinant 1. This subgroup is called the **special orthogonal group** and is denoted $SO(n)$. That $SO(n)$ is a subgroup follows from the fact that the product of matrices with determinant 1 will have determinant 1 and the inverse of a matrix with determinant 1 will have determinant 1. Note that a general element of $O(n)$ will have determinant ± 1, since $A^tA = I$ implies $1 = \det A^t \det A = (\det A)^2$. Let r denote the diagonal matrix whose first diagonal entry is -1 and all other diagonal entries are 1. Then multiplication by r just reflects through the plane where the first coordinate is 0. Note that the determinant of r is -1. If A is any orthogonal matrix whose determinant is -1, then the determinant of rA will be 1, and so rA will be in $SO(n)$. Thus every element A of $O(n)$ will either be in $SO(n)$ or rA will be in $SO(n)$. In group theory $SO(n)$ is called a **subgroup of index 2** of $O(n)$ (the quotient is the integers modulo 2).

We now generalize our discussion about $O(3)$ and give a description of $O(n)$ in terms of rotations and reflections. We first describe $O(2)$, which is somewhat special. Here there are only two things which can happen concerning the eigenvalues. Either

1. A is in $SO(2)$. There are two complex conjugate eigenvalues $\exp(i\theta)$ and $\exp(-i\theta)$. This includes as a special case when both are 1 ($\theta = 0$) or both

are $-1(\theta = \pi)$. Then by the preceding discussion there will be an orthonormal basis for $\mathbb{R}^2$ so that A just represents a rotation by θ in terms of this basis. But this will mean that A just rotates all of $\mathbb{R}^2$ by θ. This could also have been seen more directly in that the condition that A is in the orthogonal group will mean that the columns of A must be orthogonal unit vectors. The fact that $\det A = 1$ will imply that A must be of the form $A = \begin{pmatrix} \cos\theta & -\sin\theta \\ \sin\theta & \cos\theta \end{pmatrix}$, which is the matrix which represents rotation by θ.

2. A is not in $SO(2)$. Then the preceding discussion shows rA is in $SO(2)$, and so A is the composition of a rotation and a reflection. Actually, we can show that A represents a reflection. This can be done directly by showing $A = \begin{pmatrix} \cos\theta & \sin\theta \\ \sin\theta & -\cos\theta \end{pmatrix}$ and then figuring out which line A reflects through. A nicer approach is to note that A must have eigenvalues 1 and -1 (with orthogonal eigenvectors), and so A must be the reflection through the line which goes through the eigenvector with eigenvalue 1. Applying this to $A = \begin{pmatrix} \cos\theta & \sin\theta \\ \sin\theta & -\cos\theta \end{pmatrix}$ and writing $\theta = 2\phi$, then $A - I = \begin{pmatrix} \cos 2\phi - 1 & \sin 2\phi \\ \sin 2\phi & -\cos 2\phi - 1 \end{pmatrix} = 2\begin{pmatrix} -\sin^2\phi & \sin\phi\cos\phi \\ \sin\phi\cos\phi & -\cos^2\phi \end{pmatrix}$ and so an eigenvector for eigenvalue 1 is given by $(\cos\phi, \sin\phi)$; that is, A represents the reflection through the line through $(\cos\phi, \sin\phi)$.

■ ***Example 5.4.3.*** For $A = \begin{pmatrix} 1/\sqrt{5} & -2/\sqrt{5} \\ 2/\sqrt{5} & 1/\sqrt{5} \end{pmatrix}$, we see that $A = \begin{pmatrix} \cos\theta & -\sin\theta \\ \sin\theta & \cos\theta \end{pmatrix}$ with $\theta = \arctan 2 = 1.071$ radians. Thus multiplication by A rotates the plane by 1.071 radians. For the related matrix $B = \begin{pmatrix} 1/\sqrt{5} & 2/\sqrt{5} \\ 2/\sqrt{5} & -1/\sqrt{5} \end{pmatrix}$, we have $B = \begin{pmatrix} \cos\theta & \sin\theta \\ \sin\theta & -\cos\theta \end{pmatrix}$ for the same angle θ. Thus multiplication by B gives reflection through the line whose angle ϕ with the x-axis is $1.071/2 = .5535$ radians. Alternatively, we can find the eigenvector for the eigenvalue 1, which is (.8507,.5257), and multiplication by B will give the reflection through the line going through this vector; that is, the line $y = .6180x$. ■

Now let us return to $O(n)$. The eigenvalues will have to be $1, -1$ or be in conjugate pairs $\exp(i\theta), \exp(-i\theta)$. For each conjugate pair, our analysis in the 3 by 3 case shows that there will be a two-dimensional subspace so that multiplication by A will just rotate this subspace by the angle θ. For any pair of orthogonal eigenvectors for the eigenvalue -1, multiplication by A will just rotate this plane by the angle π. If the determinant is 1 (resp., -1), then the algebraic multiplicity of -1 will be even (resp., odd).

Finally, the eigenspace associated with the eigenvalue 1 will just be sent identically to itself. Thus multiplication by A will preserve the $+1$ eigenspace $E(1)$. If $A \in SO(n)$, then the orthogonal complement of $E(1)$ will be split into an orthogonal direct sum of two-dimensional subspaces, each of which will be rotated by some angle θ corresponding to eigenvalues $\exp(\pm i\theta)$. If $A \in O(n) \setminus SO(n)$, then a one-dimensional subspace with eigenvalue -1 can be separated off, and we may pair up the other eigenvectors for eigenvalue -1 to give rotations by π. It can be shown that these two-dimensional planes arising for distinct nonconjugate complex eigenvalues are mutually orthogonal. Also, when there is an algebraic multiplicity which is greater than 1 for a complex eigenvalue, we can use an orthonormal basis of the eigenspace to get a number of orthogonal planes which are rotated by the same angle. This leads to an orthogonal basis for $\mathbb{R}^n$ for which the matrix representing multiplication by A splits into 2 by 2 pieces corresponding to the complex eigenvalues and pairs of -1s. We can choose the order of the orthogonal basis so that it determines a matrix in $SO(n)$. There are numerous details to check (which are left as an exercise at the end of the chapter), but this leads to the following form of the Spectral Theorem.

THEOREM 5.4.1. Spectral Theorem for O(n). If A is in $SO(n)$, then there is a matrix B in $SO(n)$ so that

$$B^{-1}AB = \begin{pmatrix} I & 0 & \dots & 0 \\ 0 & R_{\theta_1} & \dots & 0 \\ \vdots & \vdots & \vdots & \vdots \\ 0 & 0 & \dots & R_{\theta_k} \end{pmatrix}$$

where I denotes a p by p identity block and each R_{θ_i} denotes a 2 by 2 block which is a rotation by θ_i (including $\theta = \pi$). If A is in $O(n)$ but has determinant -1, then a similar result is true with an additional -1 on the diagonal.

■ ***Example 5.4.4.*** The matrix $A = \begin{pmatrix} 1/\sqrt{2} & 0 & 1/2 & 1/2 \\ 0 & 1/\sqrt{2} & 1/2 & -1/2 \\ -1/\sqrt{2} & 0 & -1/2 & -1/2 \\ 0 & 1/\sqrt{2} & -1/2 & 1/2 \end{pmatrix}$ is orthogonal with determinant -1. It diagonalizes as $A = SDS^{-1}$ with

$$S = \begin{pmatrix} -.3536 & .8536 & .2464 - .1118i & .2464 + .1118i \\ -.1464 & .3536 & -.5949 + .2700i & -.5949 - .2700i \\ .8536 & .3536 & -.1118 - .2464i & -.1118 + .2464i \\ .3536 & .1464 & .2700 + .5949i & .2700 - .5949i \end{pmatrix}$$

$$D = \begin{pmatrix} -1 & 0 & 0 & 0 \\ 0 & 1 & 0 & 0 \\ 0 & 0 & .7071 + .7071i & 0 \\ 0 & 0 & 0 & .7071 - .7071i \end{pmatrix}$$

Then if $B = \begin{pmatrix} -.3536 & .8536 & .1118 & .2464 \\ -.1464 & .3536 & .2700 & -.5949 \\ .8536 & .3536 & -.2464 & -.111 \\ .3536 & .1464 & .5949 & .2700 \end{pmatrix}$, we have

$$B^{-1}AB = \begin{pmatrix} -1 & 0 & 0 & 0 \\ 0 & 1 & 0 & 0 \\ 0 & 0 & \cos\pi/4 & -\sin\pi/4 \\ 0 & 0 & \sin\pi/4 & \cos\pi/4 \end{pmatrix}$$ ■

Exercise 5.4.1.

In parts (a) and (b), identify whether multiplication by the given matrix represents a rotation or a reflection in $\mathbb{R}^2$. If it is a rotation, give the angle of rotation. If it is a reflection, give the line through which one reflects.

(a) $\begin{pmatrix} 3/5 & 4/5 \\ 4/5 & -3/5 \end{pmatrix}$

(b) $\begin{pmatrix} 3/5 & -4/5 \\ 4/5 & 3/5 \end{pmatrix}$

Exercise 5.4.2.

The following matrix represents a rotation in $\mathbb{R}^3$. Give the axis of rotation and the angle through which the plane perpendicular to that axis is rotated.

$$A = \begin{pmatrix} 2/3 & -1/3 & 2/3 \\ 2/3 & 2/3 & -1/3 \\ -1/3 & 2/3 & 2/3 \end{pmatrix}$$

Exercise 5.4.3.

Describe the operation of multiplying by the following matrix in terms of a reflection.

$$B = \begin{pmatrix} 1/9 & -8/9 & 4/9 \\ -8/9 & 1/9 & 4/9 \\ 4/9 & 4/9 & 7/9 \end{pmatrix}$$

Exercise 5.4.4.

Consider the matrix

$$A = \begin{pmatrix} .5772 & -.0937 & .1299 & .8008 \\ .0937 & .5772 & -.8008 & .1299 \\ .1299 & .8008 & .5772 & -.0937 \\ -.8008 & .1299 & .0937 & .5772 \end{pmatrix}$$

Find a matrix $B \in SO(4)$ so that $B^{-1}AB$ is in the form given by the Spectral Theorem for $O(n)$.

5.5. QUADRATIC FORMS: APPLICATIONS OF THE SPECTRAL THEOREM

We now consider symmetric bilinear forms and quadratic forms, which are generalizations of the dot product and length. Suppose that V is a real vector space and there is a "product" function $\mathcal{B} : V \times V \to \mathbb{R}$ satisfying

(1) $$\mathcal{B}(\mathbf{v}, \mathbf{w}) = \mathcal{B}(\mathbf{w}, \mathbf{v}) \qquad \text{symmetry}$$

(2) $$\mathcal{B}(\mathbf{v}, c\mathbf{w} + d\mathbf{x}) = c\mathcal{B}(\mathbf{v}, \mathbf{w}) + d\mathcal{B}(\mathbf{v}, \mathbf{x}) \qquad \text{bilinearity}$$

Note that by (1) there is a similar formula to (2) in terms of the first argument. Note also that if we had a third property of positive definiteness we would just be describing an inner product on V. $\mathcal{B}$ will be called a **symmetric bilinear form** on V. When we take $\mathbf{v} = \mathbf{w}$ and form $\mathcal{Q}(\mathbf{v}) = \mathcal{B}(\mathbf{v}, \mathbf{v})$, then $\mathcal{Q}$ is called a **quadratic form** on V (associated to the symmetric bilinear form $\mathcal{B}$).

■ ***Example 5.5.1.*** Symmetric bilinear forms generalize inner products, and so any inner product we have discussed earlier will give an example of a symmetric bilinear form. To get an example of a symmetric bilinear form which is not an inner product, consider the vector space $\mathbb{R}^n$, and define

$$\mathcal{B}((x_1, \ldots, x_n), (y_1, \ldots, y_n)) = x_1y_1 + x_2y_2 + \cdots + x_iy_i - x_{i+1}y_{i+1} - \cdots - x_ny_n$$

Here i is some fixed number between 1 and n. To make this example more specific we could take $n = 4$ and $i = 3$. Then the bilinear form would be

$$\mathcal{B}((x_1, x_2, x_3, x_4), (y_1, y_2, y_3, y_4)) = x_1y_1 + x_2y_2 + x_3y_3 - x_4y_4$$

Its associated quadratic form is

$$\mathcal{Q}((x_1, x_2, x_3, x_4)) = x_1^2 + x_2^2 + x_3^2 - x_4^2$$

That this is a symmetric bilinear form is easily checked from the definition. This particular symmetric bilinear form plays an important role in physics where the first three variables are interpreted as space variables and the fourth variable is interpreted as the time variable. ■

■ ***Example 5.5.2.*** A more general quadratic form can be defined in $\mathbb{R}^n$ once we start with any symmetric n by n matrix A. Then we can define the symmetric bilinear form $\mathcal{B}_A : \mathbb{R}^n \times \mathbb{R}^n \to \mathbb{R}$ by

$$\mathcal{B}_A(\mathbf{x}, \mathbf{y}) = \mathbf{x}^t A\mathbf{y}$$

In this definition $\mathbf{x}$ and $\mathbf{y}$ are regarded as column vectors. Bilinearity will follow from the linearity of matrix multiplication. Symmetry with follow from the fact that A is a symmetric matrix:

$$\mathcal{B}_A(\mathbf{y},\mathbf{x}) = \mathbf{y}^t A\mathbf{x} = \mathbf{y}^t A^t\mathbf{x} = (\mathbf{x}^t A\mathbf{y})^t = \mathbf{x}^t A\mathbf{y} = \mathbf{B}_A(\mathbf{x},\mathbf{y}) \qquad \blacksquare$$

Just as all linear transformations from $\mathbb{R}^n$ to $\mathbb{R}^m$ arise by multiplying by an m by n matrix, all bilinear forms on $\mathbb{R}^n$ correspond to the bilinear form $\mathcal{B}_A$ for the appropriate matrix A. For the last example we are just taking the diagonal matrix

$$A = \begin{pmatrix} 1 & 0 & 0 & 0 \\ 0 & 1 & 0 & 0 \\ 0 & 0 & 1 & 0 \\ 0 & 0 & 0 & -1 \end{pmatrix}$$

If we used the 2 by 2 matrix $A = \begin{pmatrix} 2 & 3 \\ 3 & 4 \end{pmatrix}$, we would get the bilinear form

$$\mathcal{B}_A(\mathbf{x},\mathbf{y}) = \mathbf{x}^t \begin{pmatrix} 2 & 3 \\ 3 & 4 \end{pmatrix} \mathbf{y} = 2x_1y_1 + 3x_1y_2 + 3x_2y_1 + 4y_1y_2$$

If $\{\mathbf{v}_1, \ldots, \mathbf{v}_n\}$ is a basis for V, then the product $\mathcal{B}(\mathbf{v},\mathbf{w})$ of any $\mathbf{v} = \sum_{i=1}^n c_i\mathbf{v}_i$ and $\mathbf{w} = \sum_{j=1}^n d_j\mathbf{v}_j$ will be determined using (2) as

$$\mathcal{B}(\mathbf{v},\mathbf{w}) = \sum_{i,j=1}^{n} c_i d_j \mathcal{B}(\mathbf{v}_i,\mathbf{v}_j)$$

In particular,

$$\mathcal{Q}(\mathbf{v}) = \sum_{i,j=1}^{n} c_i c_j \mathcal{B}(\mathbf{v}_i,\mathbf{v}_j)$$

Thus once the products of the basis elements with each other are known, this determines the products of any two elements of V. The property that a symmetric bilinear form is determined by its values on a basis is analogous to the property that a linear transformation is determined by its values on a basis. Just as in the case of a linear transformation, it allows us to associate a matrix to a symmetric bilinear form once a basis is chosen.

DEFINITION 5.5.1. Let $\mathcal{B} : V \times V \to \mathbb{R}$ be a symmetric bilinear form, and suppose $\mathcal{V} = \{\mathbf{v}_1, \ldots, \mathbf{v}_n\}$ is a basis for V. Let A be the matrix whose ij-entry is $\mathcal{B}(\mathbf{v}_i,\mathbf{v}_j)$. Then by (1), A will be a symmetric matrix. A is called the **matrix that represents the bilinear form in terms of the basis $\mathcal{V}$.**

Given the basis $\mathcal{V}$ and the matrix A, the value $\mathcal{B}(\mathbf{v},\mathbf{w})$ of the product of any other two vectors can be determined by the formula

$$\mathcal{B}(\mathbf{v},\mathbf{w}) = \sum_{i,j=1}^{n} a_{ij}c_i d_j = \mathbf{c}^t A\mathbf{d}$$

Let us examine how A will change when we choose a different basis. Suppose that $\mathcal{W} = \{\mathbf{w}_1, \ldots, \mathbf{w}_n\}$ is another basis for V. Then there is a matrix S so that $\mathbf{W} = \mathbf{V}S$; that is, $\mathbf{w}_j = \sum_{j=1}^{n} s_{kj}\mathbf{v}_k$. Then

$$b_{ij} = \mathcal{B}(\mathbf{w}_i, \mathbf{w}_j) = \mathcal{B}\left(\sum_{k=1}^{n} s_{ki}\mathbf{v}_k, \sum_{j=1}^{n} s_{mj}\mathbf{v}_m\right)$$

$$= \sum_{m,k=1}^{n} s_{ki}s_{mj}\mathcal{B}(\mathbf{v}_k, \mathbf{v}_m) = \sum_{k=1}^{n} s_{ki}\left(\sum_{m=1}^{n} a_{km}s_{mj}\right) = (\mathbf{S}^i)^t A\mathbf{S}^j$$

But this is just the ijth entry of S^tAS. Let us record these facts as a proposition.

PROPOSITION 5.5.1. Let V be a real vector space and $\mathcal{B}$ a symmetric bilinear form on V. If $\mathbf{v}_1, \ldots, \mathbf{v}_n$ is a basis for V, then the matrix that represents $\mathcal{B}$ with respect to this basis is $A = (a_{ij})$, with $a_{ij} = \mathcal{B}(\mathbf{v}_i, \mathbf{v}_j)$. If $\mathbf{v} = \sum_{i=1}^{n} c_i\mathbf{v}_i, \mathbf{w} = \sum d_i\mathbf{v}_i$, then $\mathcal{B}(\mathbf{v}, \mathbf{w}) = \mathbf{c}^tA\mathbf{d}$. If $\mathcal{W}$ is another basis and B is the matrix that represents $\mathcal{B}$ with respect to $\mathcal{W}$, and the two bases are related by $\mathbf{W} = \mathbf{V}S$, then $B = S^tAS$.

The conclusion of the proposition motivates the following definition.

DEFINITION 5.5.2. We say that A is **congruent to** B if there is an invertible matrix S so that $B = S^tAS$.

Note that if A is congruent to B, then B is congruent to A. For we have $A = T^tBT$, where $T = S^{-1}$. We leave it as an exercise to show that congruence provides an equivalence relation on symmetric matrices. The Spectral Theorem for symmetric matrices shows that every symmetric matrix is similar to a diagonal matrix. A natural question to ask is whether every symmetric matrix is also congruent to a diagonal matrix. Note that this is equivalent to asking whether there always is some basis for V so that the quadratic form $\mathcal{Q}(\mathbf{v}) = \mathcal{B}(\mathbf{v}, \mathbf{v})$ can be written as $\sum_{i=1}^{n} \alpha_i x_i^2$ when $\mathbf{v} = \sum_{i=1}^{n} x_i\mathbf{v}_i$. It turns out that this question is already answered by the Spectral Theorem for the simple reason that when S is an orthogonal matrix $S^tAS = S^{-1}AS$. Since the Spectral Theorem tells us that we can find an *orthogonal matrix* S so that $S^{-1}AS$ is diagonal, then this same matrix S will have S^tAS diagonal.

THEOREM 5.5.2. Principal Axes Theorem

Vector space version. Suppose V is a real vector space with an inner product $\langle\,,\rangle$ and $\mathcal{Q}$ is a quadratic form on V. Then there is an orthonormal basis $\mathbf{v}_1, \ldots, \mathbf{v}_n$ for V (with respect to $\langle\,,\rangle$) so that the matrix A for $\mathcal{Q}$ is diagonal.

Matrix version. Any real symmetric matrix A is congruent via an orthogonal matrix S to a diagonal matrix; that is, there is an orthogonal matrix S so that S^tAS is diagonal.

Exercise 5.5.1.

Show that congruence gives an equivalence relation on symmetric matrices.

Exercise 5.5.2.

Suppose A is a real symmetric matrix. Define a function $\mathcal{B}_A : \mathbb{R}^n \times \mathbb{R}^n \to \mathbb{R}$ by the formula $\mathcal{B}_A(\mathbf{v}, \mathbf{w}) = \mathbf{v}^t A\mathbf{w}$, where $\mathbf{v}$ and $\mathbf{w}$ are being regarded as column vectors.

(a) Show that $\mathcal{B}_A$ is a symmetric bilinear form.

(b) Show that every symmetric bilinear form on $\mathbb{R}^n$ arises this way for some symmetric matrix A.

Exercise 5.5.3.

Suppose $\mathcal{B}$ is a symmetric bilinear form and $\mathcal{Q}$ is the associated quadratic form. Show that $\mathcal{Q}$ satisfies the properties

(a) $\mathcal{Q}(r\mathbf{v}) = r^2\mathcal{Q}(\mathbf{v})$.

(b) $\mathcal{B}(\mathbf{v}, \mathbf{w}) = (1/2)(\mathcal{Q}(\mathbf{v} + \mathbf{w}) - \mathcal{Q}(\mathbf{v}) - \mathcal{Q}(\mathbf{w}))$.

Part (b) of Exercise 5.5.3 (called the **polarization formula**) shows how to recover the symmetric bilinear form from the quadratic form which it determines. For this reason we frequently talk about the quadratic form $\mathcal{Q}$ instead of the full symmetric bilinear form. Note, however, that the process of going from one to the other uses the symmetry property of the symmetric bilinear form. In the case of quadratic forms on $\mathbb{R}^n$, the quadratic form corresponds to a symmetric matrix $A : \mathcal{Q}(\mathbf{x}) = \mathbf{x}^t A\mathbf{x}$ and the bilinear form is just determined by $\mathcal{B}(\mathbf{x}, \mathbf{y}) = \mathbf{x}^t A\mathbf{y}$. For example, if we start with $\mathcal{Q}(\mathbf{x}) = 2x_1^2 + 6x_1x_2 - 3x_2^2$, then the matrix $A = \begin{pmatrix} 2 & 3 \\ 3 & -3 \end{pmatrix}$ and the corresponding symmetric bilinear form is

$$\mathcal{B}(\mathbf{x}, \mathbf{y}) = \mathbf{x}^t A\mathbf{y} = 2x_1y_1 + 3x_1y_2 + 3x_2y_1 - 3x_2y_2$$

We state as a proposition for future reference the way the symmetric bilinear form is determined from its associated quadratic form:

PROPOSITION 5.5.3. Polarization Formula. If $\mathcal{Q}$ is the associated quadratic form for a bilinear form $\mathcal{B}$, then

$$\mathcal{B}(\mathbf{v}, \mathbf{w}) = (1/2)(\mathcal{Q}(\mathbf{v} + \mathbf{w}) - \mathcal{Q}(\mathbf{v}) - \mathcal{Q}(\mathbf{w}))$$

■ ***Example 5.5.3.*** We look at another example which comes from analysis of recovering a symmetric bilinear form from its associated quadratic from via the polarization formula. Consider the vector space $\mathcal{C}([0, 1], \mathbb{R})$ of continuous functions from the interval $[0, 1]$ to the real numbers. Let

$$\mathcal{Q}(f) = \int_0^1 f(t)^2\,dt$$

Then the polarization formula gives

$$\mathcal{B}(f, g) = \frac{1}{2}\int_0^1 (f(t) + g(t))^2 - f(t)^2 - g(t)^2\,dt = \int_0^1 f(t)g(t)\,dt$$

It is easy to check that the resulting bilinear form is symmetric. ■

5.5.1. APPLICATION TO GEOMETRY

Suppose we have an equation

$$ax^2 + 2bxy + cy^2 = 1$$

in the plane. We wish to see what the solutions of this look like. To do this we first write $\mathbf{z} = \begin{pmatrix} x \\ y \end{pmatrix}$, $A = \begin{pmatrix} a & b \\ b & c \end{pmatrix}$ and check that

$$ax^2 + 2bxy + cy^2 = \mathbf{z}^t A\mathbf{z}$$

Note that this product gives us a quadratic form on $\mathbb{R}^2$ and A is just the matrix of this quadratic form with respect to the standard orthonormal basis $\mathbf{e}_1, \mathbf{e}_2$. But the Principal Axes Theorem says that there is an orthogonal matrix S so that $S^tAS = \begin{pmatrix} \alpha_1 & 0 \\ 0 & \alpha_2 \end{pmatrix} = D$. Note that the α_i are just the eigenvalues of the symmetric matrix A. We change coordinates by letting $\mathbf{w} = \begin{pmatrix} u \\ v \end{pmatrix}$ and $S\mathbf{w} = \mathbf{z}$, or, equivalently, $S\begin{pmatrix} u \\ v \end{pmatrix} = \begin{pmatrix} x \\ y \end{pmatrix}$. In these new coordinates

$$\mathbf{z}^t A\mathbf{z} = \mathbf{w}^t S^t AS\mathbf{w} = \begin{pmatrix} u & v \end{pmatrix}\begin{pmatrix} \alpha_1 & 0 \\ 0 & \alpha_2 \end{pmatrix}\begin{pmatrix} u \\ v \end{pmatrix} = \alpha_1 u^2 + \alpha_2 v^2 = 1$$

In the new coordinates, $(u, v) = (1, 0)$ corresponds to $(x, y) = S\mathbf{e}_1$ in the old coordinates. Similarly, $(u, v) = (0, 1)$ in the new coordinates corresponds to $(x, y) = S\mathbf{e}_2$ in the old coordinates. When we choose S to be in $SO(2)$, which is always possible, we see that the new coordinate axes are obtained from the old ones just by rotating using the rotation matrix S. Then there are four cases depending on α_i.

1. If both α_i are less than or equal to 0, there is no solution.

2. If one is 0 and the other is positive, then the solution consists of two parallel lines which are in the direction of the eigenvector for the 0 eigenvalue.
3. If one is positive and the other is negative, then the solutions form a hyperbola with the axes given by the eigenvectors of A which in the new coordinates is given by

$$\frac{u^2}{d^2} - \frac{v^2}{e^2} = \pm 1$$

4. If both the eigenvalues are positive, then the solutions form an ellipse with axes given by the eigenvectors of A; in the new coordinates the equation has the form

$$\frac{u^2}{d^2} + \frac{v^2}{e^2} = 1$$

This classification can be generalized to the case of an equation

$$ax^2 + bxy + cy^2 + dx + ey + f = 0$$

To handle this we first diagonalize the quadratic part and then complete the square to put it in a normal form. When one of the eigenvalues is 0, we get a parabola. The principal axes involved in the general case will come by first rotating as above and then translating to eliminate where possible the linear terms.

■ ***Example 5.5.4.*** Consider $23x^2 - 72xy + 2y^2 = 25$. Then we look at the matrix $A = \begin{pmatrix} 23 & -36 \\ -36 & 2 \end{pmatrix}$ which we diagonalize via $S = \begin{pmatrix} .8 & .6 \\ -.6 & .8 \end{pmatrix}$ to $D = \begin{pmatrix} 50 & 0 \\ 0 & -25 \end{pmatrix}$. Thus in our new coordinates, which we get by rotating the axes by $\arctan(-.6/.8) \sim -37$ degrees, the equation will be $50u^2 - 25v^2 = 25$, or $2u^2 - v^2 = 1$, which is a hyperbola. See Figure 5.1. ■

■ ***Example 5.5.5.*** Consider $x^2 + 2xy + y^2 - \sqrt{2}x + \sqrt{8}y - 1 = 0$. We first deal with the quadratic form part $x^2 + 2xy + y^2$. We look at the matrix $A = \begin{pmatrix} 1 & 1 \\ 1 & 1 \end{pmatrix}$ which is diagonalized by $S = 1/\sqrt{2}\begin{pmatrix} 1 & -1 \\ 1 & 1 \end{pmatrix}$ with $D = \begin{pmatrix} 2 & 0 \\ 0 & 0 \end{pmatrix}$. Then letting $\begin{pmatrix} x \\ y \end{pmatrix} = S\begin{pmatrix} u \\ v \end{pmatrix}$, we get the equation in the new variables to be $2u^2 + u + 3v - 1 = 0$. We then complete the square on the u terms to give $2(u + 1/4)^2 + 3v = 17/16$, which is a parabola. See Figure 5.2. ■

Exercise 5.5.4.

Sketch the graph of $7x^2 - 6xy + 7y^2 = 20$.

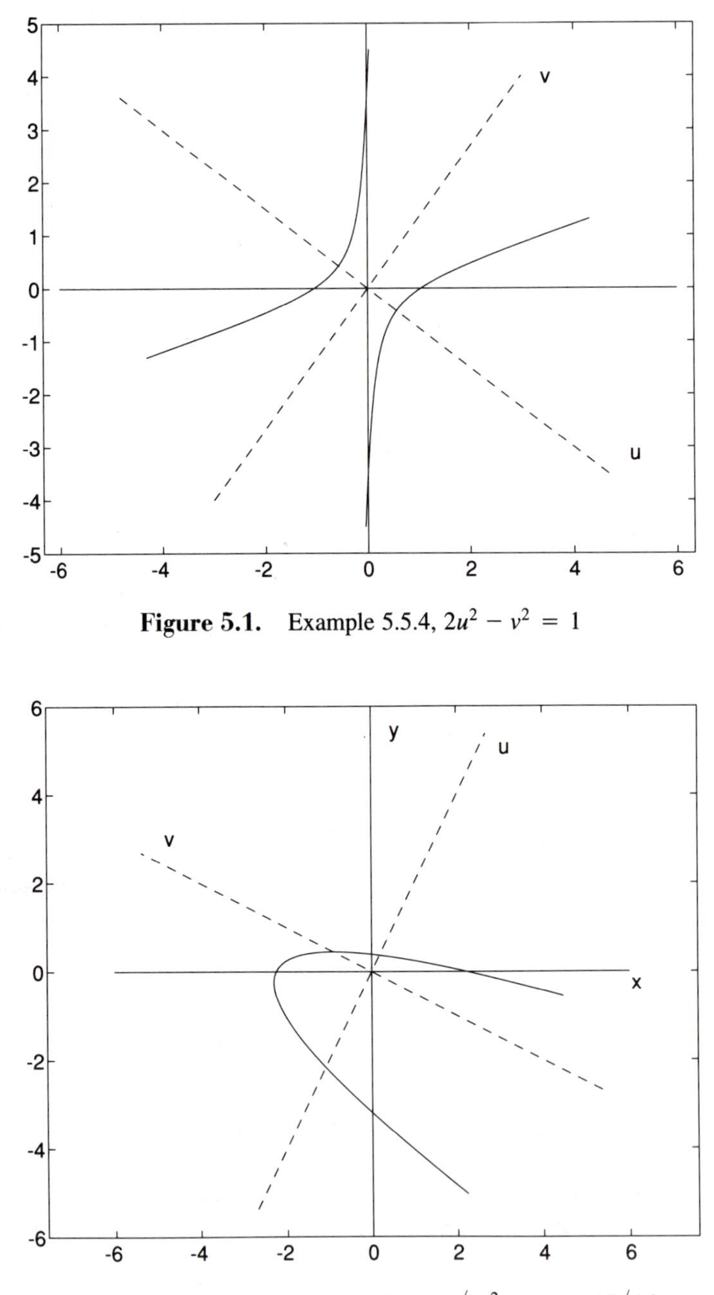

Figure 5.1. Example 5.5.4, $2u^2 - v^2 = 1$

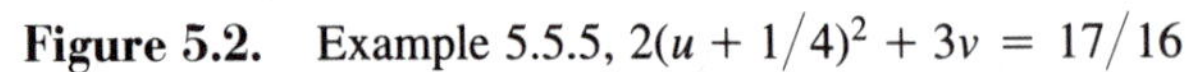

Figure 5.2. Example 5.5.5, $2(u + 1/4)^2 + 3v = 17/16$

Exercise 5.5.5.

Let $A = \begin{pmatrix} 11 & \sqrt{75} \\ \sqrt{75} & 1 \end{pmatrix}$.

(a) Find an orthogonal matrix S and a diagonal matrix D so that $S^t AS = D$.
(b) Sketch the graph of $11x^2 + 2\sqrt{75}xy + y^2 = 16$.

Exercise 5.5.6.

Sketch the graph of $2x^2 + 8xy + 8y^2 + 10\sqrt{5}y = 10$.

5.5.2. QUADRICS IN THREE DIMENSIONS

The same ideas can be applied in higher dimensions. Basically what the Principal Axes Theorem will allow us to do is to rotate and translate the axes so that in new coordinates an equation involving quadratic, linear, and constant terms can be put in a normal form. Although some degenerate cases can arise, the primary normal forms are

(1) $$\frac{x^2}{a^2} + \frac{y^2}{b^2} + \frac{z^2}{c^2} = 1 \qquad \text{ellipsoid}$$

This corresponds to three positive eigenvalues.

(2) $$\frac{x^2}{a^2} + \frac{y^2}{b^2} = \frac{z}{c} \qquad \text{elliptic paraboloid}$$

This corresponds to two positive eigenvalues and one 0 eigenvalue.

(3) $$\frac{x^2}{a^2} + \frac{y^2}{b^2} = \frac{z^2}{c^2} \qquad \text{elliptic cone}$$

This corresponds to two positive eigenvalues and one negative eigenvalue and the right-hand side being 0.

(4) $$\frac{x^2}{a^2} + \frac{y^2}{b^2} - \frac{z^2}{c^2} = 1 \qquad \text{hyperboloid of one sheet}$$

This corresponds to two positive eigenvalues and one negative eigenvalue.

(5) $$\frac{x^2}{a^2} - \frac{y^2}{b^2} - \frac{z^2}{c^2} = 1 \qquad \text{hyperboloid with two sheets}$$

This corresponds to one positive eigenvalue and two negative eigenvalues.

(6) $$\frac{x^2}{a^2} - \frac{y^2}{b^2} = \frac{z}{c} \qquad \text{hyperbolic paraboloid}$$

This corresponds to one positive, one negative, and one zero eigenvalue.

We will content ourselves with one example.

■ ***Example 5.5.6.*** Consider the quadric

$$4x^2 - 4xy + 8xz + 4y^2 + 8yz - 2z^2 = 6$$

This corresponds to the symmetric matrix

$$A = \begin{pmatrix} 4 & -2 & 4 \\ -2 & 4 & 4 \\ 4 & 4 & -2 \end{pmatrix}$$

which has diagonalizing matrix

$$S = \begin{pmatrix} -.9050 & -.1194 & -.4082 \\ .2980 & -.8629 & -.4082 \\ -.3035 & -.4911 & .8165 \end{pmatrix} \text{ giving } D = \begin{pmatrix} 6 & 0 & 0 \\ 0 & 6 & 0 \\ 0 & 0 & -6 \end{pmatrix}$$

Thus in the new variables

$$S\begin{pmatrix} u \\ v \\ w \end{pmatrix} = \begin{pmatrix} x \\ y \\ z \end{pmatrix}$$

the equation has the form

$$6u^2 + 6v^2 - 6w^2 = 6$$

which is a hyperboloid of one sheet. The new coordinate axes are along the column vectors of S. ■

Exercise 5.5.7.

Determine which one of the standard forms the following quadric is, and give a rough sketch of its graph.

$$x^2 - y^2 + 3z^2 + 2xy - 4xz + 6yz = 9$$

Hint: The eigenvalues of the matrix $\begin{pmatrix} 1 & 1 & -2 \\ 1 & -1 & 3 \\ -2 & 3 & 3 \end{pmatrix}$ are 1.3354, −3.4004, and 5.0650.

5.5.3. APPLICATION TO CALCULUS: MAXIMA AND MINIMA

Suppose we have a differentiable function $z = f(x, y)$ of two variables and we wish to find the maxima and minima of f. In particular, we are interested in relative maxima and minima, that is, those points (x_0, y_0) for which there is a small disk D about the point so that for all (x, y) in D either $f(x_0, y_0) \geq f(x, y)$—a **local maximum**—or

$f(x_0, y_0) \leq f(x, y)$—a **local minimum**. It is easy to show that a necessary condition for a local extremum is that both partial derivatives vanish at (x_0, y_0). We will assume that this holds from now on. There is a test, called the Second Derivative Test, which can sometimes determine whether (x_0, y_0) is a local extremum or is a **saddle point** (a point where no matter how small the neighborhood there will always be points in D with both larger and smaller values). We wish to explain this test in terms of a quadratic form attached to the problem, which is called the Hessian. Consider the quadratic form $\mathcal{Q}(x, y)$ which has the matrix given by the second order partial derivatives

$$H(x, y) = \begin{pmatrix} f_{11}(x, y) & f_{12}(x, y) \\ f_{21}(x, y) & f_{22}(x, y) \end{pmatrix}$$

With the assumption that the second-order partial derivatives are continuous, it is known that $f_{12} = f_{21}$, and so this will be a symmetric matrix as required. Of course it will vary from point to point, but at least it will vary continuously. From this it can be shown that its eigenvalues will also vary continuously. Now the chain rule can be combined with the Taylor's Theorem to show that for (x, y) near (x_0, y_0), if we put $\mathbf{h} = (x, y) - (x_0, y_0)$, then

$$f(x, y) - f(x_0, y_0) = \tfrac{1}{2}\mathbf{h}^t H((x_0, y_0) + t\mathbf{h})\mathbf{h}$$

for some $t, 0 \leq t \leq 1$. Thus the question of local maxima, local minima, and saddle points boils down to the question of whether this is always negative or positive or can take on both positive and negative values. This just depends on the signs of the eigenvalues of $H((x_0, y_0) + t\mathbf{h})$. But if $\mathbf{h}$ is small enough and 0 is not an eigenvalue of $H(x_0, y_0)$, then the eigenvalues of $H((x_0, y_0) + t\mathbf{h})$ will have the same signs as the eigenvalues of $H(x_0, y_0)$. Thus we get the following criterion.

PROPOSITION 5.5.4. Criterion for Local Extrema. Suppose that the first-order partials of f vanish at (x_0, y_0) and that the second-order partial derivatives of f are continuous. Let H denote the Hessian matrix of second-order partial derivatives.

1. If $H(x_0, y_0)$ has two positive eigenvalues, then f has a local minimum at (x_0, y_0).
2. If $H(x_0, y_0)$ has two negative eigenvalues, then f has a local maximum at (x_0, y_0).
3. If $H(x_0, y_0)$ has one positive and one negative eigenvalue, then f has a saddle point point at (x_0, y_0).

In calculus classes this result is usually not given in terms of the eigenvalues of the Hessian but in terms of an equivalent criterion. Before stating such a criterion let us introduce some terminology.

DEFINITION 5.5.3. We say that a quadratic form $\mathcal{Q}(\mathbf{v})$ is **positive definite** if $\mathcal{Q}(\mathbf{v}) = \mathcal{B}(\mathbf{v}, \mathbf{v}) \geq 0$ for all $\mathbf{v}$ and equals 0 iff $\mathbf{v} = \mathbf{0}$. In terms of a matrix representing $\mathcal{B}$ with respect to an orthonormal basis this just means that the eigenvalues are all positive. If we replace $\geq$ by $\leq$ we get the definition of a **negative definite** quadratic form, which corresponds to having the eigenvalues all negative. If none of the eigenvalues are 0, then we say that the quadratic form is **nondegenerate**. This includes the two preceding cases, but also includes the case of some positive and some negative eigenvalues. Saddle points will arise when the Hessian is nondegenerate but is not definite (which is called **indefinite**).

Thus the preceding criteria can be summarized by saying that if the Hessian is nondegenerate, then a minimum occurs when it is positive definite, a maximum occurs when it is negative definite, and a saddle point occurs when it is nondegenerate and indefinite. There is a simple test for when a quadratic form given by a matrix is nondegenerate, for the determinant of a matrix just gives the product of the eigenvalues. Then the quadratic form is nondegenerate iff the determinant of the associated matrix is not 0. In the 2 by 2 case it will have both positive and negative eigenvalues when the determinant is negative, but both eigenvalues will be of the same sign when the determinant is positive. To determine that sign, it suffices to evaluate the quadratic form on any one vector, that is, form $\mathbf{h}^t H\mathbf{h}$ for any $\mathbf{h}$. The $\mathbf{h}$ which is usually chosen for this test is $\mathbf{e}_1$ which just gives the $(1, 1)$ entry of the Hessian. This leads to the usual second derivative test.

PROPOSITION 5.5.5. Second Derivative Test. Suppose that $f(x, y)$ has continuous second partial derivatives and is defined on an open set in $\mathbb{R}^2$ containing (x_0, y_0) and the first-order partial derivatives vanish at (x_0, y_0). Suppose also that $DET = f_{11}f_{22} - f_{12}^2$ evaluated at (x_0, y_0).

1. If $DET > 0$ and $f_{11}(x_0, y_0) < 0$, then (x_0, y_0) is a local maximum.
2. If $DET > 0$ and $f_{11}(x_0, y_0) > 0$, then (x_0, y_0) is a local minimum.
3. If $DET < 0$, then (x_0, y_0) is a saddle point.

There are similar criteria for functions of several variables to have local maxima, local minima, and saddle points. The analysis is essentially the same. The additional complications in the criteria merely arise because it is harder to test for when the quadratic form is positive (or negative) definite or nondegenerate but indefinite. The following result is the multivariable analog of Proposition 5.5.5.

PROPOSITION 5.5.6. Second Derivative Test in $\mathbb{R}^n$. Suppose $f : \mathbb{R}^n \to \mathbb{R}$ is a function of n variables and P is a point in $\mathbb{R}^n$ where all the first derivatives vanish. Suppose also that the second-order partial derivatives are continuous. If $H(\mathbf{x})$ is the matrix whose ij-entry is the second-order partial derivative $f_{ij}(\mathbf{x})$, H is called the Hessian matrix, and it is a symmetric matrix, thus determining a quadratic form $\mathcal{Q}(\mathbf{x})$. If

$\mathcal{Q}(P)$ is positive definite (the eigenvalues are all positive), then P is a local minimum. If $\mathcal{Q}(P)$ is negative definite (the eigenvalues are all negative), then P is a local maximum. If $\mathcal{Q}(P)$ is nondegenerate but indefinite (0 is not an eigenvalue but there are positive and negative eigenvalues), then P is a saddle point.

■ ***Example 5.5.7.*** For the function $2x^2 + 6xy + y^2$, there is a critical value at the origin. The Hessian matrix (at $\mathbf{0}$) is $2\begin{pmatrix} 2 & 3 \\ 3 & 1 \end{pmatrix}$, which has determinant -28. Hence there is a saddle point at the origin. ■

■ ***Example 5.5.8.*** The function $f(x, y) = x \sin x - \cos y - xy$ has a critical point at the origin, since $f_1 = x \cos x + \sin x - y$ and $f_2 = \sin y - x$ both vanish there. When we compute the Hessian matrix, we get $f_{11} = -x \sin x + 2 \cos x - y$, $f_{12} = -1$, $f_{22} = \cos y$. Evaluating at the origin gives the matrix $\begin{pmatrix} 2 & -1 \\ -1 & 1 \end{pmatrix}$. Since this has positive determinant and $f_{11}(\mathbf{0}) > 0$, there will be a local minimum at the origin. ■

■ ***Example 5.5.9.*** Consider the two functions

$$f(x, y, z) = (1/2)(4x^2 + 4y^2 + 2z^2 + 2xy + 2xz - 2yz)$$

and

$$g(x, y, z) = 1/2(11x^2 + 11y^2 + 2z^2 + 4xy + 5xz - 5yz)$$

Each function has a critical point at the origin, and the Hessian matrices are $\begin{pmatrix} 4 & 1 & 1 \\ 1 & 4 & -1 \\ 1 & -1 & 2 \end{pmatrix}$ and $\begin{pmatrix} 11 & 4 & 5 \\ 4 & 11 & -5 \\ 5 & -5 & 2 \end{pmatrix}$. Although these matrices look very similar, the first has eigenvalues $5, 4, 1$ and so is positive definite, and the second has eigenvalues $15, 12, -3$ and so is indefinite. Thus f has a local minimum at the origin and g has a saddle point there. ■

Exercise 5.5.8.

For the following functions determine whether they have a local maximum, local minimum, or saddle point at the origin.

(a) $f(x, y) = x^2 + xy - y^2 - 1$.

(b) $f(x, y, z) = x^2 + xz + y^2 + x \sin x + 4z^2$.

Exercise 5.5.9.

Determine whether $f(x, y) = 11x^2 + 2\sqrt{75}xy + y^2$ has a local maximum, local minimum, or saddle point at the origin.

Exercise 5.5.10.

Determine whether $f(x, y, z) = x^2 + \sin yz$ has a local maximum, local minimum, or saddle point at the origin.

5.6. CHAPTER 5 EXERCISES

5.6.1. Show that if $V = \mathbb{R}^n$, then the complexification $V_{\mathbb{C}}$ is isomorphic to $\mathbb{C}^n$ as complex vector spaces; that is, there is a complex vector space isomorphism $L: \mathbb{C}^n \to V_{\mathbb{C}}$.

5.6.2. Show that if $V = \mathcal{D}(\mathbb{R}, \mathbb{R}^n)$, then the complexification $V_{\mathbb{C}}$ is isomorphic to $\mathcal{D}(\mathbb{R}, \mathbb{C}^n)$ as complex vector spaces; that is, there is a complex vector space isomorphism $L: \mathcal{D}(\mathbb{R}, \mathbb{C}^n) \to V_{\mathbb{C}}$.

5.6.3. Show that if $V = \mathcal{M}(m, n)$, then the complexification $V_{\mathbb{C}}$ is isomorphic to $\mathcal{M}(m, n, \mathbb{C})$, the complex vector space of m by n complex matrices under the usual operations of addition and complex scalar multiplication, as complex vector spaces; that is, there is a complex vector space isomorphism $L: \mathcal{M}(m, n, \mathbb{C}) \to V_{\mathbb{C}}$.

5.6.4. Show that if $L: V \to W$ is an isomorphism between real vector spaces, then $L_{\mathbb{C}}: V_{\mathbb{C}} \to W_{\mathbb{C}}$ is a complex vector space isomorphism between their complexifications.

5.6.5. (a) Show that if $V = \mathcal{F}(X, \mathbb{R})$, then the complexification $V_{\mathbb{C}}$ is isomorphic to $\mathcal{F}(X, \mathbb{C})$.
(b) Use part (a) and Exercise 5.6.4 to show that the complexification of $\mathbb{R}^n$ is isomorphic to $\mathbb{C}^n$ by showing that $\mathbb{R}^n$ is isomorphic to $\mathcal{F}(\{1, \ldots, n\}, \mathbb{R})$ and $\mathbb{C}^n$ is isomorphic to $\mathcal{F}(\{1, \ldots, n\}, \mathbb{C})$.

5.6.6. We only defined the adjoint of a linear transformation in the context of a linear transformation from an inner product space to itself.
(a) Show that if V, W are each finite-dimensional inner product spaces and $L: V \to W$ is a linear transformation, then there is a linear transformation (which is still called the adjoint) $L^*: W \to V$ which is uniquely defined by the formula

$$\langle L\mathbf{v}, \mathbf{w} \rangle = \langle \mathbf{v}, L^*\mathbf{w} \rangle$$

In this formula the inner product on W is used on the left-hand side and the inner product on V is used on the right-hand side.

(b) Show that if $\mathcal{V}$ is an orthonormal basis for V and $\mathcal{W}$ is an orthonormal basis for W, then there is the formula

$$[L^*]_{\mathcal{V}}^{\mathcal{W}} = ([L]_{\mathcal{W}}^{\mathcal{V}})^*$$

5.6.7. Consider the inner product space $\mathcal{P}^2(\mathbb{R}, \mathbb{R})$ of polynomials of degree less than or equal to 2 with inner product given by

$$\langle p, q \rangle = \int_{-1}^{1} p(t)q(t)\, dt$$

Using the notation of the previous exercise, find the adjoint of the linear transformation

$$L : \mathcal{P}^2(\mathbb{R}, \mathbb{R}) \to \mathbb{R}, L(p) = \int_{-1}^{1} p(t)\, dt$$

In Exercises 5.6.8–5.6.15 the Spectral Theorem guarantees that there is a unitary matrix S and a diagonal matrix D with $A = SDS^{-1}$. Find S and D.

5.6.8. $A = \begin{pmatrix} .5 + .5i & .5 + .5i \\ .5 + .5i & -.5 - .5i \end{pmatrix}$.

5.6.9. $A = \begin{pmatrix} i & i \\ i & i \end{pmatrix}$.

5.6.10. $A = \begin{pmatrix} (1 - i)/2 & i/\sqrt{2} \\ (1 + i)/2 & 1/\sqrt{2} \end{pmatrix}$.

5.6.11. $A = \begin{pmatrix} 1 & 1 & 4 \\ 1 & 1 & 4 \\ 4 & 4 & -2 \end{pmatrix}$.

5.6.12. $A = \begin{pmatrix} 5/6 & -1/3 & -1/6 \\ -1/3 & 1/3 & -1/3 \\ -1/6 & -1/3 & 5/6 \end{pmatrix}$.

5.6.13. $A = \begin{pmatrix} 1/\sqrt{3} & -1/\sqrt{2} & -1/\sqrt{6}i \\ 1/\sqrt{3} & 0 & 2/\sqrt{6}i \\ 1/\sqrt{3}i & 1/\sqrt{2}i & 1/\sqrt{6} \end{pmatrix}$.

5.6.14. $A = \begin{pmatrix} .5 & .5 & .5 & .5 \\ .5 & -.5 & .5 & -.5 \\ .5 & -.5 & -.5 & .5 \\ .5 & .5 & -.5 & -.5 \end{pmatrix}$.

5.6.15. $A = \begin{pmatrix} 2.5 & -.5 & 0 & -1 \\ -.5 & 2.5 & -1 & 0 \\ 0 & -1 & 2.5 & -.5 \\ -1 & 0 & -.5 & 2.5 \end{pmatrix}$.

In Exercises 5.6.16–5.6.20 write the given matrix as a linear combination of orthogonal projection matrices as given by the Spectral Theorem.

5.6.16. $\begin{pmatrix} 1 & 2 \\ 2 & 1 \end{pmatrix}$

5.6.17. $\begin{pmatrix} 0 & 1 & 1 \\ -1 & 0 & 1 \\ -1 & -1 & 0 \end{pmatrix}$

5.6.18. $\begin{pmatrix} 34/30 & -2/30 & -1/3 \\ -2/30 & 31/30 & 1/6 \\ -1/3 & 1/6 & 11/6 \end{pmatrix}$

5.6.19. $\begin{pmatrix} 1/\sqrt{2} & 1/2 & 1/2 & 0 \\ 0 & 1/2 & -1/2 & 1/\sqrt{2} \\ 1/\sqrt{2} & -1/2 & -1/2 & 0 \\ 0 & -1/2 & 1/2 & 1/\sqrt{2} \end{pmatrix}$

5.6.20. $\begin{pmatrix} 1.75 & -.25 & .25 & .25 \\ -.25 & 1.75 & .25 & .25 \\ .25 & .25 & 1.75 & -.25 \\ .25 & .25 & -.25 & 1.75 \end{pmatrix}$

5.6.21. Show that for every diagonalizable 2 by 2 real matrix A with a nonreal complex eigenvalue $a + ib$, there is a real invertible matrix S with $S^{-1}AS = \begin{pmatrix} a & b \\ -b & a \end{pmatrix}$.

5.6.22. Generalize the previous exercise to 3 by 3 diagonalizable real matrices with nonreal complex eigenvalues.

5.6.23. The Cayley–Hamilton theorem says that every matrix A satisfies its characteristic polynomial; that is, $(A - c_1 I)\dots(A - c_n I) = 0$ where $c_1, \dots, c_n$ are the eigenvalues. Verify this directly as follows.
 (a) Use Exercise 5.2.7 which says that every matrix is similar to an upper triangular matrix to reduce to the case where A is upper triangular.
 (b) Verify the Cayley–Hamilton theorem directly for upper triangular 3 by 3 matrices.
 (c) Prove the Cayley–Hamilton theorem for general upper triangular matrices.

5.6.24. This problem outlines another proof of the Cayley–Hamilton theorem (see Exercise 5.6.23).
 (a) Consider the set $S_n[x]$ of polynomials in a variable x whose coefficients are n by n matrices. Show that if A is any n by n matrix, there is an equality in $S_n[x]$ given by

$$(A - xI)\,\mathrm{adj}\,(A - xI) = C_A(x)I$$

where $C_A(x) = \det(A - xI)$ is the characteristic polynomial of A. Here $\mathrm{adj}(A - xI)$ is the adjunct matrix of $A - xI$, which will be a polynomial of degree $n - 1$ with matrix coefficients.

(b) Show that if there is an equality in $S_n[x]$, there is an equality in $\mathcal{M}(n, n)$ found by evaluating at an n by n matrix B which commutes with A.

(c) By evaluating the equality in (a) at the matrix A itself, prove the Cayley–Hamilton theorem, $C_A(A) = \mathbf{0}$.

5.6.25. Use the fact that given A, there is a unitary matrix S and an upper triangular matrix T with $A = STS^{-1}$ (see Exercise 5.2.7) to prove versions of Spectral Theorem as follows:

(a) For a Hermitian, skew Hermitian, or unitary matrix, the matrix is diagonal if it is triangular.

(b) For a normal matrix, the matrix is diagonal if it is triangular. (Hint: Use a direct calculation of the $(1, 1)$-entries of AA^* and A^*A and an inductive argument.)

(c) Show that if $A = STS^{-1}$ for unitary S, then A being Hermitian, skew Hermitian, unitary, or normal implies T is.

(d) Use (a)–(c) to prove the Spectral Theorem for these four types of matrices.

5.6.26. Let V be a vector space of dimension n and $L : V \to V$ a linear transformation. $V_1 \subset V$ is said to be an **invariant subspace for** L if $L(V_1) \subset V_1$. Suppose V_1 is an invariant subspace for L.

(a) Show that the restriction $L_1 = L|V_1$ gives a linear transformation from V_1 to V_1.

(b) Show that the eigenvalues of L_1 are a subset of the eigenvalues of L.

(c) Suppose that $V = V_1 \oplus V_2$ and both V_1 and V_2 are invariant subspaces for L. Show that if $\mathbf{v} = \mathbf{v}_1 + \mathbf{v}_2$, $\mathbf{v}_i \in V_i$, is a decomposition of an eigenvector $\mathbf{v}$ for some eigenvalue c, then each $\mathbf{v}_i$ is either $\mathbf{0}$ or an eigenvector for L (and for L_i) for the same eigenvalue.

5.6.27. Suppose that $L, M : V \to V$ are linear transformations between an n-dimensional vector space V and itself and $LM = ML$. Suppose also that L, M are each diagonalizable. Let $E(c_1) \oplus \ldots \oplus E(c_k)$ be a direct sum decomposition of V into the eigenspaces for L.

(a) Show that these eigenspaces for L are invariant subspaces for M.

(b) Use the preceding exercise to show that if $\mathbf{v}$ is an eigenvector for M, then one of its components in the direct sum decomposition is a common eigenvector for both M and L.

(c) Use part (b) and induction on the number of distinct eigenvalues of L to show that there is a basis of vectors $\mathbf{v}_1, \ldots, \mathbf{v}_n$ of common eigenvectors for L and M.

5.6.28. Suppose A, B are symmetric matrices and $AB = BA$. Use the preceding exercise to show that there is a matrix S with $S^{-1}AS$ and $S^{-1}BS$ each diagonal. Show that there is an orthogonal matrix S with this property.

5.6.29. Let

$$A = \begin{pmatrix} 8 & -1 & -1 \\ -1 & 8 & -1 \\ -1 & -1 & 8 \end{pmatrix}, \quad B = \begin{pmatrix} 1 & 2 & 0 \\ 2 & 0 & 1 \\ 0 & 1 & 2 \end{pmatrix}$$

Show that $AB = BA$ and there is an orthogonal matrix S so that $S^{-1}AS$ and $S^{-1}BS$ are each diagonal.

5.6.30. Let

$$A = \begin{pmatrix} 2 & -1 & 1 \\ -1 & 2 & -1 \\ 1 & -1 & 4 \end{pmatrix}, \quad B = \begin{pmatrix} 11 & 1 & -2 \\ 1 & 11 & 2 \\ -2 & 2 & 8 \end{pmatrix}$$

Show that there is an orthonormal basis of common eigenvectors for A, B.

5.6.31. Find the singular value decomposition and pseudoinverse of $A = \begin{pmatrix} 1 & 2 \\ 0 & 0 \\ 1 & 2 \end{pmatrix}$, and use them to find the least squares solution of $A\mathbf{x} = \begin{pmatrix} 1 \\ 1 \\ 2 \end{pmatrix}$.

5.6.32. Suppose that A is an m by n matrix whose columns are orthogonal nonzero vectors $\mathbf{v}_1, \ldots, \mathbf{v}_n$.

(a) Show that the nonzero singular values of A are $\|\mathbf{v}_1\|, \ldots, \|\mathbf{v}_n\|$ so $\Sigma_r = \operatorname{diag}(\|\mathbf{v}_1\|, \ldots, \|\mathbf{v}_n\|)$.

(b) Show that we can choose $Q_{2r} = I_n$ and $Q_{1r} = \begin{pmatrix} \frac{\mathbf{v}_1}{\|\mathbf{v}_1\|} & \cdots & \frac{\mathbf{v}_n}{\|\mathbf{v}_n\|} \end{pmatrix}$ in the singular value decomposition.

5.6.33. Let $A = \begin{pmatrix} 1 & -1 \\ 2 & 0 \\ 0 & 2 \\ 1 & 1 \end{pmatrix}$. Find the singular value decomposition and pseudoinverse of A. Use the pseudoinverse to solve the least squares problem $A\mathbf{x} = (1, 2, 1, 2)$.

5.6.34. Let $A = \begin{pmatrix} 1 & 2 & 3 \\ 3 & 2 & 1 \\ -1 & 0 & 1 \end{pmatrix}$, $\mathbf{b} = (1, 0, 0)$.

(a) Give the singular value decomposition of A.
(b) Give the pseudoinverse of A.
(c) Find the least squares solution of $A\mathbf{x} = \mathbf{b}$.

5.6.35. Find the decomposition $A = Q_{1r}\Sigma_r Q_{2r}^t$ for

$$A = \begin{pmatrix} -1/2 & -1/2 & 3/\sqrt{2} \\ 3/2 & 3/2 & -1/\sqrt{2} \\ 3/2 & 3/2 & -1/\sqrt{2} \\ -1/2 & -1/2 & 3/\sqrt{2} \end{pmatrix}$$

Use it to find the pseudoinverse for A and to solve the least squares problem $A\mathbf{x} = (1, 2, 0, -1)$.

5.6.36. Find the decomposition $A = Q_{1r}\Sigma_r Q_{2r}^t$ for

$$A = \begin{pmatrix} 1/\sqrt{2} & 0 & 1/\sqrt{2} \\ \sqrt{2} & 1/\sqrt{2} & \sqrt{2} \\ 0 & 0 & 0 \\ \sqrt{2} & -1/\sqrt{2} & \sqrt{2} \\ 0 & 1 & 0 \end{pmatrix}$$

Use it to find the pseudoinverse for A and to solve the least squares problem $A\mathbf{x} = (1,0,0,0,0)$.

5.6.37. Show that if X is the pseudoinverse of A, then

$$AXA = A, XAX = X$$

and both XA and AX are symmetric.

5.6.38. Suppose that A is a positive definite symmetric matrix. Show that the singular value decomposition $A = Q_1\Sigma Q_2^t$ coincides with the decomposition $A = QDQ^{-1}$ given by the Spectral Theorem.

5.6.39. Generalize the preceding exercise by determining the relationship between the singular value decomposition and the spectral decomposition for a symmetric matrix. Prove your assertion.

5.6.40. Show that if A is a square matrix of rank 1, then the pseudoinverse is a multiple of A^t. Describe what the multiple is in terms of the nonzero eigenvalue of A^tA.

5.6.41. Suppose that L is a linear transformation from $\mathbb{R}^3$ to $\mathbb{R}^3$ which reflects each vector $\mathbf{v}$ in the plane $x + y + z = 0$.

(a) Find an orthonormal basis $\mathbf{v}_1, \mathbf{v}_2, \mathbf{v}_3$ of $\mathbb{R}^3$ so that $\mathbf{v}_1, \mathbf{v}_2$ is a basis for the plane.

(b) Find the matrix of L with respect to this basis.

(c) Describe geometrically the adjoint L^* of L. (Hint: What are the eigenvalues and eigenvectors of L?)

(d) Find the matrix A so that $L(\mathbf{v}) = A\mathbf{v}$.

(e) Write A as a linear combination of projection matrices.

5.6.42. A symmetric matrix has eigenvalues 1, 2, 3, 4. Eigenvectors corresponding to 1, 2, 3 are $(1,1,1,1), (1,1,-1,-1)$, and $(1,-1,1,-1)$. Find the remaining eigenvector and find $A\mathbf{x}$ where $\mathbf{x} = (1,0,0,0)$.

5.6.43. In parts (a) and (b), identify whether multiplication by the given matrix represents a rotation or a reflection in $\mathbb{R}^2$. If it is a rotation, give the angle of rotation. If it is a reflection, give the line of reflection.

(a) $\begin{pmatrix} 1/2 & -\sqrt{3}/2 \\ \sqrt{3}/2 & 1/2 \end{pmatrix}$

(b) $\begin{pmatrix} 1/2 & \sqrt{3}/2 \\ \sqrt{3}/2 & -1/2 \end{pmatrix}$

5.6.44. In parts (a) and (b), identify whether multiplication by the given matrix represents a rotation or a reflection in $\mathbb{R}^2$. If it is a rotation, give the angle of rotation. If it is a reflection, give the line through which one reflects.

(a) $\begin{pmatrix} 1/3 & \sqrt{8}/3 \\ \sqrt{8}/3 & -1/3 \end{pmatrix}$

(b) $\begin{pmatrix} 1/3 & \sqrt{8}/3 \\ -\sqrt{8}/3 & 1/3 \end{pmatrix}$

For the matrices A in Exercises 5.6.45–5.6.49, find an orthogonal matrix Q so that $Q^{-1}AQ$ is in the form given by the Spectral Theorem for $O(n)$.

5.6.45. $A = \begin{pmatrix} -.2 & -.8 & -2\sqrt{2}/5 \\ -.8 & -.2 & 2\sqrt{2}/5 \\ 2\sqrt{2}/5 & -2\sqrt{2}/5 & .6 \end{pmatrix}$.

5.6.46. $\begin{pmatrix} 0 & -1 & 0 \\ -1 & 0 & 0 \\ 0 & 0 & 1 \end{pmatrix}$

5.6.47. $\begin{pmatrix} .8 & .2 & -2\sqrt{2}/5 \\ .2 & .8 & 2\sqrt{2}/5 \\ 2\sqrt{2}/5 & -2\sqrt{2}/5 & .6 \end{pmatrix}$

5.6.48. $\begin{pmatrix} 1/\sqrt{3} & 0 & 1/\sqrt{3} & 1/\sqrt{3} \\ 0 & 1/\sqrt{3} & 1/\sqrt{3} & -1/\sqrt{3} \\ 1/\sqrt{3} & 1/\sqrt{3} & -1/\sqrt{3} & 0 \\ 1/\sqrt{3} & -1/\sqrt{3} & 0 & -1/\sqrt{3} \end{pmatrix}$

5.6.49. $\begin{pmatrix} .1381 & .4714 & -.8047 & -.3333 \\ -.4714 & .8047 & .3333 & .1381 \\ -.3333 & -.1381 & .2357 & -.9024 \\ -.8047 & -.3333 & -.4310 & .2357 \end{pmatrix}$

5.6.50. Give the matrix A so that multiplication by A gives the orthogonal reflection through the plane $x - 2y + z = 0$. (Hint: First choose an orthonormal basis $\mathbf{v}_1, \mathbf{v}_2$ for this plane, and extend it to an orthonormal basis for $\mathbb{R}^3$. Then if Q denotes the orthogonal matrix with these basis vectors as columns, express $A = QDQ^{-1}$.)

5.6.51. Give the matrix A so that multiplication by A gives the orthogonal reflection through the plane $2x + y - z = 0$. (Hint: First choose an orthonormal basis $\mathbf{v}_1, \mathbf{v}_2$ for this plane, and extend it to an orthonormal basis for $\mathbb{R}^3$. Then if Q denotes the orthogonal matrix with these basis vectors as columns, express $A = QDQ^{-1}$.)

5.6.52. Give the matrix A which is a rotation of the plane $x + 2y + z = 0$ by an angle of $\pi/6$. (Hint: First find an orthonormal basis $\mathbf{q}_1, \mathbf{q}_2, \mathbf{q}_3$ for $\mathbb{R}^3$ so that $\mathbf{q}_1$ is normal to the plane and $\mathbf{q}_2, \mathbf{q}_3$ is an orthonormal basis for the plane. Determine what this rotation should do to these basis vectors.)

5.6.53. Give the matrix A which is a rotation of the plane $2x - 3y + z = 0$ by an angle of $\pi/6$. (Hint: First find an orthonormal basis $\mathbf{q}_1, \mathbf{q}_2, \mathbf{q}_3$ for $\mathbb{R}^3$ so that $\mathbf{q}_1$ is normal to the plane and $\mathbf{q}_2, \mathbf{q}_3$ is an orthonormal basis for the plane. Determine what this rotation should do to these basis vectors.)

Exercises 5.6.54–5.6.57 fill in the details in the proof of Spectral Theorem for $O(n)$.

5.6.54. Suppose that $\mathbf{c}_1 = \mathbf{s}_1 + i\mathbf{t}_1, \ldots, \mathbf{c}_k = \mathbf{s}_k + i\mathbf{t}_k$ is a orthonormal basis of eigenvectors for the complex eigenvalue $a + ib$ for the real orthogonal matrix A. Using the conjugate eigenvalue, it was shown in Section 5.4 that for any p, we must have $\langle \mathbf{s}_p, \mathbf{t}_p \rangle = 0, \|\mathbf{s}_p\| = \|\mathbf{t}_p\| = 1/\sqrt{2}$. Show that the vectors $\mathbf{s}_1, \mathbf{t}_1, \ldots, \mathbf{s}_k, \mathbf{t}_k$ are mutually orthogonal. (Hint: You will need to use the conjugate eigenvalue and eigenvectors.)

5.6.55. Suppose that $\mathbf{c}_1 = \mathbf{s}_1 + i\mathbf{t}_1, \mathbf{c}_2 = \mathbf{s}_2 + i\mathbf{t}_2$ are eigenvectors for distinct eigenvalues of a real orthogonal matrix A. Show that the vectors $\mathbf{s}_1, \mathbf{t}_1, \mathbf{s}_2, \mathbf{t}_2$ are mutually orthogonal.

5.6.56. Using the two previous exercises prove the Spectral Theorem for $O(n)$.

5.6.57. Suppose that $Q^{-1}AQ = R$ where $Q = (\mathbf{q}_1 \ \ \mathbf{q}_2)$ is an orthogonal matrix with determinant -1 and R is a 2 by 2 rotation matrix $\begin{pmatrix} \cos\theta & -\sin\theta \\ \sin\theta & \cos\theta \end{pmatrix}$.

Show that we can also write $\overline{Q}^{-1}A\overline{Q} = \overline{R}$ where $\overline{Q} = (-\mathbf{q}_1 \ \ \mathbf{q}_2)$ and $\overline{R} = \begin{pmatrix} \cos\phi & -\sin\phi \\ \sin\phi & \cos\phi \end{pmatrix}$ with $\phi = -\theta$. Use this to verify the claim that in the statement of the Spectral Theorem for $O(n)$ we can always choose the orthogonal matrix Q so that Q is in $SO(n)$.

5.6.58. Determine whether each of the following statements is true or false. Justify your answers.
(a) The eigenvalues of a real orthogonal matrix must be real numbers.
(b) If A is a 3 by 3 matrix with eigenvalues $0, -1, 3$ and eigenvectors $\mathbf{v}_1 = (1, 1, 0), \mathbf{v}_2 = (1, -1, 0), \mathbf{v}_3 = (0, 0, 1)$, then A is a symmetric matrix.
(c) If a real symmetric matrix has positive eigenvalues and $A - 3I$ has a negative eigenvalue, then A has an eigenvalue between 0 and 3.
(d) A projection matrix has only 0 and 1 as eigenvalues.

5.6.59. Let $A = \begin{pmatrix} .9 & .2 \\ .2 & .6 \end{pmatrix}$.
(a) Find an orthogonal matrix S and a diagonal matrix D with $A = SDS^{-1}$.
(b) Sketch the graph of $9x^2 + 4xy + 6y^2 = 10$.
(c) Determine whether $f(x, y) = 9x^2 + 4xy + 6y^2$ has a local maximum, local minimum, or saddle point at the origin.

5.6.60. Let $A = \begin{pmatrix} 1/2 & -\sqrt{3}/2 \\ -\sqrt{3}/2 & -1/2 \end{pmatrix}$.
(a) Find an orthogonal matrix S and a diagonal matrix D with $A = SDS^{-1}$.
(b) Sketch the graph of $x^2/2 + \sqrt{3}xy - y^2/2 = 1$.

(c) Determine whether $f(x, y) = x^2/2 + \sqrt{3}xy - y^2/2$ has a local maximum, local minimum, or saddle point at the origin.

5.6.61. Sketch the graph of the ellipse $5x^2 + 6xy + 5y^2 = 8$. Indicate the principal axes of the ellipse in your sketch.

5.6.62. Sketch the graph of $x^2 + 4xy + 3y^2 = 1$. Indicate the principal axes in your sketch.

5.6.63. Sketch the graph of $18x^2 + 8xy + 12y^2 = 10$.

5.6.64. Sketch the graph of $x^2 + 2xy + y^2 + \sqrt{2}x + 3\sqrt{2}y = 0$.

5.6.65. Sketch the graph of $2x^2 - 2xy + 2y^2 - 2\sqrt{2}x + 4\sqrt{2}y = 0$.

5.6.66. By finding the eigenvalues, determine which of the standard quadrics (ellipsoid, hyperboloid of one sheet, etc.) is given by the equation. In each case give a rough sketch of the graph.
(a) $9x^2 + 6y^2 + 6z^2 - 6xy - 6xz = 12$
(b) $9x^2 - 6xy - 6xz + 12yz = 12$
(c) $-7x^2 - 4y^2 - 4z^2 + 10xy + 10xz + 4yz = 12$

5.6.67. By finding the eigenvalues, determine which of the standard quadrics (ellipsoid, hyperboloid of one sheet, etc.) is given by the equation. In each case give a rough sketch of the graph.
(a) $17x^2 + 14y^2 + 17z^2 - 4xy - 2xz - 4yz = 36$
(b) $-x^2 + 14y^2 - z^2 - 4xy + 34xz - 4yz = 36$
(c) $-x^2 + 14y^2 - z^2 - 4xy + 34xz - 4yz = 0$

5.6.68. Determine whether $(0, 0)$ is a local maximum, local minimum, a saddle point, or none of the above for the following functions.
(a) $f(x, y) = 5x^2 - 6xy + 8y^2 + 6$
(b) $g(x, y) = x \sin y - \cos x + 4$

5.6.69. Determine whether $(0, 0)$ is a local maximum, local minimum, a saddle point, or none of the above for the following functions.
(a) $f(x, y) = 1 + \sin x \sin y$
(b) $g(x, y) = e^x \cos y - x + y^2$

5.6.70. Determine whether $(0, 0, 0)$ is a local maximum, local minimum, a saddle point, or none of the above for the following functions.
(a) $f(x, y, z) = xy \cos z + z^2$.
(b) $g(x, y, z) = x^2 + y^2 + z^2 - xy$.

5.6.71. Determine whether $(0, 0, 0, 0)$ is a local maximum, local minimum, a saddle point, or none of the above for the following function.

$$f(x, y, z, w) = x^2 + y^2 + z^2 + w^2 + xy + xz + xw + yz + yw + zw$$

CHAPTER 6

NORMAL FORMS

6.1. QUADRATIC FORMS: NORMAL FORM

This chapter will discuss further two topics we have studied. The first section will give a normal form for a quadratic form. We will show that any symmetric matrix is congruent to a diagonal matrix with $\pm 1, 0$ as the diagonal entries. We will also give a test for when a quadratic form is positive definite in terms of the ideas in Chapter 1. The second section will discuss the Jordan form of a matrix, which is a normal form for a square matrix. The Jordan form generalizes the diagonal matrix with eigenvalues on the diagonal for a diagonalizable matrix. We will then apply this to solutions of differential equations.

In Section 5.5 we discussed quadratic forms and proved the Principal Axes Theorem, which allowed us to diagonalize a quadratic form via an orthogonal matrix. Section 5.5 emphasized the aspects of quadratic forms most connected to the Principal Axes Theorem. In this section we wish to emphasize other properties of quadratic forms which don't hinge as much on the orthogonality properties of a diagonalizing matrix. Recall that a quadratic form $\mathcal{Q} : V \to \mathbb{R}$ is determined from a symmetric bilinear form $\mathcal{B} : V \times V \to \mathbb{R}$ by $\mathcal{Q}(\mathbf{v}) = \mathcal{B}(\mathbf{v}, \mathbf{v})$ and that the polarization formula allows us to recover the full symmetric bilinear form from its associated quadratic form. For this reason, we will speak mainly of the quadratic form itself, leaving the full symmetric bilinear form in the background. As a reversal on our usual practice, we will work mainly with matrices in this section and then deduce facts about general vector spaces as a consequence of our results for matrices. The quadratic form in which we are interested in this context is just $\mathcal{Q}_A(\mathbf{x}) = \mathbf{x}^t A\mathbf{x}$ for a symmetric matrix A.

We first define a diagonal matrix $D(p, n, d)$ by letting the first p diagonal entries be $+1$, the next n diagonal entries be -1, and the last d diagonal entries be 0. For example,

$$D(2,3,2) = \begin{pmatrix} 1 & 0 & 0 & 0 & 0 & 0 & 0 \\ 0 & 1 & 0 & 0 & 0 & 0 & 0 \\ 0 & 0 & -1 & 0 & 0 & 0 & 0 \\ 0 & 0 & 0 & -1 & 0 & 0 & 0 \\ 0 & 0 & 0 & 0 & -1 & 0 & 0 \\ 0 & 0 & 0 & 0 & 0 & 0 & 0 \\ 0 & 0 & 0 & 0 & 0 & 0 & 0 \end{pmatrix}$$

We start off with a consequence of the Principal Axes Theorem, which gives a normal form for quadratic forms.

THEOREM 6.1.1. Normal Form for Quadratic Forms

(a) Let A be a real symmetric n by n matrix. Then there is an invertible matrix S with $S^t AS = D(p, n, d)$.

(b) Let $\mathcal{Q} : V \to \mathbb{R}$ be a quadratic form. Then there is a basis $\mathcal{V}$ for V so that the matrix which represents Q with respect to this basis is $D(p, n, d)$. This implies that if $\mathbf{v} = \mathbf{V}\mathbf{x}$, then

$$\mathcal{Q}(\mathbf{v}) = x_1^2 + \cdots + x_p^2 - x_{p+1}^2 - \cdots - x_{p+n}^2$$

Proof. We prove only part (a). Part (b) is left as an exercise—it follows from part (a) and Proposition 5.5.1. We know from the Spectral Theorem that there is an orthogonal matrix S_1 so that $S_1^t AS_1 = D$, where D is a diagonal matrix with the eigenvalues of A down the diagonal. By permuting the columns of S (recall these columns form an orthonormal basis of eigenvectors), we may arrange that the entries of D are ordered so that the positive ones come first, then the negative ones, and finally the 0 eigenvalues. We then will have $D = \text{diag}(r_1^2, \dots, r_p^2, -s_1^2, \dots, -s_n^2, 0, \dots, 0)$, where the notation indicates we are taking the diagonal matrix with the given entries. Letting

$$S_2 = \text{diag}(1/r_1, \dots, 1/r_p, 1/s_1, \dots, 1/s_p, 1, \dots, 1)$$

we have $S_2^t DS_2 = D(p, n, d)$. If we let $S = S_1S_2$, we have $S^t AS = D(p, n, d)$. ∎

■ ***Example 6.1.1.*** Consider the matrix

$$A = (1/3)\begin{pmatrix} 1 & 5 & -4 \\ 5 & 1 & 4 \\ -4 & 4 & -8 \end{pmatrix}$$

This matrix has eigenvalues $2, -4, 0$ and an orthonormal matrix of eigenvectors

$$S_1 = \begin{pmatrix} -1/\sqrt{2} & 1/\sqrt{6} & -1/\sqrt{3} \\ -1/\sqrt{2} & -1/\sqrt{6} & 1/\sqrt{3} \\ 0 & 2/\sqrt{6} & 1/\sqrt{3} \end{pmatrix}$$

The matrix $S_2 = \operatorname{diag}(1/\sqrt{2}, 1/2, 1)$ and

$$S = S_1S_2 = \begin{pmatrix} -1/2 & 1/(2\sqrt{6}) & -1/\sqrt{3} \\ -1/2 & -1/(2\sqrt{6}) & -1/\sqrt{3} \\ 0 & 1/\sqrt{6} & 1/\sqrt{3} \end{pmatrix}$$

Then $S^tAS = D(1,1,1)$. ■

Exercise 6.1.1.

For each of the symmetric matrices A, find a matrix S so that S^tAS is in normal form.

$$\text{(a)} \begin{pmatrix} 1 & 2 \\ 2 & 3 \end{pmatrix}, \quad \text{(b)} \ (1/10)\begin{pmatrix} 6 & 42 & -10 \\ 42 & 69 & 5 \\ -10 & 5 & -35 \end{pmatrix}$$

Exercise 6.1.2.

Prove part (b) of the theorem on the normal form for quadratic forms.

Our approach for putting a quadratic form into its normal form first requires for us to solve the eigenvalue–eigenvector problem. There is a much simpler approach, which we will now explain. It is basically the routine of completing the square. We first illustrate the method with an example.

■ ***Example 6.1.2.*** Let $A = \begin{pmatrix} 2 & 1 & 1 \\ 1 & -1 & 1 \\ 1 & 1 & 2 \end{pmatrix}$ and

$$\mathcal{Q}(\mathbf{x}) = \mathbf{x}^tA\mathbf{x} = 2x_1^2 + 2x_1x_2 + 2x_1x_3 - x_2^2 + 2x_2x_3 + 2x_3^2$$

Working with $\mathcal{Q}$ we first complete the square to remove mixed terms involving the first coordinate. We write all of the terms involving x_1 as

$$2(x_1^2 + x_1x_2 + x_1x_3) = 2(x_1 + (1/2)(x_2 + x_3))^2 - (1/2)(x_2^2 + 2x_2x_3 + x_3^2)$$

Now we let $y_1 = x_1 + (1/2)(x_2 + x_3)$. Then our quadratic expression can be rewritten as $2y_1^2 - (3/2)x_2^2 + x_2x_3 + (3/2)x_3^2$. We have thus replaced terms involving the old first coordinate with a simple square term involving the new first coordinate y_1. We next look at the part of the expression which involves x_2. This is $-(3/2)x_2^2 + x_2x_3$. We complete the square on this term to rewrite it as

$$(-3/2)(x_2^2 - (2/3)x_2x_3) = (-3/2)(x_2 - (1/3)x_3)^2 + (1/6)x_3^2$$

We next introduce $y_2 = x_2 - (1/3)x_3$, $y_3 = x_3$. Then our expression becomes $2y_1^2 - (3/2)y_2^2 + (5/3)y_3^2$. Finally, if we substitute $z_1 = \sqrt{2}\,y_1$, $z_2 = \sqrt{5/3}\,y_3$, $z_3 = \sqrt{3/2}\,y_2$, we may rewrite the quadratic expression as $z_1^2 + z_2^2 - z_3^2$. In terms of matrices,

$\mathbf{y} = \begin{pmatrix} 1 & 1/2 & 1/2 \\ 0 & 1 & -1/3 \\ 0 & 0 & 1 \end{pmatrix} \mathbf{x}$, so $\mathbf{x} = \begin{pmatrix} 1 & -1/2 & -2/3 \\ 0 & 1 & 1/3 \\ 0 & 0 & 1 \end{pmatrix} \mathbf{y}$. Since

$$\mathbf{y} = \begin{pmatrix} 1/\sqrt{2} & 0 & 0 \\ 0 & 0 & \sqrt{2/3} \\ 0 & \sqrt{3/5} & 0 \end{pmatrix} \mathbf{z}$$

we can write $\mathbf{x} = S\mathbf{z}$, where $S = \begin{pmatrix} 1/\sqrt{2} & -2/\sqrt{15} & -1/\sqrt{6} \\ 0 & 1/\sqrt{15} & \sqrt{2/3} \\ 0 & \sqrt{3/5} & 0 \end{pmatrix}$. Then $S^tAS = D(2, 1, 0)$. ■

The technique used in this example can be generalized to handle a general quadratic form $\sum_{i,j=1}^{n} a_{ij}x_ix_j$ on $\mathbb{R}^n$ as follows. The process described next reduces the problem to one on $\mathbb{R}^{n-1}$ and is solved by recursively applying the algorithm.

Algorithm for Putting a Symmetric Matrix into Normal Form

1. If we have the zero matrix, then the form is already in normal form. If not, and all diagonal entries are zero, but $a_{ij} \neq 0$, then form a change of variables $x_j = y_i + y_j, y_k = x_k, k \neq j$. In the new variables, the diagonal entry $b_{ii} = 2a_{ij} \neq 0$. After making such a change, we can assume that there is a nonzero diagonal entry. If necessary, permute coordinates so the first diagonal entry is nonzero.
2. Calling the variables x_i again, complete the square to rewrite

$$a_{11}x_1^2 + \sum_{j=2}^{n} 2a_{1j}x_1x_j = a_{11}\left(x_1 + \left(\sum_{j=2}^{n}(a_{1j}/a_{11})x_j\right)\right)^2 - a_{11}\left(\sum_{j=2}^{n}(a_{1j}/a_{11})x_j\right)^2$$

When we change variables to $y_1 = x_1 + (\sum_{j=2}^{n}(a_{1j}/a_{11})x_j), y_k = x_k, k \neq 1$, the expression becomes

$$a_{11}y_1^2 + (\text{quadratic in } y_2, \ldots, y_n)$$

Thus the two steps can be applied recursively to the remaining quadratic to put it in the correct form, except for the coefficients not yet being $\pm 1, 0$.
3. As a last step, we can change a term $\pm c_i^2 y_i^2$ to $\pm z_i^2$ by the substitution $z_i = c_i y_i$ and then permute them to get in the desired order.

The algorithm can be restated in terms of how we bring the original matrix A into normal form. We are looking at an expression $\mathbf{x}^tA\mathbf{x}$ and then changing variables $\mathbf{x} = P\mathbf{y}$ so that the expression $\mathbf{x}^tA\mathbf{x} = \mathbf{y}^tP^tAP\mathbf{y} = \mathbf{y}^tD(p, n, d)\mathbf{y}$. For this to happen, we must be making $P^tAP = D(p, n, d)$. The matrix P which arises will be a composition of matrices corresponding to the individual changes P_i in coordinates. Actually, each change will be in the form of $\mathbf{y} = S_i\mathbf{x}$, so for each step P_i will arise as S_i^{-1}. The matrices which arise for the first step of the problem are just simple elementary matrices or permutation matrices. The matrices in the last step of the algorithm are just a diagonal matrix and a permutation matrix. The most interesting matrix is one from the middle step of the algorithm, which gets rid of the mixed terms for a single variable. Then $\mathbf{y} = S\mathbf{x}$, where

$$S = \begin{pmatrix} 1 & \mathbf{c} \\ \mathbf{0} & I_{n-1} \end{pmatrix}, \quad \mathbf{c} = (c_2 \quad \ldots \quad c_n)$$

and $c_j = a_{1j}/a_{11}$. Then $P = S^{-1} = \begin{pmatrix} 1 & -\mathbf{c} \\ \mathbf{0} & I_{n-1} \end{pmatrix}$.

One interesting facet about this process occurs when we never have to do the first step. Then multiplying A to get P^tAP can be thought of as two operations. The first multiplication step $A \to P^tA$ is *exactly the same step* that we do in the step in Gaussian elimination which produces the zeros beneath the leading nonzero entry in the first column. When we go from $P^tA \to P^tAP$ by multiplying on the right we are just doing the exact same column operations that multiplying by P^t on the left achieved in terms of row operations. Thus this step can be thought of as a **symmetric Gaussian elimination step.**

■ ***Example 6.1.3.*** We look at the matrix

$$A = \begin{pmatrix} 3 & 1 & -1 \\ 1 & 2 & 2 \\ -1 & 2 & 1 \end{pmatrix}$$

corresponding to the quadratic form

$$3x_1^2 + 2x_1x_2 - 2x_1x_3 + 2x_2^2 + 4x_2x_3 + x_3^2$$

that is, this is just $\mathbf{x}^tA\mathbf{x}$. We first form $P = \begin{pmatrix} 1 & -1/3 & 1/3 \\ 0 & 1 & 0 \\ 0 & 0 & 1 \end{pmatrix}$ and

$$B = P^tAP = \begin{pmatrix} 3 & 0 & 0 \\ 0 & 5/3 & 7/3 \\ 0 & 7/3 & 2/3 \end{pmatrix}$$

We next form $Q = \begin{pmatrix} 1 & 0 & 0 \\ 0 & 1 & -7/5 \\ 0 & 0 & 1 \end{pmatrix}$ and

$$C = Q^tBQ = \begin{pmatrix} 3 & 0 & 0 \\ 0 & 5/3 & 0 \\ 0 & 0 & -13/5 \end{pmatrix}$$

We then form $T = \begin{pmatrix} 1/\sqrt{3} & 0 & 0 \\ 0 & \sqrt{3/5} & 0 \\ 0 & 0 & \sqrt{5/13} \end{pmatrix}$ and

$$D = T^tCT = \begin{pmatrix} 1 & 0 & 0 \\ 0 & 1 & 0 \\ 0 & 0 & -1 \end{pmatrix} = D(2,1,0)$$

If $S = PQT$, then $S^tAS = D(2,1,0)$. ■

■ ***Example 6.1.4.*** We now look at the matrix

$$A = \begin{pmatrix} 0 & 2 & 1 \\ 2 & 1 & 0 \\ 1 & 0 & 1 \end{pmatrix}$$

Because of the first zero in the $(1,1)$-position, we first use $P = \begin{pmatrix} 0 & 1 & 0 \\ 1 & 0 & 0 \\ 0 & 0 & 1 \end{pmatrix}$ to form

$$B = P^tAP = \begin{pmatrix} 1 & 2 & 0 \\ 2 & 0 & 1 \\ 0 & 1 & 1 \end{pmatrix}$$

We then use $Q = \begin{pmatrix} 1 & -2 & 0 \\ 0 & 1 & 0 \\ 0 & 0 & 1 \end{pmatrix}$ to form

$$C = Q^tBQ = \begin{pmatrix} 1 & 0 & 0 \\ 0 & -4 & 1 \\ 0 & 1 & 1 \end{pmatrix}$$

We then use $R = \begin{pmatrix} 1 & 0 & 0 \\ 0 & 1 & 1/4 \\ 0 & 0 & 1 \end{pmatrix}$ to form

$$D = R^tCR = \begin{pmatrix} 1 & 0 & 0 \\ 0 & -4 & 0 \\ 0 & 0 & 5/4 \end{pmatrix}$$

We then use $T = \begin{pmatrix} 1 & 0 & 0 \\ 0 & 0 & 1/2 \\ 0 & \sqrt{4/5} & 0 \end{pmatrix}$ to form

$$E = T^tDT = \begin{pmatrix} 1 & 0 & 0 \\ 0 & 1 & 0 \\ 0 & 0 & -1 \end{pmatrix} = D(2,1,0)$$

Then if $S = PQRT$, we have

$$S^tAS = E = D(2,1,0)$$ ■

■ ***Example 6.1.5.*** Let

$$A = \begin{pmatrix} 0 & 1 & 1 \\ 1 & 0 & 2 \\ 1 & 2 & 0 \end{pmatrix}$$

Following the substitution $x_2 = y_1 + y_2$ in step 1 of the algorithm to get a nonzero 11-entry, we use $P = \begin{pmatrix} 1 & 0 & 0 \\ 1 & 1 & 0 \\ 0 & 0 & 1 \end{pmatrix}$ to form

$$B = P^t AP = \begin{pmatrix} 2 & 1 & 3 \\ 1 & 0 & 2 \\ 3 & 2 & 0 \end{pmatrix}$$

Using $Q = \begin{pmatrix} 1 & -1/2 & -3/2 \\ 0 & 1 & 0 \\ 0 & 0 & 1 \end{pmatrix}$, we form

$$C = Q^t BQ = \begin{pmatrix} 2 & 0 & 0 \\ 0 & -1/2 & 1/2 \\ 0 & 1/2 & -9/2 \end{pmatrix}$$

Using $R = \begin{pmatrix} 1 & 0 & 0 \\ 0 & 1 & 1 \\ 0 & 0 & 1 \end{pmatrix}$, we form

$$D = R^t CR = \begin{pmatrix} 2 & 0 & 0 \\ 0 & -1/2 & 0 \\ 0 & 0 & -4 \end{pmatrix}$$

Using $T = \begin{pmatrix} 1/\sqrt{2} & 0 & 0 \\ 0 & \sqrt{2} & 0 \\ 0 & 0 & 1/2 \end{pmatrix}$, we form

$$E = T^t DT = \begin{pmatrix} 1 & 0 & 0 \\ 0 & -1 & 0 \\ 0 & 0 & -1 \end{pmatrix} = D(1, 2, 0)$$

Letting $S = PQRT$, we have

$$S^t AS = D(1, 2, 0)$$

■

Exercise 6.1.3.

For each of the following matrices A, give S and $D(p, n, d)$ so that $S^t AS = D(p, n, d)$.

(a) $\begin{pmatrix} 21 & 12 & 3 \\ 12 & 17 & 4 \\ 3 & 4 & 2 \end{pmatrix}$

(b) $\begin{pmatrix} 0 & -1 & 0 \\ -1 & -1 & -1 \\ 0 & -1 & 0 \end{pmatrix}$

(c) $\begin{pmatrix} 0 & -1 & 0 & -1 \\ -1 & 2 & -2 & 0 \\ 0 & -2 & 1 & -1 \\ -1 & 0 & -1 & -2 \end{pmatrix}$

If the matrix A is positive (resp., negative) definite ($\mathbf{x}^t A\mathbf{x} > 0$ for $\mathbf{x} \neq \mathbf{0}$) (resp., <0), then no diagonal entry will be zero. This will persist as we continue modifying the matrix. This follows from the fact that a matrix congruent to a positive definite one is still positive definite and is left as an exercise. Thus we basically are doing the reduction in terms of the standard "symmetric Gaussian elimination moves." On the other hand, if the diagonal entries remain positive as we perform these steps, then we can make the matrix congruent to a sum of squares, and it is positive definite. This gives a test for when the matrix is positive definite. There is, of course, an analogous statement for negative definite matrices.

When we do one "symmetric Gaussian elimination" step to clear the first row and column, the result of multiplying A by P^t to get $P^t A$ has $P^t A = \begin{pmatrix} a_{11} & \mathbf{a}' \\ \mathbf{0} & B \end{pmatrix}$ where we have cleared the first column beneath the leading entry. When we multiply on the right by P, this serves only to clear the first row and doesn't change B, so $P^t AP = \begin{pmatrix} a_{11} & \mathbf{0} \\ \mathbf{0} & B \end{pmatrix}$ for the same matrix B. This means that we can effectively ignore the multiplication by the P terms on the right and then just do the Gaussian elimination on A to make it upper triangular, and then throw away the terms above the diagonal to see the final normal form. Of course this works only if we always get nonzero terms on the critical diagonal elements at each stage. This will be guaranteed if the matrix is definite.

A way of detecting whether it is definite is then suggested by this. When we do one step of Gaussian elimination by adding multiples of a row to ones beneath it, it will not change the determinant of the whole matrix. But it will also not change any of the subdeterminants of the k by k matrices in the upper left-hand corner, since they are also obtained from their predecessors by the same process. If the matrix is positive definite, then each of these subdeterminants will have to turn out to be positive. Conversely, if each one is positive, this means that the critical diagonal entries must turn out to be positive at each step, and the matrix is positive definite. This provides a computable criterion for the matrix being positive definite. There is a similar criterion for negative definite matrices by requiring these subdeterminants to start off negative and alternate in sign. We collect these observations in the following proposition.

PROPOSITION 6.1.2. Criteria for A Being Positive Definite

1. A symmetric matrix A is positive definite iff each pivot is positive when performing Gaussian elimination to reduce A to U with no row exchanges.
2. A symmetric matrix A is positive definite iff each of the subdeterminants of the submatrices in the upper left-hand corners are positive.

Exercise 6.1.4

Test whether the following matrices are positive definite, negative definite, or neither by computing subdeterminants. For each matrix, put it in its normal form by following the preceding algorithm.

$$\text{(a)} \begin{pmatrix} 2 & 1 \\ 1 & 2 \end{pmatrix}, \quad \text{(b)} \begin{pmatrix} -4 & -8 & 3 \\ -8 & -11 & 0 \\ 3 & 0 & 5 \end{pmatrix}, \quad \text{(c)} \begin{pmatrix} -14 & -16 & -3 \\ -16 & -21 & 0 \\ -3 & 0 & -5 \end{pmatrix},$$

$$\text{(d)} \begin{pmatrix} 2 & 1 & 1 & 0 \\ 1 & 2 & -1 & -1 \\ 1 & -1 & 2 & 1 \\ 0 & -1 & 1 & 2 \end{pmatrix}$$

Exercise 6.1.5.

Verify that if A is positive definite and B is congruent to A, then B is also positive definite. Use this to show that if A is positive definite, then our algorithm will need only to repeatedly use step 2 in reducing it to a diagonal matrix, and only positive entries will appear on the diagonal as claimed.

Exercise 6.1.6.

Verify that a positive definite matrix will only have positive entries on the diagonal. Give an example of a positive definite 2 by 2 matrix which has negative off-diagonal entries.

Sylvester's law of inertia generalizes the fact that if a matrix is positive definite, then any congruent matrix is also positive definite.

THEOREM 6.1.3. Sylvester's Law of Inertia For a symmetric m by m matrix A, the numbers p, n, d for which A is congruent to $D(p, n, d)$ are invariants of the congruence class of A.

Proof. Suppose A is congruent to $D(p, n, d)$. We need an invariant characterization of these numbers. We give the argument for p; the argument for n is analogous, and then the result follows for d since $d = m - p - n$. We form the quadratic form $\mathcal{Q}_A(\mathbf{x}) = \mathbf{x}^t A\mathbf{x}$. Choose S so that $S^t AS = D(p, n, d) = D$. We have $\mathcal{Q}_A(S\mathbf{e}_i) = \mathcal{Q}_D(\mathbf{e}_i) = 1, 1 \leq i \leq p$, and $\mathcal{Q}_A(S\mathbf{e}_i) \leq 0, p+1 \leq i \leq m$. The corresponding bilinear form for $\mathcal{Q}_A$ will also satisfy $\mathcal{B}_A(S\mathbf{e}_i, S\mathbf{e}_j) = 0, i \neq j$. Now let P denote the dimension of the largest dimensional subspace T on which $\mathcal{Q}_A$ is positive definite. Since $\mathcal{Q}_A$ is positive definite on the subspace spanned by $S\mathbf{e}_1, \ldots, S\mathbf{e}_p$, we know that $P \geq p$. Note that on the subspace U with basis $S\mathbf{e}_{p+1}, \ldots, S\mathbf{e}_m$, we have $\mathcal{Q}_A(\mathbf{x}) \leq 0$ for $\mathbf{x} \in U$. If $P > p$, then there would be a nonempty intersection between T and U since the sum of their dimensions would be greater than m. But this is impossible since $\mathcal{Q}_A(\mathbf{x}) > 0$ for nonzero $\mathbf{x} \in T$ and $\mathcal{Q}_A(\mathbf{x}) \leq 0$ for $\mathbf{x} \in U$. Thus we must have $P = p$.

The number p is characterized as the dimension of the largest dimensional subspace on which $\mathcal{Q}_A$ is positive definite. This dimension is an invariant of the congruence class of the matrix A. If A is congruent to B, say, $B = V^t AV$, and R is the largest dimensional subspace on which $\mathcal{Q}_B$ is positive definite, then the subspace $V(R)$ which arises by multiplying each element of R by the matrix V will be a subspace of the same dimension (since V is invertible) on which $\mathcal{Q}_A$ is positive definite. Thus the dimension of the largest positive definite dimensional subspace for $\mathcal{Q}_A$ is greater than or equal to that for $\mathcal{Q}_B$. Reversing the role of A and B (using congruence is an equivalence relation) shows that these two dimensions must be equal. We thus conclude that p is an invariant of the equivalence class. ■

Sylvester's theorem can be restated entirely in terms of quadratic forms. Suppose $(V_i, \mathcal{Q}_i)$ denotes a pair consisting of a real vector space V_i and a quadratic form $\mathcal{Q}_i$ defined on $V_i, i = 1, 2$. We say that $(V_2, \mathcal{Q}_2)$ is **isomorphic** to $(V_1, \mathcal{Q}_1)$ if there is an isomorphism $L : V_2 \rightarrow V_1$ so that $\mathcal{Q}_1(L\mathbf{v}) = \mathcal{Q}_2(\mathbf{v})$ for all $\mathbf{v} \in V_2$. If $(V_2, \mathcal{Q}_2)$ is isomorphic to $(V_1, \mathcal{Q}_1)$, then L^{-1} can be used to show that $(V_1, \mathcal{Q}_1)$ is isomorphic to $(V_2, \mathcal{Q}_2)$. In fact, isomorphism of quadratic forms provides an equivalence relation on quadratic forms. Our argument in the proof of Sylvester's theorem using the invertible matrix V can be viewed as using the fact that V provides an isomorphism between the quadratic forms $(\mathbb{R}^m, \mathcal{Q}_B)$ and $(\mathbb{R}^m, \mathcal{Q}_A)$. There is the following analog of Sylvester's theorem for quadratic forms.

THEOREM 6.1.4. Sylvester's Theorem for Quadratic Forms Let $(V, \mathcal{Q})$ denote an m-dimensional vector space V and a quadratic form $\mathcal{Q}$ on V. The dimensions p, n of the largest dimensional subspaces on which $\mathcal{Q}$ is, respectively, positive definite, negative definite are invariants of the isomorphism class of the quadratic form. There is a basis for V so that the matrix for the quadratic form with respect to this basis is $D(p, n, d)$, $d = m - p - n$.

Exercise 6.1.7.

Show that the relation of isomorphism of quadratic forms is an equivalence relation.

Exercise 6.1.8.

Prove Sylvester's theorem for quadratic forms.

6.2. JORDAN CANONICAL FORM

We now return to a discussion of the general case where a linear transformation may not be diagonalizable. Let V be a complex vector space of dimension n. We can still find a "nice" basis for V so that the matrix of L with respect to this basis is particularly simple, although as we have seen, it cannot always be chosen to be diagonal.

We first consider a special type of linear transformation.

DEFINITION 6.2.1. We say that the linear transformation N is **nilpotent** if $N^k = 0$ for some positive integer k.

Suppose N is nilpotent. Then the only eigenvalue of N is 0, for if c is an eigenvalue of N and $N\mathbf{v} = c\mathbf{v}$, then $N^k\mathbf{v} = c^k\mathbf{v}$. Thus c^k is an eigenvalue of $N^k = 0$, and so $c^k = 0$ and $c = 0$. Suppose that N is nilpotent and $N^k = 0$ but N^{k-1} is not the zero linear transformation. Then let $\mathbf{v}_k$ be a vector in V so that $N^{k-1}\mathbf{v}_k$ is *not* $\mathbf{0}$. Consider the vectors $\mathbf{v}_k, \mathbf{v}_{k-1} = N\mathbf{v}_k, \mathbf{v}_{k-2} = N\mathbf{v}_{k-1}, \cdots, \mathbf{v}_1 = N\mathbf{v}_2 = N^{k-1}\mathbf{v}_k$. First note that $\mathbf{v}_1, \cdots, \mathbf{v}_k$ forms an independent set in V. To see this, suppose $c_1\mathbf{v}_1 + \cdots + c_k\mathbf{v}_k = \mathbf{0}$. Applying N to this gives the equation $c_2\mathbf{v}_1 + \cdots + c_k\mathbf{v}_{k-1} = \mathbf{0}$. The result can then be proven by induction on the number k; the details are left as an exercise.

In the case where the dimension of V is k, this would give a basis for V and the matrix of N with respect to this basis would be

$$J_{0,k} = \begin{pmatrix} 0 & 1 & 0 & \dots & 0 \\ 0 & 0 & 1 & \dots & 0 \\ \vdots & \vdots & \vdots & \vdots & \vdots \\ 0 & 0 & \dots & 0 & 1 \\ 0 & 0 & \dots & 0 & 0 \end{pmatrix}$$

This k by k matrix has zeros except on the superdiagonal entries $a_{(i,i+1)} = 1$.

DEFINITION 6.2.2. The matrix $J_{0,k}$ is called a **small Jordan block of size k** associated to the eigenvalue 0. The set of vectors $\mathbf{v}_1, \ldots, \mathbf{v}_k$ is called a **Jordan string of length k,** the vector $\mathbf{v}_k$ is called the **generator of the string**, and the vector $\mathbf{v}_1$ is called the **end of the string.**

Note that $\mathbf{v}_1$ will be an eigenvector for N associated to the eigenvalue 0. In the general case when $k \leq \dim V$, we at least have that N sends the subspace spanned by these k vectors to itself and the matrix of the restriction of N to this subspace is given by $J_{0,k}$.

DEFINITION 6.2.3. By a **big Jordan block associated with 0** we mean a matrix J so that J can be broken up into some number p of small Jordan blocks arranged down the diagonal; that is,

$$J = \begin{pmatrix} J_{0,k_1} & 0 & 0 & \dots & 0 \\ 0 & J_{0,k_2} & 0 & \dots & 0 \\ 0 & 0 & \dots & \dots & 0 \\ 0 & 0 & \dots & 0 & J_{0,k_p} \end{pmatrix}$$

where J_{0,k_i} is a small Jordan block of size k_i associated to the eigenvalue 0.

The Jordan canonical form for a nilpotent linear transformation is just a matrix in the form of J, and the main result for nilpotent matrices is that there is a basis of V so that the matrix of N with respect to this basis is in the form of a big Jordan block associated to 0. In this case the sum of the sizes of the small Jordan blocks involved will just be $n = \dim V$. Note that the vector space V is splitting as a direct sum of p subspaces, each of which has a basis so that the matrix of N with respect to this basis is given by a small Jordan block. The problem in finding this basis becomes one of finding the generators of the Jordan strings. Next suppose L has the form $cI + N$, where N is nilpotent. If we can find a basis as was done earlier so that the matrix of N with respect to this basis is given by a big Jordan block associated to 0, then the matrix of L with respect to this basis is just given by $cI + J$. This is found from J by adding c to each diagonal entry. Note that this matrix is also composed of small Jordan blocks, where we have added c along the diagonal to each small Jordan block. A typical small Jordan block looks like

$$J_{c,k} = \begin{pmatrix} c & 1 & 0 & \dots & 0 \\ 0 & c & 1 & \dots & 0 \\ \vdots & \vdots & \vdots & \vdots & \vdots \\ 0 & 0 & \dots & c & 1 \\ 0 & 0 & \dots & 0 & c \end{pmatrix}$$

DEFINITION 6.2.4. A small Jordan block for the eigenvalue c is a matrix of the form $J_{c,k}$. A big Jordan block for the eigenvalue c is a matrix J_c so that J_c can be broken up into p small Jordan blocks arranged down the diagonal; that is,

$$J = \begin{pmatrix} J_{c,k_1} & 0 & 0 & \dots & 0 \\ 0 & J_{c,k_2} & 0 & \dots & 0 \\ 0 & 0 & \dots & \dots & 0 \\ 0 & 0 & \dots & 0 & J_{c,k_p} \end{pmatrix}$$

where J_{c,k_i} is a small Jordan block of size k_i associated to the eigenvalue c.

For a general matrix, the idea is to find all the distinct eigenvalues and to show that associated with each eigenvalue there is a collection of linearly independent vectors so that L sends the subspace spanned by these vectors to itself and the matrix of L with respect to this collection is a big Jordan block associated to this eigenvalue. It can be shown that putting all these vectors together for the distinct eigenvalues gives a basis for V and the matrix of L with respect to this basis will have big Jordan blocks associated to the distinct eigenvalues arranged down the diagonal:

$$J = \begin{pmatrix} J_1 & 0 & 0 & \dots & 0 \\ 0 & J_2 & 0 & \dots & 0 \\ 0 & 0 & \dots & \dots & 0 \\ 0 & 0 & \dots & \dots & J_k \end{pmatrix}$$

Here J_i denotes a big Jordan block associated to the ith eigenvalue.

DEFINITION 6.2.5. A matrix is said to be in **Jordan canonical form** if it is in the form of the preceding J.

Note that a basis for the eigenvectors associated to the eigenvalue c_i is just given by the vectors which lie at the ends of the various Jordan strings. The linear transformation L will be diagonalizable only when all the Jordan strings are of length 1, and then the Jordan canonical form will agree with a diagonal matrix. Here is an example of a 10 by 10 matrix which is in Jordan canonical form.

$$\begin{pmatrix} 1 & 1 & 0 & 0 & 0 & 0 & 0 & 0 & 0 & 0 \\ 0 & 1 & 1 & 0 & 0 & 0 & 0 & 0 & 0 & 0 \\ 0 & 0 & 1 & 0 & 0 & 0 & 0 & 0 & 0 & 0 \\ 0 & 0 & 0 & 1 & 1 & 0 & 0 & 0 & 0 & 0 \\ 0 & 0 & 0 & 0 & 1 & 0 & 0 & 0 & 0 & 0 \\ 0 & 0 & 0 & 0 & 0 & 2 & 0 & 0 & 0 & 0 \\ 0 & 0 & 0 & 0 & 0 & 0 & 2 & 0 & 0 & 0 \\ 0 & 0 & 0 & 0 & 0 & 0 & 0 & 4 & 1 & 0 \\ 0 & 0 & 0 & 0 & 0 & 0 & 0 & 0 & 4 & 0 \\ 0 & 0 & 0 & 0 & 0 & 0 & 0 & 0 & 0 & 4 \end{pmatrix}$$

Here the vector $\mathbf{v}_3$ is a generator of a Jordan string of length 3 associated with the eigenvalue 1 (giving us one small Jordan block). The vector $\mathbf{v}_5$ is a generator of a Jordan string of length 2 associated to the eigenvalue 1. The first 5 by 5 piece constitutes a big Jordan block associated to the eigenvalue 1. Note that the eigenspace associated to eigenvalue 1 is two dimensional with basis $\mathbf{v}_1$ and $\mathbf{v}_4$. These are just the end vectors of the two Jordan strings for the eigenvalue 1. For the eigenvalue 2 there are two Jordan strings of length 1, and the vectors $\mathbf{v}_6$, $\mathbf{v}_7$ are eigenvectors for the eigenvalue 2. These constitute a basis for the eigenspace for the eigenvalue 2. The vector $\mathbf{v}_9$ is a generator for a Jordan string of length 2 associated to the eigenvalue 4. The vector $\mathbf{v}_{10}$ gives a Jordan string of length 1, and the vectors $\mathbf{v}_8$, $\mathbf{v}_{10}$, which are the end vectors for the two Jordan strings associated to the eigenvalue 4, form a basis for the eigenspace of L associated to this eigenvalue.

We are ready to state the main theorem regarding the Jordan canonical form.

THEOREM 6.2.1. Linear Transformation Version. Let V be a complex vector space of dimension n and $L : V \to V$ a linear transformation. There is a basis for V so that the matrix of L with respect to this basis is in Jordan canonical form.

Matrix version. Any square matrix is similar (over the complex numbers) to a matrix in Jordan canonical form.

Proof. We will prove the linear transformation version. The matrix version follows by using the linear transformation which is given by multiplying by the square matrix A. We will refer to a basis as described in the theorem as a Jordan basis of V for L.

We first suppose that 0 is one of the eigenvalues. We will work by induction on the dimension of V. When the dimension of V is 1, the result is satisfied by choosing any nonzero vector as our basis. Thus suppose the result is true when the dimension is less than $n = \dim V$. Look at the range of L. First note that L sends the range of L to itself and that the dimension of the range of L is less than n by our hypothesis that 0 is an eigenvalue. Thus the induction assumption applied to the restriction of L to the range of L asserts that there is a Jordan basis $\{\mathbf{v}_i, i = 1, \dots, r\}$ of $\mathcal{R}(L)$ for $L \mid \mathcal{R}(L) : \mathcal{R}(L) \to \mathcal{R}(L)$ for which the matrix of L is in Jordan canonical form.

There will be p vectors in the Jordan basis $\{\mathbf{v}_i\}$ that are end vectors of p Jordan strings associated to the eigenvalue 0. Note it is possible $p = 0$ if 0 is not an eigenvalue for the restriction of L to $\mathcal{R}(L)$. These p vectors will be independent eigenvectors for the eigenvalue 0. Let us call these p vectors $\mathbf{z}_1, \dots, \mathbf{z}_p$ and call the generator vectors for these strings $\mathbf{w}_1, \dots, \mathbf{w}_p$. Call the length of the ith string $k(i)$. Note that all these are part of our basis $\{\mathbf{v}_i\}$ for the range of L. Since the $\mathbf{w}_j$ are in the range of L, there are vectors $\mathbf{u}_j$ so that $L\mathbf{u}_j = \mathbf{w}_j$. The $\mathbf{u}_j$ will be generators for Jordan strings of length $k(j) + 1$ with end vectors $\mathbf{z}_j$. The dimension of the null space of L is $n - r$ and we already have p independent vectors $\mathbf{z}_1, \dots, \mathbf{z}_p$ in that null space. Thus we can find $n-r-p$ vectors $\mathbf{x}_1, \dots, \mathbf{x}_{n-r-p}$ so that the $\mathbf{x}_k$ together with the $\mathbf{z}_i$ give a basis for the null space of L.

We claim that when we add the vectors $\mathbf{x}_k$ and $\mathbf{u}_j$ to the Jordan basis $\{\mathbf{v}_i\}$, this will give a Jordan basis for V so that the matrix of L with respect to this basis is in Jordan canonical form. First note that it has the correct number of vectors for a basis for V since it contains $r + p + (n - r - p) = n$ vectors. Thus it suffices to show that it spans V; that is, every vector in V is a linear combination of the original Jordan basis $\{\mathbf{v}_i\}$ for $\mathcal{R}(L)$ and the vectors $\{\mathbf{x}_k\}$ and $\{\mathbf{u}_j\}$. Let $\mathbf{v}$ be a vector in V. Then $L\mathbf{v} \in \mathcal{R}(L)$ and so can be written as a linear combination $L(\mathbf{v}) = \sum a_i\mathbf{v}_i$ of the Jordan basis for the range of L. However, each of the vectors $\mathbf{v}_i$ that is in this basis for the eigenvalue 0 that is not already a generating vector for a string is also L of one of the basis elements. The generator $\mathbf{w}_q$ is in the image of $\mathbf{u}_q$ by construction. For the vectors in strings associated to other eigenvalues, note that the vectors giving the basis for the small Jordan blocks span a subspace which is invariant under L and that the restriction of L to this subspace is invertible (since the small Jordan block is invertible). Thus the vectors in these strings are in the image of some linear combination of the basis vectors for this subspace.

Putting all this together, we get that $L\mathbf{v}$ is also L of a linear combination of our Jordan basis elements $\mathbf{v}_i$ and the $\mathbf{u}_j$: $L(\mathbf{v}) = L(\sum b_i\mathbf{v}_i + \sum c_j\mathbf{u}_j)$. Then $L(\mathbf{v} - \sum b_i\mathbf{v}_i - \sum c_j\mathbf{u}_j) = \mathbf{0}$. Thus $\mathbf{v} - \sum b_i\mathbf{v}_i - \sum c_j\mathbf{u}_j$ can be written as a linear combination of the $\mathbf{z}_k$ (which are some of our $\mathbf{v}_i$) and the $\mathbf{x}_k$. This means that the vector $\mathbf{v}$ is a linear combination of the original basis elements $\mathbf{v}_i$, the $\mathbf{u}_j$ and the $\mathbf{x}_k$. Thus these vectors give a basis for V.

By construction the matrix of L with respect to this basis will be in Jordan canonical form. It will have all the original Jordan blocks for the eigenvalues different from 0. For the eigenvalue 0 there will be $(n - r - p)$ strings of length 1 corresponding to the $\mathbf{x}_i$ and the original p strings of length $k(i)$ will now be strings of length $k(i) + 1$ with generators the $\mathbf{u}_i$. This completes the proof when L has 0 as an eigenvalue.

For a general L, suppose one of its eigenvalues is c. Then $L - cI$ will have 0 as an eigenvalue, and so there will be a basis so that the matrix of $L - cI$ is in Jordan form. Then the matrix of L will also be in Jordan form with respect to this basis (we will just be adding c to each diagonal entry). ■

The idea behind the above proof is due to A. F. Filippov and was originally published in *Moscow University Vestnik* 26 (1971), 18–19—we first saw an outline of the argument in [7]. Unfortunately, the inductive nature of this proof does not suggest an easy method to find the basis so that L is in Jordan form. However, looking at the Jordan form shows that the key is to find the generators of the Jordan strings. Also, the Jordan form shows that we can treat the distinct eigenvalues separately. The key to finding the generators for the eigenvalue c is to look at the successive null spaces of $(L - cI)^k$ for higher and higher powers of k. Alternatively, we can find the end vectors of the Jordan string and then work backward from these to find the generators. We will illustrate both methods in our examples.

■ ***Example 6.2.1.*** Let $A = \begin{pmatrix} 0 & 1 \\ -1 & 2 \end{pmatrix}$. Then the only eigenvalue is 1 with multiplicity 2. Then $A - I = \begin{pmatrix} -1 & 1 \\ -1 & 1 \end{pmatrix}$ and we see that the eigenvectors for the eigenvalue 1 are multiples of $(1, 1)$. We then compute $(A - I)^2 = \begin{pmatrix} 0 & 0 \\ 0 & 0 \end{pmatrix}$. Thus every vector in $\mathbb{R}^2$ is in the null space of $(A - I)^2$, and so we can take as the generator vector for the Jordan string of length 2 any vector that is *not* a multiple of $(1, 1)$. If we choose $\mathbf{v}_2 = \mathbf{e}_1$, then $\mathbf{v}_1 = (A - I)\mathbf{v}_2 = (-1, -1)$. Thus $S = \begin{pmatrix} -1 & 1 \\ -1 & 0 \end{pmatrix}$ and $J = \begin{pmatrix} 1 & 1 \\ 0 & 1 \end{pmatrix}$.

An alternative method is to consider the vector $(1, 1)$ to be the end vector $\mathbf{v}_1$ of a Jordan string of length 2 and then to solve $(A - I)\mathbf{v}_2 = \mathbf{v}_1$ for $\mathbf{v}_2 = (-1, 0)$ by Gaussian elimination. This would lead to the same J but $S = \begin{pmatrix} 1 & -1 \\ 1 & 0 \end{pmatrix}$. ■

Note that when $n = 2$ and the matrix is not diagonalizable, the Jordan form must be $J = \begin{pmatrix} c & 1 \\ 0 & c \end{pmatrix}$, where c is the repeated eigenvalue of algebraic multiplicity 2.

■ ***Example 6.2.2.*** Let $A = \begin{pmatrix} 3 & -2 & 5 \\ -1 & 2 & 1 \\ -1 & 1 & 0 \end{pmatrix}$. Then we compute the characteristic polynomial as $-(x-1)(x-2)^2$, so the eigenvalues are 1 and 2, and 2 has algebraic multiplicity 2. We compute $A - I = \begin{pmatrix} 2 & -2 & 5 \\ -1 & 1 & 1 \\ -1 & 1 & -1 \end{pmatrix}$ and solve for the eigenvector $\mathbf{v}_1 = (1, 1, 0)$. We then compute $A - 2I = \begin{pmatrix} 1 & -2 & 5 \\ -1 & 0 & 1 \\ -1 & 1 & -2 \end{pmatrix}$ and solve to find that the eigenspace associated to the eigenvalue 2 is one dimensional, generated by the eigenvector $(1, 3, 1)$. We then compute $(A - 2I)^2 = \begin{pmatrix} -2 & 3 & -7 \\ -2 & 3 & -7 \\ 0 & 0 & 0 \end{pmatrix}$ and find a basis for the null space given by $(3, 2, 0)$ and $(7, 0, 2)$. Since neither of these is a multiple of $(1, 3, 1)$, we could use either one as a generator for the Jordan string of length 2. We choose $\mathbf{v}_3 = (3, 2, 0)$ and compute $\mathbf{v}_2 = (A - 2I)\mathbf{v}_3 = (-1, -3, -1)$. Then the matrix

$$S = \begin{pmatrix} 1 & -1 & 3 \\ 1 & -3 & 2 \\ 0 & -1 & 0 \end{pmatrix} \quad \text{and} \quad J = \begin{pmatrix} 1 & 0 & 0 \\ 0 & 2 & 1 \\ 0 & 0 & 2 \end{pmatrix}$$

Using the alternate method from the last example, we could have set $\mathbf{v}_2 = (1, 3, 1)$ and then solved $(A - 2I)\mathbf{v}_3 = \mathbf{v}_2$ for $\mathbf{v}_3 = (-3, -2, 0)$. We get the same J as before but now $S = \begin{pmatrix} 1 & 1 & -3 \\ 1 & 3 & -2 \\ 0 & 1 & 0 \end{pmatrix}$. ■

This example gives one of the three possibilities for the Jordan form of a 3 by 3 nondiagonalizable matrix (up to order and the choice of eigenvalues). The other ones would be where there is an eigenvalue of algebraic multiplicity 3 and either two or one independent eigenvectors, giving Jordan forms

$$J = \begin{pmatrix} c & 1 & 0 \\ 0 & c & 0 \\ 0 & 0 & c \end{pmatrix} \quad \text{or} \quad J = \begin{pmatrix} c & 1 & 0 \\ 0 & c & 1 \\ 0 & 0 & c \end{pmatrix}$$

■ ***Example 6.2.3.*** Let $A = 1/2\begin{pmatrix} 3 & -1 & 1 & 1 \\ 0 & 2 & 2 & 0 \\ 0 & 0 & 2 & 2 \\ -1 & 1 & 1 & 3 \end{pmatrix}$. Then we compute the characteristic polynomial and factor it as $(x-1)^3(x-2)$. We compute $(A - 2I) =$

$1/2\begin{pmatrix} -1 & -1 & 1 & 1 \\ 0 & 1 & 2 & 0 \\ 0 & 0 & 1 & 2 \\ -1 & 1 & 1 & -1 \end{pmatrix}$ and solve for the eigenvector $\mathbf{v}_1 = (1, 1, 1, 1)$. We then compute $A - I = 1/2\begin{pmatrix} 1 & -1 & 1 & 1 \\ 0 & 0 & 2 & 0 \\ 0 & 0 & 0 & 2 \\ -1 & 1 & 1 & 1 \end{pmatrix}$ and find a one-dimensional eigenspace that has basis $(1, 1, 0, 0)$. Note that this tells us that the big Jordan block associated to the eigenvalue 1 can't have two small Jordan blocks in it since then the eigenspace would have been two dimensional. Thus there must be one Jordan string of length 3. To find it we next compute

$$(A - I)^2 = 1/2\begin{pmatrix} 0 & 0 & 0 & 2 \\ 0 & 0 & 0 & 2 \\ -1 & 1 & 1 & 1 \\ -1 & 1 & 1 & 1 \end{pmatrix}$$

which has null space with basis $(1, 1, 0, 0)$ and $(1, 0, 1, 0)$. We next compute

$$(A - I)^3 = 1/2\begin{pmatrix} -1 & 1 & 1 & 1 \\ -1 & 1 & 1 & 1 \\ -1 & 1 & 1 & 1 \\ -1 & 1 & 1 & 1 \end{pmatrix}$$

which has null space with basis the two vectors together with the vector $(1, 0, 0, 1)$. We then let $\mathbf{v}_4 = (1, 0, 0, 1)$ and find $\mathbf{v}_3 = (A - I)\mathbf{v}_4 = (1, 0, 1, 0)$ and $\mathbf{v}_2 = (A - I)\mathbf{v}_3 = (1, 1, 0, 0)$. We have

$$S = \begin{pmatrix} 1 & 1 & 1 & 1 \\ 1 & 1 & 0 & 0 \\ 1 & 0 & 1 & 0 \\ 1 & 0 & 0 & 1 \end{pmatrix} \quad \text{and} \quad J = \begin{pmatrix} 2 & 0 & 0 & 0 \\ 0 & 1 & 1 & 0 \\ 0 & 0 & 1 & 1 \\ 0 & 0 & 0 & 1 \end{pmatrix}$$

Let us now recompute the Jordan string for the eigenvalue 1 by the alternate method. We would first choose $\mathbf{v}_2 = (1, 1, 0, 0)$, which is a basis vector for the null space of $A - I$. This will be the end vector of our Jordan string of length 3. We find $\mathbf{v}_3$ by solving $(A - I)\mathbf{v}_3 = \mathbf{v}_2$ by Gaussian elimination to get $\mathbf{v}_3 = (1, 0, 1, 0)$. We next find $\mathbf{v}_4$ by solving $(A - I)\mathbf{v}_4 = \mathbf{v}_3$ to get $\mathbf{v}_4 = (1, 0, 0, 1)$. This time we get the same S and J as before. ■

Some other Jordan forms which can occur for a 4 by 4 nondiagonalizable matrix are

$$\begin{pmatrix} a & 0 & 0 & 0 \\ 0 & b & 0 & 0 \\ 0 & 0 & c & 1 \\ 0 & 0 & 0 & c \end{pmatrix}, \begin{pmatrix} a & 1 & 0 & 0 \\ 0 & a & 0 & 0 \\ 0 & 0 & b & 1 \\ 0 & 0 & 0 & b \end{pmatrix}, \begin{pmatrix} a & 0 & 0 & 0 \\ 0 & a & 0 & 0 \\ 0 & 0 & b & 1 \\ 0 & 0 & 0 & b \end{pmatrix},$$

$$\begin{pmatrix} a & 0 & 0 & 0 \\ 0 & a & 0 & 0 \\ 0 & 0 & a & 1 \\ 0 & 0 & 0 & a \end{pmatrix}, \begin{pmatrix} a & 0 & 0 & 0 \\ 0 & a & 1 & 0 \\ 0 & 0 & a & 1 \\ 0 & 0 & 0 & a \end{pmatrix}, \begin{pmatrix} a & 1 & 0 & 0 \\ 0 & a & 1 & 0 \\ 0 & 0 & a & 1 \\ 0 & 0 & 0 & a \end{pmatrix}$$

■ ***Example 6.2.4.*** We look at a more complicated example of a 6 by 6 matrix A with a single eigenvalue of algebraic multiplicity 6:

$$A = 1/2 \begin{pmatrix} 3 & 0 & -2 & 2 & 1 & 0 \\ 0 & 2 & 0 & 2 & 0 & 0 \\ 1 & 0 & 0 & 2 & 1 & 0 \\ 2 & -2 & 0 & 2 & 0 & 2 \\ 1 & 0 & -2 & 2 & 3 & 0 \\ -1 & 0 & 2 & 0 & -1 & 2 \end{pmatrix}$$

Computing the characteristic polynomial gives $(x-1)^6$, indicating that 1 is an eigenvalue of algebraic multiplicity 6. Computing the null space of $A - I$ gives a basis of $(2, 2, 1, 0, 0, 0)$, $(-1, -1, 0, 0, 1, 0)$, $(0, 1, 0, 0, 0, 1)$. Using the idea of the proof of the main theorem, we look at the range of $(A - I)$, which has basis

$$\mathbf{u}_1 = (1/2, 0, 1/2, 1, 1/2, -1/2),\ \mathbf{u}_2 = (0, 0, 0, -1, 0, 0),\ \mathbf{u}_3 = (1, 1, 1, 0, 1, 0)$$

We then find the Jordan form restricted to the range of $A - I$. We apply $A - I$ to the three vectors to get

$$A\mathbf{u}_1 = -A\mathbf{u}_2 = (1, 1, 1, 0, 1, 0),\ A\mathbf{u}_3 = (0, 0, 0, 0, 0, 0)$$

Thus the null space of the restriction of $A - I$ to the range of $A - I$ has basis

$$\mathbf{z}_1 = \mathbf{u}_1 + \mathbf{u}_2 = (1/2, 0, 1/2, 0, 1/2, -1/2),\ \mathbf{z}_2 = \mathbf{u}_3 = (1, 1, 1, 0, 1, 0)$$

These will be end vectors for Jordan strings when restricting to the range of $A - I$. By looking at $A - I$ on a basis of the range of $A - I$, we find that the vector $\mathbf{w}_2 = (0, 0, 0, 1, 0, 0)$ is a generating vector for a Jordan string of length 2 and the vector $\mathbf{z}_1$ corresponds to a vector of length 1. We thus choose

$$\mathbf{v}_2 = \mathbf{z}_1 = (1/2, 0, 1/2, 0, 1/2, -1/2),\ \mathbf{v}_4 = \mathbf{z}_2 = (1, 1, 1, 0, 1, 0),\ \mathbf{v}_5 = (0, 0, 0, 1, 0, 0)$$

We then solve $(A - I)\mathbf{v}_3 = \mathbf{v}_2, (A - I)\mathbf{v}_6 = \mathbf{v}_5$ to get

$$\mathbf{v}_3 = (1, 1, 0, 0, 0, 0),\ \mathbf{v}_6 = (0, -1, 0, 0, 0, 0)$$

Finally we extend $\mathbf{z}_1, \mathbf{z}_2$ to a basis of the null space of $A - I$ to add a final vector $\mathbf{v}_1 = (2, 2, 1, 0, 0, 0)$. Then we get a Jordan form J with $A = SJS^{-1}$,

$$J = \begin{pmatrix} 1 & 0 & 0 & 0 & 0 & 0 \\ 0 & 1 & 1 & 0 & 0 & 0 \\ 0 & 0 & 1 & 0 & 0 & 0 \\ 0 & 0 & 0 & 1 & 1 & 0 \\ 0 & 0 & 0 & 0 & 1 & 1 \\ 0 & 0 & 0 & 0 & 0 & 1 \end{pmatrix}, \quad S = \begin{pmatrix} 2 & 1/2 & 1 & 1 & 0 & 0 \\ 2 & 0 & 1 & 1 & 0 & -1 \\ 1 & 1/2 & 0 & 1 & 0 & 0 \\ 0 & 0 & 0 & 0 & 1 & 0 \\ 0 & 1/2 & 0 & 1 & 0 & 0 \\ 0 & -1/2 & 0 & 0 & 0 & 0 \end{pmatrix}$$ ■

Note that if we take a small Jordan block associated to the eigenvalue c of size k, then $(L - cI)^k$ will send all the vectors in the Jordan string to 0. Thus if k is the longest length of any Jordan string associated to c, we see that $(L - cI)^k$ will send *all* the basis vectors associated to that eigenvalue to zero. If the distinct eigenvalues are $c_1, \ldots, c_p$ and the longest Jordan string associated to c_i is of length $k(i)$, then the product $(L - c_1 I)^{k(1)} \cdots (L - c_p I)^{k(p)}$ will send all the vectors in the basis to 0. Thus this product is the zero linear transformation. Note that the factors involved are just the factors for the characteristic polynomial for L with L replacing the variable of the polynomial. The powers $k(i)$ will always be less than or equal to the algebraic multiplicities of the eigenvalues, and so this polynomial will actually divide the characteristic polynomial. It can be shown that this is the polynomial of minimal degree which L satisfies. It is called the **minimal polynomial** of the linear transformation L. When it is just the product of distinct linear factors that will mean that L is diagonalizable. Thus the minimal polynomial gives some information about the Jordan form but does not determine it entirely except in very special circumstances. Since L satisfies the minimal polynomial, and the minimal polynomial divides the characteristic polynomial we can conclude that L also satisfies the characteristic polynomial. This fact is known as the **Cayley–Hamilton theorem** and is frequently proved as one of the steps in deriving the Jordan form. Other proofs of the Cayley–Hamilton theorem were outlined in Exercises 5.6.23 and 5.6.24 of Chapter 5.

Let us return to the problem of solving the first-order system $x' = Ax$ when the matrix A is not diagonalizable. Recall that we can solve this if we can compute $\exp(At)\mathbf{v}_i$ for a basis $\mathbf{v}_i$ for $\mathbb{C}^n$. But the basis associated with the Jordan form will allow us to do this. For consider the part of the basis which corresponds to a Jordan string of length k associated to an eigenvalue c. Call these vectors $\mathbf{z}_1, \ldots, \mathbf{z}_k$. Then $(A - cI)^j \mathbf{z}_p = \mathbf{z}_{p-j}$ for $1 \leq p \leq k, p - j \geq 1$ and zero otherwise. Thus

$$\begin{aligned} \exp(At)\mathbf{z}_p &= \exp(ct)\exp[(A - cI)t]\mathbf{z}_p = \exp(ct)[I + t(A - cI) + \cdots]\mathbf{z}_p \\ &= \exp(ct)\{\mathbf{z}_p + t\mathbf{z}_{p-1} + (t^2/2)\mathbf{z}_{p-2} + \cdots + [t^{p-1}/(p-1)!]\mathbf{z}_1\} \end{aligned}$$

Let us apply this in Examples 6.2.1–6.2.3. In Example 6.2.1 with the 2 by 2 matrix we get the two solutions $\begin{pmatrix} -1 \\ -1 \end{pmatrix}\exp(t)$ and $\begin{pmatrix} 1 - t \\ -t \end{pmatrix}\exp(t)$, and the general solution is a linear combination $c_1\begin{pmatrix} -1 \\ -1 \end{pmatrix}\exp(t) + c_2\begin{pmatrix} 1 - t \\ -t \end{pmatrix}\exp(t)$. We can then solve an initial value problem by evaluating at 0 and solving for c_1, c_2 as before. This will be equivalent to solving the matrix equation $S\mathbf{c} = \mathbf{x}_0$.

For Example 6.2.2, we get one solution $\begin{pmatrix} 1 \\ 0 \\ 0 \end{pmatrix}\exp(t)$ from the eigenvalue 1. From the Jordan string of length 2 associated to the eigenvalue 2, we get the 2 solutions

$\begin{pmatrix} -1 \\ -3 \\ -1 \end{pmatrix} \exp(2t)$ and $\begin{pmatrix} 3-t \\ 2-3t \\ -t \end{pmatrix} \exp(2t)$. The general solution is a linear combination $c_1 \begin{pmatrix} 1 \\ 0 \\ 0 \end{pmatrix} \exp(t) + c_2 \begin{pmatrix} -1 \\ -3 \\ -1 \end{pmatrix} \exp(2t) + c_3 \begin{pmatrix} 3-t \\ 2-3t \\ -t \end{pmatrix} \exp(2t)$. To solve an initial value problem with $\mathbf{x}(0) = \mathbf{x}_0$, we solve for $\mathbf{c}$ by solving $S\mathbf{c} = \mathbf{x}_0$.

In Example 6.2.3, the eigenvalue 2 contributes the solution $\begin{pmatrix} 1 \\ 1 \\ 1 \\ 1 \end{pmatrix} \exp(2t)$. The Jordan string of length 3 associated to the eigenvalue 1 contributes three solutions $\begin{pmatrix} 1 \\ 1 \\ 0 \\ 0 \end{pmatrix} \exp(t), \begin{pmatrix} 1+t \\ t \\ 1 \\ 0 \end{pmatrix} \exp(t), \begin{pmatrix} 1+t+t^2/2 \\ t^2/2 \\ t \\ 1 \end{pmatrix} \exp(t)$. The general solution is a linear combination

$$c_1 \begin{pmatrix} 1 \\ 1 \\ 1 \\ 1 \end{pmatrix} \exp(2t) + c_2 \begin{pmatrix} 1 \\ 1 \\ 0 \\ 0 \end{pmatrix} \exp(t) + c_3 \begin{pmatrix} 1+t \\ t \\ 1 \\ 0 \end{pmatrix} \exp(t) + c_4 \begin{pmatrix} 1+t+t^2/2 \\ t^2/2 \\ t \\ 1 \end{pmatrix} \exp(t)$$

and an initial value problem is solved by solving $S\mathbf{c} = \mathbf{x}_0$ for $\mathbf{c}$.

Exercise 6.2.1.

Find the Jordan form for each of the following matrices. Give the matrices S, J so that $A = SJS^{-1}$. Also, give the minimal polynomial in each case.

(a) $\begin{pmatrix} 0 & 2 \\ 0 & 0 \end{pmatrix}$

(b) $\begin{pmatrix} 1 & 1 & 1 \\ 0 & 1 & 1 \\ 0 & 0 & 1 \end{pmatrix}$

(c) $\begin{pmatrix} 1 & 2 & 1 \\ 0 & 1 & 0 \\ 0 & 0 & 2 \end{pmatrix}$

(d) $\begin{pmatrix} 1 & 2 & 0 \\ 0 & 2 & 0 \\ -2 & -2 & -1 \end{pmatrix}$

(e) $\begin{pmatrix} 2 & 2 & 1 \\ 0 & 3 & 0 \\ -1 & 2 & 4 \end{pmatrix}$

Exercise 6.2.2.

For parts (a), (b), and (e) in Exercise 6.2.1, find the general solution to $\mathbf{x}' = A\mathbf{x}$. Also, solve the initial value problem, $\mathbf{x}(0) = \mathbf{e}_2$.

Exercise 6.2.3.

Fill in the details of the sketch given at the beginning of this section to show using induction that if $N^k = 0, N^{k-1}\mathbf{v}_k \neq 0$, and $\mathbf{v}_1, \ldots, \mathbf{v}_k$ are chosen by $\mathbf{v}_{k-j} = N^j\mathbf{v}_k$, then these k vectors are independent.

6.3. CHAPTER 6 EXERCISES

6.3.1. By using the criteria from Proposition 6.1.2 of positive subdeterminants, show that each of the following matrices is positive definite.

(a) $\begin{pmatrix} 2 & 1 & 1 \\ 1 & 2 & 1 \\ 1 & 1 & 2 \end{pmatrix}$

(b) $\begin{pmatrix} 2 & 1 & 1 & 1 \\ 1 & 2 & 1 & 1 \\ 1 & 1 & 2 & 1 \\ 1 & 1 & 1 & 2 \end{pmatrix}$

(c) $\begin{pmatrix} 2 & 1 & 1 & 1 & 1 \\ 1 & 2 & 1 & 1 & 1 \\ 1 & 1 & 2 & 1 & 1 \\ 1 & 1 & 1 & 2 & 1 \\ 1 & 1 & 1 & 1 & 2 \end{pmatrix}$

6.3.2. Generalize the preceding exercise by showing that the matrix whose non-diagonal entries are 1 and diagonal entries are 2 is positive definite.

6.3.3. By computing the pivots in Gaussian elimination determine whether the following matrices are positive definite.

(a) $\begin{pmatrix} 1 & 1 & 1 & 1 \\ 1 & 2 & 1 & 1 \\ 1 & 1 & 3 & 1 \\ 1 & 1 & 1 & 4 \end{pmatrix}$

(b) $\begin{pmatrix} 4 & 3 & 2 & 1 \\ 3 & 4 & 2 & 1 \\ 2 & 2 & 4 & 3 \\ 1 & 1 & 3 & 4 \end{pmatrix}$

(c) $\begin{pmatrix} 2 & 1 & 4 & 1 \\ 1 & 2 & 3 & 1 \\ 4 & 3 & 2 & 1 \\ 1 & 1 & 1 & 1 \end{pmatrix}$

6.3.4. By checking the determinants of submatrices, determine whether the following matrices are positive definite or negative definite.

(a) $\begin{pmatrix} 2 & 5 & 3 & 1 & 1 \\ 5 & 13 & 5 & 2 & 5 \\ 3 & 5 & 16 & 4 & 5 \\ 1 & 2 & 4 & 8 & 5 \\ 1 & 5 & 5 & 5 & 10 \end{pmatrix}$

(b) $\begin{pmatrix} 2 & 5 & 3 & 1 & 1 \\ 5 & 13 & 5 & 2 & 5 \\ 3 & 5 & 18 & 4 & 5 \\ 1 & 2 & 4 & 8 & 5 \\ 1 & 5 & 5 & 5 & 10 \end{pmatrix}$

(c) $\begin{pmatrix} -9 & 0 & -7 & 1 & -2 & 3 \\ 0 & -20 & -17 & -3 & -11 & 5 \\ -7 & -17 & -26 & -4 & -11 & 8 \\ 1 & -3 & -4 & -5 & -3 & 4 \\ -2 & -11 & -11 & -3 & -11 & 1 \\ 3 & 5 & 8 & 4 & 1 & -11 \end{pmatrix}$

6.3.5. Determine whether each of the following matrices is definite. Find the matrix $D(p,n,d)$ to which it is similar, and the matrix S so that $S^t AS = D(p,n,d)$.

(a) $\begin{pmatrix} 2 & 0 & 1 & 1 \\ 0 & 3 & 3 & 2 \\ 1 & 3 & 6 & 3 \\ 1 & 2 & 3 & 2 \end{pmatrix}$, (b) $\begin{pmatrix} 6 & -2 & 2 & 2 \\ -2 & 9 & 8 & 5 \\ 2 & 8 & 13 & 8 \\ 2 & 5 & 8 & 5 \end{pmatrix}$

6.3.6. Determine whether the following matrices A are definite. Find a matrix S and $D(p,n,d)$ so that $S^t AS = D(p,n,d)$.

(a) $\begin{pmatrix} 4 & 4 & -1 & 7 \\ 4 & 1 & 0 & -2 \\ -1 & 0 & 4 & -4 \\ 7 & -2 & -4 & 1 \end{pmatrix}$, (b) $\begin{pmatrix} 6 & 0 & -3 & 3 \\ 0 & 9 & 4 & 6 \\ -3 & 4 & 6 & 0 \\ 3 & 6 & 0 & 9 \end{pmatrix}$,

(c) $\begin{pmatrix} 2 & 1 & 0 & 0 & 0 & 0 & 0 & 0 \\ 1 & 2 & 1 & 0 & 0 & 0 & 0 & 0 \\ 0 & 1 & 2 & 1 & 0 & 0 & 0 & 0 \\ 0 & 0 & 1 & 2 & 1 & 0 & 0 & 0 \\ 0 & 0 & 0 & 1 & 2 & 1 & 0 & 1 \\ 0 & 0 & 0 & 0 & 1 & 2 & 1 & 0 \\ 0 & 0 & 0 & 0 & 0 & 1 & 2 & 0 \\ 0 & 0 & 0 & 0 & 1 & 0 & 0 & 2 \end{pmatrix}$

6.3.7. For a symmetric matrix A, define the **signature** of A to be $p - n$, where A is congruent to $D(p, n, d)$. Show the signature is an invariant of the congruence class of A. Show that it is also given by the number of positive eigenvalues of A minus the number of negative eigenvalues of A.

Exercises 6.3.8–6.3.12 below are related. You should use definitions and statements from earlier exercises in working later ones.

6.3.8. For a quadratic form $\mathcal{Q}$ on an m-dimensional vector space V, define its **null space** to be the set

$$\mathcal{N}(\mathcal{Q}) = \{\mathbf{v} \in V : \mathcal{B}(\mathbf{v}, \mathbf{w}) = 0 \text{ for all } \mathbf{w} \in V\}$$

(a) Show that $\mathcal{Q}$ can be represented by the matrix $D(n, p, d)$ iff $\mathcal{N}(\mathcal{Q})$ is d-dimensional.

(b) Give an example of a quadratic form so that the $\mathcal{N}(\mathcal{Q}) = \{\mathbf{0}\}$ but the dimension of the largest dimensional subspace on which $\mathcal{Q}$ is identically zero is positive.

6.3.9. A quadratic form $\mathcal{Q}$ is called **nondegenerate** if the null space $\mathcal{N}(\mathcal{Q}) = \{\mathbf{0}\}$. Show that a quadratic form is nondegenerate iff $\det A \neq 0$ for any matrix which represents A.

6.3.10. Suppose V is a real vector space of dimension n and $\mathcal{B}$ is a symmetric bilinear form on V. Define $C : V \to \mathcal{L}(V, \mathbb{R})$ by $C(\mathbf{v})(\mathbf{w}) = \mathcal{B}(\mathbf{v}, \mathbf{w})$. Here $\mathcal{L}(V, \mathbb{R})$ is the n-dimensional vector space of linear transformations from V to $\mathbb{R}$. Show that C is an isomorphism iff the associated quadratic form $\mathcal{Q}$ is nondegenerate.

6.3.11. Using the same notation as in the last problem, let $\mathcal{V} = \{\mathbf{v}_1, \ldots, \mathbf{v}_n\}$ be a basis of V, and let $\mathcal{D} = \{L_1, \ldots, L_n\}$ be the corresponding basis of $\mathcal{L}(V, \mathbb{R})$ defined by $L_i(\mathbf{v}_j) = \mathbf{0}$ if $i \neq j$ and $L_i(\mathbf{v}_i) = 1$. Show that the matrix which represents C with respect to these bases is the same as the matrix which represents the bilinear form $\mathcal{B}$ with respect to the basis $\mathcal{V}$.

6.3.12. Let $\mathcal{B}$ be a symmetric bilinear form on the n-dimensional inner product space $(V, \langle\rangle)$. Let $\mathcal{V} = \{\mathbf{v}_1, \ldots, \mathbf{v}_n\}$ be an orthonormal basis for V. Define a linear transformation $L : V \to V$ associated to $\mathcal{B}$ as follows. For $\mathbf{v} \in V$, define $L(\mathbf{v}) = \mathbf{w}$ by the formula $\langle \mathbf{u}, \mathbf{w} \rangle = \mathcal{B}(\mathbf{u}, \mathbf{v})$ for all $\mathbf{u}$.

(a) Show that this equation defines a unique linear transformation.

(b) Show that the matrix $L_{\mathcal{V}}^{\mathcal{V}}$ is the same matrix which represents $\mathcal{B}$ with respect to the matrix $\mathcal{V}$.

(c) Show that the linear transformation L is self-adjoint.
(d) Show that every self-adjoint linear transformation arises in this way for some symmetric bilinear form $\mathcal{B}$.

6.3.13. Suppose A is a positive definite m by m matrix. Show that in the decomposition $O_1 A = U$, the matrix O_1 is lower triangular. If $P = O_1^t$, show that $P^t AP$ is a diagonal matrix $\mathrm{diag}(d_1^2, \ldots, d_m^2)$. Show that if $D = \mathrm{diag}(1/d_1, \ldots, 1/d_m)$ and $S = PD$, then $S^t AS = I_m = D(m, 0, 0)$. Show this equation means $A = LL^t$ for a lower triangular matrix L. This is another derivation of the Cholesky decomposition discussed in Section 1.8.

6.3.14. Let $A = \begin{pmatrix} 2 & 2 & -1 \\ 0 & 2 & 1 \\ 0 & 0 & 4 \end{pmatrix}$.

(a) Find matrices S, J so that $S^{-1}AS = J$ is in Jordan canonical form.
(b) Solve the system of differential equations $\mathbf{x}' = A\mathbf{x}$, $\mathbf{x}(0) = (1, 1, 0)$.

6.3.15. Let $A = \begin{pmatrix} 1 & 1 & 1 & 0 \\ -1 & 1 & 0 & 1 \\ 0 & 0 & 1 & 1 \\ 0 & 0 & -1 & 1 \end{pmatrix}$.

(a) Find matrices S, J so that $S^{-1}AS = J$ is in Jordan canonical form.
(b) Solve the system of differential equations $\mathbf{x}' = A\mathbf{x}$, $\mathbf{x}(0) = (1, 1, 1, 0)$.

6.3.16. For each of the matrices A, find a matrix S so that $S^{-1}AS$ is in Jordan canonical form.

(a) $\begin{pmatrix} 1 & -1 & 0 \\ -3 & 3 & 4 \\ 2 & -2 & -2 \end{pmatrix}$

(b) $\begin{pmatrix} 1 & 0 & 1 \\ -1 & 2 & 2 \\ 1 & -1 & 0 \end{pmatrix}$

(c) $\begin{pmatrix} -10 & -2 & 20 & -26 \\ -7.5 & 1 & 8.5 & -11.5 \\ -8.5 & -2 & 19.5 & -23.5 \\ -2.5 & -1 & 7.5 & -8.5 \end{pmatrix}$

(d) $\begin{pmatrix} -5 & -2 & 14 & -20 \\ -5.25 & -1.75 & 13.5 & -21.25 \\ -1.75 & -.25 & 4.5 & -4.75 \\ 1.25 & .75 & -3.5 & 6.25 \end{pmatrix}$

6.3.17. For each of the matrices A, find a matrix S so that $S^{-1}AS$ is in Jordan canonical form.

(a) $\begin{pmatrix} 1.5 & -1.5 & -1 & -.5 \\ 2.5 & -1.5 & -2 & -2.5 \\ .5 & -.5 & -1 & -.5 \\ 0 & -1 & 1 & 1 \end{pmatrix}$

(b) $\begin{pmatrix} 0 & 0 & 1 & 1 \\ 1.5 & -.5 & -1 & -1.5 \\ -1 & 1 & 1 & 1 \\ -.5 & -.5 & 2 & 1.5 \end{pmatrix}$

(c) $\begin{pmatrix} .5 & -.5 & 1 & .5 \\ 1 & 0 & -1 & -1 \\ .5 & -.5 & 1 & -.5 \\ -1.5 & .5 & 2 & 2.5 \end{pmatrix}$

6.3.18. For each of the matrices in Exercise 6.3.16, give the general solution of the differential equation $\mathbf{x}' = A\mathbf{x}$.

6.3.19. For each of the matrices in Exercise 6.3.17, solve the differential equation $\mathbf{x}' = A\mathbf{x}$ with the initial value $\mathbf{x}(0) = (-1, 1, -1, 1)$.

6.3.20. Suppose that A is a real matrix. Show that if A has a nonreal complex eigenvalue, then the Jordan canonical form for A must have at least two big Jordan blocks.

6.3.21. Suppose $D : \mathcal{P}^4(\mathbb{R}, \mathbb{R}) \to \mathcal{P}^4(\mathbb{R}, \mathbb{R})$ is the differentiation operator, $D(p) = p'$. What are the eigenvalues of D? What are the corresponding eigenvectors? What is the Jordan canonical form for this linear transformation?

6.3.22. Without computing the Jordan form, find the characteristic polynomial and minimal polynomial for each of the following matrices by evaluating factors of the characteristic polynomial on the given matrix.

(a) $\begin{pmatrix} -4 & 1 & 8 & -1 & -3 \\ 93 & -19 & -152 & 18 & 60 \\ -27 & 6 & 45 & -5 & -18 \\ -93 & 21 & 152 & -16 & -63 \\ 0 & 0 & 0 & 0 & 1 \end{pmatrix}$

(b) $\begin{pmatrix} 1 & 0 & 0 & 2 & 0 \\ -21 & -1 & 2 & -35 & 3 \\ 6 & -1 & 1 & 10 & 0 \\ 21 & -4 & 4 & 35 & 0 \\ 0 & -2 & 2 & 0 & 1 \end{pmatrix}$

6.3.23. Give an example of a matrix whose characteristic polynomial is $(x-1)^2(x-3)^2$ and whose minimal polynomial is $(x-1)(x-3)^2$.

6.3.24. Give an example of two matrices with different Jordan canonical forms but the same minimal polynomial.

6.3.25. Suppose $N : V \to V$ is nilpotent, where $n = \dim V$. Show that $N^n = 0$. (Hint: What are the eigenvalues of N and what is the characteristic polynomial.)

Exercises 6.3.26–6.3.28 concern the notion of **generalized eigenvectors**. Suppose $L : V \to V$ is a linear transformation from a complex vector space V of dimension n to itself. If c is an eigenvalue of L, then $\mathbf{v}$ is called a **generalized eigenvector for the eigenvalue** c if $\mathbf{v} \neq \mathbf{0}$, and there is a positive integer k so that $(L - cI)^k\mathbf{v} = 0$. Denote by $E(c; k) = \mathcal{N}((L - cI)^k)$ and $GE(c) = \cup_{k \geq 1} E(c, k)$. Then $E(c; 1) = E(c)$ is the eigenspace of eigenvectors of L and $GE(c)$ is the generalized eigenspace of generalized eigenvectors.

6.3.26. (a) Show that the spaces $E(c; k)$ have to stabilize: that is, there is some k so that $E(c; k) = E(c; k + j)$ for all $j \geq 1$.

(b) Use the Jordan form for L to show that $E(c; k) = E(c; k + 1)$ implies $GE(c) = E(c; k)$.

(c) Verify the conclusion of part (b) without using the Jordan form. (Hint: Suppose we have an equality $E(c; k) = E(c; k + 1)$. Then show that $\mathbf{v} \in E(c; k + 2)$ implies $\mathbf{v} \in E(c; k + 1)$ by looking at $(L - cI)\mathbf{v}$.)

6.3.27. Suppose that $\mathbf{v}, \mathbf{w}$ are generalized eigenvectors for A corresponding to the eigenvalues a, b, where $a \neq b$. Show that $\mathbf{v}$ and $\mathbf{w}$ are independent.

6.3.28. Use the Jordan form to show that for every linear transformation $L : V \to V$ there is a basis of generalized eigenvectors.

6.3.29. Use the Jordan canonical form to show that every matrix can be approximated arbitrarily closely by a diagonalizable matrix. Here we are identifying the matrix with a point in Euclidean space of dimension n^2 with the usual distance function. (Hint: Make the eigenvalues distinct by a small change.)

APPENDIX A

SOLUTIONS TO EMBEDDED EXERCISES

A.1. SOLUTIONS TO CHAPTER 1 EXERCISES

1.1.1. (a) $(x, y) = (2.5 + 1.5y, y) = (2.5, 0) + y(1.5, 1)$

(b) $(x, y, z) = (x, 1.75 + .25z, z) = (0, 1.75, 0) + x(1, 0, 0) + z(0, .25, 1)$

(c) $(x_1, x_2, x_3, x_4, x_5) = (.2 + .8x_3 - .2x_4, x_2, x_3, x_4, x_5) = (.2, 0, 0, 0, 0)$
$+ x_2(0, 1, 0, 0, 0) + x_3(.8, 0, 1, 0, 0) + x_4(-.2, 0, 0, 1, 0) + x_5(0, 0, 0, 0, 1)$

1.2.1. 57

1.2.2. $\begin{pmatrix} 57 \\ 11 \\ 11 \end{pmatrix}$

1.2.3. $\begin{pmatrix} 3 & 9 & -2 \\ 2 & -2 & 0 \\ -1 & 1 & 3 \end{pmatrix}\begin{pmatrix} x \\ y \\ z \end{pmatrix} = \begin{pmatrix} 2 \\ 3 \\ 0 \end{pmatrix}$

1.2.4. $3x_1 + 4x_2 + 5x_3 = 1$

$x_1 + 2x_2 + x_3 = 0$

1.2.5. $A\mathbf{c} = \begin{pmatrix} 4 \\ 8 \\ 0 \end{pmatrix}$, $\quad A(\mathbf{b} + \mathbf{c}) = A\mathbf{b} + A\mathbf{c} = \begin{pmatrix} 12 \\ 12 \\ 0 \end{pmatrix}$,

$5\mathbf{b} - 3\mathbf{c} = \begin{pmatrix} -4 \\ 12 \\ -8 \end{pmatrix}$, $\quad A(5\mathbf{b} - 3\mathbf{c}) = \begin{pmatrix} 28 \\ -4 \\ 0 \end{pmatrix}$,

$$5A\mathbf{b} - 3A\mathbf{c} = \begin{pmatrix} 40 \\ 20 \\ 0 \end{pmatrix} - \begin{pmatrix} 12 \\ 24 \\ 0 \end{pmatrix} = \begin{pmatrix} 28 \\ -4 \\ 0 \end{pmatrix}$$

1.3.1. $\mathbf{x} = (1, 2, 0, 0) + c_1(-3, -2, 1, 0) + c_2(0, 0, 0, 1)$

1.3.2. There is no solution.

1.3.3. $\mathbf{x} = (0, 2, 0) + c_1(1, 0, 0) + c_2(0, -1, 1)$.

1.3.4. $\mathbf{x} = (-1, 7, 0, 2, 0) + c_1(-1, -1, 1, 0, 0) + c_2(-1, -1, 0, -1, 1)$

1.3.5. $\mathbf{x} = (2, 0, 1, 0) + c_1(0, 1, 0, 0) + c_2(-1, 0, -2, 1)$

1.3.6. $\mathbf{x} = (1, 0, 0, 2) + c_1(-2, 1, 0, 0) + c_2(-3, 0, 1, 0)$

1.3.7. $\mathbf{x} = (1, 0, 2, 1, 1, 0) + c_1(-2, 1, 0, 0, 0, 0) + c_2(-1, 0, -2, -1, -2, 1)$

1.3.8. $\mathbf{x} = (1, 0, 2, 3, 1, 0, 1, 0) + c_1(1, 1, 0, 0, 0, 0, 0, 0)$
$\quad + c_2(0, 0, -2, -3, -1, 1, 0, 0) + c_3(-1, 0, -1, 0, -2, 0, 0, 1)$

1.4.1. $O(2, 1; -2)$, $Om(2; 1/2)$, $O(1, 2; 1)$ $\left(\begin{array}{cc|c} 1 & 0 & 1 \\ 0 & 1 & -1 \end{array}\right)$, $x_1 = 1, x_2 = -1$

1.4.2. $O(2, 1; 1)$, $Om(2; 1/3)$, $O(1, 2; -2)$, $Om(1; 1/2)$

$$\left(\begin{array}{ccc|c} 1 & 2 & 0 & 10/3 \\ 0 & 0 & 1 & -4/3 \end{array}\right), \quad \mathbf{x} = \begin{pmatrix} 10/3 \\ 0 \\ -4/3 \end{pmatrix} + c_1 \begin{pmatrix} -2 \\ 1 \\ 0 \end{pmatrix}$$

1.4.3. $O(2, 1; -2)$, $Op(2, 3)$, $Om(3; 1/2)$, $O(2, 3; -2)$, $O(1, 3; -3)$,

$Om(2, 1/2)$, $O(1, 2; -2)$, $\left(\begin{array}{ccc|c} 1 & 0 & 0 & 1 \\ 0 & 1 & 0 & 1 \\ 0 & 0 & 1 & 0 \end{array}\right)$, $x_1 = 1, x_2 = 1, x_3 = 0$

1.4.4. $x_1 = 1, x_2 = 0, x_3 = 1$.

1.4.5. $O(2, 1; 2)$, $O(3, 1; -2)$, $O(3, 2; -1)$, $Om(3; 1/3)$, $O(2, 3; -2)$,

$O(1, 2; -1)$, $Om(1; 1/2)$, $\begin{pmatrix} 1 & 0 & 2 & 0 \\ 0 & 1 & -1 & 0 \\ 0 & 0 & 0 & 1 \end{pmatrix}$

1.4.6. $O(2, 1; 1)$, $O(3, 2; -1)$, $O(2, 3; -1)$, $O(1, 3; -1)$,

$O(1, 2; 1)$, $Om(1; 1/2)$, $\begin{pmatrix} 1 & 0 & 0 & 3 & 0 \\ 0 & 0 & 1 & 2 & 0 \\ 0 & 0 & 0 & 0 & 1 \end{pmatrix}$

1.4.7. $O(2, 1; 1)$, $O(4, 1; -1)$, $O(3, 2; -2)$, $O(4, 3; 1)$, $O(3, 4; -1)$, $O(2, 4; -1)$,
$O(1, 4; -1)$, $Om(3; 1/2)$, $O(2, 3; -1)$, $Om(2; 1/2)$, $O(1, 2; -3)$

$$\begin{pmatrix} 1 & 0 & 0 & .5 & 0 & 0 & -1 \\ 0 & 1 & 0 & .5 & 0 & 0 & 0 \\ 0 & 0 & 0 & 0 & 1 & 0 & 0 \\ 0 & 0 & 0 & 0 & 0 & 1 & 1 \end{pmatrix}$$

1.4.8. $\mathbf{x} = (-.5, -.5, 0, 0, 0, 3, 0) + c_1(0, 0, 1, 0, 0, 0, 0) + c_2(-.5, -.5, 0, 1, 0, 0, 0) + c_3(1, 0, 0, 0, 0, -1, 1)$

1.4.9. There is a solution iff $-3b_1 + b_2 + b_3 = 0$. If so, the system reduces to

$$\left(\begin{array}{ccccc|c} 1 & 0 & -.5 & -.5 & -1 & .5b_1 - .5b_2 \\ 0 & 1 & 2 & 1 & 3 & b_2 \end{array}\right)$$

The general solution is $\mathbf{x} =$

$(.5b_1 - .5b_2, b_2, 0, 0, 0) + c_1(.5, -2, 1, 0, 0) + c_2(.5, -1, 0, 1, 0) + c_3(1, -3, 0, 0, 1)$

1.5.1. $BA = \begin{pmatrix} 16 & 9 & 0 \\ 8 & 1 & 2 \end{pmatrix}$, $BC = \begin{pmatrix} 9 & 14 & 19 & 24 \\ 5 & 8 & 11 & 14 \end{pmatrix}$, $AC = \begin{pmatrix} 3 & 3 & 3 & 3 \\ 2 & 2 & 2 & 2 \\ 6 & 9 & 12 & 15 \end{pmatrix}$

1.5.2. (a) $(5 \quad 8 \quad 11 \quad 14) = 1\,(1 \quad 1 \quad 1 \quad 1) + 2\,(1 \quad 2 \quad 3 \quad 4) + 1\,(2 \quad 3 \quad 4 \quad 5)$

(b) $\begin{pmatrix} 14 \\ 8 \end{pmatrix} = 1\begin{pmatrix} 1 \\ 1 \end{pmatrix} + 2\begin{pmatrix} 2 \\ 2 \end{pmatrix} + 3\begin{pmatrix} 3 \\ 1 \end{pmatrix}$

1.5.3. (a) $\begin{pmatrix} 1 & 0 \\ 0 & -1 \end{pmatrix}$

(b) $\begin{pmatrix} \cos 45^\circ & -\sin 45^\circ \\ \sin 45^\circ & \cos 45^\circ \end{pmatrix} = \begin{pmatrix} 1/\sqrt{2} & -1/\sqrt{2} \\ 1/\sqrt{2} & 1/\sqrt{2} \end{pmatrix}$

1.5.4. $E(2, 1; -1/2)$, $E(3, 1; -2)$, $E(3, 2; 12)$, $D(3; 1/3)$, $E(2, 3; -1/2)$, $E(1, 3; -1)$, $D(2; -2)$, $E(1, 2; 1)$, $D(1; 1/2)$

1.5.5. $O_1 = \begin{pmatrix} 1 & 0 & 0 \\ -1 & 1 & 0 \\ -1 & -1 & 1 \end{pmatrix}$. Equivalent equations are $U\mathbf{x} = \begin{pmatrix} 1 \\ 0 \\ 0 \end{pmatrix}$ and $U\mathbf{x} = \begin{pmatrix} 1 \\ 0 \\ -2 \end{pmatrix}$, where $U = \begin{pmatrix} 1 & 1 & 1 & 1 \\ 0 & 1 & 2 & 3 \\ 0 & 0 & 0 & 0 \end{pmatrix}$. The first has a solution but the second doesn't.

1.5.6. $O = \begin{pmatrix} 1/2 & 1 & -1/2 \\ -3/2 & 0 & 1/2 \\ 1 & -1/2 & 0 \end{pmatrix}$, $R = \begin{pmatrix} 1 & 0 & 0 & 0 \\ 0 & 1 & 0 & 1 \\ 0 & 0 & 1 & 0 \end{pmatrix}$, $O\mathbf{d} = \begin{pmatrix} 1 \\ 1 \\ -1 \end{pmatrix}$

The solution is $(1, 1, -1, 0) + c_1(0, -1, 0, 1)$.

1.5.7. The ij-entry of $(AB)C$ is $(AB)_iC^j$ and the ij-entry of $A(BC)$ is $A_i(BC)^j$.

$$\begin{aligned} (AB)_iC^j &= \Big(\sum_{k=1}^{n} a_{ik}b_{k1}\Big)c_{1j} + \Big(\sum_{k=1}^{n} a_{ik}b_{k2}\Big)c_{2j} + \cdots + \Big(\sum_{k=1}^{n} a_{ik}b_{kn}\Big)c_{nj} \\ &= \sum_{k,p=1}^{n} (a_{ik}b_{kp})c_{pj} = a_{i1}\Big(\sum_{p=1}^{n} b_{1p}c_{pj}\Big) + \cdots + a_{in}\Big(\sum_{p=1}^{n} b_{np}c_{pj}\Big) = A_i(BC)^j. \end{aligned}$$

1.5.8. If A and B are lower triangular matrices, then $a_{ij} = b_{ij} = 0$ for $i < j$. Fix $i < j$. Then $(AB)_{ij} = a_{i1}b_{1j} + \cdots + a_{in}b_{nj}$. For $k \leq i$, we have $k < j$ and so $b_{kj} = 0$. When $k > i$, we have $a_{ik} = 0$. Thus each product in the sum is a product of 0 and another number, so the sum is zero. To compute the diagonal element $(AB)_{ii}$, the same argument shows that the only nonzero term is $a_{ii}b_{ii}$, which will be 1 if both of these are 1.

1.6.1. (a) $\begin{pmatrix} 1 & 0 & 0 \\ -2 & 1 & 0 \\ 0 & 0 & 1 \end{pmatrix}$ (b) $\begin{pmatrix} 1 & 0 & 0 \\ 0 & 1/3 & 0 \\ 0 & 0 & 1 \end{pmatrix}$ (c) $\begin{pmatrix} 0 & 1 & 0 \\ 1 & 0 & 0 \\ 0 & 0 & 1 \end{pmatrix}$

1.6.2. $(AB)^{-1} = \begin{pmatrix} 1 & 0 & 0 \\ -2/3 & 1/3 & 0 \\ 0 & 0 & 1 \end{pmatrix}$, $(ABC)^{-1} = \begin{pmatrix} -2/3 & 1/3 & 0 \\ 1 & 0 & 0 \\ 0 & 0 & 1 \end{pmatrix}$

1.6.3. $A^{-1} = \begin{pmatrix} -1 & -3 & 2 \\ 1/2 & 1 & -1/2 \\ -1 & -1 & 1 \end{pmatrix}$. The solution is $\mathbf{x} = A^{-1}\mathbf{b} = \begin{pmatrix} 1 \\ 1/2 \\ 0 \end{pmatrix}$.

1.6.4. A, C, E are invertible

1.6.5. $Om(3;.2)$, $O(2,3;-1)$, $O(1,3;-1)$, $Om(2;1/3)$, $O(1,2;-1)$

$$A^{-1} = \begin{pmatrix} 1 & -1/3 & -2/15 \\ 0 & 1/3 & -1/15 \\ 0 & 0 & 1/5 \end{pmatrix}$$

$O(3,1;-3)$, $O(3,2;-1)$, $O(2,3;-1)$, $O(1,3;-1)$, $Om(2;1/2)$, $O(1,2;-3)$

$$C^{-1} = \begin{pmatrix} -1/2 & -2 & 1/2 \\ 3/2 & 1 & -1/2 \\ -3 & -1 & 1 \end{pmatrix}$$

1.6.6. $A^t = \begin{pmatrix} 1 & 2 & 2 \\ 3 & 4 & 1 \\ 5 & 6 & 3 \end{pmatrix}$, $B^t = \begin{pmatrix} 3 & 2 & 1 \\ 2 & 1 & 4 \end{pmatrix}$, $AB = \begin{pmatrix} 14 & 25 \\ 20 & 32 \\ 11 & 17 \end{pmatrix}$

$$(AB)^t = \begin{pmatrix} 14 & 20 & 11 \\ 25 & 32 & 17 \end{pmatrix} = B^t A^t$$

1.6.7. $P = \begin{pmatrix} \mathbf{e}_{c(1)} \\ \vdots \\ \mathbf{e}_{c(n)} \end{pmatrix}$, $P^t = (\mathbf{e}_{c(1)} \quad \ldots \quad \mathbf{e}_{c(n)})$, $(PP^t)_{ij} = \langle \mathbf{e}_{c(i)}, \mathbf{e}_{c(j)} \rangle = I_{ij}$

$P^t P = I$ since a solution to $A\mathbf{x} = I$ also satisfies $\mathbf{x}A = I$ for square $\mathbf{x}$. Alternatively, P^t is also a permutation matrix, so we can apply the first argument to it.

1.6.8. If $A^t = A^{-1}$, then

$$\langle A\mathbf{x}, A\mathbf{y} \rangle = \mathbf{x}^t A^t A\mathbf{y} = \mathbf{x}^t I \mathbf{y} = \mathbf{x}^t \mathbf{y} = \langle \mathbf{x}, \mathbf{y} \rangle.$$

Conversely, if $\langle A\mathbf{x}, A\mathbf{y} \rangle = \langle \mathbf{x}, \mathbf{y} \rangle$ for all $\mathbf{x}, \mathbf{y}$, then choosing $\mathbf{x} = \mathbf{e}_i$ and $\mathbf{y} = \mathbf{e}_j$, then $\langle \mathbf{e}_i, \mathbf{e}_j \rangle = (A^t A)_{ij}$, so $A^t A = I$ and $A^t = A^{-1}$.

1.6.9. O_1 and O both have the same last row $(-1 \quad -1 \quad 1)$, so we get $(-1,-1,1)$ as a solution. The general solution is $c(-1,-1,1)$.

1.6.10. The condition that $\mathbf{c}$ is perpendicular to these three vectors can be rewritten as

$$\begin{pmatrix} 1 & 1 & 2 \\ 2 & 3 & 5 \\ 2 & 1 & 3 \end{pmatrix} \mathbf{c} = \begin{pmatrix} 0 \\ 0 \\ 0 \end{pmatrix},$$

which is the same as the equation $A'\mathbf{c} = \mathbf{0}$. Its solutions are just multiples of $(-1,-1,1)$.

1.7.1. (a) -18 (b) -18 (c) 9 (d) 2

1.7.2. (a) 1 (b) 1 (c) -2 (d) 24

1.7.3. (a) -9 (b) -3

1.7.4. In our proof we used the corresponding properties of the determinant.
Multilinearity:

$$\begin{aligned} &D(A_1,\ldots,aA_i' + bA_i'',\ldots,A_n) = \det(AB)/\det(B) \\ &= \det(A_1B,\ldots,aA_i'B + bA_i''B,\ldots,A_n)/\det(B) \\ &= (a\det(A_1B,\ldots,A_i'B,\ldots,A_nB) + b\det(A_1B,\ldots,A_i''B,\ldots,A_nB))/\det(B) \\ &= aD(A_1,\ldots,A_i',\ldots,A_n) + bD(A_1,\ldots,A_i'',\ldots,A_n) \end{aligned}$$

Alternating property:

$$\begin{aligned} D(A_1,\ldots,A_j,\ldots A_i,\ldots,A_n) &= \det(A_1B,\ldots,A_jB,\ldots,A_iB,\ldots,A_n)/\det(B) \\ &= -\det(A_1B,\ldots,A_iB,\ldots,A_jB,\ldots,A_n)/\det(B) \\ &= -D(A_1,\ldots,A_i,\ldots A_j,\ldots,A_n) \end{aligned}$$

Normalization:

$$D(I) = \det(B)/\det(B) = 1$$

1.7.5. (a) $2(2) - 2(7) = -10$ (b) $-2(7) - 2(-2) = -10$

1.7.6. $\operatorname{adj} A = \begin{pmatrix} 2 & -3 & 2 & -7 \\ 2 & 2 & -8 & -2 \\ -4 & 6 & -4 & 4 \\ 0 & -10 & 10 & 0 \end{pmatrix}, \ A^{-1} = \begin{pmatrix} -.2 & .3 & -.2 & .7 \\ -.2 & -.2 & .8 & .2 \\ .4 & -.6 & .4 & -.4 \\ 0 & 1 & -1 & 0 \end{pmatrix}$

1.7.7. $x_i = (1/\det A)(\operatorname{adj} A)_i\mathbf{b} = (1/\det A)(b_1A_{1i} + \cdots + b_nA_{ni}) = \det B(i)/\det A$, where we are expanding $\det B(i)$ by the ith column. The solution is $\mathbf{x} = (1.7, 3.2, -.4, -2)$.

1.7.8. [1 2 3] even $\rightarrow 2\cdot 1\cdot 3 = 6$, [1 3 2] odd $\rightarrow 2\cdot 4\cdot 1 = 8$, [2 1 3] odd $\rightarrow 1\cdot 2\cdot 3 = 6$, [2 3 1] even $\rightarrow 1\cdot 4\cdot 0 = 0$, [3 1 2] even $\rightarrow 3\cdot 2\cdot 1 = 6$, [3 2 1] odd $\rightarrow 3\cdot 1\cdot 0 = 0$; thus $\det A = 6 - 8 - 6 + 0 + 6 - 0 = -2$.

1.7.9. (a) $\mu(\sigma) = 3$ and we can use 3 flips
(b) $\mu(\sigma) = 9$ and we can use 3 flips

1.7.10. 1/2

1.7.11. 2

1.7.12. The determinant is $-r^2 \sin\phi$ and its absolute value is $r^2 \sin\phi$. Thus to find the volume enclosed by the sphere of radius 2 we integrate the function 1 over the ball of radius 2, which is then transformed by the change of variables formula to the integral of $r^2 \sin\phi$ over a rectangle where $0 \le r \le 2, 0 \le \theta \le 2\pi, 0 \le \phi \le \pi$. This integral is $64\pi/3$.

1.8.1. $L = \begin{pmatrix} 1 & 0 & 0 \\ 0.5 & 1 & 0 \\ 0.5 & -0.2 & 1 \end{pmatrix}$, $U = \begin{pmatrix} 2 & 1 & 1 \\ 0 & 2.5 & -0.5 \\ 0 & 0 & 3.4 \end{pmatrix}$, $\mathbf{c} = \begin{pmatrix} 4 \\ 3 \\ -3 \end{pmatrix}$, $\mathbf{x} = \begin{pmatrix} 2 \\ 1 \\ -1 \end{pmatrix}$

1.8.2. $\begin{pmatrix} 1.4142 & .7071 & .7071 \\ 0 & 1.5811 & -.3162 \\ 0 & 0 & 1.8439 \end{pmatrix}$

1.8.3. $L = \begin{pmatrix} .25 & 1 & 0 \\ .5 & .6667 & 1 \\ 1 & 0 & 0 \end{pmatrix}$, $U = \begin{pmatrix} 4 & 2 & 0 \\ 0 & 1.5 & 1 \\ 0 & 0 & 3.3333 \end{pmatrix}$

The solution is $\mathbf{x} = (1, 1, -3)$.

1.9.1. $(I-C)^{-1} = \begin{pmatrix} 2.9487 & 1.6026 & .7692 \\ 1.2821 & 2.4359 & .7692 \\ .8974 & .7051 & 1.5385 \end{pmatrix}$ and $\mathbf{p} = (I-C)^{-1}\mathbf{d} = \begin{pmatrix} 468.4615 \\ 368.4615 \\ 236.9231 \end{pmatrix}$.

1.9.2. (a) The new equation is just the negative of the sum of the other three equations from the other three nodes.

(b) The equation coming from the outer loop $5I_1 + 4I_2 - 2I_6 = 10$ is the sum of the three equations coming from the inner loops.

1.9.3. The left hand side of the matrix is unchanged but the right hand vector becomes $(0, 0, 0, 15, -5, 0)$. The current

$$\mathbf{I} = (1.4642, .1997, 1.2646, -.7404, -.5241, -.9401)$$

1.9.4. The equations are

$$\begin{pmatrix} -1 & 0 & 0 & 1 & 0 \\ 1 & -1 & 0 & 0 & -1 \\ 0 & 1 & -1 & 0 & 0 \\ 0 & 1 & 1 & 0 & -1 \\ 1 & 0 & 0 & 1 & 1 \end{pmatrix} \mathbf{I} = \begin{pmatrix} 0 \\ 0 \\ 0 \\ 10 \\ 10 \end{pmatrix},$$

with solution $(5, 5, 5, 5, 0)$.

1.9.5. The equations are

$$\begin{pmatrix} -1 & 0 & 0 & 1 & -1 & 0 & 0 & 0 \\ 1 & -1 & 0 & 0 & 0 & -1 & 0 & 0 \\ 0 & 1 & -1 & 0 & 0 & 0 & -1 & 0 \\ 0 & 0 & 1 & -1 & 0 & 0 & 0 & -1 \\ 1 & 0 & 0 & 0 & -1 & 1 & 0 & 0 \\ 0 & 1 & 0 & 0 & 0 & -1 & 1 & 0 \\ 0 & 0 & 1 & 0 & 0 & 0 & -1 & 1 \\ 0 & 0 & 0 & 1 & 1 & 0 & 0 & -1 \end{pmatrix} \mathbf{I} = \begin{pmatrix} 0 \\ 0 \\ 0 \\ 0 \\ 1 \\ 1 \\ 1 \\ 1 \end{pmatrix}$$

This has solution $\mathbf{I} = (1, 1, 1, 1, 0, 0, 0, 0)$. From the symmetry of the circuit with respect to rotation, we must have the same currents in branches one through four, and the same current in branches five through eight. That the last four are 0 uses the fact that their sum is 0 by the equation at node 5. That the first four are each 1 uses the fact that their sum is 4 by the voltage equation on the outer loop.

A.2. SOLUTIONS TO CHAPTER 2 EXERCISES

2.1.1. (a) is a subspace; the rest are not. (d) and (e) don't contain the zero vector. (b) and (c) are closed under scalar multiplication but are not closed under addition.

2.1.2. (a) $\mathcal{N}(A)$, (b) $\mathcal{R}(A)$ where in both cases $A = \begin{pmatrix} 3 & 2 & -1 & 1 \\ 2 & -3 & 1 & -5 \\ 1 & 1 & 4 & -7 \end{pmatrix}$

2.1.3. (a) $\text{span}((.6724, -.6379, 1.7414, 1))$
(b) $\text{span}((3, 2, 1), (2, -3, 1), (-1, 1, 4), (1, -5, -7))$

2.1.4. (b), (c), (d) are subspaces. The others are not as they fail to contain the zero vector.

2.1.5. (a) and (b) are subspaces; (c) and (d) are not as they fail to contain the zero matrix.

2.1.6. The closure property under scalar multiplication holds for (a) and (b) but not for (c) and (d). The closure property under addition holds for (c) and (d) but not for (a) and (b). They all satisfy both forms of associativity and multiplication by 1 sends a vector to itself. They fail to satisfy either distributive axiom. The addition is not commutative for any of them. For (a), (c), (d) there is an additive identity, which is the identity matrix, but there is no identity for (b). For (c) and (d) the inverse property is satisfied, but it is not satisfied for (a) and (b).

2.1.7. Each element of $\text{span}(\mathbf{v}_1, \ldots, \mathbf{v}_k)$ is a linear combination $a_1\mathbf{v}_1 + \cdots + a_k\mathbf{v}_k$, so in particular it is a linear combination $a_1\mathbf{v}_1 + \cdots + a_m\mathbf{v}_m$ where the numbers $a_i = 0$ when $i > k$. For an example when they are equal, take $k = 1$ and $m = 2$ with $\mathbf{v}_1 = \mathbf{e}_1$ and $\mathbf{v}_2 = 2\mathbf{e}_1$. For an example when they are different, take $k = 1$ and $m = 2$ with $\mathbf{v}_1 = \mathbf{e}_1$ and $\mathbf{v}_2 = \mathbf{e}_2$.

2.2.1. (a) it is not in the subspace
(b) $\mathbf{v} = \mathbf{v}_1 + 2\mathbf{v}_2 - \mathbf{v}_3$

2.2.2. (a) dependent; (b) dependent; (c) independent; (d) independent

2.2.3. (a) independent; (b) dependent

2.2.4. They span S since a general matrix in S is of the form

$$\begin{pmatrix} x & y \\ y & z \end{pmatrix} = x\begin{pmatrix} 1 & 0 \\ 0 & 0 \end{pmatrix} + y\begin{pmatrix} 0 & 1 \\ 1 & 0 \end{pmatrix} + z\begin{pmatrix} 0 & 0 \\ 0 & 1 \end{pmatrix}.$$

They are also independent since

$$a\begin{pmatrix}1 & 0\\0 & 0\end{pmatrix} + b\begin{pmatrix}0 & 1\\1 & 0\end{pmatrix} + c\begin{pmatrix}0 & 0\\0 & 1\end{pmatrix} = \begin{pmatrix}0 & 0\\0 & 0\end{pmatrix}$$

means $\begin{pmatrix}a & b\\b & c\end{pmatrix} = \begin{pmatrix}0 & 0\\0 & 0\end{pmatrix}$ and so $a = b = c = 0$. Thus they are a basis.

2.2.5. $(x, y, z) = z(1, 1, 1) + (y - z)(1, 1, 0) + (x - y)(1, 0, 0)$.

2.2.6. This follows from the fact that two vectors spanning $\mathbb{R}^3$ would imply the standard basis vectors $\mathbf{e}_1, \mathbf{e}_2, \mathbf{e}_3$ are dependent by Proposition 2.2.

2.2.7. Since $(3, 1, 2) = (1, 2, 1) + (2, -1, 1)$, this subspace is spanned by the two vectors $(1, 2, 1), (2, -1, 1)$. Since neither of these is a multiple of the other, they are independent. Thus the subspace involved is a plane. Its normal vector is given by the cross product of these two vectors, which is $(3, 1, -5)$. Thus the equation of the plane is $3x + y - 5z = 0$.

2.2.8. A basis is given by

$$\begin{pmatrix}0 & 1 & 0\\-1 & 0 & 0\\0 & 0 & 0\end{pmatrix}, \begin{pmatrix}0 & 0 & 1\\0 & 0 & 0\\-1 & 0 & 0\end{pmatrix}, \begin{pmatrix}0 & 0 & 0\\0 & 0 & 1\\0 & -1 & 0\end{pmatrix}.$$

2.2.9. We argue on the dimension of the subspace S. Since $S \subset \mathbb{R}^3$, $\dim S \leq 3$. If $\dim S = 0$, then the subspace is $\{\mathbf{0}\}$. If $\dim S = 1$ and $\mathbf{v}$ is a basis, then $S = \{c\mathbf{v}\}$ is just the line through the origin and $\mathbf{v}$. If $\dim S = 2$ with basis $\mathbf{v}_1, \mathbf{v}_2$, then S is the plane through the origin with normal vector $\mathbf{v}_1 \times \mathbf{v}_2$. If $\dim S = 3$, then $S = \mathbb{R}^3$ since a subspace with the same dimension as the whole space must equal the whole space.

2.3.1. Verification for $k = 1$: $L(a_1\mathbf{v}_1) = a_1L(\mathbf{v}_1)$ by the linearity property under scalar multiplication. Suppose that the assertion is true for $k < n$ and let $k = n$. Then $L(a_1\mathbf{v}_1 + \cdots + a_n\mathbf{v}_n) = L(a_1\mathbf{v}_1) + L(a_2\mathbf{v}_2 + \cdots + a_n\mathbf{v}_n) = a_1L(\mathbf{v}_1) + a_2L(\mathbf{v}_2) + \cdots + a_nL(\mathbf{v}_n)$, where we are using the induction hypothesis to rewrite the second term.

2.3.2. We just need to see that it is closed under taking linear combinations. Suppose that L and M are linear transformations and a, b are real numbers. Then

$$\begin{aligned}(aL + bM)(c\mathbf{v} + d\mathbf{w}) &= aL(c\mathbf{v} + d\mathbf{w}) + bM(c\mathbf{v} + d\mathbf{w})\\ &= acL(\mathbf{v}) + adL(\mathbf{w}) + bcM(\mathbf{v}) + bdM(\mathbf{w})\\ &= c(aL + bM)(\mathbf{v}) + d(aL + bM)(\mathbf{w})\end{aligned}$$

so $aL + bM$ is a linear transformation.

2.3.3. (a), (b), (e), (f), (g), (h) are linear transformations and the others are not.

2.3.4. (a) $(2, 3, 5) = 2(1, 0, 1) + 3(0, 1, 1)$; $L(2, 3, 5) = (2, 5, 6, 11)$
(b) $(x, y, z) = x(1, 0, 1) + y(0, 1, 1)$; $L(x, y, z) = (x, 2x + y, 3x, 4x + y)$

2.3.5. (a) $(10, 11, 12) = -2(1, 2, 3) + 3(4, 5, 6)$; $L(10, 11, 12) = (-2, -1, -6, -5)$
(b) $(x, y, z) = (-5x/3 + 4y/3)(1, 2, 3) + (2x/3 - y/3)(4, 5, 6)$;
$L(x, y, z) = (-5x/3 + 4y/3, -8x/3 + 7y/3, -5x + 4y, -6x + 5y)$

2.4.1. L sends the standard basis for $\mathbb{R}^3$ to the basis from Exercise 2.2.4 for S, so it is an isomorphism.

2.4.2. Suppose $c_1\mathbf{v}_1 + \cdots + c_n\mathbf{v}_n = \mathbf{0}$. Applying L to each side gives $\mathbf{0} = L(\mathbf{0}) = L(c_1\mathbf{v}_1 + \cdots + c_n\mathbf{v}_n) = c_1\mathbf{w}_1 + \cdots + c_n\mathbf{w}_n$. Now the independence of the $\mathbf{w}_i$ implies $c_1 = \cdots = c_n = 0$.

2.4.3. No, since a line is of dimension 1 and a plane is of dimension 2.

2.4.4. The matrices A and B exist since the $\mathbf{w}_i$ span V and so the $\mathbf{v}_j$ can be written in terms of them. Since the $\mathbf{v}_p$ span V, $\mathbf{w}_m$ can be written in terms of them. Let us write these statements as $(\mathbf{v}_1 \quad \ldots \quad \mathbf{v}_n) = (\mathbf{w}_1 \quad \ldots \quad \mathbf{w}_n)\, A$ and $(\mathbf{w}_1 \quad \ldots \quad \mathbf{w}_n) = (\mathbf{v}_1 \quad \ldots \quad \mathbf{v}_n)\, B$. Then we get

$$(\mathbf{v}_1 \quad \ldots \quad \mathbf{v}_n) = (\mathbf{v}_1 \quad \ldots \quad \mathbf{v}_n)\, AB.$$

Since $\mathbf{v}_1, \ldots, \mathbf{v}_n$ is a basis, each vector can be expressed as a linear combination of the basis elements in exactly one way. Thus we get $AB = I$. Similarly, we can show $BA = I$.

2.4.5. First suppose that L is $1-1$ and $c_1\mathbf{w}_1 + \cdots + c_n\mathbf{w}_n = \mathbf{0}$. Use $c_1\mathbf{w}_1 + \cdots + c_n\mathbf{w}_n = L(c_1\mathbf{v}_1 + \cdots + c_n\mathbf{v}_n)$ and L $1-1$ to see that $c_1\mathbf{v}_1 + \cdots + c_n\mathbf{v}_n = \mathbf{0}$. Since the $\mathbf{v}_i$ are independent, we get that each $c_i = 0$ and so the $\mathbf{w}_i$ are independent. Next suppose that L is onto. Then every vector $\mathbf{w} \in W$ is equal to $L(\mathbf{v})$ for some $\mathbf{v} \in V$. Since the $\mathbf{v}_i$ span V, we can write $\mathbf{v} = c_1\mathbf{v}_1 + \cdots + c_n\mathbf{v}_n$. Thus $\mathbf{w} = L(c_1\mathbf{v}_1 + \cdots + c_n\mathbf{v}_n) = c_1\mathbf{w}_1 + \cdots + c_n\mathbf{w}_n$, and so the $\mathbf{w}_i$ span W.

2.4.6. Let $\mathbf{s}_1, \ldots, \mathbf{s}_k$ be a basis of S. Since these are a basis, they are independent. On the other hand, these are $k = \dim S = \dim T$ independent vectors in T, so Proposition 3.5 implies they are a basis of T.

2.5.1. $\mathcal{N}(A) : (-2/3, 1/3, 1)$; $\mathcal{N}(A') : (-2, 1, 1)$;
$\mathcal{R}(A) : (1, 2, 0), (2, 1, 3)$; $\mathcal{R}(A') : (1, 2, 0), (0, -3, 1)$

2.5.2. $\mathcal{N}(A) : (1, -1, 1, 0), (0, -1, 0, 1)$; $\mathcal{N}(A') : (-3, -2, 1)$;
$\mathcal{R}(A) : (1, 0, 3), (0, 1, 2)$; $\mathcal{R}(A') : (1, 0, -1, 0), (0, 1, 1, 1)$

2.5.3. $\mathcal{N}(A) : (1/2, -5/4, 1/4, 1, 0), (-1/2, -1/4, 1/4, 0, 1)$;
$\mathcal{N}(A') : (-2, -1, 1, 1)$; $\mathcal{R}(A) : (1, 0, 2, 0), (2, 1, 1, 4), (0, 1, 1, 0)$;
$\mathcal{R}(A') : (1, 2, 0, 2, 1), (0, 1, 1, 1, 0), (0, 0, 4, -1, -1)$

2.5.4. $\mathcal{N}(A) : (-1, 0, 1, 0, 0, 0), (-1/3, 1/3, 0, 1, 0, 0), (-4/3, 2/3, 0, 0, 1, 0), (-2/3, 1/3, 0, 0, 0, 1)$; $\mathcal{N}(A') : (-1, -1, 1, 0), (1, -1, 0, 1)$;
$\mathcal{R}(A) : (1, 2, 3, 1), (-1, 1, 0, 2)$;
$\mathcal{R}(A') : (1, -1, 1, 0, 2, 1), (0, 3, 0, 1, -2, -1)$.

2.5.5. A basis for $\mathcal{R}(A)$ is given by $\sin t, \cos t$. A basis for $\mathcal{N}(L)$ is given by $(-1/\sqrt{2}, -1/\sqrt{2}, 1)$.

2.5.6. 1

2.5.7. Note that L sends a basis to a basis, so it is an isomorphism. We can verify that $L(\mathbf{e}_1)$, $L(\mathbf{e}_2)$, $L(\mathbf{e}_3)$, $L(\mathbf{e}_4)$ is a basis by showing either that these vectors are independent or that they span the 2 by 2 matrices. To see that they are independent, assume that

$$c_1\begin{pmatrix}1 & 0\\ 0 & 0\end{pmatrix} + c_2\begin{pmatrix}1 & 1\\ 0 & 0\end{pmatrix} + c_3\begin{pmatrix}1 & 1\\ 1 & 0\end{pmatrix} + c_4\begin{pmatrix}1 & 1\\ 1 & 1\end{pmatrix} = \begin{pmatrix}0 & 0\\ 0 & 0\end{pmatrix}.$$

This gives the matrix equation $A\mathbf{c} = \mathbf{0}$, where $A = \begin{pmatrix}1 & 1 & 1 & 1\\ 0 & 1 & 1 & 1\\ 0 & 0 & 1 & 1\\ 0 & 0 & 0 & 1\end{pmatrix}$. This only has the solution $\mathbf{c} = \mathbf{0}$, proving independence.

2.5.8. If $\mathbf{w}_1, \mathbf{w}_2 \in \mathcal{R}(L)$ and $a, b \in \mathbb{R}$, then we must show that $a\mathbf{w}_1 + b\mathbf{w}_2 \in \mathcal{R}(L)$. But $\mathbf{w}_1, \mathbf{w}_2 \in \mathcal{R}(L)$ means that there exist $\mathbf{v}_1, \mathbf{v}_2$ with $L(\mathbf{v}_1) = \mathbf{w}_1$, $L(\mathbf{v}_2) = \mathbf{w}_2$. Then $L(a\mathbf{v}_1 + b\mathbf{v}_2) = aL(\mathbf{v}_1) + bL(\mathbf{v}_2) = a\mathbf{w}_1 + b\mathbf{w}_2$ and so $a\mathbf{w}_1 + b\mathbf{w}_2 \in \mathcal{R}(L)$. The range of L is closed under forming linear combinations and is thus a subspace. If $\mathbf{v}_1, \mathbf{v}_2 \in \mathcal{N}(L)$ and $a, b \in \mathbb{R}$, then we must show $a\mathbf{v}_1 + b\mathbf{v}_2$ is in $\mathcal{N}(L)$ to verify that $\mathcal{N}(L)$ is closed under forming linear combinations and is thus a subspace. But $L(\mathbf{v}_1) = L(\mathbf{v}_2) = \mathbf{0}$ implies that $L(a\mathbf{v}_1 + b\mathbf{v}_2) = aL(\mathbf{v}_1) + bL(\mathbf{v}_2) = a\mathbf{0} + b\mathbf{0} = \mathbf{0} + \mathbf{0} = \mathbf{0}$ and so $a\mathbf{v}_1 + b\mathbf{v}_2 \in \mathcal{N}(L)$.

2.5.9. (a) The perpendicular vector is $(-1, -2, 1)$. The matrix A is $(-1 \quad -2 \quad 1)$.

(b) $B = \begin{pmatrix}0 & 1 & 0\\ 2 & 0 & 1\end{pmatrix}$

2.5.10. (a) $(1, 0, -1, 1), (2, 1, 1, 0)$

(b) $(1, -3, 1, 0), (-1, 2, 0, 1)$

(c) $\begin{pmatrix}1 & -3 & 1 & 0\\ -1 & 2 & 0 & 1\end{pmatrix}$

2.6.1. They are independent since neither is a multiple of the other. The only essential property needed for $\mathbf{v}$ is that it is not a linear combination of the two given vectors. We can take $\mathbf{v} = (1, 0, 0)$.

2.6.2. A basis for S is given by the 6 matrices

$$\begin{pmatrix}1 & 0 & 0\\ 0 & 0 & 0\\ 0 & 0 & 0\end{pmatrix}, \begin{pmatrix}0 & 1 & 0\\ 1 & 0 & 0\\ 0 & 0 & 0\end{pmatrix}, \begin{pmatrix}0 & 0 & 1\\ 0 & 0 & 0\\ 1 & 0 & 0\end{pmatrix}, \begin{pmatrix}0 & 0 & 0\\ 0 & 1 & 0\\ 0 & 0 & 0\end{pmatrix}, \begin{pmatrix}0 & 0 & 0\\ 0 & 0 & 1\\ 0 & 1 & 0\end{pmatrix}, \begin{pmatrix}0 & 0 & 0\\ 0 & 0 & 0\\ 0 & 0 & 1\end{pmatrix}.$$

A basis for T is given by the six matrices

$$\begin{pmatrix}1 & 0 & 0\\ 0 & 0 & 0\\ 0 & 0 & 0\end{pmatrix}, \begin{pmatrix}0 & 1 & 0\\ 0 & 0 & 0\\ 0 & 0 & 0\end{pmatrix}, \begin{pmatrix}0 & 0 & 1\\ 0 & 0 & 0\\ 0 & 0 & 0\end{pmatrix}, \begin{pmatrix}0 & 0 & 0\\ 0 & 1 & 0\\ 0 & 0 & 0\end{pmatrix}, \begin{pmatrix}0 & 0 & 0\\ 0 & 0 & 1\\ 0 & 0 & 0\end{pmatrix}, \begin{pmatrix}0 & 0 & 0\\ 0 & 0 & 0\\ 0 & 0 & 1\end{pmatrix}.$$

A basis for $S \cap T$ is given by the three matrices

$$\begin{pmatrix}1 & 0 & 0\\ 0 & 0 & 0\\ 0 & 0 & 0\end{pmatrix}, \begin{pmatrix}0 & 0 & 0\\ 0 & 1 & 0\\ 0 & 0 & 0\end{pmatrix}, \begin{pmatrix}0 & 0 & 0\\ 0 & 0 & 0\\ 0 & 0 & 1\end{pmatrix}.$$

A basis for $S + T = M(3, 3)$ is given by the 9 matrices $N(i, j)$, which have all entries 0 except in the ij-position, where the entry is 1.

2.6.3. We show that $S \cap T = \{\mathbf{0}\}$ by checking that $c(-1, 1, 2)$ satisfies the equation of the plane iff $c = 0$, i.e. the intersection is the zero vector. Then $S + T$ will be of dimension 3, so it must be $\mathbb{R}^3$. Since there is a direct sum splitting, each vector in $\mathbb{R}^3 = S + T$ can be written uniquely as the sum of a vector in S and a vector in T.

2.6.4. A basis for $S \cap T$ is $(3, 1, -1)$. Since $\dim S \cap T = 1$, we have $\dim S + T = 3$ and so $S + T = \mathbb{R}^3$.

2.6.5. A basis for $S \cap T$ is $1 + x^3$. Since $\dim S + T = 4$, we have $S + T = \mathcal{P}^3(\mathbb{R}, \mathbb{R})$, so we can take the standard basis $1, x, x^2, x^3$.

2.7.1. The jth entry of $(\mathbf{V}C)D$ is

$$(\mathbf{V}C)D^j = \sum_{k=1}^{n} d_{kj}(\sum_{i=1}^{n} c_{ik}\mathbf{v}_i) = \sum_{i=1}^{n}(\sum_{k=1}^{n} c_{ik}d_{kj})\mathbf{v}_i.$$

Since the ij-entry of CD is $\sum_{k=1}^{n} c_{ik}d_{kj}$, this is also the jth entry of $\mathbf{V}(CD)$.

2.7.2. The statement $L(\mathbf{U}) = L(\mathbf{V})C$ just means that $L(\mathbf{u}_j) = \sum_{i=1}^{n} c_{ij}L(\mathbf{v}_i)$. This follows immediately from the fact that L is a linear transformation and so sends linear combinations to linear combinations together with $\mathbf{u}_j = \sum_{i=1}^{n} c_{ij}\mathbf{v}_i$.

2.7.3. Following the hint, we have $L(\mathbf{V}) = \mathbf{W}A$, $M(\mathbf{W}) = \mathbf{V}B$, where A and B are inverses. Then $ML(\mathbf{V}) = M(\mathbf{V}A) = M(\mathbf{V})A = \mathbf{V}BA = \mathbf{V}$. Thus ML is the identity. Similarly we can show $LM(\mathbf{W}) = \mathbf{W}$ and so LM is the identity. Thus L is an isomorphism.

2.7.4. $\begin{pmatrix} 0 & 0 & 0 \\ 1 & 0 & 0 \\ 0 & 1/2 & 0 \\ 0 & 0 & 1/3 \end{pmatrix}$

2.7.5. $\begin{pmatrix} 1 & 0 \\ 0 & -1 \end{pmatrix}, \begin{pmatrix} 3/5 & 4/5 \\ 4/5 & -3/5 \end{pmatrix}$

2.7.6. $\mathbf{U} = \mathbf{V}\mathcal{T}_{\mathcal{V}}^{\mathcal{U}}, \mathbf{V} = \mathbf{U}\mathcal{T}_{\mathcal{U}}^{\mathcal{V}}$ imply $\mathbf{U} = \mathbf{U}\mathcal{T}_{\mathcal{U}}^{\mathcal{V}}\mathcal{T}_{\mathcal{V}}^{\mathcal{U}}$. Since the coefficients with respect to a basis are uniquely determined, this means that $I = \mathcal{T}_{\mathcal{U}}^{\mathcal{V}}\mathcal{T}_{\mathcal{V}}^{\mathcal{U}}$, and so $\mathcal{T}_{\mathcal{U}}^{\mathcal{V}} = (\mathcal{T}_{\mathcal{V}}^{\mathcal{U}})^{-1}$.

2.7.7. (a) $\begin{pmatrix} -1/3 & 2/3 & 1/3 \\ 2/3 & -7/3 & 1/3 \\ 1/3 & 1/3 & -1/3 \end{pmatrix}$

(b) same as part (a)

2.7.8. $\begin{pmatrix} -1/3 & 2/3 \\ 2/3 & -1/3 \end{pmatrix}$

2.7.9. $\begin{pmatrix} 1 & 0 & 0 \\ 0 & 1 & 0 \\ 0 & 0 & -1 \end{pmatrix}, \begin{pmatrix} 1/3 & -2/3 & 2/3 \\ -2/3 & 1/3 & 2/3 \\ 2/3 & 2/3 & 1/3 \end{pmatrix}$

2.7.10. $\begin{pmatrix} 1 & 0 & 0 \\ 0 & 0 & -1 \\ 0 & 1 & 0 \end{pmatrix}, \begin{pmatrix} .3333 & .9107 & -.2440 \\ -.2440 & .3333 & .9107 \\ .9107 & -.2440 & .3333 \end{pmatrix}$

2.7.11. The easiest way to do this problem is to first find $L_{\mathcal{S}}^{\mathcal{S}}$ and then find the other representatives of L from it.

(a) $L_{\mathcal{B}}^{\mathcal{B}} = \begin{pmatrix} -1 & -1 & -1 & -1 \\ 0 & 0 & -1 & -1 \\ 0 & 0 & 1 & -1 \\ 0 & 0 & 0 & 2 \end{pmatrix}.$

(b) $\mathcal{T}_{\mathcal{S}}^{\mathcal{B}} = \begin{pmatrix} 1 & 1 & 1 & 1 \\ 0 & 1 & 1 & 1 \\ 0 & 0 & 1 & 1 \\ 0 & 0 & 0 & 1 \end{pmatrix}.$

(c) The answer is $L_{\mathcal{S}}^{\mathcal{S}} = \begin{pmatrix} -1 & 0 & 0 & 0 \\ 0 & 0 & 0 & 0 \\ 0 & 0 & 1 & 0 \\ 0 & 0 & 0 & 2 \end{pmatrix}$. In the second computation we use

$$\mathcal{T}_{\mathcal{B}}^{\mathcal{S}} = (\mathcal{T}_{\mathcal{S}}^{\mathcal{B}})^{-1} = \begin{pmatrix} 1 & -1 & 0 & 0 \\ 0 & 1 & -1 & 0 \\ 0 & 0 & 1 & -1 \\ 0 & 0 & 0 & 1 \end{pmatrix}.$$

(d) $L_{\mathcal{S}}^{\mathcal{B}} = \begin{pmatrix} -1 & -1 & -1 & -1 \\ 0 & 0 & 0 & 0 \\ 0 & 0 & 1 & 1 \\ 0 & 0 & 0 & 2 \end{pmatrix}.$

2.7.12. reflexive: $A = IAI$ implies A is similar to A.
symmetric: $B = C^{-1}AC$ implies $A = CBC^{-1} = (C^{-1})^{-1}B(C^{-1})$
transitive: $C = E^{-1}BE$, $B = D^{-1}AD$ implies

$$C = E^{-1}D^{-1}ADE = (DE)^{-1}A(DE)$$

2.7.13. $L(1) = -1; L(x) = 0; L(x^2) = x^2; L(x^3) = 2x^3$, so

$$[L] = \begin{pmatrix} -1 & 0 & 0 & 0 \\ 0 & 0 & 0 & 0 \\ 0 & 0 & 1 & 0 \\ 0 & 0 & 0 & 2 \end{pmatrix}.$$

Thus $\mathcal{N}([L]) = \text{span}(\mathbf{e}_2), \mathcal{R}([L]) = \text{span}(\mathbf{e}_1, \mathbf{e}_3, \mathbf{e}_4)$, and so a basis for $\mathcal{N}(L)$ is given by x, and a basis for $\mathcal{R}(L)$ is given by $1, x^2, x^3$.

2.7.14. (a) $[L] = \begin{pmatrix} 3 & 0 \\ 0 & 2 \end{pmatrix}, [x + 1] = \begin{pmatrix} 1 \\ 1 \end{pmatrix}.$

(b) The solution of $[L]\mathbf{c} = [x+1]$ is $= (1/3, 1/2)$ so the polynomial solution to the differential equation is $1/3 + x/2$.

2.8.1. (a) $C_d = \begin{pmatrix} 1 & 0 & 1 & 1 & 0 \\ 0 & 0 & 1 & 0 & 1 \\ 1 & 1 & 0 & 0 & 0 \\ 0 & 1 & 1 & 0 & 0 \\ 2 & 0 & 0 & 0 & 0 \end{pmatrix}, C_g = \begin{pmatrix} 1 & 0 & 2 & 1 & 2 \\ 0 & 0 & 2 & 1 & 1 \\ 2 & 2 & 0 & 1 & 0 \\ 1 & 1 & 1 & 0 & 0 \\ 2 & 1 & 0 & 0 & 0 \end{pmatrix}$

(b) $C_d = \begin{pmatrix} 0 & 1 & 0 & 0 & 0 & 0 & 0 & 0 & 0 \\ 0 & 0 & 1 & 0 & 0 & 0 & 0 & 1 & 1 \\ 0 & 0 & 0 & 1 & 1 & 0 & 0 & 0 & 0 \\ 0 & 0 & 0 & 0 & 0 & 0 & 0 & 1 & 0 \\ 0 & 0 & 0 & 0 & 0 & 0 & 0 & 1 & 0 \\ 0 & 0 & 0 & 0 & 0 & 0 & 0 & 1 & 0 \\ 0 & 0 & 0 & 0 & 0 & 0 & 0 & 1 & 0 \\ 0 & 0 & 0 & 0 & 0 & 0 & 0 & 0 & 0 \\ 0 & 0 & 0 & 0 & 0 & 1 & 1 & 0 & 0 \end{pmatrix}, C_g = \begin{pmatrix} 0 & 1 & 0 & 0 & 0 & 0 & 0 & 0 & 0 \\ 1 & 0 & 1 & 0 & 0 & 0 & 0 & 1 & 1 \\ 0 & 1 & 0 & 1 & 1 & 0 & 0 & 0 & 0 \\ 0 & 0 & 1 & 0 & 0 & 0 & 0 & 1 & 0 \\ 0 & 0 & 1 & 0 & 0 & 0 & 0 & 1 & 0 \\ 0 & 0 & 0 & 0 & 0 & 0 & 0 & 1 & 1 \\ 0 & 0 & 0 & 0 & 0 & 0 & 0 & 1 & 1 \\ 0 & 1 & 0 & 1 & 1 & 1 & 1 & 0 & 0 \\ 0 & 1 & 0 & 0 & 0 & 1 & 1 & 0 & 0 \end{pmatrix}$

2.8.2. (a) $\begin{pmatrix} 0 & 1 & 0 & 1 & 0 \\ 1 & 0 & 0 & 0 & 2 \\ 0 & 0 & 0 & 1 & 1 \\ 1 & 0 & 1 & 0 & 1 \\ 0 & 2 & 1 & 1 & 0 \end{pmatrix}$

(b) $\begin{pmatrix} 0 & 1 & 1 & 1 & 1 & 0 & 0 \\ 1 & 0 & 0 & 0 & 0 & 0 & 0 \\ 1 & 0 & 0 & 0 & 0 & 0 & 1 \\ 1 & 0 & 0 & 0 & 0 & 0 & 0 \\ 1 & 0 & 0 & 0 & 0 & 0 & 1 \\ 0 & 0 & 0 & 0 & 0 & 1 & 0 \\ 0 & 0 & 1 & 0 & 1 & 0 & 0 \end{pmatrix}$

2.8.3.

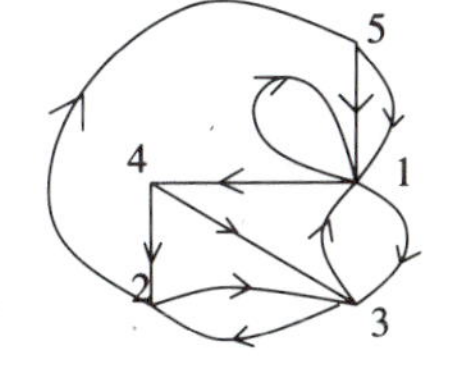

Figure A.1. Solution to 2.8.3

2.8.4.

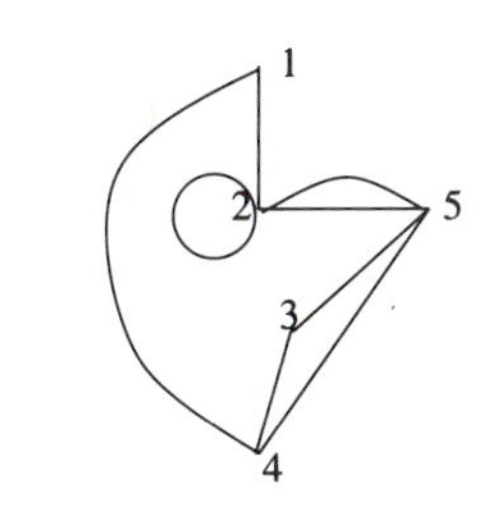

Figure A.2. Solution to 2.8.4

2.8.5. The number of paths of length less than or equal to p is the sum of the number of paths of length i, where i ranges from 1 to p. Using Proposition 2.8.1, this number is given by the ij entry of the sum $C + C^2 + \cdots + C^p$.

2.8.6. (a) The graph is not connected. There are 3 paths of length less than or equal to 5 which join the first and second vertices.

(b) It is connected. There are 10 paths of length less than or equal to 5 joining the first and second vertices.

2.8.7. The null space has basis $(0,1,0,1,0,0,0,0), (1,0,1,0,1,1,1,1)$, so it is not connected.

2.8.8. The rows of L are perpendicular to the columns of N, so the range of N is in the null space of L. Since they have the same dimension $V - 1$, they are equal.

2.8.9. The null space of $\overline{N}$ is $\{\mathbf{0}\}$ since $\overline{N}$ is of rank $V - 1$. The inclusion $\mathcal{N}(\overline{N}) \subset \mathcal{N}(\overline{N}^t R^{-1} \overline{N})$ holds since $\overline{N}\mathbf{y} = \mathbf{0}$ implies $(\overline{N}^t R^{-1} \overline{N})\mathbf{y} = \overline{N}^t R^{-1}(\overline{N}\mathbf{y}) = \mathbf{0}$. To get the other inclusion suppose $\overline{N}^t R^{-1} \overline{N}\mathbf{y} = \mathbf{0}$. Following the hint, we multiply by $\mathbf{y}^t$ to get $\mathbf{y}^t \overline{N}^t R^{-1} \overline{N}\mathbf{y} = 0$. If $\overline{N}\mathbf{y} = (a_1, \cdots, a_k)$ and the diagonal entries of R^{-1} are $d_1, \cdots, d_k$, then we have $d_1 a_1^2 + \cdots + d_k a_k^2 = 0$. Since the d_i are positive, this gives $\overline{N}\mathbf{y} = \mathbf{a} = \mathbf{0}$.

A.3. SOLUTIONS TO CHAPTER 3 EXERCISES

3.1.1. The ij-th entry of $A^t A$ is just $\mathbf{v}_i \cdot \mathbf{v}_j$. Thus the basis is orthogonal iff the nondiagonal entries of $A^t A$ are zero, and is orthonormal iff in addition the diagonal entries are 1. The rows are othogonal iff the columns of A^t are orthogonal iff $AA^t = D$. The matrix is an orthogonal matrix iff $A^t = A^{-1}$ which means that $A^t A = AA^t = I$, which is equivalent by the above to the rows and columns of A each forming orthonormal bases.

3.1.2. $\mathbf{q}_1 = (-.7071, 0, .7071), \mathbf{q}_2 = (.4082, -.8165, .4082)$

$$Q = \begin{pmatrix} -.7071 & .4082 \\ 0 & -.8165 \\ .7071 & .4082 \end{pmatrix}, R = \begin{pmatrix} 1.4142 & .7071 \\ 0 & 1.2247 \end{pmatrix}$$

The third vector is $\mathbf{q}_3 = (.5774, .5774, .5774)$.

3.1.3. $$Q = \begin{pmatrix} .7071 & .4082 \\ 0 & .8165 \\ .7071 & -.4082 \end{pmatrix}, R = \begin{pmatrix} 1.4142 & 2.1213 \\ 0 & 1.2247 \end{pmatrix}$$

3.1.4. $$Q = \begin{pmatrix} .5774 & .7001 & .4201 \\ 0 & .4201 & -.7001 \\ .5774 & -.1400 & .5601 \\ -.5774 & .5601 & .1400 \end{pmatrix}, R = \begin{pmatrix} 1.7321 & .5774 & 0 \\ 0 & 2.3805 & 1.6803 \\ 0 & 0 & .4201 \end{pmatrix}$$

3.1.5. $2(1,0,1) + 4/3(1,1,-1) + 1/3(-1,2,1)$

3.1.6. $(1 + x)$ and $(1/4 - 7/4x + x^2), 2 + x^2 = 7/4(1 + x) + (1/4 - 7/4 + x^2)$

3.1.7. $x^3 - 3x/5$

3.2.1. Bases are: $\mathcal{N}(A) : (0, -1, 0, 1), (1, -2, 1, 0)$; $\mathcal{R}(A^t) : (1, 0, -1, 0), (0, 1, 2, 1)$. We can check that the first two vectors are orthogonal to the last two. $\mathcal{N}(A^t) : (2, -1, 1)$; $\mathcal{R}(A) : (1, 2, 0), (0, 1, 1)$. We can check that the first vector is orthogonal to the last two. $A\,(1 \quad 0 \quad -1 \quad 0)^t = (2 \quad 2 \quad -2)^t, A\,(0 \quad 1 \quad 2 \quad 0)^t = (-2 \quad 1 \quad -3)^t$. Since these are two independent vectors in the 2 dimensional $\mathcal{R}(A)$, they are a basis.

3.2.2. Bases are:

$$\mathcal{N}(A) : (-1, 0, 1, 0, 0), (1, -1, 0, 1, 0), (-1, 0, 0, 0, 1);$$
$$\mathcal{R}(A^t) : (1, 1, 1, 0, 1), (0, 1, 0, 1, 0).$$

We check that the first three vectors are orthogonal to the last two.

$$\mathcal{N}(A^t) : (-2, -1, 1, 0), (-3, -2, 0, 1);$$
$$\mathcal{R}(A) : (1, 0, 2, 3), (1, 1, 3, 5).$$

We checks that the first two vectors are orthogonal to the last two.

$$A\,(1 \quad 1 \quad 1 \quad 0 \quad 1)^t = (4 \quad 1 \quad 9 \quad 14)^t, A\,(0 \quad 1 \quad 0 \quad 1 \quad 0)^t = (1 \quad 2 \quad 4 \quad 7)^t.$$

Since these are two independent vectors in the 2 dimensional $\mathcal{R}(A)$, they are a basis.

3.2.3. $(0, -1, 0, 1, 0), (1, -2, 2, 0, 1)$

3.2.4. For the polynomial $a + bx + cx^2 + dx^3$ to be orthogonal to 1 and x leads to the equations $2(a + c/3) = 0, 2(b/3 + d/5) = 0$. Taking two independent solutions of these equations gives the polynomials $x^2 - 1/3$ and $x^3 - 3x/5$, which are the next two monic orthogonal polynomials.

3.2.5. By a dimension count $\dim S^\perp = p$. Then $\mathbf{t}_1, \dots, \mathbf{t}_p$ being orthogonal to each of the vectors $\mathbf{s}_i$ implies that they are orthogonal to the subspace S spanned by the $\mathbf{s}_i$. They form p independent (since they are an orthonormal collection of vectors) vectors in a p dimensional vector space, and so they are a basis for $S^\perp$.

3.3.1. The ij-entry of AB is $\sum_{k=1}^n a_{ik}b_{kj}$. The ij-entry of $A^k B_k$ is $a_{ik}b_{kj}$, thus the ij-entry of the sum is $\sum_{k=1}^n a_{ik}b_{kj}$.

3.3.2. $\begin{pmatrix} 4/5 & 2/5 & 0 \\ 2/5 & 1/5 & 0 \\ 0 & 0 & 0 \end{pmatrix}$

3.3.3. $\begin{pmatrix} 1/5 & -2/5 & 0 \\ -2/5 & 4/5 & 0 \\ 0 & 0 & 1 \end{pmatrix}$

3.3.4. $\mathbf{x} = (1/2, 0), P = \begin{pmatrix} 1/2 & 0 & 1/2 \\ 0 & 1 & 0 \\ 1/2 & 0 & 1/2 \end{pmatrix}$

3.4.1. normal equation: $\begin{pmatrix} 5 & 1 \\ 1 & 3 \end{pmatrix}\mathbf{x} = \begin{pmatrix} -1 \\ 1 \end{pmatrix}$, solution $\mathbf{x} = (.2857, .4286)$

$$P = \begin{pmatrix} .7143 & -.1429 & .4286 \\ -.1429 & .9286 & .2143 \\ .4286 & .2143 & .3571 \end{pmatrix}$$

3.4.2. normal equation: $\begin{pmatrix} 2 & 1 & 0 \\ 1 & 2 & 3 \\ 0 & 3 & 6 \end{pmatrix}\mathbf{x} = \begin{pmatrix} 2 \\ 2 \\ 2 \end{pmatrix}$, $P = \begin{pmatrix} 2/3 & 1/3 & 1/3 \\ 1/3 & 2/3 & -1/3 \\ 1/3 & -1/3 & 2/3 \end{pmatrix}$

The solution of normal equation is $\mathbf{x} = (2/3, 2/3, 0) + c(1, -2, 1)$; the solution in the row space is (.7778, .4444, .1111). The vector in the range of A which is closest to $\mathbf{b}$ is $A\mathbf{x} = (4/3, 2/3, 2/3)$.

3.5.1. $2.5 + .1x$

3.5.2. 11/4

3.5.3. $1.25 + 1.35x - .25x^2$, $24 - 34.8333x + 16x^2 - 2.1667x^3$

3.5.4. $1.1752 + 1.1036x$; MATLAB gives $1.1844 + 1.1089x$ for the data fitting problem.

3.5.5. $-.0857 + .8571x$; MATLAB gives $-.0941 + .8982x^2$ for the data fitting problem.

3.5.6. The Fourier series is $\pi/2 + (4/\pi)(\cos x + 1/3 \cos 3x + 1/5 \cos 5x + \ldots)$. The closest function in T_2 is thus $\pi/2 + (4/\pi)\cos x$.

3.6.1. $(4/3\sqrt{10}, 4/3\sqrt{10})$

3.6.2. $Q_1 = \begin{pmatrix} .4472 & .8944 \\ .8944 & -.4472 \end{pmatrix}$, $\Sigma = \begin{pmatrix} 8.366 & 0 & 0 \\ 0 & 0 & 0 \end{pmatrix}$, $Q_2 = \begin{pmatrix} .2673 & -.9636 & .0027 \\ .5345 & .4460 & .8325 \\ .8018 & .2239 & .5541 \end{pmatrix}$

3.6.3. $A^+ = \begin{pmatrix} .0357 & .0714 & .1071 \\ .0357 & .0714 & .1071 \end{pmatrix}$. The solution is $\begin{pmatrix} .2142 \\ .2142 \end{pmatrix}$

A.4. SOLUTIONS TO CHAPTER 4 EXERCISES

4.1.1. $S = \begin{pmatrix} 1 & 1 \\ 0 & -1 \end{pmatrix}$, $D = \begin{pmatrix} 1 & 0 \\ 0 & -1 \end{pmatrix}$

4.1.2. $S = \begin{pmatrix} 2 & 1 \\ 1 & -1 \end{pmatrix}$, $D = \begin{pmatrix} 2 & 0 \\ 0 & -1 \end{pmatrix}$

4.1.3. 12 is an eigenvalue of algebraic multiplicity 2; the eigenspace is one dimensional with basis (2, 1) and so the matrix is not diagonalizable.

4.1.4. $S = \begin{pmatrix} 1 & 1 \\ i & -i \end{pmatrix}$, $D = \begin{pmatrix} 2+i & 0 \\ 0 & 2-i \end{pmatrix}$

4.1.5. $S = \begin{pmatrix} 1 & 0 & -1 \\ 0 & 1 & 0 \\ 1 & 0 & 1 \end{pmatrix}$, $D = \begin{pmatrix} 1 & 0 & 0 \\ 0 & 2 & 0 \\ 0 & 0 & 3 \end{pmatrix}$

4.2.1. If $B\mathbf{v} = \lambda\mathbf{v}$, then $A\mathbf{v} = cB\mathbf{v} = (c\lambda)\mathbf{v}$, so $\mathbf{v}$ is an eigenvector for eigenvalue $c\lambda$.

4.2.2. Let A be an n by n real matrix. Our argument will be given under the additional assumption that all eigenvalues are real; if there are complex eigenvalues, substitute $\mathbb{C}^n$ for $\mathbb{R}^n$ in the argument. Suppose the distinct eigenvalues are $\lambda_1, \ldots, \lambda_k$ with algebraic multiplicity = geometric multiplicity = $n_1, \ldots, n_k$. For each eigenvalue λ_i, choose a basis $\mathbf{v}_{i1}, \ldots, \mathbf{v}_{in_i}$ for the eigenspace $E(\lambda_i)$. We claim $\mathbf{v}_{11}, \ldots, \mathbf{v}_{1n_1}, \ldots, \mathbf{v}_{k1}, \ldots, \mathbf{v}_{kn_k}$ is a basis of eigenvectors for $\mathbb{R}^n$. Since the sum of the algebraic multiplicities is n, there will be n of them. Thus we only need show that they are independent. Suppose $\sum_{i,j} c_{ij}\mathbf{v}_{ij} = 0$. Let $\mathbf{w}_i = \sum_j c_{ij}\mathbf{v}_{ij}$. Then $\mathbf{w}_1 + \cdots + \mathbf{w}_k = 0$. By the proof of Lemma 1.1, we must have $\mathbf{w}_i = 0, i = 1, \ldots, k$. But then the independence of the $\mathbf{v}_{ij}$ for fixed i implies all $c_{ij} = 0$.

4.2.3. $\operatorname{tr} AB = \sum_{i=1}^{n} \sum_{k=1}^{n} a_{ik}b_{ki} = \sum_{k=1}^{n} \sum_{i=1}^{n} b_{ki}a_{ik} = \operatorname{tr} BA$.

4.2.4. The eigenvalues of A are 2, 1, with corresponding eigenvectors $(1, 1), (1, 3)$. The algebraic multiplicity and geometric multiplicity of each eigenvalue is 1. Thus A is diagonalizable. For B, 1 is an eigenvalue with algebraic multiplicity 2, but geometric multiplicity 1. The eigenspace $E(1)$ consists of multiples of $(1, 3)$. Thus B is not diagonalizable. For C there are two distinct eigenvalues, 1 and 3. The eigenvalue 3 has algebraic and geometric multiplicity 1, with eigenvector $(0, 1, 1)$. The eigenvalue 1 has algebraic multiplicity 2, but geometric multiplicity 1, with eigenvectors all multiples of $(1, 0, 1)$. Thus C is not diagonalizable.

4.2.5. For A, the trace is 3 and the determinant is 2. Thus the eigenvalues are $1, 2$. For B the trace is 2 and the determinant is -2. This means that for the other two eigenvalues except 1, their sum is 1 and their product is -2. Thus the other two eigenvalues are $-1, 2$.

4.2.6. If $\mathbf{n}$ is the normal vector to the plane, then L will send $\mathbf{n}$ to $\mathbf{0}$, so $\mathbf{n}$ is an eigenvector for the eigenvalue 0. On the other hand, if $\mathbf{v}_1, \mathbf{v}_2$ form a basis of the plane, then L will send each of them to themselves, so they are eigenvectors for the eigenvalue 1. Thus the geometric multiplicity of 1 is at least 2. Using the fact that the algebraic multiplicities of all the eigenvalues must add up to 3, this means that 0 is an eigenvalue of geometric multiplicity 1 and 1 is an eigenvalue of geometric multiplicity 2. Thus the characteristic polynomial is $-x(x-1)^2$. The required basis is $(1, 0, 1), (0, 1, 1), (1, 1, -1)$. If we form the matrix S with these eigenvectors as the columns, then $A = SDS^{-1}$.

Thus $A = \begin{pmatrix} 1/3 & -2/3 & 2/3 \\ -2/3 & 1/3 & 2/3 \\ 2/3 & 2/3 & 1/3 \end{pmatrix}$.

4.3.1. All of our verifications will depend on the fact that the real numbers satisfy the corresponding properties.We first verify that the complex numbers form an abelian group under addition.

(1) Associativity under addition:

$$\begin{aligned}((a+ib)+(c+id))+(e+if) &= ((a+c)+i(b+d))+(e+if)\\ &= ((a+c)+e)+i((b+d)+f))\\ (a+ib)+((c+id)+(e+if)) &= (a+ib)+((c+e)+i(d+f))\\ &= a+(c+e)+i(b+(d+f))\end{aligned}$$

Associativity then follows from associativity of real addition.
(2) Commutativity under addition:

$$\begin{aligned}(a+ib)+(c+id) &= (a+c)+i(b+d)\\ &= (c+a)+i(d+b)\\ &= (c+id)+(a+ib)\end{aligned}$$

(3) Existence of additive identity ($= 0 + i0$):

$$(a+ib)+(0+i0) = (a+0)+i(b+0) = a+ib$$

(4) Existence of additive inverses ($= -a - ib$):

$$(a+ib)+((-a)+i(-b)) = (a-a)+i(b-b) = 0+i0$$

We now verify the complex numbers form an abelian group under multiplication.
(1) Associativity under multiplication:

$$\begin{aligned}((a+ib)(c+id))(e+if) &= ((ac-bd)+i(ad+bc))(e+if)\\ &= (ace-bde-bcf-adf)+i(bce+ade+acf-bdf)\\ &= (a+ib)((ce-df)+i(cf+de)) = (a+ib)((c+id)(e+if))\end{aligned}$$

(2) Commutativity under multiplication:

$$\begin{aligned}(a+ib)(c+id) &= (ac-bd)+i(ad+bc)\\ &= (ca-db)+i(da+cb)\\ &= (c+id)(a+ib)\end{aligned}$$

(3) Existence of multiplicative identity ($= 1 + i0$) :

$$(a+ib)(1+i0) = a+ib$$

(4) Existence of multiplicative inverses ($= (a - ib)/(a^2 + b^2)$) :

$$(a+ib)(a-ib)/(a^2+b^2) = (a^2+b^2)/(a^2+b^2)+i0 = 1+i0$$

We verify the distributive law.

$$\begin{aligned}(a+ib)((c+id)+(e+if)) &= (a+ib)((c+e)+i(d+f))\\ &= (a(c+e)-b(d+f))+i(a(d+f)+b(c+e))\\ &= (ac+ae-bd-bf)+i(ad+af+bc+be)\\ &= ((ac-bd)+i(ad+bc))+i((ae-bf)+i(af+be))\\ &= ((a+ib)(c+id)+(a+ib)(e+if)\end{aligned}$$

4.3.2 We let $z = a + ib$, $w = c + id$.

$$\begin{aligned}\overline{(a+ib)+(c+id)} &= \overline{(a+c)+i(b+d)} = (a+c)-i(b+d)\\ &= (a-ib)+(c-id) = \overline{a+ib}+\overline{c+id}\\ \overline{(a+ib)(c+id)} &= \overline{(ac-bd)+i(ad+bc)} = (ac-bd)-i(ad+bc)\\ &= (a-ib)(c-id) = \overline{a+ib}\,\overline{c+id}\end{aligned}$$

4.3.3. The checks are exactly the same for showing $\mathbb{R}^n$ is a vector space. The identity is $(0,\ldots,0)$ and the inverse of $(z_1,\ldots,z_n)$ is $(-z_1,\ldots,-z_n)$. Since operations are done coordinatewise, the verifications hinge on those for $\mathbb{C}$ being a field. In general, if F is a field then F^n forms a vector space.

4.3.4. The ith entry for $\overline{A\mathbf{v}}$ is $\overline{\sum_{j=1}^n a_{ij}v_j} = \sum_{j=1}^n \overline{a_{ij}v_j} = \sum_{j=1}^n \bar{a}_{ij}\bar{v}_j$, which is the ith entry for $\bar{A}\bar{\mathbf{v}}$.

4.3.5. $\langle \mathbf{v}, \mathbf{w}\rangle = \overline{v_1}w_1 + \cdots + \overline{v_n}w_n = \overline{\overline{w_1}v_1 + \cdots + \overline{w_n}v_n} = \overline{\langle \mathbf{w}, \mathbf{v}\rangle}$

4.3.6. (a) Hermitian (b) skew Hermitian (c) unitary and Hermitian (d) Hermitian (e) unitary

4.3.7. If A and B are Hermitian, then $A^* = A$, $B^* = B$ so $(A+B)^* = A^* + B^* = A + B$; hence $A + B$ is Hermitian. The sum of two skew Hermitian matrices is skew Hermitian, but the sum of two unitary matrices is not necessarily unitary.

4.3.8. If A is Hermitian, then $(iA)^* = -iA^* = -iA$ so iA is skew Hermitian.

4.3.9. As suggested in the hint, let $\mathbf{x} = \mathbf{e}_i$, $\mathbf{y} = \mathbf{e}_j$. Then

$$\overline{a_{ji}} = \langle A\mathbf{e}_i, \mathbf{e}_j\rangle = \langle \mathbf{e}_i, B\mathbf{e}_j\rangle = b_{ij}.$$

For the converse,

$$\langle A\mathbf{x}, \mathbf{y}\rangle = \mathbf{x}^* A^* \mathbf{y} = \mathbf{x}^* B\mathbf{y} = \langle \mathbf{x}, B\mathbf{y}\rangle.$$

4.4.1. All three parts have the same $S = \begin{pmatrix} 0 & 1 & 1\\ 1 & 0 & 1\\ 1 & 1 & 1\end{pmatrix}$. We give D in terms of the diagonal entries.

(a) 1,2,4; It tends to infinity along the line through $(1, 1, 1)$.

(b) .1,.5,1; It tends to $\begin{pmatrix} 1 & 1 & -1\\ 1 & 1 & -1\\ 1 & 1 & -1\end{pmatrix}$.

(c) .5,.1,.1; It tends to zero along the line through $(0, 1, 1)$.

4.4.2. The limiting distribution is $(.2308, .3077, .4615)$. The probability that we are in state 1 after 20 steps is .2308.

4.4.3. The limiting distribution of populations will be $1/3$ in A and $2/3$ in B. This means 10 million in A and 20 million in B. After 10 years there will be 12,529,400 in A, 17,046,900 in B and 423,700 in C.

4.4.4. (2) is just a restatement that the sum of the entries is 1. (3) then becomes the statement that each column of T is a probability vector. For (1) note that

$$\mathbf{r}(T\mathbf{x}) = (\mathbf{r}T)\mathbf{x} = \mathbf{r}\mathbf{x} = 1$$

4.4.5. Under the hypotheses, it was shown that there is a one dimensional subspace of eigenvectors for T, which consists of multiples of a unique probability vector $\mathbf{p}$. Moreover, all of the other eigenvectors have length less than 1. The columns of T^k are just $T^k\mathbf{e}_j$, and these are all probability vectors. Each column must tend to a multiple of $\mathbf{p}$, and since they are all probability eigenvectors they all tend to $\mathbf{p}$. Positivity follows from $T^k\mathbf{p} = \mathbf{p}$ having positive entries when T^k is positive.

4.4.6. 1.2247

4.4.7. For $\mathbf{x}_0 = (1, 0)$, the vectors alternate between $(1, 0)$ and $(0, 1)$. For initial vector $(1, 1)$, the vectors are constantly $(1, 1)$.

4.4.8. The limit exists since there is a largest eigenvalue of algebraic multiplicity 1. The limiting value is the normalized eigenvector $(-1/\sqrt{5}, 0, -2/\sqrt{5})$.

4.4.9. $A = \begin{pmatrix} 1 & 2 \\ 1 & 0 \end{pmatrix}$; the limiting ratio is 2.

4.4.10. $A = \begin{pmatrix} 1 & 2 & 3 \\ 1 & 0 & 0 \\ 0 & 1 & 0 \end{pmatrix}$; the limiting ratio is 2.3744.

4.4.11. $A = \begin{pmatrix} 1 & 0 & 0 & 1 \\ 1 & 0 & 0 & 0 \\ 0 & 1 & 0 & 0 \\ 0 & 0 & 1 & 0 \end{pmatrix}$; the limiting ratio is 1.3803.

4.4.12. $A = \begin{pmatrix} 0 & 1 & 0 & 1 \\ 1 & 0 & 0 & 0 \\ 0 & 1 & 0 & 0 \\ 0 & 0 & 1 & 0 \end{pmatrix}$; there is no limiting ratio.

4.4.13. The power method gives 16.1168 as the largest eigenvalue and (.2320, .5253, .8187) as the corresponding eigenvector. The characteristic polynomial is $-x^3 + 15x^2 + 18x = -x(x^2 - 15x + 18)$. Thus the eigenvalues are $0, \frac{15}{2} \pm \sqrt{153}$ which are $0, -1.1168, 16.1168$. The eigenvector for 16.1168 is the one found by the power method.

4.4.14. The largest eigenvalue is 8, with eigenvector $(-.7071, 0, -.7071, 0)$.

4.4.15. The power method fails to converge. The problem arises since the largest eigenvalues are the complex pair $4.7250 \pm .3332i$ and they are of the same magnitude.

4.4.16. One cube root of A is $\begin{pmatrix} -1/2 & 3/2 \\ 3/2 & -1/2 \end{pmatrix}$.

4.4.17. The four square roots are

$$\pm\begin{pmatrix} 4 - i & -4 + 2i \\ 2 - i & -2 + 2i \end{pmatrix}, \pm\begin{pmatrix} -4 - i & 4 + 2i \\ -2 - i & 2 + 2i \end{pmatrix}.$$

4.4.18. Suppose $\begin{pmatrix} a & b \\ c & d \end{pmatrix}^2 = \begin{pmatrix} 0 & 1 \\ 0 & 0 \end{pmatrix}$. Then we have equations

$$a^2 + bc = 0, bc + d^2 = 0, b(a + d) = 1, c(a + d) = 0.$$

From the third equation we get $b \neq 0, a + d \neq 0$. From the fourth equation this gives $c = 0$. Substituting into the first two gives $a = d = 0$. But this contradicts $a + d \neq 0$. Hence there are no solutions.

4.4.19. $\begin{pmatrix} -6.6533 & 14.0424 \\ -7.0212 & 14.4102 \end{pmatrix}$

4.5.1. $c_1 \exp(7t) \begin{pmatrix} 2 \\ 1 \end{pmatrix} + c_2 \exp(-5t) \begin{pmatrix} 2 \\ -1 \end{pmatrix}$

4.5.2. $\exp(-2t) \begin{pmatrix} 2\cos t - 4\sin t \\ -\cos t - 3\sin t \end{pmatrix}$

4.5.3. $\exp(2t) \begin{pmatrix} 1 \\ -2 \\ -1 \end{pmatrix}$

4.5.4. $\frac{-1}{3} e^{-2t} + \frac{4}{3} e^t$

4.5.5. $\frac{3}{2} e^t - \frac{2}{3} e^{2t} + \frac{1}{6} e^{-t}$

4.5.6. $\cos t$

4.5.7. $\frac{-3}{2} \cos \sqrt{3}t - \frac{\sqrt{3}}{6} \sin \sqrt{3}t + \frac{9}{2} \cos t - \frac{1}{2} \sin t$

4.5.8. $e^{2t} - te^{2t}$

4.5.9. $-\frac{3}{25} \cos 2t - \frac{4}{25} \sin 2t + \frac{28}{25} e^t - \frac{4}{5} te^t$

4.5.10. $-2e^t + 3te^t - 2t^2 e^t$

4.5.11. e^{2t}

A.5. SOLUTIONS TO CHAPTER 5 EXERCISES

5.1.1. That $V_{\mathbb{C}}$ forms an abelian group under addition follows directly from the fact that V does and the addition operation is coordinate wise. For the associative property,

$$\begin{aligned} ((a + ib)(c + id))(\mathbf{v}_1, \mathbf{v}_2) &= ((ac - bd) + i(bc + ad))(\mathbf{v}_1, \mathbf{v}_2) \\ &= ((ac - bd)\mathbf{v}_1 - (bc + ad)\mathbf{v}_2, (ac - bd)\mathbf{v}_2 + (bc + ad)\mathbf{v}_1) \end{aligned}$$

$$\begin{aligned}(a+ib)((c+id)(\mathbf{v}_1,\mathbf{v}_2)) &= (a+ib)(c\mathbf{v}_1-d\mathbf{v}_2,c\mathbf{v}_2+d\mathbf{v}_1)\\ &= ((ac-bd)\mathbf{v}_1-(bc+ad)\mathbf{v}_2,(bc+ad)\mathbf{v}_1+(ac-bd)\mathbf{v}_2).\end{aligned}$$

The distributive property is given by

$$\begin{aligned}(a+ib)(\mathbf{v}_1+\mathbf{w}_1,\mathbf{v}_2+\mathbf{w}_2) &= (a(\mathbf{v}_1+\mathbf{w}_1)-b(\mathbf{v}_2+\mathbf{w}_2),a(\mathbf{v}_2+\mathbf{w}_2)+b(\mathbf{v}_1+\mathbf{w}_1))\\ &= (a\mathbf{v}_1-b\mathbf{v}_2,a\mathbf{v}_2+b\mathbf{v}_1)+(a\mathbf{w}_1-b\mathbf{w}_2,a\mathbf{w}_2+b\mathbf{w}_1)\\ &= (a+ib)(\mathbf{v}_1,\mathbf{v}_2)+(a+ib)(\mathbf{w}_1,\mathbf{w}_2)\\ ((a+ib)+(c+id))(\mathbf{v}_1,\mathbf{v}_2) &= ((a+c)+i(b+d))(\mathbf{v}_1,\mathbf{v}_2)\\ &= ((a\mathbf{v}_1-b\mathbf{v}_2)+(c\mathbf{v}_1-d\mathbf{v}_2),(b\mathbf{v}_1+a\mathbf{v}_2)+(d\mathbf{v}_1+c\mathbf{v}_2))\\ &= (a+ib)(\mathbf{v}_1,\mathbf{v}_2)+(c+id)(\mathbf{v}_1,\mathbf{v}_2).\end{aligned}$$

Finally, $1(\mathbf{v}_1,\mathbf{v}_2)=(\mathbf{v}_1,\mathbf{v}_2)$.

5.1.2. $$\begin{aligned}\overline{(a+ib)(\mathbf{w}_1,\mathbf{w}_2)} &= \overline{(a\mathbf{w}_1-b\mathbf{w}_2,b\mathbf{w}_1+a\mathbf{w}_2)} = (a\mathbf{w}_1-b\mathbf{w}_2,-b\mathbf{w}_1-a\mathbf{w}_2)\\ &= (a-ib)(\mathbf{w}_1,-\mathbf{w}_2) = \overline{a+ib}\,\overline{(\mathbf{w}_1,\mathbf{w}_2)}\end{aligned}$$

$$\begin{aligned}\overline{(\mathbf{w}_1,\mathbf{w}_2)+(\mathbf{y}_1,\mathbf{y}_2)} &= \overline{(\mathbf{w}_1+\mathbf{y}_1,\mathbf{w}_2+\mathbf{y}_2)} = (\mathbf{w}_1+\mathbf{y}_1,-\mathbf{w}_2-\mathbf{y}_2)\\ &= (\mathbf{w}_1,-\mathbf{w}_2)+(\mathbf{y}_1,-\mathbf{y}_2) = \overline{(\mathbf{w}_1,\mathbf{w}_2)}+\overline{(\mathbf{y}_1,\mathbf{y}_2)}\end{aligned}$$

5.1.3. $$\begin{aligned}L_{\mathbb{C}}((a+ib)(\mathbf{v}_1,\mathbf{v}_2)) &= L_{\mathbb{C}}(a\mathbf{v}_1-b\mathbf{v}_2,a\mathbf{v}_2+b\mathbf{v}_1)\\ &= (L(a\mathbf{v}_1-b\mathbf{v}_2),L(a\mathbf{v}_2+b\mathbf{v}_1)) = (aL(\mathbf{v}_1)-bL(\mathbf{v}_2),aL(\mathbf{v}_2)+bL(\mathbf{v}_2))\\ &= (a+ib)(L(\mathbf{v}_1),L(\mathbf{v}_2)) = (a+ib)L_{\mathbb{C}}(\mathbf{v}_1,\mathbf{v}_2).\end{aligned}$$

$$\begin{aligned}L_{\mathbb{C}}((\mathbf{w}_1,\mathbf{w}_2)+(\mathbf{y}_1,\mathbf{y}_2)) &= L_{\mathbb{C}}(\mathbf{w}_1+\mathbf{y}_1,\mathbf{w}_2+\mathbf{y}_2)\\ &= (L(\mathbf{w}_1+\mathbf{y}_1),L(\mathbf{w}_2+\mathbf{y}_2)) = (L(\mathbf{w}_1)+L(\mathbf{y}_1),L(\mathbf{w}_2)+L(\mathbf{y}_2))\\ &= (L(\mathbf{w}_1),L(\mathbf{w}_2))+(L(\mathbf{y}_1),L(\mathbf{y}_2)) = L_{\mathbb{C}}(\mathbf{w}_1,\mathbf{w}_2)+L_{\mathbb{C}}(\mathbf{y}_1,\mathbf{y}_2)\end{aligned}$$

5.1.4. As vectors in $V_{\mathbb{C}}$, these are ordered pairs $(\mathbf{v}_1,0),\ldots,(\mathbf{v}_n,0)$. Let $(\mathbf{w},\mathbf{y})\in V_{\mathbb{C}}$. Then there are real constants $a_1,\ldots,a_n,b_1,\ldots,b_n$ so that

$$\mathbf{w}=a_1\mathbf{v}_1+\cdots+a_n\mathbf{v}_n,\mathbf{y}=b_1\mathbf{v}_1+\cdots+b_n\mathbf{v}_n$$

Then

$$(a_1+ib_1)(\mathbf{v}_1,0)+\cdots+(a_n+ib_n)(\mathbf{v}_n,0)=(\mathbf{w},\mathbf{y})$$

and so these vectors span $V_{\mathbb{C}}$. To see that they are independent, suppose

$$(a_1+ib_1)(\mathbf{v}_1,0)+\cdots+(a_n+ib_n)(\mathbf{v}_n,0)=(\mathbf{0},\mathbf{0})$$

Then this gives

$$a_1\mathbf{v}_1+\cdots+a_n\mathbf{v}_n=\mathbf{0},b_1\mathbf{v}_1+\cdots+b_n\mathbf{v}_n=\mathbf{0}$$

The independence of the $\mathbf{v}_i$ then implies that all a_i,b_i are 0. That A represents L means that $L(\mathbf{v}_j)=\sum_{i=1}^n a_{ij}\mathbf{v}_i$. Then

$$L_{\mathbb{C}}(\mathbf{v}_j,\mathbf{0})=(L(\mathbf{v}_j),\mathbf{0})=(\sum_{i=1}^n a_{ij}\mathbf{v}_i,\mathbf{0})=\sum_{i=1}^n a_{ij}(\mathbf{v}_i,\mathbf{0})$$

which implies that this same matrix still represents $L_{\mathbb{C}}$.

5.1.5. (a) Suppose L is unitary, and $\mathbf{v}_1,\ldots,\mathbf{v}_n$ is an orthonormal basis. Then if A is the matrix which represents L with respect to this basis, L^* is represented by A^*. Then $L^*L=I$ implies that the identity is represented by

$A^*A = I$, which says that A is unitary. Conversely, suppose L is represented by A and we know that A is unitary. Then L^* is represented by A^*, and so L^*L is represented by $A^*A = I$, which implies that $L^*L = I$.

The solutions of (b), (c) follow the same pattern as above, where one goes from $L \pm L^* = 0$ to $A \pm A^* = 0$ and back.

(d) Choose the basis $\mathbf{v}_1 = \mathbf{e}_1, \mathbf{v}_2 = 2\mathbf{e}_2$ for $\mathbb{C}$, which is not an orthonormal basis. Let L be the linear transformation given by multiplication by the matrix $\begin{pmatrix} 0 & -1 \\ 1 & 0 \end{pmatrix}$. This linear transformation is both unitary and skew Hermitian, by the criteria given above. However, the matrix which represents it with respect to the above matrix is $\begin{pmatrix} 0 & -2 \\ 1/2 & 0 \end{pmatrix}$, which is neither unitary nor skew Hermitian. For the Hermitian counterexample, we can take multiplication by $\begin{pmatrix} 0 & 1 \\ 1 & 0 \end{pmatrix}$.

5.2.1. (a) $S = \begin{pmatrix} 1/\sqrt{2} & -1/\sqrt{6} & 1/\sqrt{3} \\ 1/\sqrt{2} & 1/\sqrt{6} & -1/\sqrt{3} \\ 0 & 2/\sqrt{6} & 1/\sqrt{3} \end{pmatrix}, D = \begin{pmatrix} 3 & 0 & 0 \\ 0 & 9 & 0 \\ 0 & 0 & 6 \end{pmatrix}$

(b) $S = \begin{pmatrix} .6455 & .0573 - .6430i & -1/\sqrt{6} \\ -.1291 + .6325i & -.6414 + .0725i & -1/\sqrt{6} \\ .2582 + .3162i & -.2921 - .2852i & 2/\sqrt{6} \end{pmatrix}$,

$$D = \begin{pmatrix} \sqrt{6}i & 0 & 0 \\ 0 & -\sqrt{6}i & 0 \\ 0 & 0 & 0 \end{pmatrix}$$

(c) $S = \begin{pmatrix} 1/\sqrt{3} & .4133 - .4316i & -.3510 + .4316i \\ 1/\sqrt{3} & -.5106 - .1580i & -.1824 - .5896i \\ 1/\sqrt{3} & .0973 + .5896i & 7.5334 + .1580i \end{pmatrix}$,

$$D = \begin{pmatrix} 0 & 0 & 0 \\ 0 & -21 & 0 \\ 0 & 0 & 21 \end{pmatrix}$$

(d) $S = \begin{pmatrix} -0.7071 & -0.7071 & 0 \\ 0.7071i & -0.7071i & 0 \\ 0 & 0 & 1.0000 \end{pmatrix}$,

$$D = \begin{pmatrix} 0.7071 + 0.7071i & 0 & 0 \\ 0 & 0.7071 - 0.7071i & 0 \\ 0 & 0 & 1 \end{pmatrix}$$

5.2.2. (a) $A = 3P_1 + 9P_2 + 6P_3$, where

$$P_1 = \begin{pmatrix} 1/2 & 1/2 & 0 \\ 1/2 & 1/2 & 0 \\ 0 & 0 & 0 \end{pmatrix}, P_2 = \begin{pmatrix} 1/6 & -1/6 & -1/3 \\ -1/6 & 1/6 & 1/3 \\ -1/3 & 1/3 & 2/3 \end{pmatrix}, P_3 = \begin{pmatrix} 1/3 & -1/3 & 1/3 \\ -1/3 & 1/3 & -1/3 \\ 1/3 & -1/3 & 1/3 \end{pmatrix}$$

(b) $A = \sqrt{6}iP_1 - \sqrt{6}iP_2$, where

$$P_1 = \begin{pmatrix} .4167 & -.0833 - .4082i & .1667 - .2041i \\ -.0833 + .4082i & .4167 & .1667 + .2041i \\ .1667 + .2041i & .1667 - .2041i & .1667 \end{pmatrix}$$

and $P_2 = \overline{P_1}$.

(c) $A = -21P_1 + 21P_2$, where

$$P_1 = \begin{pmatrix} .3571 & -.1429 + .2857i & -.2143 - .2857i \\ -.1429 - .2857i & .2857 & -.1429 + .2857i \\ -.2143 + .2857i & -.1429 - .2857i & .3571 \end{pmatrix}$$

and

$$P_2 = \begin{pmatrix} .3095 & -.1905 - .2857i & -.1190 + .2857i \\ -.1905 + .2857i & .3810 & -.1905 - .2857i \\ -.1190 - .2857i & -.1905 + .2857i & .3095 \end{pmatrix}.$$

(d) $A = (.7071 + .7071i)P_1 + (.7071 - .7071i)P_2 + P_3$, where

$$P_1 = \begin{pmatrix} 1/2 & i/2 & 0 \\ -i/2 & 1/2 & 0 \\ 0 & 0 & 0 \end{pmatrix}, P_2 = \begin{pmatrix} 1/2 & -i/2 & 0 \\ i/2 & 1/2 & 0 \\ 0 & 0 & 0 \end{pmatrix}, P_3 = \begin{pmatrix} 0 & 0 & 0 \\ 0 & 0 & 0 \\ 0 & 0 & 1 \end{pmatrix}.$$

5.2.3. $(A+A^*)^* = A^*+A = A+A^*$ so $A+A^*$ is Hermitian. $(A-A^*)^* = A^*-A = -(A-A^*)$ so $A-A^*$ is skew Hermitian. $A = 1/2(A+A^*)+1/2(A-A^*)$. Since a Hermitian matrix and a skew Hermitian matrix are both linear combinations of projection matrices, then A is a linear combination of projection matrices.

5.2.4. $\begin{pmatrix} 1+i & 1-i \\ 1-i & 1+i \end{pmatrix}$

5.2.5. (1) implies (2): Suppose that L is normal. Then $LL^* = L^*L$ and thus $LL^*\mathbf{v} = L^*L\mathbf{v}$ for all $\mathbf{v} \in V$. Hence $\langle L^*L\mathbf{v}, \mathbf{w}\rangle = \langle LL^*\mathbf{v}, \mathbf{w}\rangle$ for all $\mathbf{v}, \mathbf{w} \in V$.
(2) implies (3): If (2) holds for all vectors $\mathbf{v}, \mathbf{w} \in V$, then in particular it holds when we choose $\mathbf{v}, \mathbf{w}$ out of a basis.
(3) is equivalent to (4): This just uses

$$\langle L\mathbf{v}_i, L\mathbf{v}_j\rangle = \langle L^*L\mathbf{v}_i, \mathbf{v}_j\rangle, \langle L^*\mathbf{v}_i, L^*\mathbf{v}_j\rangle = \langle LL^*\mathbf{v}_i, \mathbf{v}_j\rangle.$$

(4) implies (1): It suffices to show that LL^* and L^*L agree on some basis of V. Choose the basis so that (4) holds. Then using the equivalence of (3) and (4), we see that the inner products of $LL^*\mathbf{v}_i$ and $LL^*\mathbf{v}_i$ with the basis vectors $\mathbf{v}_1, \ldots, \mathbf{v}_n$ are all the same. Thus the difference $LL^*\mathbf{v}_i - LL^*\mathbf{v}_i$ has inner product 0 with every basis vector. Hence it must have inner product 0 with every vector in V.

But the only vector with this property is $\mathbf{0}$. Thus $LL^*\mathbf{v}_i = L^*L\mathbf{v}_i$ for every basis vector, and hence for all vectors, implying (1).

5.2.6. (a) $L(\mathbf{v}_i) = c_i\mathbf{v}_i$. Thus

$$\overline{c_i} = \langle c_i\mathbf{v}_i, \mathbf{v}_i\rangle = \langle L\mathbf{v}_i, \mathbf{v}_i\rangle = \langle \mathbf{v}_i, L^*\mathbf{v}_i\rangle.$$

Also,

$$0 = \langle c_j\mathbf{v}_j, \mathbf{v}_i\rangle = \langle L\mathbf{v}_j, \mathbf{v}_i\rangle = \langle \mathbf{v}_j, L^*\mathbf{v}_i\rangle$$

for $j \neq i$. But

$$L^*\mathbf{v}_i = \sum_{j=1}^{n}\langle \mathbf{v}_j, L^*\mathbf{v}_i\rangle\mathbf{v}_j = \overline{c_i}\mathbf{v}_i.$$

(b) By part (4) of Exercise 5.2.5. we just have to find a basis so that $\langle L\mathbf{v}_i, L\mathbf{v}_j\rangle = \langle L^*\mathbf{v}_i, L^*\mathbf{v}_j\rangle$. We choose the basis of eigenvectors. Then

$$\langle L\mathbf{v}_i, L\mathbf{v}_j\rangle = \langle c_i\mathbf{v}_i, c_j\mathbf{v}_j\rangle = \overline{c_i}c_j\delta_{ij} = \langle \overline{c_i}\mathbf{v}_i, \overline{c_j}\mathbf{v}_j\rangle = \langle L^*\mathbf{v}_i, L^*\mathbf{v}_j\rangle.$$

5.2.7. For $n = 1$ the matrix is already triangular; we can take $S = (1)$. Following the outline, we assume it is true when the size of the matrix is less than k and prove it for $n = k$. We then find an eigenvalue μ and corresponding eigenvector $\mathbf{v}$, which we can normalize to be a unit vector. Then renaming $\mathbf{v}$ as $\mathbf{v}_1$, we extend it to an orthonormal basis $\mathbf{v}_1, \ldots, \mathbf{v}_n$ for $\mathbb{C}^n$. These vectors then form the columns of a unitary matrix S_1. Since $\mathbf{v}_1$ is an eigenvector, we have $AS_1 = S_1\begin{pmatrix}\mu & D\\ \mathbf{0} & C\end{pmatrix}$. We rename this last matrix B and have $S_1^{-1}AS_1 = B$. Then by the induction assumption there is a unitary matrix U with $U^{-1}CU = T_1$, where T_1 is triangular. If $S_2 = \begin{pmatrix}1 & 0\\ 0 & U\end{pmatrix}$, then S_2 is unitary since U is. Also, we compute that $S_2^{-1}BS_2 = \begin{pmatrix}\mu & D\\ 0 & T_1\end{pmatrix} = T$, which is triangular. If $S = S_1S_2$, then

$$S^{-1}AS = S_2^{-1}(S_1^{-1}AS_1)S_2 = S_2^{-1}BS_2 = T.$$

5.3.1. $Q_1 = \begin{pmatrix}1/\sqrt{14} & 2/\sqrt{5} & 3/\sqrt{70}\\ 2/\sqrt{14} & -1/\sqrt{5} & 6/\sqrt{70}\\ 3/\sqrt{14} & 0 & -5/\sqrt{70}\end{pmatrix}$, $Q_2 = \begin{pmatrix}1/\sqrt{5} & -2/\sqrt{5}\\ 2/\sqrt{5} & 1/\sqrt{5}\end{pmatrix}$,

$$\Sigma = \begin{pmatrix}\sqrt{70} & 0\\ 0 & 0\\ 0 & 0\end{pmatrix}, \quad A^+ = (1/70)\begin{pmatrix}1 & 2 & 3\\ 2 & 4 & 6\end{pmatrix}$$

5.3.2. $Q_1 = \begin{pmatrix}2/\sqrt{5} & 1/\sqrt{5}\\ 1/\sqrt{5} & 2/\sqrt{5}\end{pmatrix}$, $Q_2 = \begin{pmatrix}2/\sqrt{30} & -1/\sqrt{5} & 2/\sqrt{6}\\ 5/\sqrt{30} & 0 & -1/\sqrt{6}\\ 1/\sqrt{30} & 2/\sqrt{5} & 1/\sqrt{6}\end{pmatrix}$,

$$\Sigma = \begin{pmatrix} \sqrt{6} & 0 & 0 \\ 0 & 1 & 0 \end{pmatrix}, \ A^+ = \begin{pmatrix} 1/3 & -1/3 \\ 1/3 & 1/6 \\ -1/3 & 5/6 \end{pmatrix}$$

5.3.3. $A^+ = \begin{pmatrix} .1382 & .1789 & -.1707 & -.0325 \\ -.1301 & .0081 & .2195 & .0894 \\ .0081 & .1870 & .0488 & .0569 \end{pmatrix}$ and the solution to the least squares problem is $(.1138, .1870, .3008)$.

5.4.1. Part (a) is a reflection through $y = x/2$. Part (b) is a rotation of .9273 radians.

5.4.2. The axis goes through $(1/\sqrt{3}, 1/\sqrt{3}, 1/\sqrt{3})$ and the angle is $\pi/3$ radians.

5.4.3. We are reflecting through the plane $2x + 2y - z = 0$.

5.4.4. $B = \begin{pmatrix} .5 & -.5 & .5 & .5 \\ .5 & .5 & -.5 & .5 \\ .5 & -.5 & -.5 & -.5 \\ .5 & .5 & .5 & -.5 \end{pmatrix}$,

$$B^{-1}AB = \begin{pmatrix} .7071 & .7071 & 0 & 0 \\ -.7071 & .7071 & 0 & 70 \\ 0 & 0 & .4472 & -.8944 \\ 0 & 0 & .8944 & .4472 \end{pmatrix}$$

5.5.1. Reflexivity: $A = I^t AI$

Symmetry: If B is congruent to A, then there is an invertible matrix S with $B = S^t AS$. Then $A = (S^{-1})^t BS^{-1}$, so A is congruent to B.

Transitivity: Suppose that B is congruent to A and C is congruent to B. Then there are invertible matrices S, T with $B = S^t AS, C = T^t BT$. Then

$$C = T^t(S^t AS)T = (ST)^t A(ST),$$

which means that C is congruent to A.

5.5.2. Symmetry:

$$b(\mathbf{v}, \mathbf{w}) = \mathbf{v}^t A\mathbf{w} = (\mathbf{v}^t A\mathbf{w})^t = \mathbf{w}^t A^t \mathbf{v} = \mathbf{w}^t A\mathbf{v} = b(\mathbf{w}, \mathbf{v})$$

Bilinearity:

$$b(\mathbf{v}, c\mathbf{w} + d\mathbf{x}) = \mathbf{v}^t A(c\mathbf{w} + d\mathbf{x}) = c\mathbf{v}^t A\mathbf{w} + d\mathbf{v}^t A\mathbf{x} = cb(\mathbf{v}, \mathbf{w}) + db(\mathbf{v}, \mathbf{x})$$

5.5.3. (a) $q(r\mathbf{v}) = b(r\mathbf{v}, r\mathbf{v}) = r^2 b(\mathbf{v}, \mathbf{v}) = r^2 q(\mathbf{v})$.

(b) $(1/2)(q(\mathbf{v} + \mathbf{w}) - q(\mathbf{v}) - q(\mathbf{w}))$

$= 1/2(b(\mathbf{v} + \mathbf{w}, \mathbf{v} + \mathbf{w}) - b(\mathbf{v}, \mathbf{v}) - b(\mathbf{w}, \mathbf{w}))$

$= (1/2)(b(\mathbf{v}, \mathbf{v}) + b(\mathbf{v}, \mathbf{w}) + b(\mathbf{w}, \mathbf{v}) + b(\mathbf{w}, \mathbf{w}) - b(\mathbf{v}, \mathbf{v}) - b(\mathbf{w}, \mathbf{w}))$

$= (1/2)(b(\mathbf{v}, \mathbf{w}) + b(\mathbf{w}, \mathbf{v})) = b(\mathbf{v}, \mathbf{w})$

5.5.4.

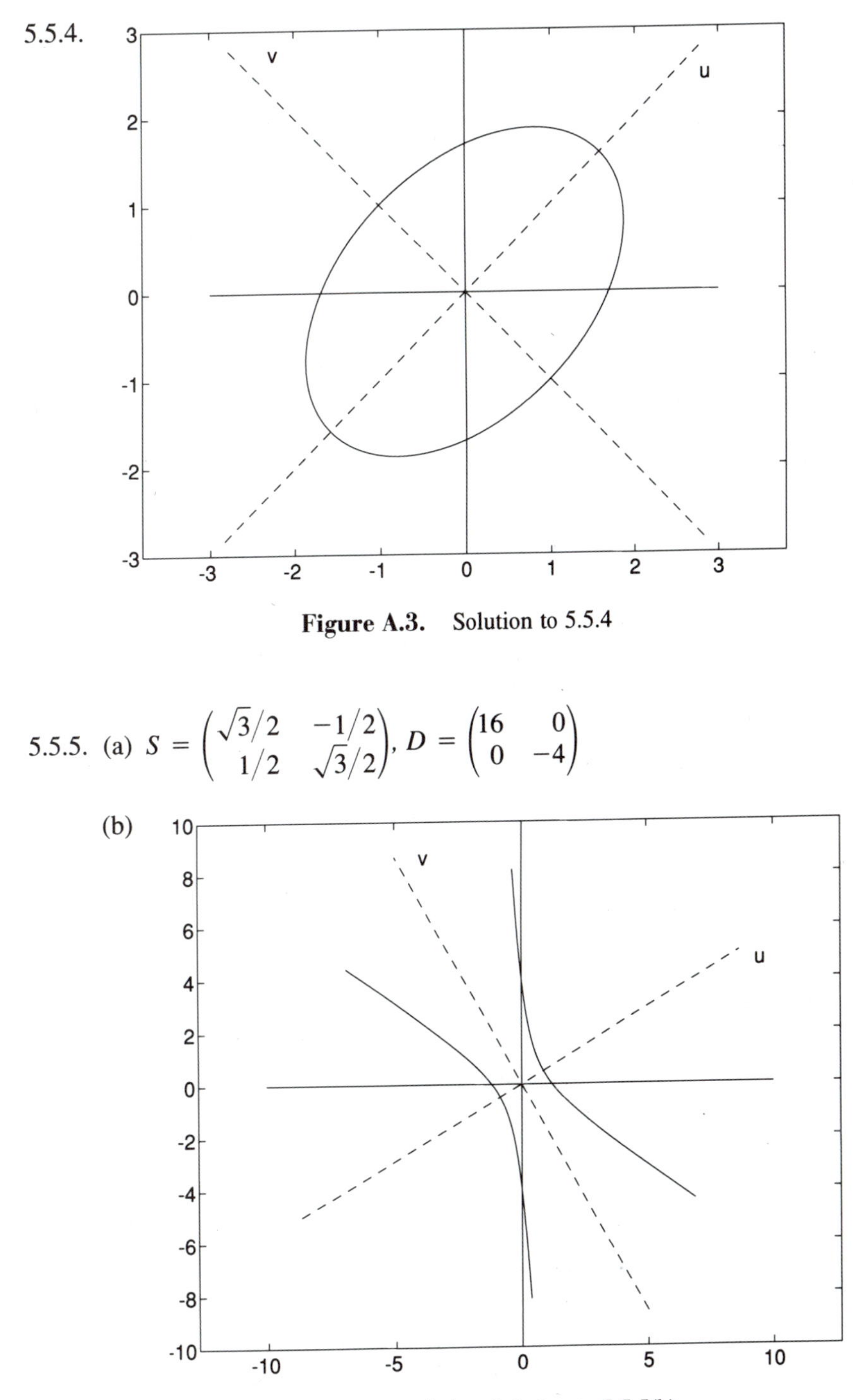

Figure A.3. Solution to 5.5.4

5.5.5. (a) $S = \begin{pmatrix} \sqrt{3}/2 & -1/2 \\ 1/2 & \sqrt{3}/2 \end{pmatrix}$, $D = \begin{pmatrix} 16 & 0 \\ 0 & -4 \end{pmatrix}$

(b)

Figure A.4. Solution to 5.5.5(b)

5.5.6.

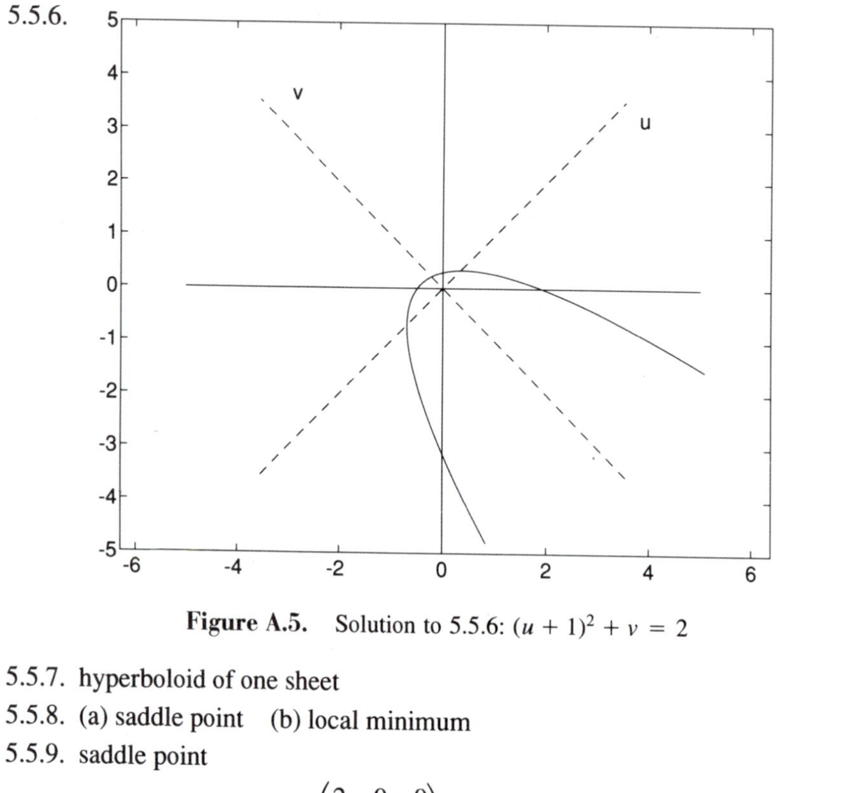

Figure A.5. Solution to 5.5.6: $(u + 1)^2 + v = 2$

5.5.7. hyperboloid of one sheet

5.5.8. (a) saddle point (b) local minimum

5.5.9. saddle point

5.5.10. The Hessian matrix is $\begin{pmatrix} 2 & 0 & 0 \\ 0 & 0 & 1 \\ 0 & 1 & 0 \end{pmatrix}$, so the eigenvalues are $2, -1, 1$ and there is a saddle point.

A.6. SOLUTIONS TO CHAPTER 6 EXERCISES

6.1.1. (a) $S = \begin{pmatrix} .2554 & 1.7508 \\ .4133 & -1.0820 \end{pmatrix}, D = D(1, 1, 0)$.

(b) $S = \begin{pmatrix} .1491 & .8165 & .1826 \\ .2981 & -.4082 & -.0913 \\ 0 & -.4082 & .4564 \end{pmatrix}, D = D(1, 2, 0)$.

6.1.2. Choose an arbitrary basis $\mathcal{W}$ so that the matrix which represents $\mathcal{Q}$ with respect to this basis is A. By part (a) there is a matrix S with $D(p, n, d) = S^t AS$. Choose a new basis $\mathcal{V}$ so that the matrix S is the transition matrix from $\mathcal{V}$ to $\mathcal{W}$, i.e. $\mathbf{V} = \mathbf{W}S$. Then Proposition 6.5.1 implies that the matrix which represents $\mathcal{Q}$ with respect to $\mathcal{W}$ is $D(p, n, d)$. The last statement follows from $\mathcal{Q}(\mathbf{v}) = \mathbf{x}^t D(p, n, d)\mathbf{x}$.

6.1.3. (a) $S = \begin{pmatrix} 2 & 4 & 1 \\ 4 & 1 & 0 \\ 1 & 0 & 1 \end{pmatrix}$ and $D(3, 0, 0)$.

(b) $S = \begin{pmatrix} 0 & 1 & -1 \\ -1 & 1 & 0 \\ 1 & -1 & 1 \end{pmatrix}$ and $D(1, 1, 1)$.

(c) $S = \begin{pmatrix} 1 & 1 & 0 & 0 \\ -1 & 0 & 1 & -1 \\ -1 & -1 & 1 & -1 \\ 0 & 0 & 0 & 1 \end{pmatrix}$ and $D(2, 2, 0)$.

6.1.4. (a) A is positive definite, with $S = \begin{pmatrix} .7071 & -.4082 \\ 0 & .8165 \end{pmatrix}$.

(b) A is indefinite, with $S = \begin{pmatrix} -7.3790 & -.8944 & .5000 \\ 5.3666 & .4472 & 0 \\ 4.4721 & 0 & 0 \end{pmatrix}$ and $S^tAS =$ $D(1, 2, 0)$.

(c) A is negative definite, with $S = \begin{pmatrix} .2673 & -.6937 & -10.2199 \\ 0 & .6070 & 7.7866 \\ 0 & 0 & 6.1644 \end{pmatrix}$.

(d) A has normal form $D(3, 0, 1)$ (called positive semi–definite) with

$$S = \begin{pmatrix} .7071 & -.4082 & -.2887 & -1 \\ 0 & .8165 & .5774 & 1 \\ 0 & 0 & 0 & 1 \\ 0 & 0 & .8660 & 0 \end{pmatrix}.$$

6.1.5. A being positive definite means $\mathbf{x}^tA\mathbf{x} > 0$ for all $\mathbf{x} \neq 0$. If B is congruent to A, then there is an invertible matrix S with $B = S^tAS$. Then $\mathbf{y}^tB\mathbf{y} = (S\mathbf{y})^tA(S\mathbf{y}) > 0$ for all $\mathbf{y}$, showing B is also positive definite. Since the entries on the diagonal are $\mathbf{e}_i^tA\mathbf{e}_i$, for a matrix which is positive definite, they will be positive. During the execution of the algorithm, each matrix obtained along the way will be congruent to the original matrix and so will be positive definite if the original matrix is. This means that we only have to use step (2) for a positive definite matrix in reducing it to a diagonal matrix with positive entries on the diagonal since at each step the matrix will have a positive entry as the next pivot.

6.1.6. This follows from $\mathbf{e}_i^tA\mathbf{e}_i = a_{ij}$ and A positive definite means $\mathbf{x}^tA\mathbf{x} > 0$ for all $\mathbf{x} \neq \mathbf{0}$. An example would be $\begin{pmatrix} 2 & -1 \\ -1 & 2 \end{pmatrix}$.

6.1.7. The identity provides an isomorphism between (V, q) and itself. If $L : (V_2, q_2) \to (V_1, q_1)$ is an isomorphism, then $L^{-1} : (V_1, q_1) \to (V_2, q_2)$ is also an isomorphism. Finally, $L : (V_2, q_2) \to (V_1, q_1)$, $M : (V_3, q_3) \to (V_2, q_2)$ being isomorphisms imply that $LM : (V_3, q_3) \to (V_1, q_1)$ is an isomorphism.

6.1.8. We give the argument for p being an invariant; the other cases are similar. Let P_2 be a subspace of V_2 of maximal dimension p_2 on which q_2 is positive definite. Then q_1 will be positive definite on the subspace $L(P_2)$, where $L : (V_2, q_2) \to (V_1, q_1)$ is an isomorphism. Thus the dimension p_1 of the maximal positive definite subspace for (V_1, q_1) must satisfy $p_2 \leq p_1$. Reversing the roles of the two spaces and using L^{-1} similarly gives $p_1 \leq p_2$, hence p is an invariant of the isomorphism class. The second statement follows from the fact that there is a basis so that the matrix is $D(p', n', d')$, and with this basis we see as in the proof of Sylvester's law of inertia for symmetric matrices that the dimension of the maximal positive definite subspace must be p', hence $p' = p$, with similar results for n, d.

6.2.1. (a) $S = \begin{pmatrix} 2 & 0 \\ 0 & 1 \end{pmatrix}$, $J = \begin{pmatrix} 0 & 1 \\ 0 & 0 \end{pmatrix}$

(b) $S = \begin{pmatrix} 1 & 1 & 0 \\ 0 & 1 & 0 \\ 0 & 0 & 1 \end{pmatrix}$, $J = \begin{pmatrix} 1 & 1 & 0 \\ 0 & 1 & 1 \\ 0 & 0 & 1 \end{pmatrix}$

(c) $S = \begin{pmatrix} 2 & 0 & 1 \\ 0 & 1 & 0 \\ 0 & 0 & 1 \end{pmatrix}$, $J = \begin{pmatrix} 1 & 1 & 0 \\ 0 & 1 & 0 \\ 0 & 0 & 2 \end{pmatrix}$.

(d) $S = \begin{pmatrix} 1 & 0 & 2 \\ 0 & 0 & 1 \\ -1 & 1 & -2 \end{pmatrix}$, $J = \begin{pmatrix} 1 & 0 & 0 \\ 0 & -1 & 0 \\ 0 & 0 & 2 \end{pmatrix}$

(e) $S = \begin{pmatrix} -1 & 1 & 2 \\ 0 & 0 & 1 \\ -1 & 0 & 0 \end{pmatrix}$, $J = \begin{pmatrix} 3 & 1 & 0 \\ 0 & 3 & 0 \\ 0 & 0 & 3 \end{pmatrix}$

6.2.2. (a) $c_1 \begin{pmatrix} 1 \\ 0 \end{pmatrix} + c_2 \begin{pmatrix} 2t \\ 1 \end{pmatrix}$; IVP: $\begin{pmatrix} 2t \\ 1 \end{pmatrix}$

(b) $c_1 \exp(t) \begin{pmatrix} 1 \\ 0 \\ 0 \end{pmatrix} + c_2 \exp(t) \begin{pmatrix} 1+t \\ 1 \\ 0 \end{pmatrix} + c_3 \exp(t) \begin{pmatrix} t^2/2 + t \\ t \\ 1 \end{pmatrix}$

IVP: $-\exp(t) \begin{pmatrix} 1 \\ 0 \\ 0 \end{pmatrix} + \exp(t) \begin{pmatrix} 1+t \\ 1 \\ 0 \end{pmatrix}$

(e) $c_1 \exp(3t)\begin{pmatrix} -1 \\ 0 \\ 1 \end{pmatrix} + c_2 \exp(3t)\begin{pmatrix} 1-t \\ 0 \\ -t \end{pmatrix} + c_3 \exp(3t)\begin{pmatrix} 2 \\ 1 \\ 0 \end{pmatrix}$

IVP: $-2\exp(3t)\begin{pmatrix} 1-t \\ 0 \\ -t \end{pmatrix} + \exp(3t)\begin{pmatrix} 2 \\ 1 \\ 0 \end{pmatrix}$

6.2.3. We do this by induction, starting with $k = 1$. Then there is only one nonzero vector $\mathbf{v}_1$, so this is an independent set. We assume that this is true for $k < n$, and suppose $k = n$. Suppose that we have a relation

$$c_1\mathbf{v}_1 + \cdots + c_n\mathbf{v}_n = \mathbf{0}.$$

Then apply N to this equation to get

$$c_2\mathbf{v}_1 + \cdots + c_n\mathbf{v}_{n-1} = \mathbf{0}.$$

By the induction hypothesis, $c_2 = \cdots = c_n = 0$. Returning to our original equation, this gives $c_1\mathbf{v}_1 = \mathbf{0}$, which implies $c_1 = 0$, completing the induction step.

BIBLIOGRAPHY

[1] R. Bellman. *Introduction to Matrix Analysis.* New York: McGraw-Hill, 1970.

[2] W. E. Boyce and R. C. DiPrima. *Elementary Differential Equations and Boundary Value Problems.* New York: John Wiley, 1976.

[3] G. Golub and C. van Loan. *Matrix Computations.* Baltimore, MD: John Hopkins, 1987.

[4] G. E. Forsythe and C. B Moler, *Computer Solution of Linear Algebraic Systems.* Englewood Cliffs, NJ: Prentice Hall, 1967.

[5] F. R. Gantmacher. *Applications of the Theory of Matrices.* New York: Chelsea, 1959.

[6] W. Rudin. *Real and Complex Analysis.* New York: McGraw-Hill, 1966.

[7] G. Strang, *Linear Algebra and its Applications.* San Diego, CA: Harcourt Brace Jovanovich, 1988.

INDEX